BURNS

Burns can result from heat or chemicals. To treat extensive thermal burns, have the victim lie down, place the cleanest available cloth material over all burned body areas to exclude air, and call an ambulance or physician to obtain emergency professional treatment with utmost speed. If the victim is conscious and can swallow, give him water, tea, coffee, or other nonalcoholic beverages.

To treat small thermal burns soak a sterile gauze pad or clean cloth in a baking soda solution (24–30 g of sodium bicarbonate in 1 liter of water, lukewarm if possible), place the pad over the burn, and bandage loosely. Do not disturb or open blisters. If the skin is not broken, immerse burned part in clean cold water or apply clean ice to relieve pain.

To treat chemical burns, flush with water immediately. Remove all clothing that has been contaminated (best in a shower or stream of water). If the burned area is extensive, have the victim lie down and call a physician at once. If the victim is conscious and can swallow, give him plenty of nonalcoholic beverages to drink. Do not apply ointments, grease, sodium bicarbonate, or other substances to extensive burns. All burns except those where the skin is reddened in a small area only should be seen by a physician.

Every burn, either thermal or chemical, can be complicated by shock. The signs of shock are a pale face, shallow breathing, cold or clammy skin with beads of perspiration on the forehead and palms of hands, and chills or complaint of chilled feeling. Frequently shock is accompanied by nausea or vomiting. In shock resulting from burns the liquid part of the blood flows into the burned areas; there may not be enough blood volume left to keep the brain, heart, and other organs functioning normally.

To treat shock, keep the victim lying down (preferably with feet elevated), keep his airway open, keep him warm if the weather is cold, and give water or other nonalcoholic beverage if he can swallow. Never give alcoholic beverages. Always call a physician if the victim appears to be suffering from shock.

SAFETY

The material on these pages is no more than a brief presentation of safety information for the organic chemistry laboratory. The authors believe that safety should be a topic for early discussion in any laboratory course in organic chemistry. Potential hazards of each experiment should be carefully pointed out throughout the course. Books on laboratory safety, such as *CRC Handbook of Laboratory Safety* (2nd ed.), N. V. Steere, ed., Chemical Rubber Co., Cleveland, Ohio, 1971, should be available in every university or college chemical laboratory. Careful study of such books is recommended.

Laboratory Practice of Organic Chemistry

Laboratory Practice of Organic Chemistry

FIFTH EDITION

Thomas L. Jacobs
University of California, Los Angeles
William E. Truce
Purdue University
G. Ross Robertson
Late of the University of California, Los Angeles

Macmillan Publishing Co., Inc.
New York
Collier Macmillan Publishers
London

Printed in the United States of America

Macmillan Publishing Co., Inc.
866 Third Avenue, New York, New York 10022

Collier-Macmillan Canada, Ltd.

Library of Congress Cataloging in Publication Data

Jacobs, Thomas Lloyd, (date)
Laboratory practice of organic chemistry.

First–3d ed. by G. R. Robertson, 4th ed. by G. R. Robertson and T. L. Jacobs.
1. Chemistry, Organic—Laboratory manuals.
I. Truce, William E., joint author. II. Robertson, George Ross, (date) joint author. III. Robertson, George Ross, (date) Laboratory practice of organic chemistry. IV. Title.
QD261.J27 1974 547'.0028 73-7349
ISBN 0-02-360050-0

Printing: 2345678 Year: 567890

Preface to the Fifth Edition

The laboratory practice of organic chemistry has undergone major changes during the past fifteen years. The most significant of these has been the widespread application of spectroscopic methods which have become the prime avenue of approach to structure determination. Spectroscopy is also used to guide the search for optimum conditions for a given synthesis and to permit precise control of a reaction. Other major changes have been the widespread application of vapor phase chromatography and the carrying out of experiments on a semimicro or micro scale as a routine practice (made possible in part by vapor phase chromatography).

These changes are reflected in the fifth edition of this laboratory manual. A new chapter on spectroscopic techniques has been added, and spectra for starting materials and products are presented for several experiments. A number of problems involving infrared or nuclear magnetic resonance spectra are included, the section on qualitative analysis takes into account spectroscopic methods, and literature references are given to experiments involving spectroscopic techniques that are suitable for undergraduate laboratory courses. More space is devoted to vapor phase chromatography and small-scale experiments are more plentiful.

It is recognized, however, that spectroscopic methods only supplement the older laboratory techniques. The practicing chemist must still learn to prepare a substantial amount of an organic compound in good yield, purify it by distillation or crystallization, and determine its melting point or boiling point. Furthermore, spectroscopic equipment is expensive and may be unavailable for the beginning organic laboratory. Actual determination of spectra is rather routine and in many industrial laboratories is done by specialists. The authors believe that the interpretation of spectra is an important part of a beginning course in organic chemistry, but that it must be an integral part of the whole course. Most textbooks of organic chemistry now cover spectroscopy and there seems no reason for a lengthy presentation in a laboratory manual.

As in previous editions, the special feature that distinguishes this work from other manuals of organic chemistry is an extensive but informal treatment of the principles underlying laboratory manipulations. Part 1 consists of this material and of a description of laboratory technique. Part 2 includes directions for synthetic experiments, an introduction to qualitative organic analysis, and references to the literature describing optional experiments suitable for beginning or intermediate students. Explanations of principles are thus kept distinct from directions for experiments. This separation is desirable because each principle is involved in several experiments, and an instructor seldom uses every experiment since more experiments are described than are needed for a year course.

To be sure, many of the chemical principles discussed in Part 1 may be found in abstract form in courses of physical chemistry, which, unfortunately, many students never reach. Most of these concepts are best illustrated in the laboratory of organic chemistry. The authors feel that an adequate treatment of such fundamental material, physical as well as chemical, would go far to remove the stigma of "cookbook" so often attached to elementary literature in this field. For students contemplating careers in medicine or agriculture, as well as for chemists, the understanding of such principles should have more lasting value than performing and memorizing additional organic reactions.

In the fifth edition, discussion of theory and general technique has been extensively revised. The chapter on solvents and solubility has been expanded to include solvent effects on organic reactions, and dipolar aprotic solvents are discussed briefly. Some little-used experiments have been deleted. Emphasis continues to be on acquisition of synthetic skill in the organic laboratory.

There is continuing need for improvement of the writing skills of scientists. The authors believe that this is more likely to occur if constant attention is given to the problem throughout a student's undergraduate career. The chapter "Writing a Research Report" is designed to offer practice toward this goal. Such a project arouses a surprising amount of enthusiasm if the instructor supplies a like share of enthusiastic interest himself. The chapter on "The Literature of Organic Chemistry" has been brought up to date and should assist in this project.

In several of the discussions of experiments, suggestions of reaction mechanism have been incorporated, though without great elaboration. Representations of reactions are often left incomplete, leaving the student to balance the equations. Molar or equivalent concentrations of reagents are often specified instead of simple metric units, in order to direct attention to the basis of chemical calculation.

The authors wish to thank colleagues, and also correspondents from outside institutions, for valuable suggestions.

It has been unfortunate that illness and death prevented Professor Robertson from participating in the present revision of this manual. He was the sole author of the first three editions and the major contributor to the fourth. The first edition pioneered clear presentation of physical chemical principles with the practical laboratory techniques and synthetic experiments for which they formed the basis. The plan continues to be the unique feature of the manual. This edition of *Laboratory Practice of Organic Chemistry* is dedicated to the memory of Professor Robertson.

THOMAS L. JACOBS
WILLIAM E. TRUCE

Contents

Laboratory Practice of Organic Chemistry

Part 1

Theory and General Technique

Chapter 1

Introduction

In elementary inorganic chemistry a student gets the impression that manufacture of a chemical product, such as potassium bromide or copper sulfide, is merely a question of mixing together the proper reagents in the proportion called for by chemical arithmetic. The law of conservation of mass, if nothing else, would insist that one get the full yield called for by theory, regardless of the operator's skill. From organic chemistry one soon learns that the problems of synthesis run far beyond the arithmetic of ordinary chemical equations. Two or more competing reactions usually ensue, leading to several products. Both the main reaction and undesirable side reactions are influenced, often very markedly, by changes in temperature, concentration, solvent, and acidity. Catalysts can frequently be found that favor one reaction. Exposure to the air may be significant. The skill of the experimenter is often decisive.

A main objective of the laboratory portion of a course in organic chemistry is an appreciation of the importance of choice of reaction conditions. Selection of the optimum conditions for a desired process depends upon sound knowledge of the physical and chemical principles underlying each reaction, and understanding of the principles underlying laboratory technique. Part 1 of this volume is therefore devoted to a critical discussion of both the theory and general practice of laboratory methods. Part 2 contains practical experimental examples illustrating these principles and emphasizing the importance of understanding reactions for the choice of procedures.

THE SIGNIFICANCE OF LABORATORY WORKMANSHIP

The laboratory portion of an organic chemistry course should seek to increase understanding of the principles of organic chemistry, to give substance and reality to the subject by direct contact with organic compounds, and to develop in the student competence in carrying out experiments with these compounds. Foremost among the requirements for competent laboratory work is a thorough understanding of the experiment to be undertaken. No procedure should be started without detailed study of the experiment and a careful plan for carrying it out. In many laboratories a student is not permitted to withdraw necessary supplies from stock or to start an experiment until he has demonstrated in an oral or written quiz that he understands what he is about to do. Such careful study and planning are essential for the development of good laboratory technique.

Also of prime importance is appreciation of safety in the laboratory. Not only must the student learn the rules, but still more important, he must also understand the principles and the experience that have brought forth these rules. Hazards of solvent fires as well as strictly chemical risks should be studied and thoroughly comprehended before a laboratory course is begun. Each experiment should be considered from the standpoint of safety before it is undertaken. Specific discussion of some of the more important hazards of organic laboratory work will be found on the inside of the front cover of this book.

The experiments in Part 2 have been carefully tested and represent workable procedures for many standard organic transformations. As the student advances, he should become so familiar with the standard laboratory procedures and with the principles of organic reactions that such detailed directions are no longer necessary. Most experimental material in the original literature is given with far less detail. The advanced student must be able to devise satisfactory procedures on his own.

The scale of an experiment depends upon the objective sought. Much information can be obtained more rapidly by semimicro or micro experiments than by experiments on a larger scale. However, it is often desirable to prepare considerable quantities of organic compounds in the laboratory (as starting materials, for biological testing, etc.). Moreover, a sequence of reactions must often be carried out, and sufficient starting material taken to ensure workable amounts in later stages. The experiments in Part 2 are designed to offer experience in all these aspects of organic laboratory work.

Lastly, it is an open secret that the laboratory instructor in organic chemistry often identifies a good workman by discovering clean apparatus, neatly assembled, on a clean laboratory desk. This does not mean the total absence of all extraneous matter not actually in use. Evidence of slovenly workmanship appears not so much from the presence of a recently used match or a few drops of water splashed during current operations. It is yesterday's or last week's dust, dirt, or disorder which is more characteristic and thus inexcusable.

As knowledge of the relationship between properties and structure of organic compounds has increased, the organic chemist has turned more and more to physical methods for structure determination. Spectroscopic methods have become increasingly important. Nuclear magnetic resonance (nmr) spectroscopy has now been applied routinely to organic compounds for over a decade and is of importance for following the course of organic reactions as well as for structure determination. Infrared spectroscopy continues to be highly useful, and Raman spectra can now be obtained almost as readily as infrared. Mass spectroscopy is of use not only for molecular-weight determination but for structural work as well. Increasing use is made of isotopic marking for mechanism studies and mass spectra are essential in such work. The range of ultraviolet spectra has been increased and such spectra are now more useful. X-ray crystallography has advanced to such a point that the most complicated organic structures may sometimes be more rapidly attacked by this method than by any other.

A brief introduction to these methods, particularly infrared and nmr spectroscopy, is presented in Chapter 13, which also includes a number of references on spectroscopic methods.

Chapter **2**

General Laboratory Requirements

Note: The following paragraphs outline sound practices often adhered to in organic laboratory courses. It is expected that the instructor in charge will announce the regulations to be followed, which may differ from those below.

1. Notebooks. Records of laboratory work should be kept in ink in a bound notebook about $8-8\frac{1}{2} \times 10-11$ in. in size. Looseleaf notebooks are not acceptable in industrial or research practice and should not be permitted in academic laboratory courses. Books made from square, cross-sectional paper printed with pale blue or green lines are convenient for graphs, sketches, and tables. The pages should be numbered and a few pages left at the beginning for a table of contents. Each experiment or entry should be dated.

Success in laboratory work in organic chemistry requires thorough knowledge of the experiment to be performed and careful planning for efficient utilization of time. Notebook entries should reflect such planning. If the experiment calls for records of weights, volumes, temperatures, etc., a tabular arrangement prepared in advance is often most convenient. Data should be recorded directly in the notebook, not on scraps of paper.

Notebook entries should include the book or journal from which the directions were obtained (with proper page reference), equations for the reactions, and a record of all starting chemicals and reagents (including supplier, constants as given on the bottle or in the catalog, and quantities in molar and metric units). It is not necessary to copy extensive directions for experiments. If an experiment detailed in this manual is involved, the description can be quite brief, but changes in the procedure or unexpected occurrences should be noted carefully. Times (for addition, reflux, etc.) should be recorded. If the experiment is interrupted between laboratory periods, a precise indication of the point in the experiment and the length of time of the interruption is necessary. If directions are taken from other manuals or from other literature, the procedure should be given in sufficient detail to be intelligible to a chemist. Actual yields of products (crude and purified), theoretical yields, and percentage yields should be recorded (see Chapter 3). Finally, conclusions and comments (including suggestions for improvements) should be given.

The worker should never tear out any sheet carrying unsatisfactory data; instead such data should be annotated with the appropriate critical comments. If the experiment did not go well, enter what information you have about the difficulties.

The keeping of a course notebook is training for the recording of research results or of any experimental results obtained on an industrial chemical job. It is essential that such records be accurate and reliable—they may have to stand court inspection by experienced lawyers bent on proving them unreliable.

2. Reports and Products. Most chemists enjoy carrying out experimental work far more than preparing a report on what they have done. However, results are of little value if they are not reported to those who may use or are interested in them. The principle extends to reports on beginning experiments. The form of report will vary in different laboratories, but the student would be wise to realize that concise, attractive presentation of results is one of the best ways to establish the quality of his work. In beginning laboratory work the notebook and bottled product commonly are presented to the laboratory instructor for approval. A brief oral quiz may be part of the report. More often a brief written report giving the student's name, the experiment number and title, equation, yield, yield calculations, and physical constants of the product will be required with the bottled product. Experiments involving a sequence of reactions or an investigation disclosing results not in the literature will require longer reports (see Chapter 15).

Products should be stored in neatly labeled bottles or sample vials. Wide-mouth vessels are used for solids and narrow-mouth for liquids. The label should be written neatly in ink, should be firmly and attractively stuck onto the container, and should give at least the name of the product, the boiling point or melting point, the yield, the weight of the empty container (tare weight), and the name of the student. Usually empty containers are weighed with stoppers, but the practice specified by the instructor should be followed. Do not use preliminary pencil-scrawled labels at any time. The standard of workmanship in bottling and labeling is merely that required in any first-class pharmacy. A neat label might appear as shown on page 8.

tert-BUTYL CHLORIDE
bp 52°
Yield 37.5 g Tare 41 g
ANN D. FISHER

Products should be carefully dried and placed in dry containers. Solids that stick to the walls of the container or liquids that are cloudy are judged to be wet and will usually be downgraded.

CLEANING AND DRYING GLASSWARE

3. Use of Solvents. Glass vessels are most easily cleaned *immediately after use*. Since it is a highly oxidized material, glass is preferentially wetted by water rather than the various oily and tarry constituents accompanying the water on the surface of a newly soiled vessel. Simple rinsing with water may eliminate the organic matter at this stage. If, through delay, the water evaporates, the tar will reach the actual glass surface and adhere tightly. It is well to develop the habit of washing each piece of apparatus as soon as it is emptied.

As aids in cleaning, certain high-grade washing powders are available. These contain fine abrasives that aid in removal of tarry matter but are not coarse or sharp enough to scratch laboratory glassware. Some workers favor mixtures of the water softener (and detergent) trisodium phosphate and soap powder. Others prefer synthetic detergents.

Tarry matter is often handled to advantage with the aid of a small amount of acetone, which has the special merit of being an excellent solvent at room temperature. In view of the cost, one should use only a few milliliters of this solvent in one application. Keep acetone away from open flames.

4. Solvent–Abrasive Combination. Acetone and an abrasive washing powder may be combined in economical fashion as a thin paste rubbed upon a tarry surface, as of a soiled beaker, or in troublesome cases even upon the hands of the worker. Perhaps a teaspoonful of powder and a few milliliters of acetone will be adequate. The tar is at least loosened, if not completely dissolved, and is promptly scattered in the mealy mass left upon evaporation of the volatile acetone. Cautious addition of a very small amount of water now emulsifies impurities, leaving, finally, an easy job of rinsing.

5. Acid Cleaners. Use of strong inorganic acids as cleaning agents involves serious hazards of fire, personal injury, and destruction of clothing. Nitric acid is particularly dangerous. The practice, now fairly rare, of maintaining a large heated evaporating dish of concentrated nitric acid or sulfuric acid–chromic anhydride in the hood for the cleaning of glass equipment causes costly damage to flues, fans, and motors of the hood system. Such heated acid baths are dangerous and have caused many laboratory accidents; they should not be permitted.

6. Troublesome Cleansing Problems. Limited use of hot cleaning solution in small volume for removal of resistant dark polymeric material that may contain elemental carbon, is sometimes necessary. Such deposits, sometimes formed by scorching residual organic matter over a free flame, are treated by placing about a gram of sodium or potassium dichromate and about 5 ml of concentrated sulfuric acid in the flask, heating *(in the hood!)* over a quiet bare blue flame, spreading both flame and reagent evenly until the acid fumes strongly. Care must be taken to heat gently and to keep the mouth of the flask pointed away from the face. Safety glasses or goggles should be worn in the laboratory at all times but are especially important here. The flask is allowed to cool after the heating, the acid is poured out, and the flask rinsed with water.

7. Drying of Glassware. Much time is saved by washing glassware immediately after using, thus taking advantage of spontaneous drying overnight. When prompt drying is required, the clean, wet glass vessel is well drained of water and treated with one or two 10-ml portions of acetone; or a larger portion of "wash acetone" is used to rinse the vessel and the acetone is returned to the special bottle. Following careful drainage, air is sucked or blown out of the vessel with the aid of a clean glass tube attached to an aspirator, piped vacuum service, or compressed-air cock. **Warning:** *Compressed air often contains oil droplets, and the stream must then be passed through a suitable cleaning-drying train.* A drying tube filled with loosely packed absorbent cotton is satisfactory if the cotton is renewed fairly frequently.

8. Cork Rolling and Boring. Use of standard-taper glassware has decreased use of corks, but the following information is still valuable and skill in manipulation of corks worthwhile. Preparation of a cork to fit a test tube or other piece of equipment involves choice of a cork of ample size to fit, about as shown in Fig. 1(a). The cork is then rolled gently in a rotary iron cork press until it is softened and reduced in diameter to fit, as shown in Fig. 1(b).

A hole is easily bored in a cork with a sharp cork borer. The borer is sharpened by rotation against the knife, A, of the sharpener (Fig. 2) while it is pressed gently against the center core, B. Care is taken not to press too hard at C lest the cutting edge be nicked or at D lest the edge be burst.

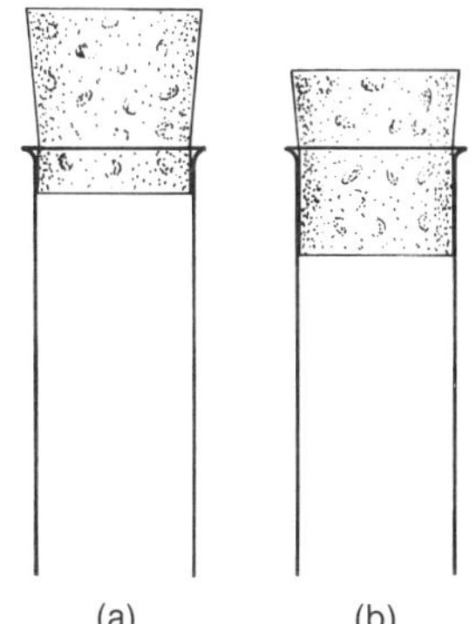

Fig. 1. Fitting of Corks.

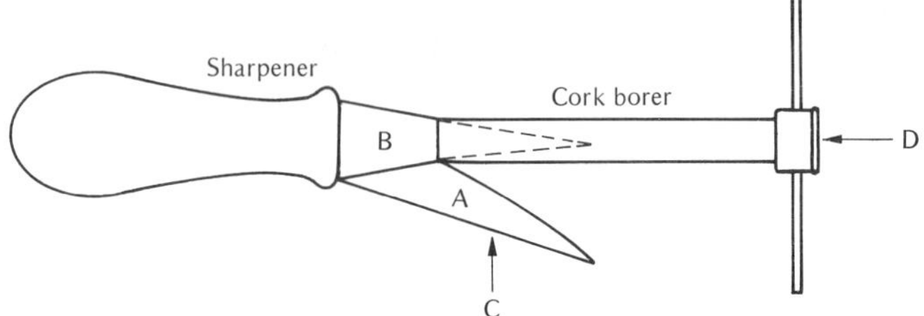

Fig. 2. Sharpening a Cork Borer.

Choose a borer of very slightly smaller diameter than the tube to be fitted. Cut the hole with continuous rotary motion of the borer, starting from the small end of the cork and carefully centering the hole if a one-hole cork is needed. The borer is a cutter, not a punch; do not force it to burst its way through.

As the borer advances through the cork, examine the tool from the side several times to see that it is cutting a straight hole without slant. If much resistance is encountered, remove the borer, eject the jammed cork mass from the borer, and return the tool to place. After the borer is about halfway through, withdraw it and eliminate any piece of cork that may come out with the borer. There are now three different methods of completing the task, as favored by various workers:

(a) Start a hole from the other end of the cork. If a conscious effort is made to keep the borer aligned with the hole in the first end as the cork and borer are rotated, the two holes meet cleanly in the center.
(b) Holding the cork in the open between thumb and fingers, continue the boring from the same side as started. A very sharp borer is required and the boring must take place with much rotation and only gentle pressure, or a ragged, torn aperture will be produced on the top side of the cork. This will result in a much less satisfactory seal and the cork may leak around the aperture.
(c) Stand the half-bored cork upon an old (waste) cork and continue the boring with much rotation until the borer has gone into the waste cork.

Warning: *Do not press the borer into the desk or other hard objects. If the cutting edge of a brass borer should be driven even once firmly against hardwood, it will require a special machine-shop job to restore its usefulness.*

The thermometer or tube that is to be installed should fit the hole in the cork snugly but not so tightly that there is danger of breaking the glass piece. To obtain such a fit one must often enlarge a hole slghtly with the rat-tail file. Take care that such filing is applied uniformly, without side swing that would produce a poorly fitting conical or oval hole.

9. Boring Rubber Stoppers. Unfortunately, rubber stoppers of the best "pure-gum" quality are the most difficult of all stoppers to bore. A lubricant such as dilute sodium hydroxide solution is needed. Alcohol, glycerol, soap solution, or even plain water is also helpful. The borer must be given the keenest cutting edge possible and should be about one size larger than one used with cork. The technique of rotary cutting instead of pushing is of greater importance than ever.

Stopper-boring machines are available in many laboratories and make boring of rubber stoppers easier. These machines may be used for corks, but it is often not worth the time to go to the machine.

10. Removal of Tubes and Thermometers. Many painful accidents have come from struggles to remove glass tubes from rubber stoppers. Figure 3 shows a convenient and safe method of removing such a tube by slipping the smallest possible cork borer over the tube. The borer should not be any sharper than necessary and should be lubricated with soap solution or glycerol so that it will work its way into the passage without starting a new cut.

A device, sometimes called a "hand saver," which operates on the same principle, is available commercially to remove or insert glass tubing, etc., from or into rubber stoppers.

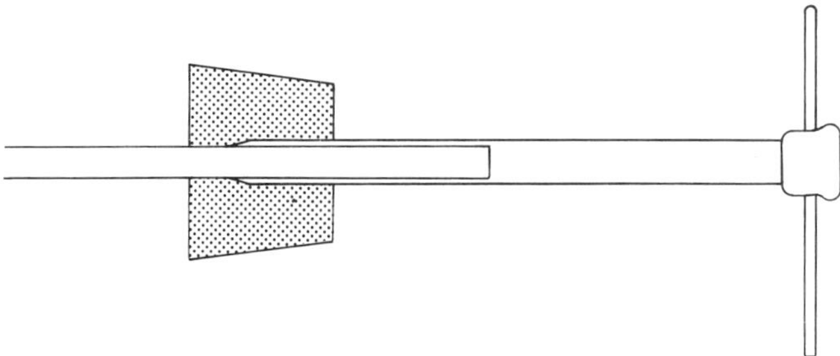

Fig. 3. Removal of Tube from Rubber Stopper.

Chapter 3

Calculation of Quantities of Materials

11. Specification of Quantities of Reactants. An overall, *balanced* equation is required before one may prescribe the correct quantities of reactants for a given organic synthesis. In some syntheses precisely the molar proportions indicated in such an equation are used. For example in the coupling of diazotized aniline with β-naphthylamine, tarry by-products are formed if either reactant is in excess.

$$C_6H_5NH_2 + HCl \longrightarrow C_6H_5NH_3Cl$$

$$C_6H_5NH_3Cl + NaNO_2 + HCl \longrightarrow C_6H_5N_2Cl + 2H_2O + NaCl$$

$$C_6H_5N_2Cl + C_{10}H_9N + CH_3CO_2Na \longrightarrow C_{16}H_{13}N_3 + NaCl + CH_3CO_2H$$

$$\underset{\text{Aniline}}{C_6H_5NH_2} + \underset{\beta\text{-Naphthylamine}}{C_{10}H_9N} + NaNO_2 + CH_3CO_2Na + 2HCl \longrightarrow \underset{\text{Phenylazo-}\beta\text{-naphthylamine}}{C_{16}H_{13}N_3} + 2NaCl + CH_3CO_2H + 2H_2O$$

The directions specify 0.03 mole each of aniline, β-naphthylamine, and sodium nitrite, just as the overall equation requires. Hydrochloric acid (15 ml of 6 *N* or 0.09 equivalent) and sodium acetate (0.07 mole) are in excess. Thus 4.3 g (0.03 mole) of β-naphthylamine and 2.07 g (0.03 mole) of sodium nitrite, both solids, are weighed out on a simple balance. One may also weigh the aniline (2.79 g, 0.03 mole), a liquid, but it is often more convenient to measure the volume. Aniline has a density of 1.022, so 2.79 g is 2.73 ml (volume = weight/density), a volume easily measured with a graduated pipet or less accurately in a small graduated cylinder.

It is unusual for a synthetic chemist to choose the exact proportions called for by theory. Reasons for variation include the following:

(a) At equilibrium there may be considerable of the starting materials present. Increase in concentration of one of the reagents will improve

the yield of product and is often desirable, provided the excess is readily removed at the end of the reaction and is inexpensive or easily recovered.

(b) If one starting material is harder to separate from the desired product than any other, an excess of other starting materials may be used to make sure that the difficultly separable material is all consumed.

(c) One reagent may not only participate in the chemical change but also be a useful solvent for the reaction; it is then used in considerable excess provided it is easily separated from product.

(d) One reagent may be quite volatile and at least some excess may be desirable to insure adequate amounts to react with the other starting materials. Often such a reagent is very easily removed at the end of the reaction, so considerable excess will be useful, as indicated in *(a)*, *(b)*, or *(c)*.

(e) If one reagent is offensive in character, it may be preferable to choose conditions so that it is all consumed.

12. Calculation of Yield. The "theoretical yield" of an organic synthesis is the amount of product that could theoretically be obtained from the amounts of starting materials taken, assuming that the reaction goes to completion as indicated by the balanced equation ignoring side reactions. For example, the synthesis of fluorescein (§342) involves reaction of two molecules of resorcinol with one of phthalic anhydride:

$$\underset{\text{Phthalic anhydride}}{C_6H_4(CO)_2O} + \underset{\text{Resorcinol}}{2C_6H_4(OH)_2} \longrightarrow \underset{\text{Fluorescein}}{C_{20}H_{12}O_5} + 2H_2O$$

The directions specify 0.05 mole of phthalic anhydride and 0.1 mole of resorcinol. The theoretical yield of fluorescein is thus 0.05 mole or 16.6 g, the maximum amount of this product that could be obtained from the amounts of starting materials specified, assuming that the product was formed and isolated quantitatively. In practice a good laboratory worker isolated 13 g of fluorescein (the actual yield) following these directions. From these data, the percentage yield is calculated:

$$\text{Percentage yield} = \frac{\text{Actual yield} \times 100}{\text{Theoretical yield}} = \frac{13.00}{16.6} = 78.3\%$$

The more usual case in which some of the starting materials are in excess requires determination of the limiting starting material and calculation of yields based on it. For example, the preparation of 1,1-diphenylethanol from phenylmagnesium bromide and ethyl acetate involves preparation of the Grignard reagent from bromobenzene and addition of ethyl acetate followed by hydrolysis with dilute acid. Directions specify 34.5 g of bromobenzene, 5.5 g of magnesium, and 8.8 g of ethyl acetate. The equations are

$$2C_6H_5Br + 2Mg \longrightarrow 2C_6H_5MgBr$$

$$2C_6H_5MgBr + CH_3CO_2C_2H_5 \longrightarrow (C_6H_5)_2C(OMgBr)CH_3 + C_2H_5OMgBr$$

$$(C_6H_5)_2C(OMgBr)CH_3 + C_2H_5OMgBr + HCl \xrightarrow{H_2O} (C_6H_5)_2C(OH)CH_3 + C_2H_5OH + 2Mg^{++} + 2Br^- + 2Cl^-$$

$$\underset{\substack{34.5\text{ g}\\0.22\text{ mole}}}{2C_6H_5Br} + \underset{\substack{5.5\text{ g}\\0.23\text{ g. at.}}}{2Mg} + \underset{\substack{8.8\text{ g}\\0.1\text{ mole}}}{CH_3CO_2C_2H_5} + 2HCl \longrightarrow \underset{\substack{14.0\text{ g}\\0.076\text{ mole}}}{(C_6H_5)_2C(OH)CH_3} + C_2H_5OH + 2Mg^{++} + 2Cl^- + 2Br^-$$

} Weights and moles of reactants and product

The equation indicates that 0.1 mole of ethyl acetate will react with 0.2 mole of bromobenzene and 0.2 gram-atom of magnesium; ethyl acetate is clearly the limiting reactant. Since the theoretical yield of l, l-diphenylethanol is 0.l mole or 19.8 g.

$$\text{Percentage yield} = \frac{14.0 \times 100}{19.8} = \frac{0.076 \times 100}{0.1} = 76\%$$

13. Change of Scale of Operations. If one does not need so large an amount of product as that predicted for certain practical directions, it is, of course, easy to scale down the respective quantities of reagents. Beginners are cautioned not to cut down the time of reaction in the same proportion. Most reaction rates depend on concentrations; if these remain the same, the yield will be the same. On the other hand, many organic reactions are carried out by adding one reagent to the solution at such a rate that the heat evolved does not force the reaction mixture out of the flask. Under these circumstances a small-scale experiment may often be completed much more rapidly. Conversely, "scaling up" sometimes involves hazards of runaway reaction. If directions call for a reaction of 10 g of X and 20 g of Y, a change to 1000 g of X and 2000 g of Y might cause so much evolution of heat that the reaction would become uncontrollable without special temperature regulation.

BALANCING OXIDATION-REDUCTION EQUATIONS

14. Oxidation State. The schemes well known in elementary chemistry that involve positive and negative valence numbers (oxidation state) or electron transfer are readily adapted to organic chemistry, with only slight added complication. For example, iron and chlorine are compared with carbon. The statements that iron in the form Fe^{3+} has a valence number, or oxidation state, of +3 and that chlorine in Cl^- has the valence number −1 are commonplace. With carbon, however, the bonds directly attached to that element are normally not ionic, and one cannot speak so confidently of a plus or minus number. It is nevertheless convenient to assign a positive or negative unit value to each valence of the carbon atom, according to the element to which that carbon bond is attached. For example, all four bonds to carbon in the methane molecule, CH_4, are attached to the electropositive element hydrogen. The carbon atom is therefore designated as of valence number, or oxidation state, of −4. In the opposite extreme the carbon atom CO_2 holds electronegative oxygen at all points, and thus for equation purposes is marked as +4.

Organic compounds with more than one carbon atom in the molecule are similarly rated. Acetic acid, for example, is of intermediate state. One carbon atom in CH_3CO_2H has a valence number of −3; the other, +3. In such a calculation the covalent bond between the two carbon atoms, being attached neither to hydrogen nor to oxygen, is considered as of zero value. The total oxidation state, or valence number, of acetic acid is thus

$$-3 + 3 = 0.$$

This illustrates the versatility of carbon (unlike simple metallic ions) in having part "positive" and part "negative" bonds. Even a single carbon atom, as in methyl alcohol, may have both positive and negative valence numbers, in this case

$$+1 - 3 = -2.$$

15. Oxidation of a Secondary Alcohol. Conversion of cyclohexanol into adipic acid is chosen to illustrate one of the common methods of calculation, that of "oxidation state." Workers preferring the "electron-transfer" device should note §16. In the present case the oxidation is effected with aid of potassium permanganate in the presence of sulfuric acid.

First in consideration are the experimental facts. Cyclohexanol, the reducing agent, is attacked by permanganate ion (the oxidizing agent) and adipic acid (the oxidized product) appears. One more of the four major factors, the reduced product, must be determined; this is manganous ion, Mn^{++}, presumably detected by qualitative test. It is now convenient to set down these four quantities graphically in systematic order.

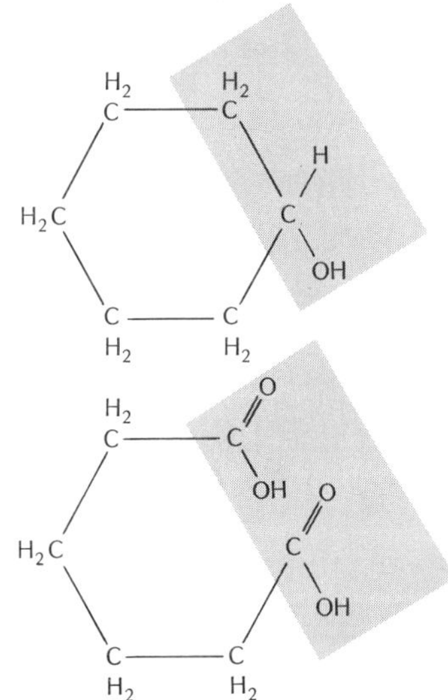

It is evident that a large part of the cyclohexanol molecule took no part in this reaction and therefore may be excluded from consideration. Grey panels identify the portions of the molecules involved in oxidation. Only the *change* in oxidation state is important. Inside the upper panel the oxidation state of carbon is

$$-2 + 0 = -2.$$

Inside the lower,

$$+3 + 3 = +6.$$

A change from −2 to +6 has occurred. This means eight electrons have been withdrawn, or oxidation to the extent of eight equivalents has been effected. Something must have been reduced and in equivalent amount, namely, permanganate.

The transformation of permanganate into manganous ion (well known in strictly inorganic chemistry) causes this oxidation in the organic material. Since heptavalent manganese (MnO_4^-) is reduced to bivalent (Mn^{++}), five electrons have been absorbed. The oxidation state of manganese has fallen from +7 to +2, and thus five equivalents of reduction have been effected.

Since 8 does not match 5, it will be obviously necessary to use 8 moles of permanganate to oxidize only 5 of cyclohexanol. Determination of this 8:5 ratio is thus the main task in the calculation. An incomplete statement of reaction, but one showing correct proportions, includes the 5 moles of cyclohexanol and 8 moles of permanganate plus whatever else is obvious at the moment:

$$5C_6H_{11}OH + 8KMnO_4 \longrightarrow 5C_4H_8(CO_2H)_2 + 8K^+ + 8Mn^{++}$$

Since the last two terms indicate need for negative ions to balance 24 plus-charges, 12 molecules of H_2SO_4 are required.

$$5C_6H_{11}OH + 8KMnO_4 + 12H_2SO_4 \longrightarrow 5C_4H_8(CO_2H)_2 + 8K^+ + 8Mn^{++} + 12SO_4^{--}$$

If the change in oxidation state has now been correctly represented and proper balance of ionic charges has been effected, nothing remains but to add water and to assemble ions into molecules if that be desired. Simple inventory of H and O atoms reveals the requirement of 17 moles of water as product.

$$5C_6H_{11}OH + 8KMnO_4 + 12H_2SO_4 \longrightarrow 5C_4H_8(CO_2H)_2 + 4K_2SO_4 + 8MnSO_4 + 17H_2O$$

16. Method of Half-reactions. Although the method of oxidation number, as illustrated in §15 is relatively easy to learn and use, many instructors regard it only as a practical trick or device that has no sound physical basis. The varying positive or negative count is not a real property of the carbon atoms themselves. Preference is noted for a system that recognizes the transfer of electrons from the material oxidized (reducing agent) to the material being reduced (oxidizing agent). Such a concept leads to a method often employed in studies of analytical chemistry, where the accounting of oxidation and reduction is expressed in separate equations dealing with so-called half-reactions.

For illustration, the oxidation of ethyl alcohol to acetic acid is chosen. The experimental facts are first determined:

Potassium permanganate + Ethyl alcohol + Sulfuric acid (dil.) ⟶ Acetic acid + Manganous ion

Separate equations are now written, one to show the main reaction of the oxidizing agent and the other of the reducing agent. In the first equation, permanganate ion (MnO_4^-) is the essential reagent:

$$\underset{\text{Oxidized form}}{MnO_4^-} + \underset{\text{Protons}}{8H^+} \longrightarrow \underset{\text{Reduced form}}{Mn^{++}} + \underset{\text{Water}}{4H_2O} \qquad (1)$$

To balance equation (1) it is necessary, as in most such reactions, to furnish the proper quantity of hydrogen ions, or protons, to take care of the oxygen no longer needed after the "oxidized form," or oxidizing agent, was reduced. In the present case, sulfuric acid provided the required protons, and water was thus formed. Protons and water thus constitute two essential accessory materials required in an equation for such a transformation.

An expression for the relation of alcohol to acetic acid may now be written in a form analogous to equation (1):

$$\underset{\text{Oxidized form}}{CH_3CO_2H} + \underset{\text{Protons}}{4H^+} \longleftarrow \underset{\text{Reduced form}}{CH_3CH_2OH} + \underset{\text{Water}}{H_2O} \qquad (2)$$

While (2) follows (1) in pattern, item for item, the actual chemical reaction goes in the *opposite direction,* as shown by the arrow. Consolidation of equations (1) and (2) would therefore require that (2) be written in reverse order:

$$CH_3CH_2OH + H_2O \longrightarrow CH_3CO_2H + 4H^+ \qquad (3)$$

Although equations (1) and (3) are balanced with respect to numbers of atoms of each element involved, neither is correct with respect to ionic charges. Equation (1) needs five additional negative charges, or electrons, on the left side, while (3) needs four electrons on the right side. Equations for the half-reactions are therefore completed, as (4) and (5), by adding electrons (e^-):

$$MnO_4^- + 8H^+ + 5e^- \longrightarrow Mn^{++} + 4H_2O \qquad (4)$$

$$CH_3CH_2OH + H_2O \longrightarrow CH_3CO_2H + 4H^+ + 4e^- \qquad (5)$$

It is now necessary to add (4) to (5) algebraically so that the "e" terms will disappear, since the number of electrons involved in the oxidation half-reaction must be equal to the number of electrons in the reduction half-reaction. The terms of (4) are thus multiplied by 4 and those of (5) by (5), and the addition is completed:

$$4MnO_4^- + 32H^+ + 20e^- \longrightarrow 4Mn^{++} + 16H_2O \qquad (6)$$

$$5CH_3CH_2OH + 5H_2O \longrightarrow 5CH_3CO_2H + 20H^+ + 20e^- \qquad (7)$$

$$4MnO_4^- + 5CH_3CH_2OH + 12H^+ \longrightarrow 4Mn^{++} + 5CH_3CO_2H + 11H_2O \qquad (8)$$

Equation (8) is a completely balanced ionic equation that reports in full the reactions of oxidation and reduction. It does not, however, list all the actual commercial reagents and products involved. Since potassium permanganate and sulfuric acid were actually used, equation (8) is written as follows for laboratory use:

$$4KMnO_4 + 5CH_3CH_2OH + 6H_2SO_4 \longrightarrow 4K^+ + 4Mn^{++} + 6SO_4^{--} + 5CH_3CO_2H + 11H_2O \qquad (9)$$

Questions

1. 5-Methyl-5-nonanol was prepared from ethyl acetate and *n*-butyl bromide by a Grignard reaction essentially as described in Experiment 14. From 22 g of ethyl acetate, 75 g of *n*-butyl bromide, and 13 g of magnesium was obtained 24 g of 5-methyl-5-nonanol. Calculate the theoretical yield and the percentage yield of the product. Which starting material is limiting?

2. From chlorination of 360 g of ethane were isolated 410 g of ethyl chloride and 250 g of ethylene dichloride as the only products obtained as pure compounds. Calculate the percentage yield of each product and the overall percentage yield for the process.

3. If four reagents were required for a synthesis, how would you decide which of the four should serve as basis for calculation of theoretical yield?

4. Bromoform was prepared from 20 ml of acetone and 8 ml of bromine by the haloform reaction (§295). The yield was 10 g. Calculate the theoretical yield and the percentage yield.

5. Oxidation of ethylbenzene with aqueous potassium permanganate gives potassium benzoate, manganese dioxide, potassium carbonate, potassium hydroxide, and water. Write a balanced equation for this reaction. Calculate the weight of potassium permanganate needed to oxidize 53 g of ethyl benzene. If 340 g of potassium permanganate is used and the yield of benzoic acid after acidification is 40 g, calculate the theoretical yield and the percentage yield.

6. When aqueous potassium permanganate is mixed with ethyl alcohol in the absence of sulfuric acid, acetate ion and manganese dioxide are the main products. What proportions of permanganate and alcohol are required? Assuming that 0.2 *M* aqueous potassium permanganate is used, what would be the approximate pH of the final solution (neglect the volume of alcohol)?

7. When *n*-butyl alcohol is allowed to react in the cold with sodium dichromate and sulfuric acid, butyl butyrate is produced. Calculate the number of moles of dichromate theoretically required to convert 1 mole of alcohol to the maximum yield of ester.

8. In the presence of acid how many gram-atoms of zinc are required in theory to convert 1 mole of nitrobenzene to 1 mole of aniline?

9. Ferric chloride converts *p*-aminophenol to benzoquinone; ferrous ion and ammonium ion are also produced. Write the equation for the reaction and calculate the amount of ferric chloride required by theory to react with 218 g of *p*-aminophenol.

10. When a green plant converts carbon dioxide from air into glucose, what is the overall oxidation-reduction equation?

11. Write the equation for reduction of cyclohexanone by $LiAlH_4$ in which all four hydrogens are used for the reduction. Show the products before water is added to the reaction mixture, and also the reaction when

water is added as the final step. Calculate the weight of $LiAlH_4$ that would be required to reduce 49 g of cyclohexanone. What is the theoretical yield of cyclohexanol? If the actual yield is 45 g, what is the percentage yield?

12. Write equations for the reduction of propionic acid to *n*-propyl alcohol by $LiAlH_4$. Remember that the fast reaction of $LiAlH_4$ with an acid is libertation of H_2 and formation of a salt ($CH_3CH_2CO_2Li$). Before addition of water the products are lithium and aluminum propoxides. Addition of water forms the free alcohol. Write the equations for the various steps and the overall equation. What is the theoretical amount of $LiAlH_4$ required to reduce 14.8 g of propionic acid. If 6 g of $LiAlH_4$ was used and 11 g of *n*-propyl alcohol isolated, calculate the theoretical and percentage yields of the alcohol.

Chapter 4

Conduct of the Reaction

Important considerations for carrying out a chemical reaction include (1) temperature control, (2) proper choice of apparatus, and (3) choice of solvent. The first two of these are discussed below; choice of solvent is discussed in Chapter 11.

TEMPERATURE CONTROL

17. Thermometers. Temperature is one of the most important factors controlling a chemical reaction. This fact leads to a critical scrutiny of laboratory thermometers. Precautions to be used in handling chemical thermometers, and directions for calibration are given in Experiment 1, Part 2.

Errors found in thermometers are sometimes due to carelessness or haste in manufacture, but such poor technique is not the only reason for incorrect readings. The glass-bulb section of a thermometer, which holds nearly all the mercury, tends to shrink with age and elevate the mercury column to high and, of course, erroneous levels.

18. Stem Correction. Normally the manufacturer expects the thermometer to be completely immersed in the medium whose temperature is to be measured. In ordinary operations of organic chemistry, however, a considerable part of the mercury column protrudes from the heated zone. This part is therefore at too low a temperature and thus too short. A simple mathematical correction is required to show how much longer this outer column would be if it were actually inside that zone, as the manufacturer presumably intended in his process of graduation "under total immersion."

The protruding portion of a thermometer column may be likened to a steel bar measured in cool weather. Wishing to know how much the bar would expand on a hot day, the engineer must know (1) its length,

(2) the number of degrees of elevation in temperature, and (3) the coefficient of expansion of steel. For a thermometer, the analogous formula yields the desired correction:

$$N(T - t)0.000154$$

in which N is the length of that part of the mercury column that protrudes from the heated zone. This length is most conveniently measured in degrees Celsius, since a scale of degrees is on the stem. T is the temperature of the medium being tested, and t is the average temperature of the protruding column; therefore $T - t$ is the deficiency in temperature calling for correction. Finally, 0.000154 is a constant, namely, the linear coefficient of expansion of mercury in a glass tube of the kind normally used in thermometers.

N is, of course, directly measured—not precisely, but with a rough practical estimate as to the border line between the heated medium and the cooler outside atmosphere. T may be taken, without serious error in this kind of calculation, as the uncorrected thermometer reading. The term t is determined approximately by hanging a second thermometer outside the heated medium but with bulb alongside the protruding side stem of the main thermometer. The following example is illustrative:

Example. A certain preparation of phenol boils at 180° (uncor.). The temperature of the air above the distilling flash is 30°, shown by the second thermometer. The bottom of the cork stands at the 95° graduation.

$$N = 180 - 95 = 85$$
$$T - t = 180 - 30 = 150$$

and

$$85 \times 150 \times 0.000154 = 1.9635$$

or approximately 2° correction. The boiling point of the phenol preparation is thus recorded, with sufficient accuracy for ordinary laboratory work, as $180 + 2 = 182°$ (cor.).

Some commercial thermometers are graduated so that they show the correct temperature only when they are immersed in a heated zone, such as a liquid bath, for a short distance only. A special mark is etched on the lower part of the stem to indicate the boundary between the heated zone and the outer atmosphere. A thermometer for 3-in. immersion is often used. Such thermometers are convenient but not ideal in theory at least, owing to uncertainty about the exterior temperature.

19. Control by Refluxing. A convenient method of maintaining a constant temperature for a reaction is to choose a solvent that boils at the desired temperature and to reflux the reaction mixture. Heat is ap-

[§19] plied to cause gentle boiling, and the solvent vapor is recondensed in a condenser, arranged as shown in Fig. 1, and returned to the flask. A condenser so arranged is called a reflux condenser. A solvent may be unnecessary if one of the reactants boils at a convenient temperature. Care must be taken to avoid too rapid boiling, which may cause the condenser to flood; liquid solvent may then be forced out of the top of the condenser by the rising vapor producing loss, damage and, with flammable liquids, fire hazard. Refluxing permits very long heating periods when standard-taper glass joints are used; with ordinary glassware assembled with corks, students are warned that escape of vapors around a bored cork may be surprisingly great unless the cork is carefully bored. The end of the condenser should project about 2 cm below the bottom of the cork to minimize contact of hot vapors with the cork; if it projects too far below, flooding of the condenser occurs more easily. In most student laboratories

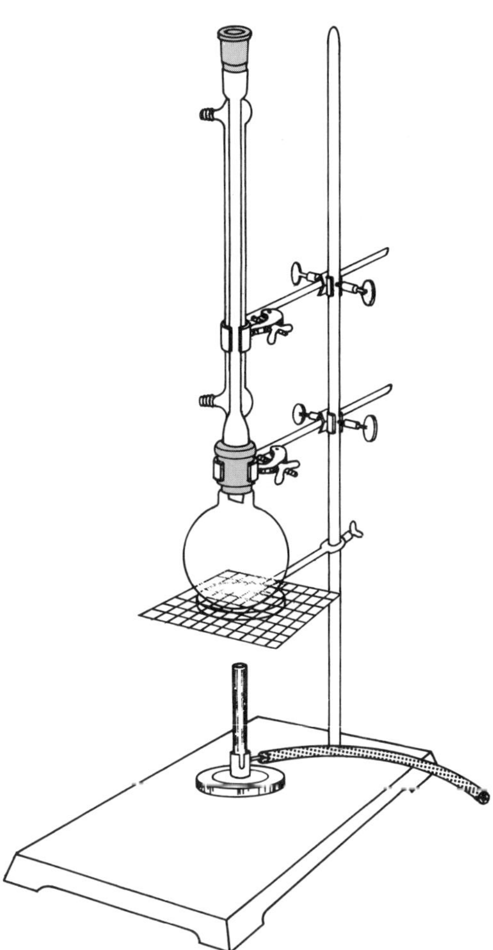

Fig. 1. Use of Reflux Condenser.

the refluxing operation is not allowed to continue overnight or while the laboratory is unattended.

Warning: *If unattended refluxing is permitted, the rubber tubing must be wired onto the condenser and water faucet, and the end of the exit hose must be placed in the drain with care. The most common cause of floods and water damage in chemical laboratories is improper attachment of rubber tubing to condensers.*

Avoid chance constrictions in the rubber tubing; these make the tubing flop about so that it may slip off the connection or the end may come out of the drain. Students are also warned of the treacherous behavior of some water faucets. Slow swelling of certain kinds of washers in faucets often causes delayed, automatic shutoff of the small current of water that enters the bottom of the condenser jacket. The reaction may be spoiled by loss of solvent from the top of the inadequately cooled condenser, and with flammable solvents there may be a serious fire hazard. Check the flow of water through the condenser 15 min after starting the reaction, and be sure the flow is stabilized if the reaction is to be left unattended. Care in turning off water at the conclusion of laboratory work is essential. Liquids boiling above 160° may be refluxed under an air-cooled condenser. This does not mean a water-cooled condenser with the water jacket empty, since this outer jacket then acts as an insulator and decreases the efficiency of the condenser. An air-cooled condenser is illustrated in Fig. 3, §62.

A wire gauze placed as shown in Fig. 1 helps to prevent local superheating (which sometimes leads to partial decomposition of the reaction mixture) and prevents development of strains in the glass. In some student laboratories open flames are no longer permitted for safety reasons, and heating is done electrically with a heating mantle (§21), hot plate, or heating bath (§20). Heating with these is usually more uniform.

Refluxing may lead to special trouble from bumping (§64) if a second phase, either solid or liquid, separates from the reaction mixture. Boiling stones should always be used to assist smooth refluxing, and a heating bath or mantle may help to reduce bumping.

20. Heating Baths. Steam or boiling water is often used to keep a laboratory vessel at constant temperature near 100°. Copper steam cones or baths are commercially available. An excellent aluminum alloy steam bath is shown in Fig. 2. In the absence of such special devices, steam (preferably from piped laboratory service) is led into a pan of water through a piece of $\frac{5}{16}$-in. copper tubing about 20 in. (50 cm) long, bent in the form of a loop, and largely immersed in the water. The entering steam condenses quietly in the copper tube without the disturbing noise produced when steam is passed into water through poorly conducting material (rubber or glass).

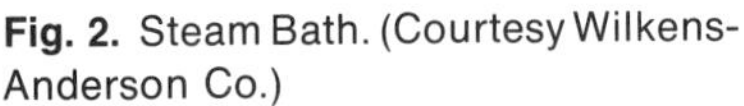

Fig. 2. Steam Bath. (Courtesy Wilkens-Anderson Co.)

For temperatures above the region of 95°, heating baths containing various other materials are useful. These give even heating but remain at a constant temperature only if the heat supplied is carefully regulated. Fusible metal alloys (e.g., Wood's metal, mp 71°, composition 50% bismuth, 25% lead, 12.5% tin, and 12.5% cadmium) are good but costly and so heavy that special care must be used in manipulating them (a small "lab jack" such as is pictured in Fig. 3, or a small hydraulic jack is recommended, with electrical heating). Wood's metal oxidizes rapidly above 300°, and vapors of lead present some hazard unless ventilation is excellent. Oil of low vapor pressure is an economical substitute. Vegetable oils such as cottonseed, sesame, rapeseed, or peanut oil are excellent from the standpoint of low vapor pressure and minimum fire hazard. Such oil should be of hydrogenated type such as household shortening, which is semisolid when cool and thus conveniently stored. Such soft material has the advantage that a flask is easily pushed into place in it at room temperature. Some workers prefer "hard-hydrogenated" oil, with melting point above 60°, because the softer material tends to flow slowly and may spill on bench or locker during storage if the containing pan becomes tipped. Hydrogenated vegetable oils will not spontaneously ignite below 300°, whereas paraffin and other petroleum products will flash and cause dangerous conflagration much below that point. Students are warned never to mix any paraffiin, motor grease, or other hydrocarbon with any hydrogenated vegetable oil used in an oil bath. Baths of hydrogenated oil smoke badly above 225°.

Silicone oils, which are dimethylsiloxane polymers, are sometimes used for heating baths. One such oil, Dow Corning 550 Fluid, has a serviceable temperature range of −40 to +230° in an open system or to 315° (its flash point) in a closed system. Such oils have low volatility and good heat transfer characteristics, but are relatively expensive ($5–$7/lb).

Fusible salt mixtures are sometimes used for temperatures above 300° *but are extremely hazardous* at these temperatures. A mixture containing 40% sodium nitrite, 7% sodium nitrate, and 53% potassium nitrate melts at 142° and has a useful range of 150–500°. The eutectic mixture of 54.4% of potassium nitrate, 18.2% sodium nitrate, and 27.3% lithium nitrate melts at 119° but attacks glass slowly and has a more powerful oxidizing action. Its use is not recommended. Dangerous burns result

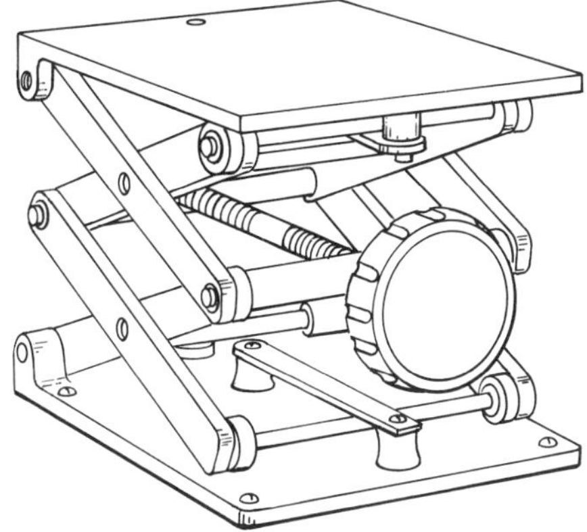

Fig. 3. Laboratory Jack.

if molten salt strikes the skin; protective clothing should be worn, and the bath should be designed to minimize any chance of spilling or splashing. Metal containers should be used and equipment removed from the bath before it solidifies.

In laboratory directions in early literature, sand baths are often specified for heating. Older glassware was more easily cracked by application of heat, but with borosilicate glass (Pyrex, Kimax, etc.) there is less need for special precautions and a wire gauze over a flame is satisfactory. A sand bath is relatively unsatisfactory as a replacement for a liquid heating bath because its heat conductivity is poor, the temperature is not easily controlled, and a flask is not readily replaced in hot sand if it has been removed (e.g., to relieve overheating).

21. Heating Mantles. Electrical resistance wire units, embedded in fireproof glass cloth "mantles," may replace heating baths for heating reaction mixtures. They are expensive, but very convenient and are available to fit flasks too large to heat easily in any other way (12 liters and larger). Such heating mantles are unsuitable for many distillations because the walls of the flask may be overheated when the liquid level drops below the top of the mantle; false temperature readings and increased decomposition in the distilling flask often result. Occasionally the vapors of an organic compound are dangerously explosive if overheated, and some explosions during distillation or reflux have been attributed to the use of heating mantles.

22. Baths for Low Temperature. Finely crushed ice in water is convenient for temperatures around 0–5°. A mixture of 23.1% salt and 76.9% ice can give a temperature of −22.4°, but in practice a salt-ice bath is useful for the range −15° to −5°. For lower temperatures solid carbon dioxide (dry ice) is often used. The dry ice is powdered and cautiously added to acetone in a Dewar vessel; rapid initial addition produces a rush of gas that often carries acetone over the rim of the vessel. Acetone serves as a heat-transfer medium; other organic liquids such as alcohol or trichloroethylene are sometimes used. The sublimation temperature of carbon dioxide is −78.5°, and temperatures somewhat below this are attained in freshly prepared baths (with acetone these are around −86°). Water condenses in such baths fairly rapidly, and in practice temperatures are not this low. Intermediate temperatures are obtained with slush baths consisting of low-melting liquids that have been partially frozen. Such baths are often prepared by mixing the liquid with liquid nitrogen or powdered dry ice. Several such baths and references to more detailed information are given by A. M. Phipps and D. N. Hume [*J. Chem. Ed.*, **45**, 664 (1968)]. For lower temperatures liquid nitrogen (bp −196°) may be used.

Warning: *Liquid oxygen boils at −183° so that oxygen may condense from the air in containers cooled in liquid nitrogen and open to the air; a mixture of liquid oxygen and an organic material may explode.*

Students are warned that mercury thermometers cannot be used below −38° because mercury is a solid below −38.87°. For lower temperatures, thermometers filled with organic liquids are often used (e.g., toluene to −90°, *n*-pentane to −130°); the liquid contains a dye to make it more easily visible. Such thermometers require far longer than mercury-filled thermometers to reach equilibrium because the fluids are less heat conducting and more viscous.

LABORATORY APPARATUS

Most organic laboratory operations are conducted in glass apparatus. Reactions are usually carried out in solution, frequently at the boiling point of the solvent or with stirring, in round-bottom flasks of borosilicate glass. This glass has a far lower coefficient of thermal expansion than ordinary "soft" glass previously used for laboratory glassware, and easily withstands moderately rapid heating or cooling. It is attacked by alkali more rapidly than ordinary glass.

23. Standard-Taper Apparatus. Almost all advanced laboratory work in organic chemistry is now conducted in apparatus provided with standard taper ground-glass joints as shown in Figs. 1 and 5. Such apparatus is now fairly common in instructional laboratories in spite of the high initial cost. The advantages in speed of assembly of equipment and avoidance of loss and contamination of materials are great.

Two kinds of standard-taper joints are in general use. The tapered joints shown in Fig. 5 are more common and have less tendency to leak than spherical joints (shown in Fig. 5 as a stirrer seal). Tapered joints have a 1:10 taper (1 mm taper for 10 mm length) and are specified[1] by two numbers (e.g., 24/40) that indicate maximum diameter and length, respectively, in millimeters. Three types differing in length have been designated long, medium, and short (only the first two are common). Many laboratories standardize on a few sizes for most equipment; 24/40 is probably the most frequently used. In the medium type the corresponding size is 24/25. The center neck of a three-neck flask is usually of larger size for convenience. Spherical or "ball and socket" joints are similarly designated by two numbers (e.g., 18/9) indicating radius of curvature and inside diameter of tubing, respectively. These joints permit flexible connections.

Carefully made joints are vaporproof without lubrication. Some workers prefer to lubricate joints with stopcock grease to prevent freezing, but this is necessary only if the joint comes in contact with decomposing organic material, which tends to cement the ground surfaces together. Lubrication may lead to contamination of the reaction mixture by the grease, which may be extracted by solvent. Teflon sleeves are now available that fit smoothly over the inner part of a tapered joint and effectively prevent both freezing and leaking. If standard-taper equipment is to be used under reduced pressure, lubrication of the joints or use of sleeves is necessary. At high temperatures Teflon decomposes to give very toxic, volatile products.

24. Flasks. One-neck flasks (Fig. 1) are satisfactory for most work where stirring is not required. A liquid may be added slowly to a reaction mixture through the reflux condenser from a dropping funnel. A three-neck flask (Fig. 4 or Fig. 5) is used when the reaction mixture is to be stirred mechanically. This flask is more convenient when slow addition of a liquid is necessary. Two-neck and four-neck flasks can be obtained but are rarely used.

25. Dropping Funnels. Dropwise addition of liquids to a reaction mixture is accomplished with a cylindrical dropping funnel. These may be graduated to permit more accurate regulation of rate of addition. Globe- or pear-shaped separatory funnels may be used as dropping funnels but are often difficult to mount because their greater diameter interfers with the stirring assembly. It is also more difficult to estimate addition rate with these.

26. Condensers. Figure 4 pictures the West condenser. The ordinary Liebig condenser is similar but has a wider water jacket that makes less efficient use of cooling water. A spiral condenser in which the cooling water passes through an internal glass spiral (Fig. 5) is very efficient

[1]Bureau of Standards Commercial Standard CS 21-39.

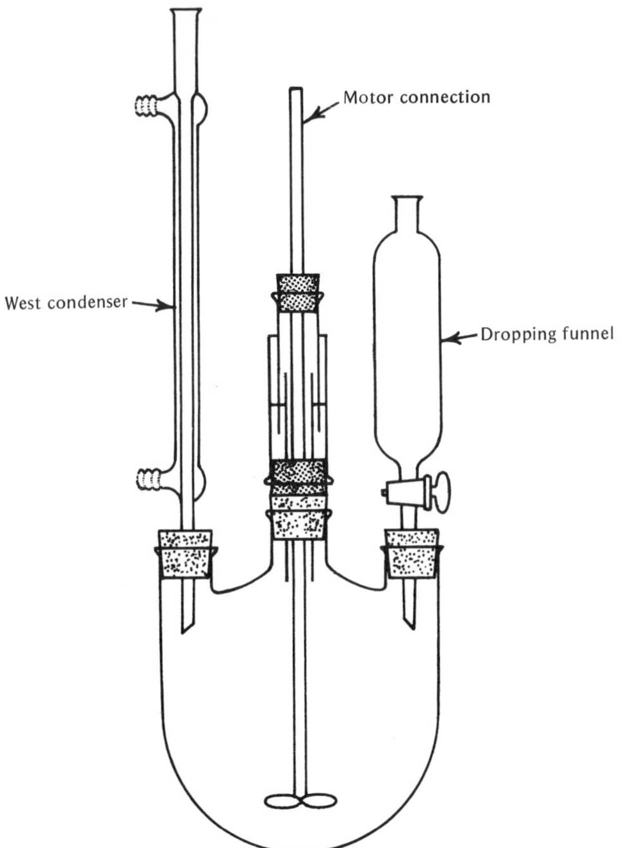

Fig. 4. Apparatus for Synthesis.

and compact. For greater efficiency the spiral is sometimes made of metal. The Graham condenser, a modified Liebig with a spiral central tube for the vapors, is unsatisfactory for reflux operations because it floods readily and a portion of the liquid may be blown out the top; it is seldom used. Other styles of condensers are shown in books on laboratory technique (see the References).

When very volatile materials must be refluxed, it may be convenient to cool the circulating water by passage through a copper spiral immersed in ice or by circulation of lower-melting liquids (brine, aqueous alcohol, etc.) through the condenser. A cold-finger condenser (Fig. 6) filled with dry ice and acetone, or even liquid nitrogen, is used frequently.

27. Assembly of Glass Apparatus. Each piece of glassware should be supported not only by rod or ring stand but also by clamp for the individual piece, as shown in Fig. 1. A three-neck flask requires a separate clamp for each added piece. Thus, in Fig. 4 two clamps attached to a

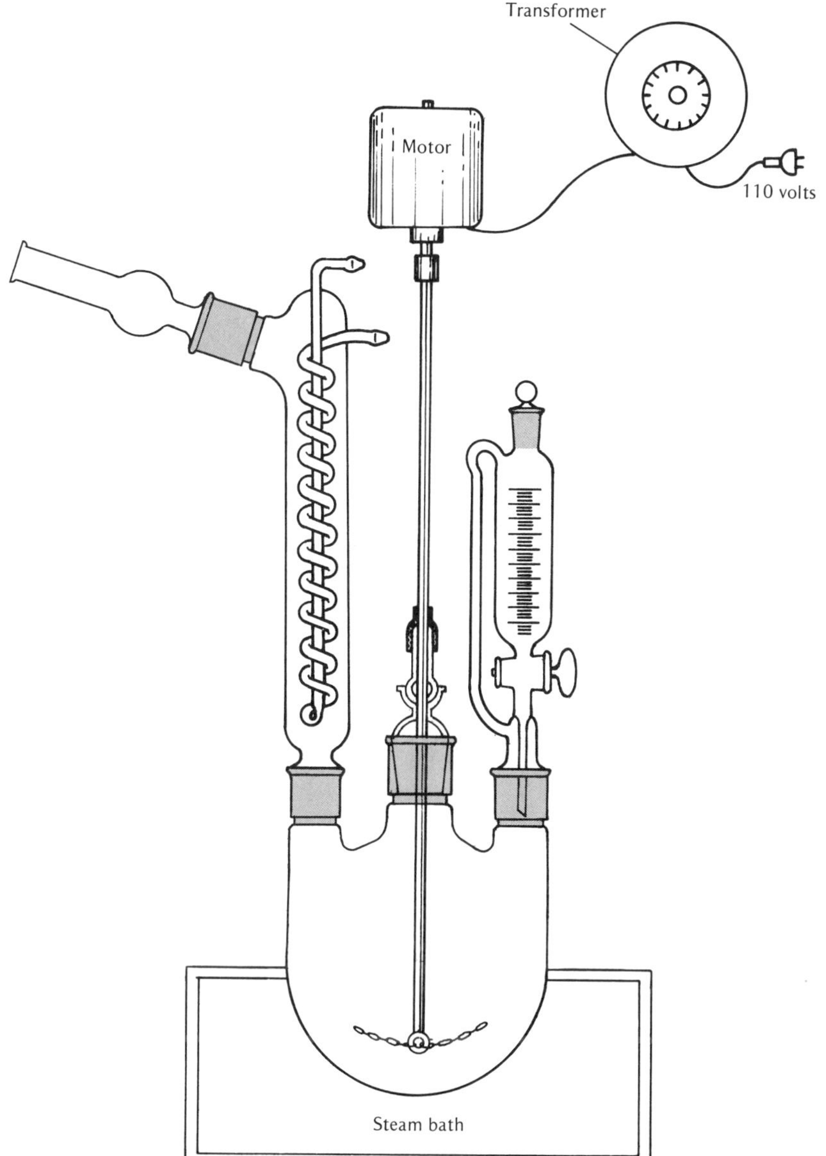

Fig. 5. Apparatus for Synthesis.

rod directly behind the stirrer shaft extend forward to clamp the center neck of the flask and the mercury seal; clamps also extend radially from the same rod to the condenser and to the dropping funnel. With standard-taper glassware these side clamps are sometimes omitted when smaller dropping funnels or condensers are used.

The jaws of clamps holding glassware should be protected with short pieces of rubber tubing. Asbestos paper or cloth should replace rubber if the glass is expected to become hot enough at the clamped spot to melt or scorch the rubber.

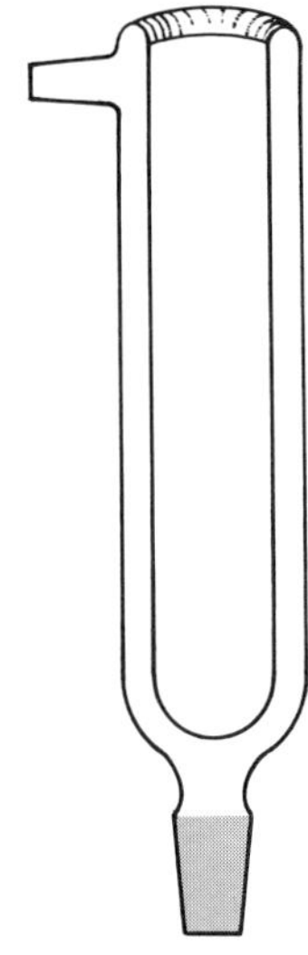

Fig. 6. Cold-Finger Apparatus.

28. Stirrers. Many heterogeneous reactions require stirring because good contact must be maintained between phases. Rapid mixing is also necessary when a reagent is added or when temperature control must be facilitated. Small electric motors with speed reduced by suitable small gears, a pulley arrangement, or a cone drive are usually employed to drive a stirrer, which may be a simple bent rod, a glass propeller, a Teflon paddle attached to a glass rod by a swivel connection, or a more elaborate device as required by the experiment or dictated by the preference of the experimenter. An efficient and easily constructed wire stirrer designed by E. B. Hershberg [*Ind. Eng. Chem., Anal. Ed.*, **8**, 313 (1936); OS–CV, 2, 117] is pictured in Fig. 5; 18- or 20-gage nichrome wire is often used to construct this stirrer, but tantalum is sometimes preferred if nichrome is attacked by the reaction mixture or has an unwanted catalytic action on the reaction.

Magnetic stirring is very convenient and is commonly used unless the reaction mixture is too viscous or requires more vigorous mixing. The stirrer is a small Teflon-coated magnet; it is rotated by another magnet placed beneath the flask and mounted on a small motor. Rate of rotation may be controlled. When it is necessary both to heat and to stir a mixture, the magnetic stirrer can be operated through a heating mantle or electrically heated bath, but at reduced torque. Convenient combinations of hot plate with magnetic stirrer can be purchased. Magnetic stirring avoids the problem of seals discussed in §29.

The References at the end of the chapter describe many stirring devices including some that are air driven or equipped with explosion-proof motors for safety.

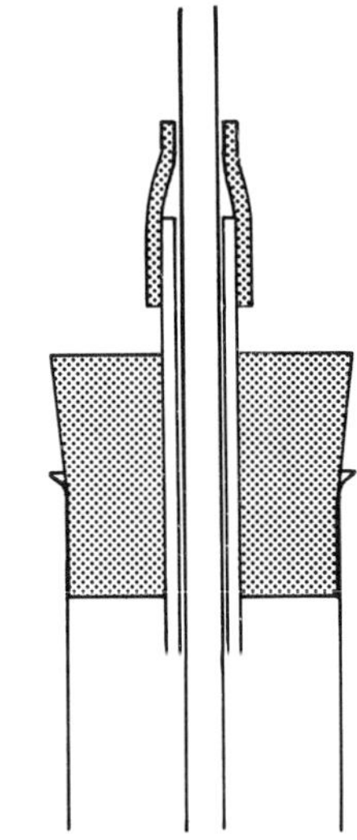

Fig. 7. Rubber-Sleeve Seal.

29. Seals for Stirrers. The necessity for preventing escape of vapors around a stirrer shaft leads to special devices permitting motion without vapor leakage. The mercury seal shown in Fig. 4 is the time-honored solution to the problem but is somewhat troublesome to assemble and introduces some hazard of mercury-vapor toxicity.

The lubricated rubber-sleeve device shown in Fig. 7 is much simpler. Care is taken that the sleeve in which the stirrer shaft rotates is a very close fit without being so small as to grasp the shaft and retard or stop the rotation. Glycerol or lubricating oil is used, according to the solubility problem at hand (preferably glycerol in most cases).

A small spherical joint mounted as shown in Fig. 5 is an especially convenient seal. The joint is lubricated with glycerol or sometimes with stopcock grease (the latter dissolves more readily in organic solvents and may contaminate the reaction mixture to a greater extent).

30. Reactions on Semimicro or Micro Scale. It is increasingly common in organic chemistry to conduct reactions on a semimicro or micro scale. The amounts of material designated by these terms is somewhat vague. A reasonable suggestion (Cheronis in Weissberger, *Technique of Organic Chemistry,* **6,** 8) differentiates between analytical procedures (where quantities up to 10 mg are considered as micro, and 10–25 mg as semimicro) and preparative procedures (where quantities up to 50 mg are classed as micro and 100–2000 mg as semimicro).

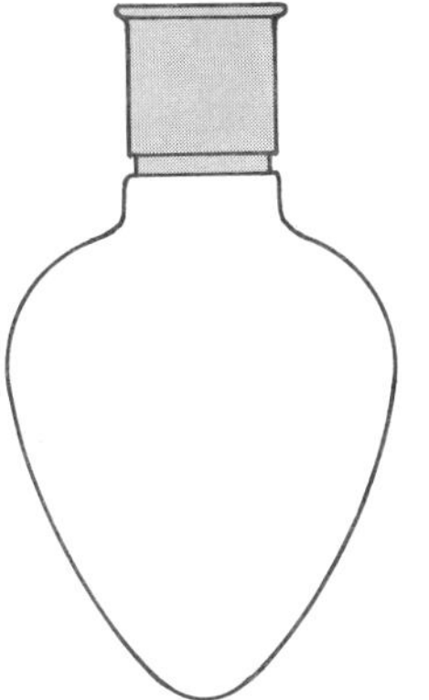

Fig. 8. Pear-Shaped Flask.

Small-scale work is both faster and less expensive in many instances, although it requires greater care than macro-scale work both in measurement of quantities and in manipulation. The development of apparatus and procedures for micro-scale experimentation has made possible the investigation of many naturally occurring substances. However, if a sequence of reactions is to be carried out on a relatively available starting material, judgment is required to choose a large enough amount for the initial steps so that the final steps can be carried out conveniently. Thus a beginning laboratory course in organic chemistry should provide experience with work on a variety of scales.

For most semimicro preparative work, the glassware required is not greatly different, except in size, from that used in macro-scale experiments. Pear-shaped flasks (Fig. 8) are usually preferred to round-bottom test tubes. A small cold-finger condenser (Fig. 9) is often useful. Hirsch funnels (Fig. 1, §96) are preferred to Büchner funnels for collecting precipitates. Separation of two liquid phases may be accomplished with medicine droppers having tips drawn out longer than ordinary.

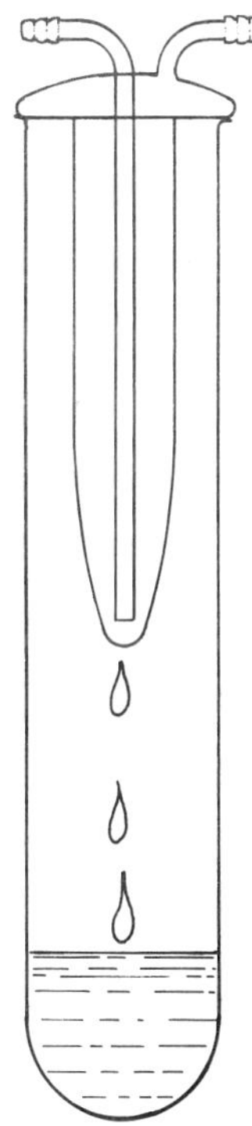

Fig. 9. Cold-Finger Apparatus.

For advanced work, many kinds of special apparatus have been devised. A discussion of these is beyond the scope of the present manual, but *Micro and Semimicro Methods* by N. D. Cheronis (Weissberger, *Technique of Organic Chemistry,* **6**) gives details. Standard-taper apparatus for small-scale work is available.

Several of the experiments in this manual illustrate semimicro procedures.

31. Specialized Apparatus. The apparatus described in this manual is mainly of the kind that has wide general applicability in organic chemistry. There is increasing use of more specialized equipment in all phases of laboratory investigation. Many experiments are carried out in inert atmospheres, and increasing use is made of steel equipment for reactions under pressure. High-speed stirring and high dilution require special apparatus. Photochemical investigations require sources of ultraviolet light and glassware transparent to such light. Laboratory glassware opaque to all but red light is also available for work with light-sensitive compounds. Pyrolytic reactions are now often used; equipment for vapor-phase investigations at controlled high temperatures has been greatly improved. Special equipment for catalytic hydrogenation and other catalytic reactions, for ozonization, and for other experiments that cannot be conducted in conventional apparatus, is necessary for advanced work. Increasing attention is paid to the hazards of work with highly toxic or explosive compounds.

A discussion of these more advanced aspects of the conduct of reactions is beyond the scope of this manual. The References include excellent discussions.

Questions

1. State two reasons why stem correction is more important in the distillation of high-boiling than of low-boiling liquids.

2. Calculate the thermometric stem correction for a liquid being distilled under the following conditions: bottom of the cork at the −20° mark; thermometer reading, 120°; outside air temperature, 40°.

3. A liquid distills under reduced pressure at 10°. The thermometer is placed so that the bottom of the cork is at −20° and the temperature of the air outside is 30°. Calculate the stem correction.

4. A thermometer was correctly graduated for partial immersion, with the mark for immersion set at −10°. Suppose the instrument were completely immersed in pure liquid naphthalene, boiling at 760-torr pressure. What would be the reading of the thermometer? See §570.

5. Suppose you had purchased a new standard thermometer with calibration certificate. Why would it be desirable to redetermine the ice point of this thermometer on future occasions when this instrument is to be used as standard of comparison? How would you use the new ice-point value so determined?

References

(See Chapter 14 for complete titles.)

Temperature Measurement: Weissberger, *Technique of Organic Chemistry*, **1** (3d ed.), Pt. 1, 259.

General Laboratory Practice: Weissberger (various volumes in both *Technique of Organic Chemistry* and *Techniques of Chemistry*); Müller–Houben–Weyl, **1;** Wiberg, 191; Bates and Schaefer; Pasto and Johnson; Vogel (1956); Vogel (1966).

Chapter 5

Distillation

32. Distillation of a Pure Substance. The technique of distillation comprises two operations: vaporization and recondensation. Ordinarily two distinct pieces of apparatus are assembled for these successive operations, a boiler and a condenser. For a single liquid one may use a simple distilling flask attached to a suitable condenser (§62.) The product, usually liquid, is called the *distillate*. Distillation is the most commonly used method for purifying liquids.

33. Vapor Pressure. Molecules of a liquid are constantly escaping from the surface into the vapor state and molecules in the vapor recondensing into the liquid. The molecules in the vapor bombard objects in contact with the vapor and thus exert pressure. In a closed vessel the rate of recondensation soon equals the rate of escape. The pressure exerted by the vapor when this equilibrium prevails is the vapor pressure of the liquid; it is determined by the temperature and the concentration of the substance in the vapor, and can be measured by a suitable gage.

34. Temperature–Vapor Pressure Relationship. As temperature rises, the kinetic energy of the molecules in the liquid increases and more molecules escape into the vapor. A new equilibrium is soon established with higher concentration in the vapor and higher vapor pressure. There is a fixed vapor pressure for each temperature at which any given substance can exist in the liquid state. Vapor-pressure tables have been compiled for many compounds at many temperatures (e.g., see the Appendix, §577). Figure 1 shows a typical plot of change in vapor pressure with temperature. Vapor pressures are usually expressed in millimeters of mercury or torr (1 torr = 1 mm of mercury[1]) because mercury

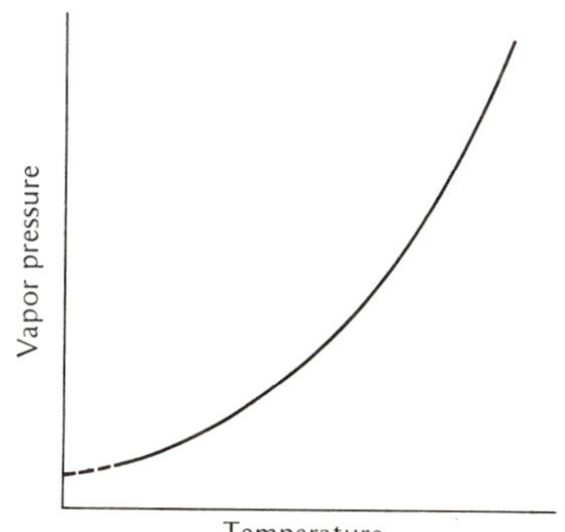

Fig. 1. Vapor Pressures of a Liquid.

[1]A standard atmosphere is 1,013,250 dynes/cm^2; by definition it is 760 torr. Thus 1 torr is equivalent to 1 mm of mercury within one part in 7×10^6. The unit is named for E. Torricelli, the Florentine scientist who discovered the properties of the barometer in 1643.

is the most common material used in vapor-pressure-measuring devices (manometers, see §70).

The rate at which vapor pressure changes with temperature is given by the Clapeyron equation:

$$\frac{dp}{dt} = \frac{\Delta H_v}{T \Delta V}$$

where ΔH_v is the heat of vaporization of the given liquid, T is the absolute temperature and ΔV is the increase in volume upon vaporization. A useful approximate equation is obtained by neglecting the volume of liquid, which is small compared with the volume of vapor, and assuming that the vapor behaves as an ideal gas so that V may be replaced by RT/p. Then

$$\frac{dp}{p} = \frac{\Delta H_v dT}{RT^2}$$

Integration of this equation gives

$$\log p = \frac{-\Delta H_v}{2.3RT} + \text{constant} \quad \text{or} \quad \log \frac{P_1}{P_2} = \frac{-\Delta H_v}{2.3R}\left(\frac{1}{T_2} - \frac{1}{T_1}\right)$$

This equation indicates that vapor pressure increases exponentially with temperature or that, to the extent that the assumptions hold and ΔH_v is constant, a plot of log p against $1/T$ will be linear.

35. Boiling Point. A liquid in an open flask at a temperature below its boiling point exerts vapor pressure characteristic of the particular liquid but less than the barometric pressure. The vapor above the liquid will contain a mixture of molecules of nitrogen, oxygen, etc. (from air), along with molecules of the liquid substance, such that the total pressure is equal to the barometric pressure. For example, diethyl ether at 20° has a vapor pressure of 442 torr (Appendix, §569). If the barometer reads 750 torr, the air must exert a pressure of 308 torr. As the temperature is raised, vapor is driven out of the flask and the partial pressure of ether increases; at 30° it is 647 torr and at 34.5° it reaches 750. At this temperature ether boils and all of the air is driven from the flask, leaving only ether molecules in the vapor. Further heat causes vaporization of more ether and ether vapor pours out of the flask. The temperature will not rise further until all of the ether has been vaporized. The boiling point is defined as the temperature at which the vapor pressure of a liquid is equal to the external pressure. It is often recorded at 760 torr (standard pressure), but in many instances it is recorded at lower pressures as well. For stable liquids an accuracy of ±0.01° can be obtained if special apparatus and precautions are taken, but in ordinary laboratory practice an accuracy of 0.5–1° is usually acceptable. Near sea level it is seldom necessary to correct the boiling point for differences in barometric pressure because change with pressure is small in this region. Many laboratories, however, are located at altitudes well above sea level where boiling points are significantly below the "normal" values found in handbooks. Tables showing in adequate detail the boiling points over the range 570–780 torr can seldom be found. A graphic method is convenient for estimating values in this range. Instead of plotting log p against $1/T$, as indicated by the Clapeyron equation, it is more convenient to use semilog paper so that one may plot p as the ordinate without reference to a log table. Arithmetical work is simplified by plotting t as abscissa instead of $1/T$. The relationship is nearly linear in the region of interest. Figure 2 presents such a graph on which common substances may be plotted. Line ABC represents such a plot for n-heptane. A and B are the boiling points at 760 torr and 600 torr, respectively, as recorded in the table inside the back cover of this book. In a laboratory at 4250-ft altitude, where the barometric reading is 650 torr, the boiling point is read from the graph as 93.2° (point C). For compounds not listed in the table the References at the end of this chapter may give the information needed.

36. Distillation Under Reduced Pressure. Organic compounds vary widely in their sensitivity to heat, but all decompose or react in some way if heated enough. In many instances some chemical change occurs if the compound is heated to its boiling point at atmospheric pressure. Distillation remains the simplest method for purification of most organic liquids; when they are unstable at normal boiling point, it is often convenient to distill at reduced pressure, which, of course, lowers the boiling point. For example, benzyl cyanide, prepared as described in §208, suffers slight decomposition at its normal boiling point, 234°, but distills as a clear colorless compound at 30 torr where it boils at about 128°. Most compounds that boil above 150° at atmospheric pressure should be distilled at reduced pressure. The technique for such distillation is described in Chapter 6.

37. Estimation of Boiling Point at Reduced Pressure. It is easy to find the boiling points of organic compounds at normal pressure in the literature, but such data for lower pressures are often unavailable and may be much less accurate (§66). It is thus important to be able to estimate boiling points at lower pressures from those at 760 torr. The following two very rough approximations are useful:

1. An organic compound boiling in the range 200–300° at 760 torr boils about 90–125° lower at 25 torr.
2. Below 25 torr the boiling point is lowered 10–15° each time the pressure is halved.

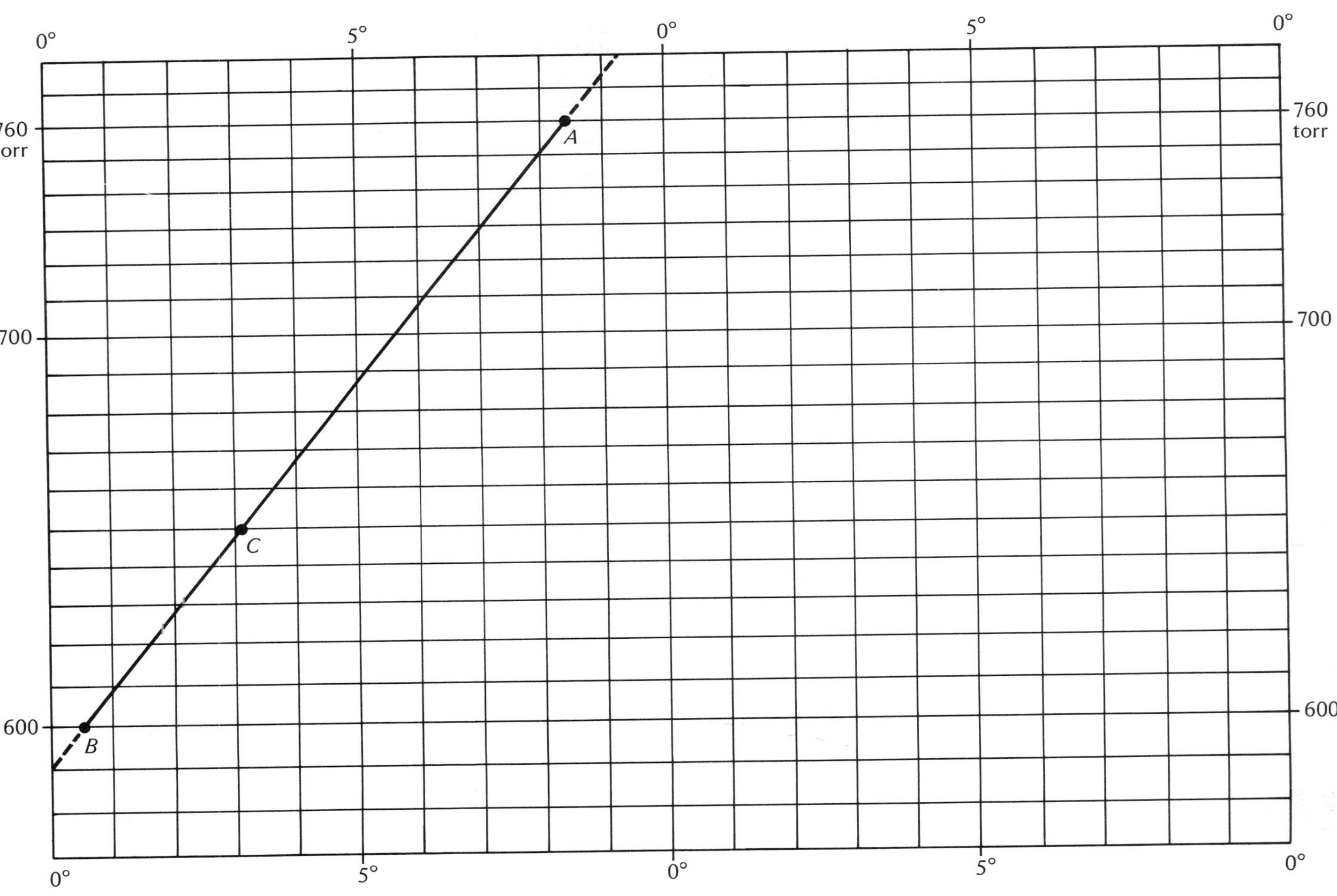

Fig. 2. Boiling Points at Higher Altitudes.

Often more accurate estimates are needed. The Clapeyron equation can be used (integrated form), provided the assumptions are valid and the molar heat of vaporization is known and is constant over the temperature range involved. In practice more accurate estimates can be made with empirical relations that have been derived (Weissberger, *Technique of Organic Chemistry,* **1** (3d ed.), Pt. 1, 471–503). Organic compounds can be grouped into types for which the vapor pressure–temperature relationships are closely similar. R. R. Dreisbach [*Pressure-Volume-Temperature Relationships of Organic Compounds,* Handbook Publishers, Inc. (1952), now available from McGraw-Hill, New York] has prepared tables that give quite accurate boiling points at various reduced pressures for each type of compound when the normal boiling point is known. The table on page 39 was prepared from the Dreisbach tables and gives rounded-off boiling points at 760 and 10 torr for several types of compounds. The problem of interpolation or extrapolation of a boiling point at intermediate pressures or pressures between 1 and 10 torr from values at these two pressures is simple. Figure 3 is a graphic blank form that can be used for this purpose. For example, 1,3,5-triisopropylbenzene boils at 235° at 760 torr, but no boiling points at lower pressure have been recorded. A student wishes to distill this compound at 25 torr. On the line "alkylbenzenes" 235° corresponds to approximately 100° at 10 torr. These values are plotted on Fig. 3 and connected by a straight line; interpolation then gives 121° as the estimated boiling point at 25 torr. An interesting modification of such plots has been devised [G. R. Robertson, *J. Chem. Ed.,* **27,** 341 (1950)]; this is more accurate for hydrophilic compounds such as glycols or alcohols.

Various other methods are available for estimation of boiling points at reduced pressure. D. R. Stull [*Ind. Eng. Chem.,* **39,** 517 (1947)] has published a table of vapor pressures at 1, 5, 10, 20, 40, 60, 100, 200, 400, and 760 torr of about 1200 pure organic compounds. A table prepared by G. G. Schlessinger covering mainly these same compounds now appears in the *Handbook of Chemistry and Physics,* R. C. Weast, ed., The Chemical Rubber Co., Cleveland, Ohio; this table gives constants A and B for the modified Clapeyron equation:

$$\log_{10} p = -0.2185 \frac{A}{K} + B$$

where p is in torr, A is the molar heat of vaporization, and K is temperature in degrees Kelvin. This handbook also has a short section on correction of boiling points to standard pressure by H. B. Hass and R. F. Newton; a formula involving the entropy of vaporization at 760 torr and a graph permitting estimation of these entropies are used. Calculation of vapor pressures at temperatures lower than the normal boiling point is also covered.

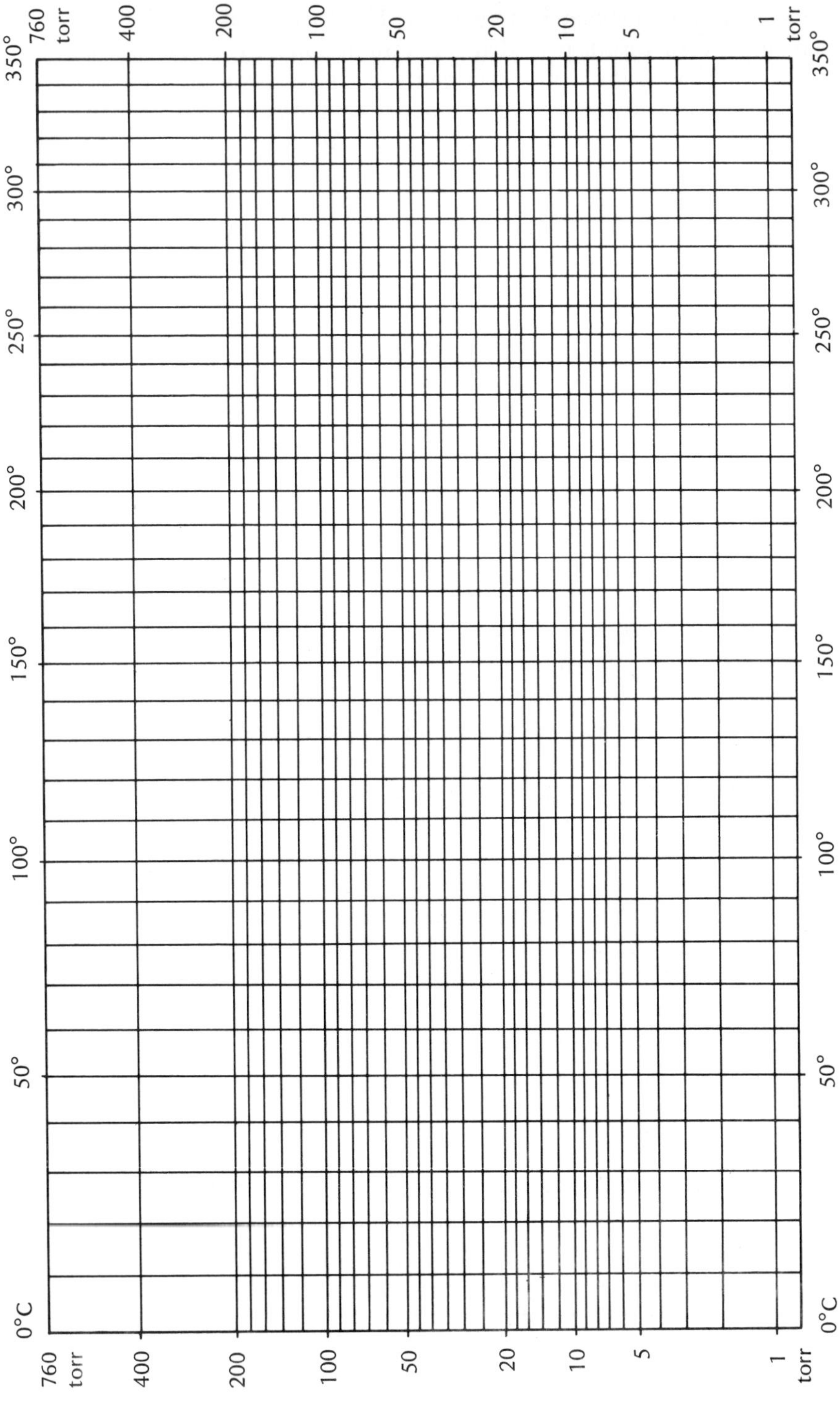

Fig. 3. Chart for Plotting Boiling Points at Reduced Pressure.

Boiling Points of Type Compounds at 10-torr Pressure[a]

Boiling Point (°C) at 760 torr	*120*	*130*	*140*	*150*	*160*	*170*	*180*	*190*	*200*	*210*	*220*	*230*	*240*	*250*	*260*	*270*	*280*	*290*	*300*	*310*	*320*	*330*
Alkanes, alkenes	13	20	28	35	43	51	59	67	75	83	91	99	107	115	123	131	140	148	156	165	173	182
Aliphatic halides	13	21	28	36	43	51	59	67	74	82	90	98	106	114	122	130	138	146	154	162	—	—
Aliphatic alcohols	29	37	44	52	59	67	74	82	89	97	104	112	119	127	134	142	149	157	164	172	179	187
Aliphatic ketones	15	22	30	37	44	52	59	66	74	81	88	96	103	110	118	129	133	140	148	155	163	170
Aliphatic acids	19	29	39	50	58	66	74	82	90	98	106	114	123	131	139	148	156	164	173	—	—	—
Aliphatic ethers	15	22	30	37	45	53	60	68	76	83	91	99	107	115	123	131	139	147	—	—	—	—
Aliphatic esters	18	25	33	40	48	56	64	71	79	87	95	103	—	—	—	—	—	—	—	—	—	—
Aliphatic nitriles	14	21	29	36	44	51	59	66	74	81	89	97	104	112	119	127	135	—	—	—	—	—
Aliphatic amines, primary	15	23	31	39	47	55	63	69	77	85	93	101	109	118	126	134	142	151	159	167	176	184
Alkylbenzenes, -cyclohexanes	13	21	28	35	43	51	58	66	73	81	89	97	104	112	120	128	136	144	152	160	—	—
Naphthalenes	—	—	—	—	—	—	—	—	—	82	91	99	108	116	125	134	143	152	161	170	179	188
Nitrobenzenes	—	—	—	—	—	—	—	—	—	82	90	97	105	113	120	128	136	144	151	159	167	—
Alkylphenols	—	—	—	—	—	—	67	75	83	90	98	106	113	121	129	137	145	152	160	168	176	184
Aromatic aldehydes	—	—	—	—	—	51	60	69	78	88	98	108	117	126	137	148	158	169	—	—	—	—
Aromatic and aliphatic–aromatic ketones	—	—	—	—	—	—	—	—	76	84	92	99	107	115	122	130	138	146	153	161	169	—
Aromatic amines	—	—	—	—	—	—	62	70	78	85	93	101	109	117	124	132	140	148	—	—	—	—
Aliphatic–aromatic ethers	—	—	—	37	45	53	60	68	75	83	91	98	106	114	122	129	137	145	153	161	—	—
Cyclopentanes	9	19	26	34	42	49	57	64	72	80	88	96	103	111	119	127	135	—	—	—	—	—
Nitroparaffins	15	23	31	39	46	55	62	70	78	86	94	103	111	118	126	135	143	151	159	—	—	—

[a]Values rounded off from data in *Pressure-Volume-Temperature Relationships of Organic Compounds* by R. R. Dreisbach; with permission of Handbook Publishers, Inc., Sandusky, Ohio.

It is often convenient to use a nomograph that permits rapid graphic estimation of vapor pressure–temperature data [Weissberger, *Technique of Organic Chemistry,* **1** (3d ed.), Pt. 1, 471–503; S. B. Lippincott and M. M. Lyman, *Ind. Eng. Chem.,* **38,** 320 (1946); F. E. E. Germann and O. S. Knight, *ibid,* **26,** 467 (1934)]. Even more convenient is a vapor pressure slide rule devised by F. T. Miles [*Ind. Eng. Chem.,* **35,** 1052 (1943)].

DISTILLATION OF MIXTURES

38. Effect of Nonvolatile Impurities. The continuous curve of Fig. 4 represents the vapor pressures of pure water. Suppose now that a small amount of the nonvolatile substance sugar is dissolved in the water. Sugar molecules interfere with the escape of water vapor; therefore the vapor pressure at any given temperature will be below normal. A second vapor-pressure curve, resembling the first, will fall just below the curve for pure water. This is indicated by the dashed curve in Fig. 4.

At a given pressure, such as 760 torr, the boiling point originally at A (100°) is elevated to B. Doubling of the sugar/water ratio in very dilute solution would double the boiling-point elevation. Such changes are accounted for by Raoult's law: *The partial vapor pressure of a component in a solution varies directly with its mole fraction.*

Expressed more formally,

$$P_A = PN_A$$

in which P is the vapor pressure of a pure liquid substance A, P_A is the partial vapor pressure of the substance in a solution obeying Raoult's law, and N_A is the mole fraction of the species A in the solution.

39. Mole Fraction and Mole Percentage. It is convenient to express the composition of solutions in mole fractions or mole percentages rather than weight percentages when dealing with vapor pressure. If the mole fractions in a two-component solution are N_A and N_B,

$$N_A = \frac{\text{(number of) moles of A}}{\text{moles of A} + \text{moles of B}} \quad \text{and} \quad N_A + N_B = 1$$

Mole percentage is simply mole fraction × 100. Mole percentage (or mole percent) of A = $100N_A$ and mole % A + mole % B = 100. For example, a propanol–water solution was made by dissolving 120 g of propanol (molecular weight 60) in 144 g of water (molecular weight 18), total 264 g. The composition in weight percentage is (120/264)100 = 45.4% propanol and (144/264)100 = 54.6% water. The solution contains 120/60 = 2 moles of propanol and 144/18 = 8 moles of water, total 10 moles. The mole fraction of propanol is 2/10 = 1/5 = 0.2 (mole percentage 20), and the mole fraction of water is 8/10 = 4/5 = 0.8 (mole per-

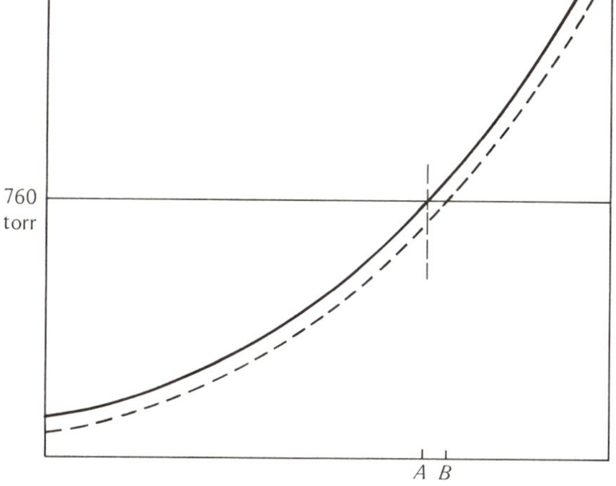

Fig. 4. Vapor Pressures of Water and of Sugar Solution.

centage 80). *Please note that any percentage figure cited in this book is for weight percent unless specified to the contrary.*

40. Application of Raoult's Law. Starting with 1000 moles of pure water at its boiling point 100°, one may now replace 1 mole of this solvent with 342 g, or 1 mole, of sugar. The fraction of solvent is reduced exactly 0.1 mole %, or to 99.9 mole % of its former value.

As Raoult's law would predict, and as actual experiment shows, the vapor pressure of the liquid falls to 99.9% of its former value, that is, from 760 to 759.24 torr. Increase of the fraction of sugar to 0.2 mole % (2 moles of sugar to 998 moles of water) will then lower the aqueous vapor pressure to 99.8% of 760, or 758.48 torr, and so on.

Such regularity of behavior occurs in almost any solution containing very low concentrations of the nonvolatile solute. As the solution becomes more concentrated, substantial deviations from the rule appear, although the general trend persists in at least rough compliance with the Raoult concept.

At these low concentrations, boiling point rises with similar regularity as vapor pressure falls. This enables one to reverse the experiment. It is possible to estimate mole fraction, and thus molecular weight, given a reasonably accurate determination of elevation of boiling point. In very dilute solutions the boiling point of water rises at the rate of about 0.3° per mole % of solute, under the assumption that the latter is not ionized.

On the basis of the foregoing discussion, it would not be satisfactory to determine the boiling point of a solvent by placing the thermometer in a solution of some foreign substance in that solvent, even if that foreign substance were nonvolatile. If, however, the thermometer is elevated to a point near the delivery sidestem of a distilling flask, it may be possible in a slow distillation to allow pure recondensed liquid to flow back over the thermometer bulb. The liquid, at equilibrium with the slowly rising vapor, will then be at the true boiling point, even if (nonvolatile) solute is present below. It is thus the standard practice to mount thermometers near the sidestem in distilling flasks.

IDEAL SOLUTIONS

41. Mixtures of Two Liquid Substances. If two liquid substances are chemically similar but do not react with each other, they will probably be miscible in all proportions and their mixtures will obey Raoult's law. Such "ideal solutions" are well illustrated by mixtures of the two inert paraffin hydrocarbons *n*-pentane and *n*-heptane.

The behavior of these two hydrocarbons is shown in the following problems.

Problems

1. What is the vapor pressure of a mixture of 108 g of *n*-pentane and 350 g of *n*-heptane at room temperature (20°)?

$$\begin{aligned} \text{108 g of } n\text{-pentane (mol. wt. 72)} &= 1.5 \text{ moles} \\ \text{350 g of } n\text{-heptane (mol. wt. 100)} &= \underline{3.5 \text{ moles}} \\ \text{Mixture} &= 5.0 \text{ moles} \end{aligned}$$

According to Raoult's law the partial pressure of *n*-pentane in the mixture will be 1.5/5.0, or 30%, of the vapor pressure of pure *n*-pentane at 20°, since 30% of the molecules are of that species. Similarly the partial vapor pressure of *n*-heptane will be 3.5/5.0, or 70% of the corresponding value for pure *n*-heptane. (See tabular data in §577.)

$$\begin{aligned} 0.30 \times 420 \text{ torr} &= 126 \text{ torr} \\ 0.70 \times 36 \text{ torr} &= \underline{25.2 \text{ torr}} \\ \text{Vapor pressure of mix} &= 151.2 \text{ torr} \end{aligned}$$

2. What is the initial boiling point of a mixture of 36 g of *n*-pentane and 150 g of *n*-heptane in a laboratory where the barometric reading is 756 torr?

A mathematician would expect to solve this problem by use of equations for the two curves in Fig. 5. In the absence of such equations, a "cut-and-try" method is chosen.

First the mole fraction is calculated; in this case $\frac{1}{4}$ pentane, $\frac{3}{4}$ heptane. According to Raoult's law, $\frac{1}{4}p_1$ (pentane) + $\frac{3}{4}p_2$ (heptane) must equal 756 torr. A rapid survey of the vapor-pressure table, with one or two possible wrong guesses, suggests that the values at 70° are promising, and these values are accordingly tested, and it is found that

$$\tfrac{1}{4}(2119) + \tfrac{3}{4}(302) = 756.25$$

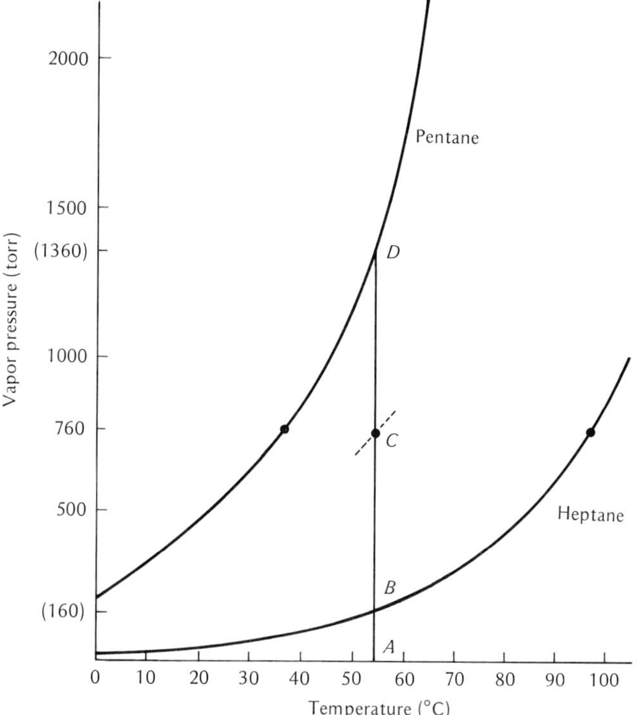

Fig. 5. Vapor Pressures of Mixtures of *n*-Pentane and *n*-Heptane.

This result, 756.25, is so close to the 756 torr specified in the problem that one may immediately conclude, with negligible error, that the boiling point of the mixture in question is 70°.

Such a neat coincidence is rare. More often the desired result is not a simple multiple of 10 and thus cannot be spotted directly in the customary table given in 10° intervals. One then resorts to the method of plotting on cross-sectional paper, as in problem 3.

3. What is the normal boiling point of a mixture of 144 g of pentane and 200 g of heptane?

The vapor pressures of this mixture are calculated by the method of problem 1 for several 10° intervals in the region where the boiling point is expected. As soon as two or more values next above 760 torr and two or more below 760 torr are found, the vapor-pressure curve in the region of interest is plotted, as in Fig. 6. In this case the temperatures 40°, 50°, 60°, and 70° are plotted against 483, 667, 907, and 1211 torr, respectively. A smooth curve through the four points, marked in Fig. 6 by small circles, shows an intersection with the horizontal line for 760 torr at 54°. The boiling point is therefore 54°, as marked at C in Fig. 5.

4. What mixture of pentane and heptane will boil at 60°? Let $x =$ the mole per cent of pentane, and $100 - x =$ the mole per cent of heptane.

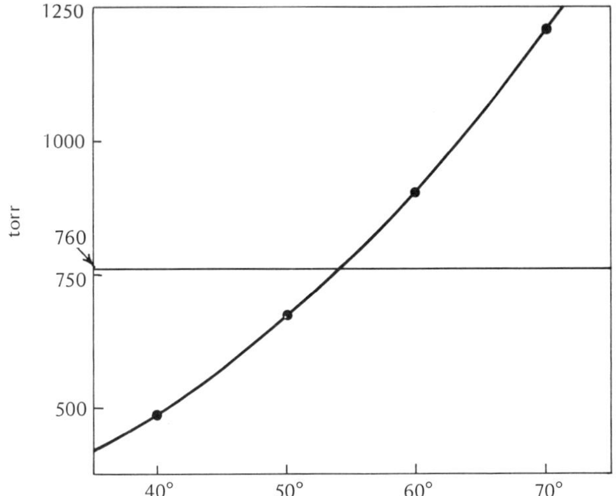

Fig. 6. Determination of Boiling Point.

According to data from §577 and Raoult's law, the partial vapor pressure of pentane will be $x\%$ of 1605 torr, the vapor pressure of pure pentane at 60°. Similarly, the partial pressure of heptane will be $(100-x)\%$ of 209 torr. The sum of these partial pressures must be 760 torr if the mixture is to be at its boiling point.

$$\left(\frac{x}{100} \times 1605\right) + \left(\frac{100-x}{100} \times 209\right) = 760$$

$$x = 39.5 \text{ (mole \% pentane)}$$
$$100 - x = 60.5 \text{ (mole \% heptane)}$$

If the proportions by weight are desired, it is convenient to calculate the weight of a total of 100 moles of the mixture:

39.5 moles ×	72 (mol. wt. pentane) =	2844 g pentane
60.5 moles ×	100 (mol. wt. heptane) =	6050 g heptane
100 moles mixture	=	8894 g

It is evident from Fig. 5 that the boiling point of a 50 mole % mixture lies closer to the boiling point of pentane than its mere mole fraction would suggest. In cases where the boiling points of the two components are still farther apart, the tendency of the boiling point of a mixture to be relatively low is more noticeable.

This behavior in part explains the fact that ether (bp 35°) is easily separated from aniline (bp 184°) by distillation.

5. What would be the molar composition of the first yield of vapor that would rise when the mixture cited in problem 4 above is boiled at 60°? By Raoult's law,

$$39.5\% \text{ (1605 torr)} = 634 \text{ torr}$$

the partial pressure of pentane in the boiling liquid, and

$$60.5\% \text{ (209 torr)} = 126 \text{ torr}$$

the partial pressure of heptane in that liquid.

Since the liquid in question is at equilibrium with the vapor produced in the boiling process, the partial pressures of the two hydrocarbons in the vapor will also be 634 and 126 torr, respectively. But Avogadro's law states that the partial pressures of the components in a mixture of gases measure the relative number of molecules of each species, respectively, in unit volume of the mixture. A simple illustration is seen in dry air at sea level (760 torr). The partial pressure of oxygen in such air is approximately 0.21 atm, or about 159 torr; of nitrogen, 0.78 atm, or about 593 torr; of minor constituents, 0.01 atm, or about 8 torr. Therefore air contains 21 molecules of oxygen for each 78 of nitrogen, or 159 oxygen to 593 nitrogen.

Similarly, the hydrocarbon vapor will contain 634 molecules of pentane for every 126 molecules of heptane, total 760 molecules. Dividing 634 by 760 and multiplying by 100, one determines the mole % of pentane, which is 83.4.

42. Boiling Point–Composition Diagram. Similar calculations give the compositions of mixtures boiling at 50°, 70°, 80°, and 90° as well as the composition of the vapor in equilibrium with each. These values are shown in the following table:

Mixtures of Pentane and Heptane

Mixture	*Boiling Point, °C*	*Mole % Pentane in Liquid*	*Mole % Pentane in Vapor*
A	50	59	92
B	60	39.5	83.4
C	70	25	70
D	80	14	52
E	90	7	27

Figure 7 is a plot of these values along with the boiling point of pure pentane (P) and pure heptane (H). The solid curve passing through PABCDEH represents the boiling points of all mixtures of pentane and heptane. The dashed curve represents vapor compositions (PA′B′C′D′E′H). The horizontal lines A′–A, B′–B, etc., are tie lines which represent the compositions (A′, B′, . . .) in equilibrium with the corresponding liquid (A, B, . . .) at each temperature. Notice that if a mixture of any composition is heated to boiling, the vapor coming off is richer in the more volatile component (in this case pentane) than the

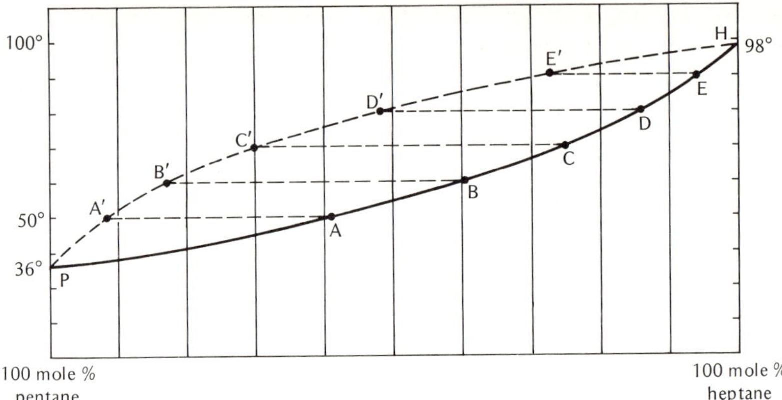

Fig. 7. Boiling Point-Composition Diagram.

starting liquid. This is the fundamental fact which makes separation by distillation possible.

43. Change During Distillation. It is clear that a vapor of which 83.4 mole % is pentane cannot be withdrawn from a liquid containing only 39.5 mole % pentane without immediate disturbance of the conditions initially present in the boiler. With loss of pentane the boiling point rises, and the distillate becomes increasingly richer in heptane; but the distillate being produced at any given moment is always richer in pentane than is the boiler liquid existing just at that moment. The temperature, at first just 60°, rises steadily and approaches 98° as the boiler goes dry.

Two practical conclusions may be drawn at once. First, a rising temperature of boiling means that *at least two* substances are present. That is, the liquid is certainly not a pure substance—or if originally pure, it has decomposed at least partially. Second, a "straight" distillation does not give a satisfactory separation of two components. Two improvements of method are possible: *(a)* to distill repeatedly into a succession of receivers, as described in §50, or *(b)* to employ a fractionating column (§51).

44. Boiling Range. Distillation is frequently the final step in purification of an organic liquid, and in practice it is difficult to obtain a product of such high purity that the boiling point is precisely constant as recorded on an ordinary laboratory thermometer. For most purposes a relatively pure compound that has a boiling range of 0.5–1.0° is acceptable. *The actual boiling range should be reported.* Students are warned that even when a commercial catalog lists an organic liquid as boiling at a single temperature (e.g., ethylbenzene, bp 136°) it is unlikely that the material supplied will actually boil this sharply on redistillation or have a purity >99%; the purity may be much less. In recent years firms supplying chemicals have improved the reliability of their service; a boiling range or other indication of purity is often given. In spite of this a practicing chemist should always be skeptical of the purity and identity of his starting materials unless he has established these himself.

Problems on Molecular Weight Relations

1. A 3% solution of glucose in water boils at 100.09°. What would be the boiling point of a 4% solution of glycerol, under the assumption that glycerol, like glucose, has practically no vapor pressure under the conditions of this experiment?

2. The molecular elevation of the boiling point of a solvent is usually given in degrees Celsius for 1 mole of solute dissolved in 1000 g of solvent. For water this value is 0.512° (at standard pressure). In §40 a rough value for the constant in different units is given as 0.3° rise in boiling point for each mole percent of solute. Calculate a more precise value for the constant in terms of mole percent.

3. A solution of 20 g of a compound in 1 liter of water boils at 100.17° at standard pressure. What is the molecular weight of the compound?

4. The molecular elevation of the boiling point of benzene is 2.53° for 1 mole of solute in 1000 g of benzene. The boiling point of benzene is 80.099°. A solution of 28.5 g of an unknown aromatic hydrocarbon in 1000 g of benzene boiled at 80.415°. What is the molecular weight of the compound?

5. When pure lactic acid, $CH_3CHOHCO_2H$, is distilled, the boiling point rises steadily. What kind of chemical reaction might be responsible for this unorthodox behavior?

Problems on Distillation

See §577 on vapor pressure for necessary data.

1. What is the vapor pressure of a mixture of 18 g of *n*-pentane, 15 g of *n*-heptane, and 11.4 g of *n*-octane at 30°?

2. What is the boiling point of a 15 mole % solution of *n*-pentane in *n*-octane at standard pressure? What would be the boiling point of this solution at an elevation of 1000 ft above sea level where the barometer reading was 740 torr?

3. When a mixture of 2 moles of bromobenzene and 1 mole of *p*-dibromobenzene was distilled under reduced pressure the first drop of distillate came over at 90°.

(a) What was the pressure in the distilling flask?

(b) What was the mole fraction of bromobenzene in the first drop of distillate that came over?

(c) What was the percentage composition by weight of this first drop?

(d) Qualitatively what change would take place in the composition of the residual liquid in the distilling flask as the distillation proceeded?

4. What mixture of *p*-dibromobenzene and α-bromonaphthalene will boil at 130° at 30.6 torr?

5. Calculate (a) the mole fraction of *n*-octane in that mixture of *n*-octane and *n*-heptane which boils at 100° and (b) the weight percentage of *n*-octane in the first vapor coming from this boiling mixture.

6. Students frequently report that the boiling point fell at some point during a distillation. Do you think this possible? Why?

NONIDEAL SOLUTIONS

45. Deviations from Raoult's Law. It is relatively unusual for a solution to display ideal behavior. Many mixtures of liquids have greater vapor pressures than would be expected from Raoult's law and are said to show a positive deviation. With other mixtures a negative deviation is encountered.

46. Positive Deviation. In this case two liquid substances seem to have a tendency to escape from each other's presence. In extreme cases they part company and separate as two layers. This mutual "dislike" leads to abnormally high combined vapor pressures in mixtures. The boiling point of such a mixture is depressed, even below the normal boiling point of the more volatile of the two components.

47. Aqueous Ethyl Alcohol. The system ethyl alcohol–water is the best-known case of positive deviation. At one composition (95.57%) aqueous ethyl alcohol boils at 78.15°, a temperature slightly below the boiling point of pure ethyl alcohol. For convenience in oral explanation, this percentage is rounded off to "96%" in the following paragraphs.

Although the boiling point of 96% alcohol is only 0.15° below 78.3°, the boiling point of pure alcohol, this small difference is enough to make much trouble in practical manufacture of the pure substance. The phenomenon makes it impossible to produce pure ("absolute") alcohol by any kind of fractional distillation of mixtures of alcohol and water alone. As a result, commercial ethyl alcohol, as used for extracts, medicines, beverages, etc., contains not over 95.57% of the essential component and usually slightly less.

48. Distillation. In Fig. 8 it is seen that 50% (aqueous) alcohol (A) will yield a distillate (A′) richer in alcohol, just as one would expect in a case complying with Raoult's law. On the other hand, 99% alcohol (B) yields a distillate (B′) richer in water—richer in the *less volatile* component. If 96% alcohol is distilled, vapor of unchanged (96%) composition

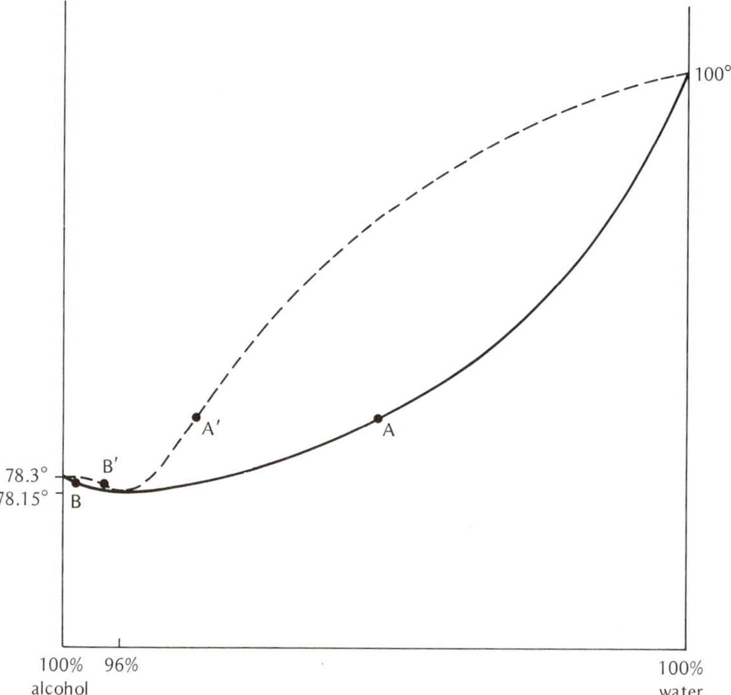

Fig. 8. Boiling Point–Composition Diagram for Ethyl Alcohol and Water.

comes over. In such a case the entire quantity of liquid could be distilled without change in temperature or composition. Such a mixture is commonly called a "constant-boiling mixture," "azeotropic mixture," or, more briefly, "azeotrope." If it consists of but two components, it is often called a "binary mixture."

Repeated partial distillation of any mixture of alcohol and water other than the azeotrope itself tends to give, as a distillate, a mixture nearer to the 96% value than does the original. Eventually the azeotropic composition is reached. If a fractionating column is included in the apparatus, the azeotropic composition is merely attained more rapidly. One cannot reach the 100% mark. To prepare anhydrous alcohol, the water must be removed chemically (for example, with calcium oxide) or by addition of a substance that will form an even lower-boiling azeotrope with the water. The latter method is used commercially by addition of benzene to form a three-component azeotropic mixture (a ternary mixture) that boils at 64.9°. The tables on page 50 give several binary mixtures that occur with common compounds and two ternary mixtures, including the water–ethanol–benzene azeotrope. Many more can be found in the References.

Binary Mixtures of Minimum (Constant) Boiling Point

Components	*Boiling Point, °C*	*Composition, %*	*Constant Boiling Point, °C*
Benzene	80.2	66.7	71.9
Isopropyl alcohol	82.5	33.3	
tert-Butyl alcohol	82.8	11.7	79.9
Water	100	88.3	
Toluene	111	50	85.8
Formic acid	100.8	50	
Methyl alcohol	64.7	44	62.3
Ethyl acetate	77.1	56	
Ethyl alcohol	78.3	20	65.7
n-Butyl chloride	78.1	80	

Ternary Mixtures of Minimum (Constant) Boiling Point

Components	*Boiling Point, °C*	*Composition, %*	*Constant Boiling Point, °C*
Water	100	7.4	64.9
Ethanol	78.3	18.5	
Benzene	80.2	74.1	
Methyl formate	31.9	52	16.95
Ethyl bromide	38.4	5	
Isopentane	27.95	43	

49. Negative Deviation. Where two liquid substances are not "hostile" to each other, they may cling together, at least loosely. Vapor pressures of mixtures are low, and boiling points high. As pictured in Fig. 9, acetone and chloroform yield a constant-boiling mixture at maximum temperature.

Distillation of an acetone–chloroform mixture of any composition other than the 20–80 azeotrope will cause gradual elimination from the system of whichever component, acetone or chloroform, happens to exist in excess of the azeotropic fraction, 20 or 80 as the case may be. Thus the residue in the boiler will eventually become 20–80, provided the boiler does not first run dry.

It is presumed that the hydrogen in chloroform, known to be somewhat labile, or loosely held, coordinates to some extent with unshared electrons on the ketone oxygen in acetone. One may then assume at any given time that part of the acetone and chloroform is tied up in rela-

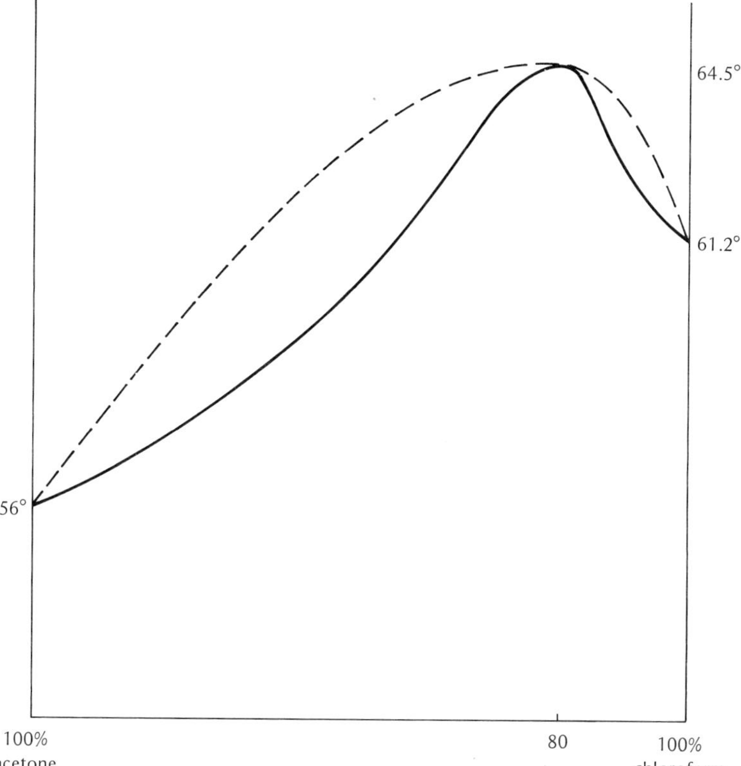

Fig. 9. Negative Deviation.

tively large complexes that would not exert vapor pressures such as those of the free pure components. On this assumption the effective percentages of free acetone and chloroform are not really 20 and 80 but some lower numbers not precisely known. Following Raoult's law, the total vapor pressure would fall short of the sum based on simple theory. Unduly low total vapor pressure would then necessitate abnormally high boiling point.

A limited number of other molecular pairs yield maximum-boiling azeotropes. Many of these depend on partial union of a more or less volatile weak base with a somewhat volatile weak acid to form a small amount of nonvolatile salt.

Negative deviation is well known in the field of common inorganic acids. For example, hydrochloric, hydrobromic, hydriodic, and nitric acids form azeotropes with water, including common "concentrated" (68%) nitric acid. This sometimes leads inexperienced workers to have insufficient respect for the less common, really concentrated, and dangerous fuming nitric acid; see §579.

Several common azeotropic mixtures of maximum boiling point are given in the following table.

Mixtures of Maximum Boiling Point

Components	*Boiling Point, °C*	*Composition, %*	*Constant Boiling Point, °C*
Acetic acid	118.1	31.3	162.
Triethylamine	89.4	68.7	
Chloroform	61.2	22	64.5
Methyl acetate	57.1	78	
Formic acid	101	77	107.1
Water	100	23	
Methyl ether	−23.7	39	−2.
Hydrogen chloride	−83.7	61	
Phenol	181.5	42	186.22
Aniline	184.4	58	

Problems

See tables in §§48 and 49.

1. Suppose a 50% solution of benzene in isopropyl alcohol were completely distilled. Describe in a general way, without exact calculation, how the composition of the distillate would change during the entire process.

2. Suppose that a solution of 10 g of toluene and 90 g of formic acid were distilled until just 50 g of distillate had been collected. In which vessel would the greater part of the toluene be, the boiler or receiver, at this point?

3. A mixture of equal weights of *tert*-butyl alcohol and water is completely distilled from a simple boiler without fractionating column or equivalent.

(*a*) State in what way—without exact calculation—the composition of the first milliliter of distillate would differ from the original mixture.

(*b*) How would the last milliliter differ from the original mixture?

4. Suppose a mixture of equal weights of chloroform and methyl acetate were distilled until 95% of the material had passed over into the receiver. What would be the composition of the liquid left in the boiler at this point?

FRACTIONAL DISTILLATION

50. Multiple Receivers. In early days, before development of modern fractionating columns, it was customary to separate two liquid compounds by repeated distillation into a succession of receivers. Since the principles thus employed are also the basis of column fractionation, it may be best to begin with an imaginary experiment employing such principles. A mixture of 500 ml of X, bp 70°, and 500 ml of Y, bp 120°, with no azeotropic phenomena involved, is distilled into five receivers.

First Distillation

Fraction	A_1	B_1	C_1	D_1	E_1
Boiling range, °C	80–86	86–93	93–101	101–108	108–115
Yield, ml	200	200	200	200	200

Variations in temperature and air currents in the laboratory might lead to slightly different temperatures and yields. The schedule is merely illustrative.

Fraction A_1 is now returned to the boiler. A series of redistillations now ensues in which each distillate, returned to the boiler, is thereby divided into portions that go into different receivers. Every time any distillate is partially redistilled, the new distillate is more like X than the liquid in the boiler, while the residue, passed along to later receivers, becomes more like Y. The scheme for distilling, pouring back, and redistilling becomes too intricate for ready comprehension in an abstract discussion like the current paragraph. The reader is therefore referred to §179, where the practical directions may be examined and (still better) tried in the laboratory. In the present discussion there is presented only the record of quantities and boiling temperatures at four more stages in the process, of which the last stage is the successful completion of fractionation.

Distillation No. 2

Fraction	A_2	B_2	C_2	D_2	E_2
Boiling Range, °C	75–79	79–89	89–104	104–113	113–117
Yield, ml	325	135	100	125	315

Distillation No. 3

Fraction	A_3	B_3	C_3	D_3	E_3
Boiling Range, °C	73–76	76–88	88–104	104–115	115–118
Yield, ml	360	110	70	105	355

Distillation No. 4

Fraction	A_4	B_4	C_4	D_4	E_4
Boiling Range, °C	71–73	73–86	86–104	104–117	117–119
Yield, ml	410	75	40	70	405

Distillation No. 5

Fraction	A_5	B_5	C_5	D_5	E_5
Boiling Range, °C	70–71	71–85	85–105	105–119	119–120
Yield, ml	475	20	10	20	475

Fractions A_5 and E_5 are fairly pure X and Y, respectively, and B_5, C_5, and D_5 may be discarded unless further material of the same kind is to be fractionated.

51. Use of Fractionating Column. A mixture of benzene (bp 80.6°) and toluene (bp 111°), which is an almost ideal solution, is first distilled through a simple tube without packing, as shown in Fig. 11. Following the notation in Fig. 10 (corresponding to that of Fig. 11) liquid A delivers a small amount of vapor A′ at the bottom of the column. Vapor A′ rises, strikes the walls of the relatively cool column, and starts to recondense. Such initial recondensation would mean the appearance of a small amount of whatever liquid mixture is normally at equilibrium with A′—which, according to Fig. 10, means liquid mixture A. The few drops of recondensed A fall back into the boiler. Some of the vapor still remains, however.

The operation of producing A from A′ must of necessity strip vapor A′ of a relatively large part of its toluene content, leaving the residual vapor richer in benzene. This residue is, of course, reduced in quantity; it is also cooler than A′. Its position on a chart like Fig. 10 must be lower than A′ (because cooler) and to the left of A′ (because richer in benzene). It is thus shown as B′. Of course B′ is just another of the various vapor mixtures that may be at equilibrium with liquid mixtures, which we mark as A, B, C, D, etc., in Fig. 10.

Vapor B′ rises to a higher level in the column, and again a small amount of recondensation occurs. Another quantity of condensate, this time of approximately B composition, runs back into the boiler. Vapor C′ re-

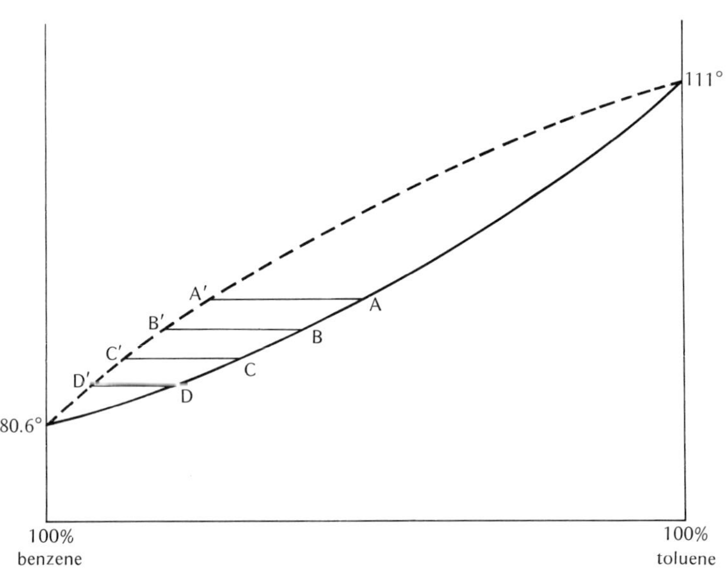

Fig. 10. Boiling Point-Composition Diagram.

mains as a residue. In like manner C′ becomes D′. If only enough vapor survives, a final residue, which is almost pure benzene, reaches the sidestem and the condenser. The yield is so small and the chance of complete interruption of distillation so great that the process is regarded as impractical. "Packing" must be introduced into the column to effect not only a more rapid but also a more distinct separation of the two components.

The column of Fig. 11 is now filled with glass beads or other appropriate "packing," and the distillation repeated.

Liquid mixture A produces vapor A′, A′ again produces a little A and changes to B′; B′ again produces a little *B*, all as before with a simple tube. But this time *B* does not immediately drop back into the boiler. It is detained in the lower part of the bead column, where it has time to stand in contact with a new supply of rising vapor A′. But B is not, and cannot be, at equilibrium with A′.

Whenever a liquid (such as B) and a vapor (such as A′), each made up of the same components but not at equilibrium with each other, are nevertheless brought together, there will be a transfer of material from one phase to the other in an attempt to reach equilibrium. In this case B contains too much benzene to be at equilibrium with A′. It will give up part of its benzene to the rising vapor A′. Similarly, A′, too rich in toluene to be at equilibrium with B, will give up toluene to phase B. Furthermore A′ is slightly cooled, B somewhat warmed.

Figure 12 shows the change. A′ becomes X′ and B becomes X. The exchange of material and leveling off in temperature cease. Liquid X falls into the boiler.

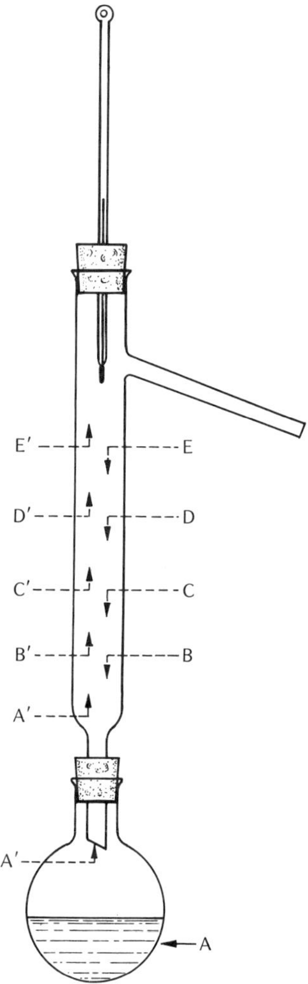

Fig. 11. Fractional Distillation.

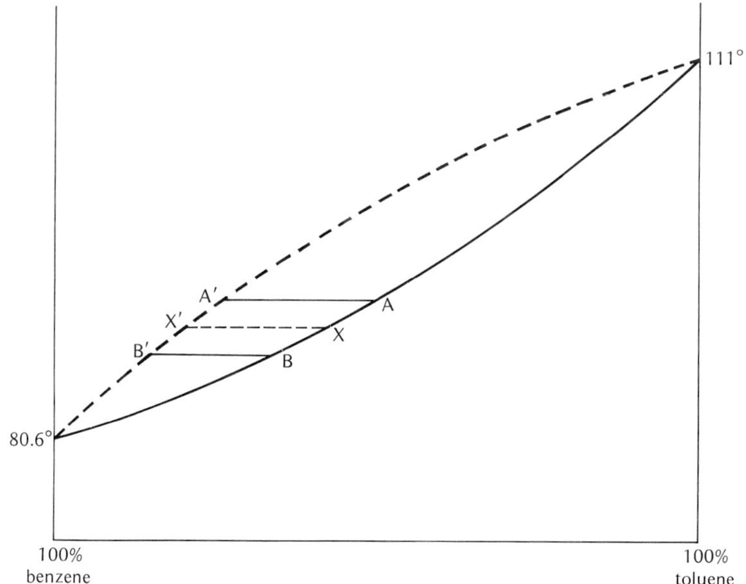

Fig. 12. Boiling Point–Composition Diagram for Distillation with a Column Containing Beads.

One might now proceed with explanations showing how vapor X′ may be encouraged to become B′, C′, etc. The important principle, however, is seen in the fact that vapor in the column has been enriched in benzene by mere holding of a falling mass of liquid, together with this vapor, in a zone of mechanical obstructions. If now a column of 3 ft length or (better yet) 50–100 ft length, as in industry, be employed, this constant enrichment of A′ in benzene finally yields at the sidestem of the column a vapor that is almost pure benzene.

52. Temperature Change. It has often been said that a fractionating column is in effect a succession of stills, each producing distillate that is then redistilled. This analogy is roughly true, as can be seen by recalling vapor A′ and liquid B into discussion, this time from the thermal standpoint. A′ is (relatively) hot. As does an oil bath, it causes liquid B to boil. Similarly, liquid B is (relatively) cool. As does a condenser, it receives upon its own surface additional distillate rich in toluene. The process of fractionation is thus a succession of numerous distillations that do not yet reach the main receiver.

53. Holdup of Liquid. Although the beads perform valuable service in separation of components, they introduce a serious difficulty by holding up an appreciable mass of condensate. The boiler goes dry prematurely, and there is no further opportunity to fractionate the mass of condensate still held between the beads. Industrially, such holdup is of little importance, since plenty of material lies in reserve. There is thus no objection to a very long column, running to the height of a four-story building or standing out in the weather like a cylindrical tower. Sharp fractional "cuts" are thus practical. In the laboratory one must resort to special designs to get comparable results.

To avoid holdup in the laboratory, the Vigreux design (Fig. 8, §75) is popular. In other models the general idea of a spiral passageway through the column seems to be of considerable value. Spinning band columns (§76) are especially efficient.

54. Distillation Curves. The comparative efficiency of distillation with and without a column is illustrated by curves showing the boiling points (as measured at the top of the column) at different times during a distillation. Figure 13 gives an example for distillation of a mixture of methanol and water (2:1) in which boiling point is plotted against weight of distillate. Curve *a* is that obtained by distillation from a simple flask; curve *b* is the result from slow distillation with a highly efficient column of low holdup. With the latter the temperature of distillation was constant at 65°, the boiling point of methanol, until nearly all of the alcohol had distilled. As the next few drops of distillate came over, the temperature rose rapidly to 100°; from there to the end of the distillation, pure water distilled and the temperature remained constant at 100°.

In practice, the sudden change from such a temperature as 65° to a much higher value like 100° may not proceed smoothly unless heat is at once applied more intensely. A low adjustment of the burner, adequate for the low-temperature part of the process, may not suffice to distill water. Thus the sudden exhaustion of the volatile substance may cause a suspension of distillation. Cool air surges back from the condenser, causing the thermometer to behave erratically and indicate temperature values that have no significance in the experiment.

An example of such comparison in distillation is given in the acetone experiment (§178).

When two liquids show a slight positive deviation from Raoult's law, not sufficient to cause an azeotropic minimum boiling point, as illustrated in the methyl alcohol and acetone examples mentioned above, it is comparatively easy to perform a distillation of the type pictured in §75. Contrariwise, irregular cases where the vapor and liquid curves lie closer together present difficult problems in fractional distillation.

55. "Theoretical Plate" Value. It is convenient to have a unit for measurement of efficiency of a fractionating column. Such a unit, long in general use, has been given the peculiar name "theoretical plate," a term related to the mechanical design of certain industrial columns. For example, a "5-theoretical-plate" column would be regarded as inefficient, whereas a 60-plate device would be a valuable research tool. The accurate determination of the plate value of a column is a specialty of chemical engineering, details of which are out of place here. A simple ideal example may be devised, however, to show the general meaning of the numerical plate value. For purposes of test on a certain fractionating column, one chooses a mixture of two liquids whose boiling

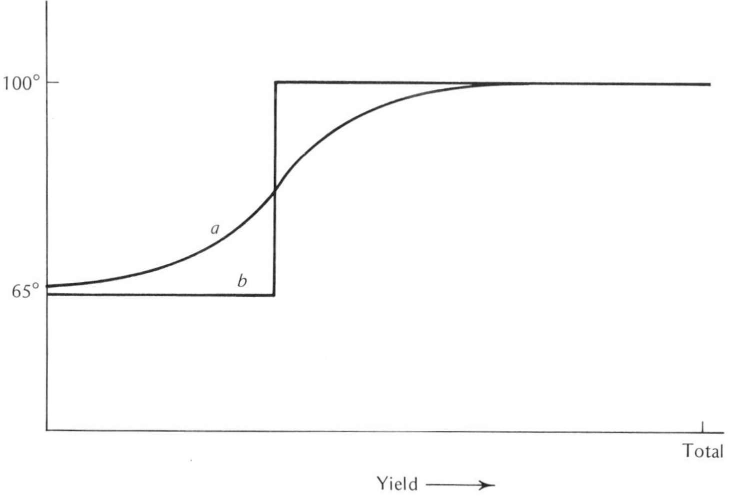

Fig. 13. Distillation Curve for Distillation of a Mixture of Methyl Alcohol and Water.

points are near together. For this example let these be substances A (bp 80°) and B (bp 82°). A mixture of 30% A, 70%B is arbitrarily chosen. Separation by fractional distillation is purposely intended to be difficult. A comparison is now made between the behavior of a sample of the 30–70 mixture in the column under test with that of a second sample distilled from a simple flask without column.

Sample 1 is carefully distilled through the column at fractionating equilibrium. The first output in the receiver is found to contain 42% of A. Although this is a mediocre achievement, it is somewhat better than the next result, obtained by distillation of the second 30–70 sample from a simple flask. The latter process is found to yield a product testing only 33% A.

The next operation is to redistill the 33% distillate from the simple flask; the result is a product about 36% A. A second redistillation, this time of the 36% product, yields 39% A, and the third and final redistillation gives 42% A. Since it required three redistillations to get a result equal to that gotten in one operation through the column, the latter is given a rating of three theoretical plates. Incidentally, chemical engineers employ graphical shortcuts to get such information, especially for columns of much higher plate rating, but this feature is omitted here.

The hypothetical column just described, though with low rating, might suffice for elementary laboratory work on easy separations, such as acetone (56°) from water (100°). Some simple columns, such as small Vigreux devices, do not have even a three-plate rating.

In laboratory research work, it is desirable to have high theoretical-plate rating without excessive height of column. This leads to the special characterization "HETP," or height equivalent to a theoretical plate. Thus a 30-plate column only 60 cm long has an HETP value of only 2 cm. Since it has small liquid capacity and minimum holdup, it is better for some research purposes than a 60-plate column 3 m high, with HETP of 5 cm.

The theoretical-plate rating assumes operation of the column at liquid-vapor equilibrium, with almost total reflux. As a result, a research column rated at 75 or 100 plates may have an output of less than 1 ml of distillate per hour. Faster operation would be possible, but fractionation would "break down" and the column would then behave as though it were of much lower theoretical-plate rating. Accordingly, maintenance of high plate value requires special temperature control.

FRACTIONATION OF NONIDEAL SOLUTIONS

56. Minimum-Boiling Mixtures. If a low-boiling azeotrope is distilled through a column, no change in composition occurs, either in boiler, column, or condenser. On the other hand, if the mixture of these particular components does not have the azeotropic composition, the column causes more or less change in composition. It is then customary to state that "fractionation occurs."

For example, if 100 g of aqueous ethyl alcohol (50%) should be distilled through an ideal, highly efficient column, 96% alcohol would be steadily received in the condenser until the boiler was completely stripped of alcohol. The temperature at the top of the column would remain constant at 78.15° during the distillation. Approximately 2 g of water would, by this time, have come over with the 50 g of alcohol, with yield of about 52 g of the azeotrope. Further attempt to distill would, after momentary interruption, cause about 48 g of water to come over at 100° (see Fig. 14).

57. Maximum-Boiling Mixtures. Mixtures in which negative deviation from Raoult's law produces maximum-boiling azeotropes and which are not at the azeotropic composition can also be separated into the azeotropic mixture and one of the pure components. This is illustrated in Fig. 15, which should be examined along with the corresponding boiling point–composition diagram of Fig. 9, §49.

Suppose a mixture of chloroform (40 g) and acetone (40 g) is distilled through an ideal fractionating column. At first pure acetone passes over until just 30 g has left the boiler. The balance is of the azeotropic composition, acetone (10 g) and chloroform (40 g). Fractionation ceases, and the entire residue, which is the azeotrope, will distill uniformly at 64.7° to the end.

58. Fractionation Under Increased Pressure. The great development in recent years of special petroleum products has made it desirable to conduct fractionation at pressures above 760 mm. Modern steel kettles and fractionating columns are thus useful over a wide range of pressure

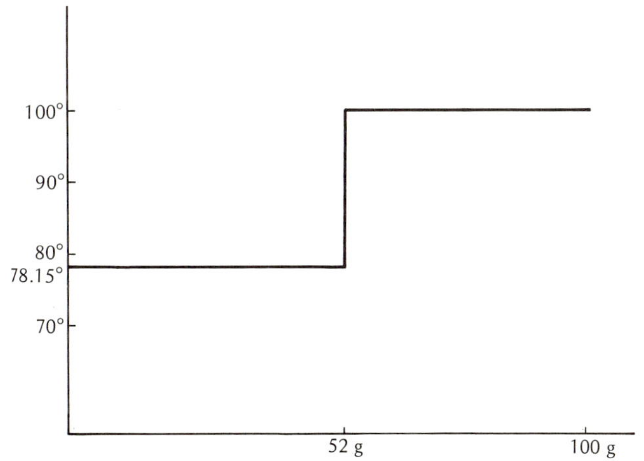

Fig. 14. Distillation Curve for Distillation of 100 g of a 50% Solution of Ethyl Alcohol Through an Efficient Column.

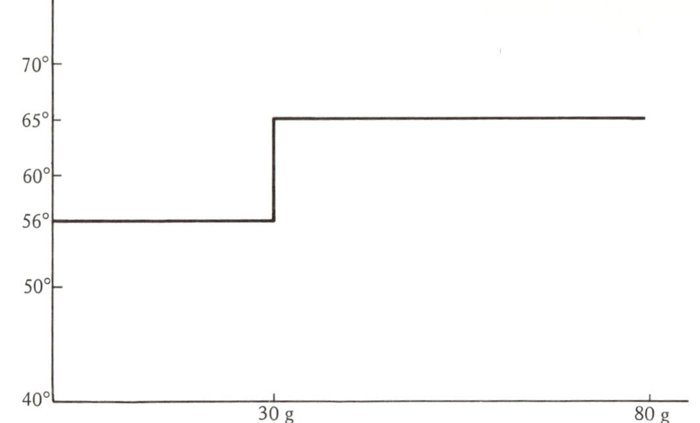

Fig. 15. Distillation Curve for Distillation of an Acetone–Chloroform Mixture Through a Column.

without change of essential principles. For example, the separation of the C_4 hydrocarbons, including butadiene, from compounds of less or greater carbon count, is a routine operation in the modern petrochemical industry. Fractionation under increased pressure, however, is less convenient in the laboratory and is seldom used.

MIXTURES OF TWO LIQUID PHASES

59. Codistillation; Steam Distillation. When a substance has so high a boiling point that decomposition is feared under simple distillation, the technique of reduced pressure may, of course, be adopted. Much simpler, especially if the material is viscous and tarry, is the device of codistillation, in which the desired compound is sent over in company with a lower-boiling liquid. Usually the new companion liquid is not miscible with the main desired product. Most often it is water, and the process is then known as "steam distillation."

The very fact that neither of the two liquids dissolves appreciably in the other is evidence at once that they do not obey Raoult's law. The situation may be regarded as an extreme case of "positive deviation" in which the combined vapor pressures are abnormally high. An example is seen in the distillation of a mixture of iodobenzene and water, two mutually insoluble liquids whose vapor pressures are represented in Figs. 16 and 17.

Iodobenzene and water behave independently of each other, even though in the same vessel. Each exerts its own full vapor pressure in addition to that of the other component. Necessarily the combined vapor pressures of the two will reach 760 torr before the temperature reaches 100°. To find the boiling point of the mixture, invert Fig. 16 and place

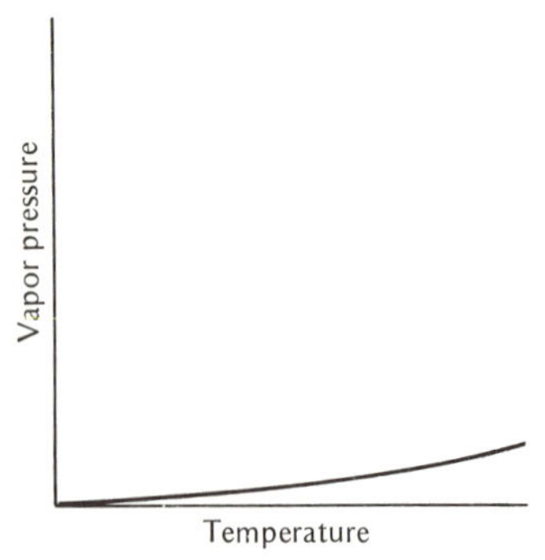

Fig. 16. Vapor Pressures of Iodobenzene.

it directly below Fig. 17, using the same horizontal or temperature axis, as in Fig. 18.

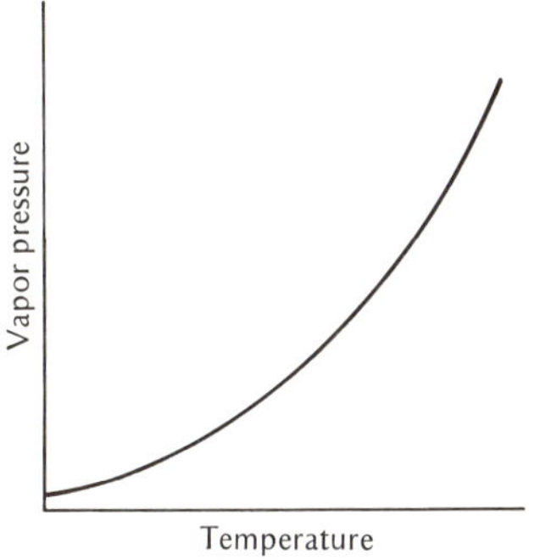

Fig. 17. Vapor Pressures of Water.

The line *DE* is graphically equal to 760-torr pressure. For the sake of simplicity, move *DE* along to the left and downward until it just fits between the curves for iodobenzene and water, as at *AC*. The intersection, *X*, marks the point where the sum of the vapor pressure of water (*AX*, 714 torr) and that of iodobenzene (*XC*, 46 torr) is 760 torr. This condition happens to occur at a temperature of 98.25°. The boiling point of this two-phase system is thus 98.25°. It does not matter how much of each liquid is present in the mixture. The above figures apply as long as both kinds of material are present.

If the iodobenzene is now replaced with a relatively nonvolatile substance like bromonaphthalene, the boiling point of the mixture will be nearer to 100°. The higher the normal boiling point of the organic component, the more closely the temperature of steam distillation approaches 100°. A low-boiling liquid, on the contrary, will boil together with water at a lower temperature. For example, a mixture of toluene (bp 111°) and water (bp 100°) boils at 85°. *In general the boiling point of a mixture of nonmiscible liquids is below the normal boiling point of the more volatile component.*

The variation of boiling point noted in cases of this type causes difficulty in the identification of certain organic substances when they are first distilled from crude reaction mixtures. For example, when *n*-butyl bromide is distilled from a mixture of sulfuric acid, butyl alcohol, water, and organic salts, as described in §198, the boiling point of the mixture proves to be low. It gives no definite clue to the composition of the distillate prior to the purification of the desired product.

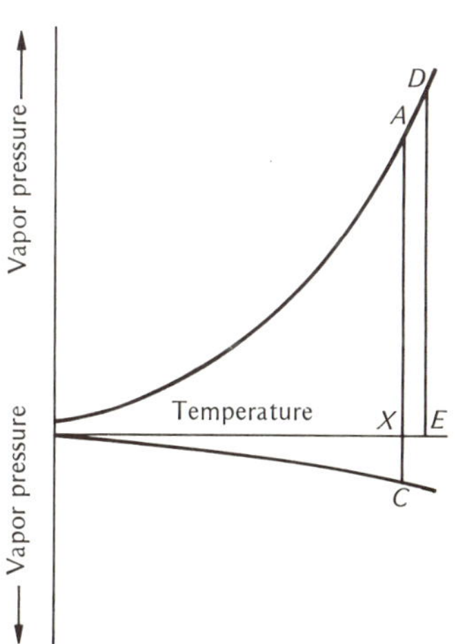

Fig. 18. Vapor Pressures of Iodobenzene and Water.

60. Efficiency of Two-Phase ("Steam") Distillation. The lines *AX* and *XC* in Fig. 18 measure the relative yields of water and iodobenzene, respectively, as collected in the receiver. In this case there will be 46 moles of iodobenzene obtained for every 714 moles of water. The truth of this simple deduction hinges upon Avogadro's law. Barring slight deviations from the simple gas laws, each molecule, regardless of size or kind, exerts the same individual vapor pressure—or partial vapor pressure—after escaping from the boiler. Thus the partial vapor pressures of 46 and 714 torr simply measure, respectively, the relative numbers of each species of molecule passing to the condenser.

The ratio of 46 to 714 cited above at first seems unfavorable in the attempt to obtain a large yield of iodobenzene in a short time. When one notes, however, that 46 moles of iodobenzene weigh 9.38 kg, whereas 714 moles of water weigh only 12.85 kg, the contrast is seen to be not great. The distillation is efficient, as over 42% of the distillate will be of the desired iodobenzene.

In practice, it is not customary to boil a mixture of water and the desired substance. As indicated in the practical directions for conducting the process (§79) steam is usually generated in a separate vessel. No change in principle is involved, however, The passage of a current of steam into an organic liquid at once introduces liquid water. The turbulent mixture of water, organic compound, and mixed vapors may be regarded as the practical equivalent of the contents of a simple flask where two liquids are being boiled together.

Actually the practice of using an outside source of steam is preferable. Direct heating of a glass vessel containing the two liquids might cause decomposition of the organic component. It might cause bumping. When the process is being used to purify a tarry preparation, direct heating might cause fracture of the flask at a point where the tar would stick and become scorched.

61. Special Cases. If two liquids are partially miscible, their mixture distills at constant minimum-boiling value. The situation is intermediate between the typical "ideal" distillation and common steam distillation of an insoluble oil.

Still another variation—distillation in superheated steam—does not follow ordinary Raoult's law predictions. The hot steam merely sweeps out the vapor of a refractory liquid, which does not yield readily to ordinary steam distillation below 100°. Such a process may be highly useful.

Problems

1. Why is the advice commonly given to conduct fractional distillation *slowly?*

2. Why is a large reflux ratio considered desirable for efficient fractionation with a column?

3. What does the position of the sidestem on a distilling flask have to do with the efficiency of fractional distillation from such a vessel, there being, of course, no column present?

4. Assuming that only a mediocre fractionating column were available, which working pressure would permit the sharper separation of bromobenzene from *p*-dibromobenzene in an equimolal mixture, 760 torr or reduced pressure in the region of 30–50 torr? Why?

5. Suppose you wished to remove the dangerously volatile and flammable butane and pentane from gasoline, leaving behind a maximum residue of the safer hexane, heptane, etc., for use in cleaning clothing. Which would be less wasteful of the desirable higher hydrocarbons: **(a)** distillation at ordinary pressure to remove the volatile substances or **(b)** blowing air through the gasoline to expedite evaporation of those substances? Explain in terms of vapor pressure–temperature curves.

6. Suppose you admitted, very slowly, a mixture of benzene (bp 80°) and toluene (bp 111°) at a point halfway up on an ideal fractionating column mounted over the customary boiler. Predict the general behavior of the apparatus, considering the possibility of getting a steady output of the pure substances. Where would you tap off the two products?

7. Draw a possible boiling point–composition curve for a mixture of two components whose slight positive deviation from Raoult's law is insufficient to make possible an azeotrope but nevertheless makes separation by distillation easier to perform than in the case of pentane–heptane.

8. Suppose you distilled a mixture of 50 g of formic acid and 50 g of water through an ideal fractionating column at a constant rate. Sketch approximately the curve that you would expect to obtain, plotting time elapsed against boiling points.

9. Answer a problem like problem 8 in which carbon tetrachloride and methyl alcohol are substituted for the formic acid and water.

10. Make a reasonable approximate estimate of the boiling point of each of the following mixtures. Figures chosen should be consistent with each other as well as with general principles:

(*a*) One mole of *n*-propyl bromide and one of water.
(*b*) One mole of *n*-propyl bromide and three of water.
(*c*) One mole of α-bromonaphthalene and one of water.
(*d*) One mole of bromobenzene and one of water.
(*e*) One mole of ethyl bromide and one of n-propyl bromide.

11. Suppose a mixture of 10 moles of *p*-dibromobenzene and 90 moles of water were distilled until 1 mole of the former had passed into the receiver. How much water would simultaneously be distilled?

12. A quantity of *n*-heptane contained 0.5% of suspended water. What fraction of the hydrocarbon must be distilled before the residue becomes dry?

13. Draw vapor-pressure curves to show how the yield of α-bromonaphthalene from steam distillation could be improved, in relation to water simultaneously distilled, by use of superheated steam.

References

(See Chapter 14 for complete titles.)

Distillation: Weissberger, *Technique of Organic Chemistry,* **4** (2nd ed.); E. W. Berg, *Physical and Chemical Methods of Separation,* ch. 2, McGraw-Hill, New York, 1963; M. Van Winkle, *Distillation,* McGraw-Hill, New York, 1968; E. Krell, *Handbook of Laboratory Distillation,* American Elsevier, New York, 1963; F. J. Zuiderweg, *Laboratory Manual of Batch Distillation,* Wiley–Interscience, New York, 1957.

Azeotropic Mixtures: Weissberger, *Technique of Organic Chemistry,* **4** (2nd ed.), 423; *Azeotropic Data* (*Advances in Chemistry Series,*

No. 6), American Chemical Society, Washington, D.C., 1952, and supplement (*Advances in Chemistry Series,* No. 35), 1962; W. Swietoslawski, *Azeotropy and Polyazeotropy,* Pergamon Press, Elmsford, N.Y., 1963.

Chapter 6

Technique of Distillation

62. Ordinary Distillation. Figures 1 and 2 picture apparatus for distillation at atmospheric pressure; observe the following precautions for assembly and use. Both flask and condenser are attached to vertical rods or ring stands (which stand directly behind them) by clamps pointing directly out from each rod. In Fig. 1 the clamp holding the distilling flask is attached above, not below, the sidearm (in Fig. 2 both the flask and the distilling head may be clamped, but more often only one). The condenser is clamped about equidistant from each end and is precisely aligned with the sidearm of the flask. A simple procedure to assure such alignment is to place the flask in position first, making sure that it is at the proper height to accommodate the Bunsen burner or whatever heating device is to be employed and that it stands vertically. The thermometer bulb is centered in the neck of the flask low enough to make sure that it is completely bathed in vapor but not lower (see §40). Next mount the condenser so that it is not yet attached to the sidearm, but is clamped near the top, points directly at the sidearm, and is cocentric with it. Finally loosen the clamp enough to slide the condenser up to attach to the sidestem, and retighten the clamp. In Fig. 1 the sidestem should pass well through the cork as shown to minimize contact of liquid or vapor with cork. The clamp jaws should be covered to protect flask and condenser jacket from breakage. Short pieces of rubber tubing are often used for this purpose, but these are unsatisfactory on clamps holding flasks that will be heated much above 100°. Clamps are available with asbestos or heat-resistant plastic coatings on the jaws. Attach the curved adapter so that the distillate is delivered directly into the receiver. If the adapter is omitted and the distillate drops through the air into the receiver, increased escape of vapors may cause fire or toxicity hazard. A receiver placed as in Fig. 3 avoids this risk, but the arrangement is less secure. Even when the flask is supported on a cork ring or block of wood with center depression as shown, accidental loss of distillate is more likely to occur. Do not use a book to support the receiver.

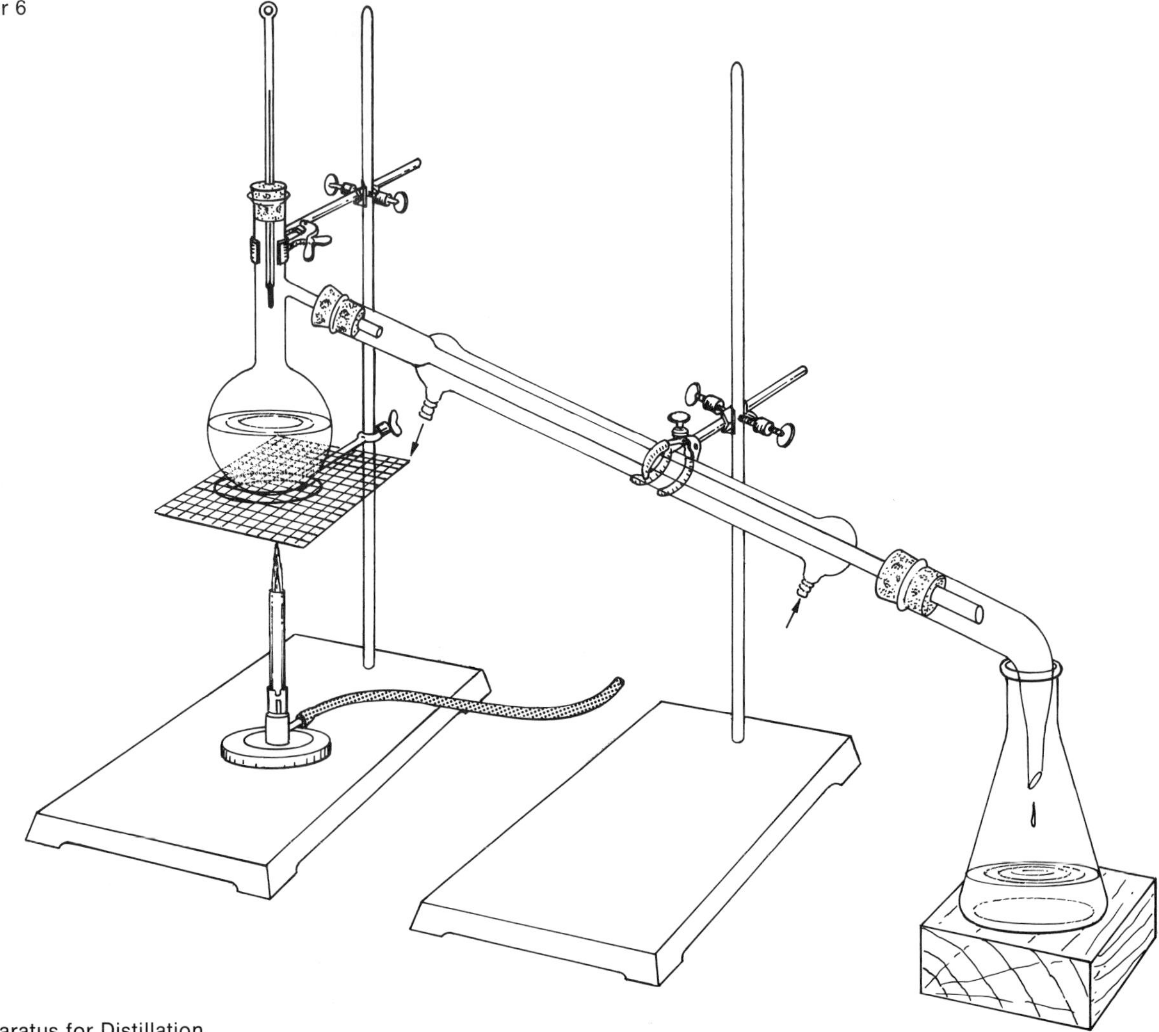

Fig. 1. Apparatus for Distillation.

Figure 3 shows the use of an air condenser, which is preferred for material boiling above 150°.

63. Fire Hazards in Distillation. Several common organic liquids are so dangerously flammable and so frequently used as solvents that some instructors ban all open flames (as of Bunsen burners) from beginning organic laboratories. Most organic compounds will burn, and flammable, highly volatile compounds must be handled with full attention to fire hazard. Beginning organic chemistry laboratories, often crowded, are especially dangerous unless students and instructors are aware of the

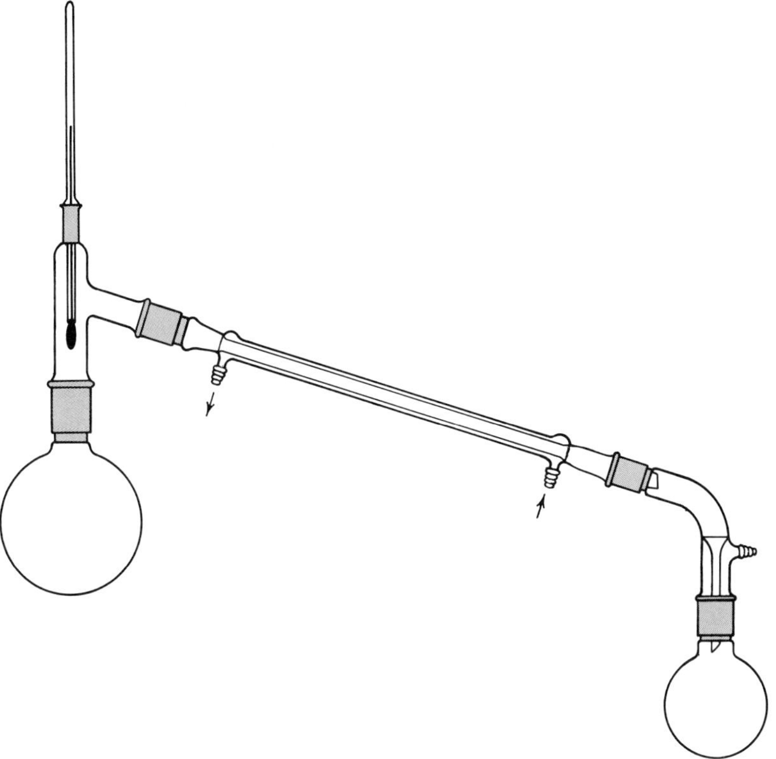

Fig. 2. Standard-Taper Apparatus for Ordinary Distillation.

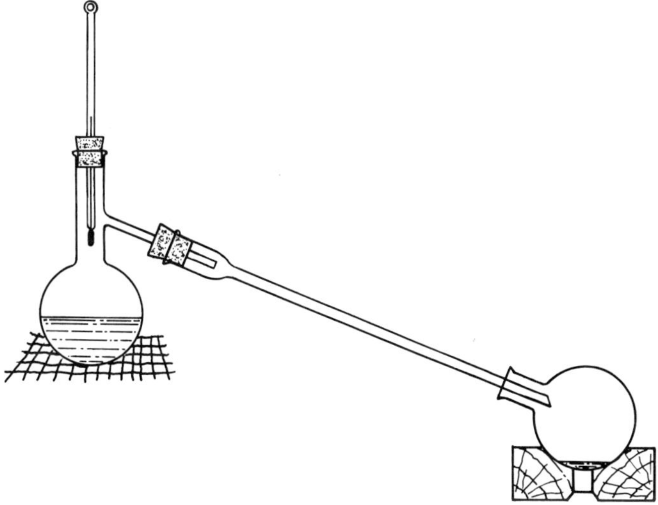

Fig. 3. Ordinary Distillation with Air Condenser. Note the placement of receiver when no adapter is used.

problems presented by each compound and take all necessary precautions.

Diethyl ether is the most common such liquid.

Warning: *No flames are permissible under or near distilling equipment when ether is being distilled or is being used.*

Even a flame as much as 10 ft distant may ignite ether if a continuous table top lies between the flame and the exposed ether (especially if a gentle draft carries the ether vapor toward the flame). Steam or hot water baths (not flame-heated), or electric hot plates with covered heating elements that do not allow contact of ether vapor with the bare element, are suitable for ether distillation. Ether vapor is often imperfectly condensed during distillation when an ordinary water-cooled condenser is used. The standard-taper apparatus shown in Fig. 2 is suitable for distillation of ether; the receiver may be cooled in ice or a piece of rubber tubing running to the floor may be attached to the side tube of the adapter for added safety. The receiver arrangement shown in Fig. 1 is sometimes used if cotton is placed in the neck of the receiver and distillation is slow. With this apparatus it is better to use a filter flask as a receiver and attach it to the adapter with a rubber stopper. A piece of rubber tubing leading to the floor may be attached to the side tube of the filter flask. Ether that has stood for some time in contact with air may contain peroxides that under certain circumstances are dangerously explosive (see §116).

Petroleum ether (bp 30–60°) is as dangerous as diethyl ether and must be handled similarly. **Ligroin** (various boiling ranges are available) must also be handled with care; the danger decreases as the boiling range increases, but precautions are always necessary to avoid fires with these hydrocarbon solvents.

Carbon disulfide (bp 48°) is an especially dangerous liquid because it may be ignited by contact with the hot surface of a "safe" hot plate or the surface of a hot steam radiator. It is also fairly toxic. Fortunately, it is not often needed in elementary work. If carbon disulfide must be distilled, warm water (60–80°) brought from a distant source of heat is used to heat the flask. The receiver should be protected as for ether distillation.

Benzene (bp 80°), **ethanol** (bp 78°), **methanol** (bp 65°), **acetone** (bp 56°), **diisopropyl ether** (bp 68°), and other flammable organic liquids boiling in the range 55–90° are less dangerous than those discussed earlier but should be distilled with suitable precautions to avoid contact of vapors with open flames. A steam bath is adequate to distill these liquids unless they are mixed with less volatile substances (as when they are used for extractions). One then uses the steam bath to distill around 90% of the solvent, and the remainder can be removed quite safely using more vigorous sources of heat (even the Bunsen burner). In crowded laboratories it is best to use protected receivers as for ether.

Heat for distillation of solvents with boiling points above 100° is best supplied by electrically heated oil baths, hot plates, or heating mantles. If a Bunsen burner is used, a wire gauze should be placed under the distilling flask and care taken to apply heat moderately. If a bare glass flask containing organic material is heated above 100° by direct application of a gas flame, tarry matter may separate and adhere to the glass surface. Such adhesion may cause even a borosilicate glass flask to break, and tars baked on in this way are often difficult to remove. Suitable precautions must also be taken to prevent the vapor of any flammable solvent from coming in contact with a free flame.

64. Bumping. The boiling of a liquid does not proceed smoothly unless adsorbed air or other gas is present on the inner surface of the boiler. There is often insufficient adsorbed gas on a polished glass surface to produce a smooth, continuous flow of bubbles during boiling. The liquid "bumps," alternately ejecting excessive amounts of vapor, possibly rushing over into the receiver or blowing a spray of impure liquid into the condenser, and then lying quiet in the boiler. The presence of mineral salts dissolved in the boiling liquid often aggravates the situation. When no vacuum pump is involved in common distillation, the introduction of certain solid objects into the boiler furnishes a steady stream of minute bubbles that initiate smooth and continuous boiling. Bits of unglazed clay plate called "boiling stones" or "chips" are commonly employed. Scraps of platinum (a metal noted for its ability to adsorb gases), carborundum, and anthracite coal have been used successfully, and more recently, polytetrafluoroethylene (Teflon polymer) chips.

> **Warning:** *If a heated liquid does not boil steadily, beware of the mistake of dropping the stone into the already hot liquid, which may be above its normal boiling point. A sudden outburst of vapor and spray may throw acid or flaming liquid upon the hand of the worker. When the forgetful operator realizes that he needs a boiling stone, he should wait 2 or 3 min after the burner is removed before applying the remedy.*

DISTILLATION UNDER REDUCED PRESSURE

65. Safety Precautions. If apparatus for distillation is to be evacuated, there arises at once the hazard of collapse. For example, the external pressure on a large flask might exceed one-half ton.

The apparatus shown in Fig. 4 is traditional for reduced-pressure distillation. It has the virtue of minimum number of rubber-stopper connections. Standard-taper equipment (§23) is now more commonly used. Figure 5 shows one such assembly, which has the following special features:

1. The two-neck Claisen head is separate from the flask below, permitting interchange with flasks of several sizes. The capillary, which must

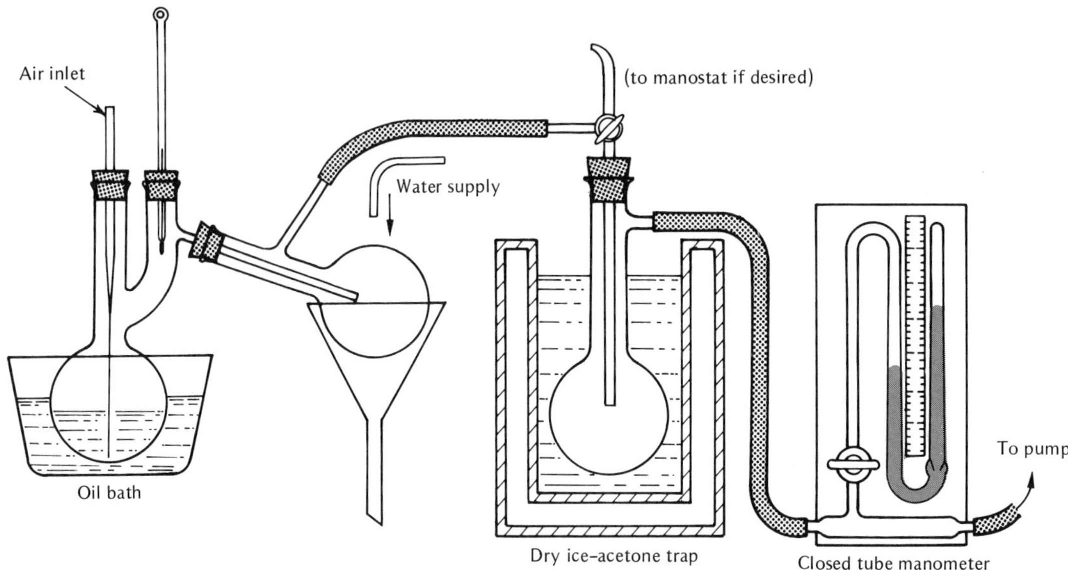

Fig. 4. Apparatus for Reduced-Pressure Distillation.

often be replaced, is attached by rubber. The thermometer may also be attached in the same manner without difficulty.

2. A special glass connector unit between condenser and receiver provides angle adjustment where required. It also has a connecting tube for the manometer and pump.
3. The gas burner is eliminated. An electric immersion heater directly heats a fusible-metal bath—or, more economically, a vegetable-oil bath. Electric current for this device comes from a hand-operated autotransformer connected to 110-volt domestic service and providing power at 0–110 volts, instantly adjustable at the will of the operator.
4. Replacing the standard ring stand is a laboratory jack (see Fig. 3, §20). This device is convenient and much safer than a fixed stand when it is necessary to lower a hot bath.

The apparatus shown in Fig. 2 may also be used for distillation under reduced pressure if the ordinary distilling head is replaced by the Claisen head shown in Fig. 5. The flask should be heated by an oil bath as shown in Figs. 4 and 5.

66. Bumping Problem at Low Pressures. The process of evacuation of distillation apparatus unfortunately removes the adsorbed air not only from the inner surface of the distilling flask but also from any boiling stones considered useful. Bumping then becomes serious. This lack of air

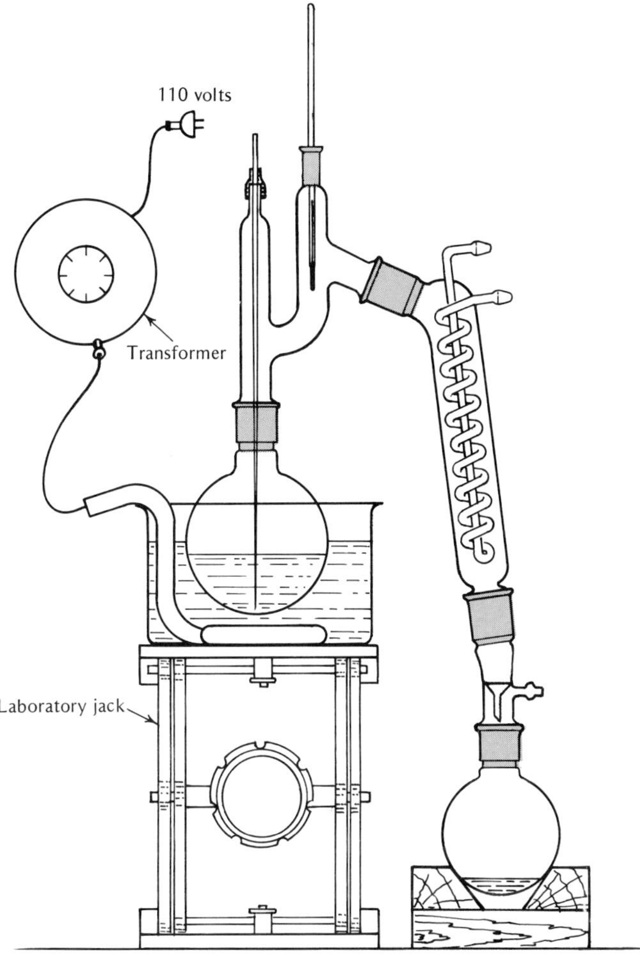

Fig. 5. Special Apparatus for Reduced-Pressure Distillation.

is compensated by introduction of a special, though minute, stream of air from the outside. A capillary tube mounted in the central neck of the flask provides the regulated stream.

A piece of common glass tubing, whose size is chosen merely because it snugly fits the hole in the stopper, is heated in the gas flame and drawn out to a very fine capillary. The opening through the capillary is so small that leakage of air into the flask will be negligible compared with removal of air by the vacuum pump or aspirator. The capillary tube is tested by trying to blow bubbles in a liquid of low viscosity like ether. If the compound under distillation is sensitive to oxidation, nitrogen instead of air is admitted to the capillary.

Under reduced pressure, the bumping nuisance is accentuated by simple operation of the gas laws, as illustrated in the following example:

Example. Suppose the benzyl cyanide cited in §36 is distilled at 760 torr. Each cubic millimeter of this liquid, arbitrarily taken as a portion being vaporized at a given moment, expands about 350-fold, producing 0.35 ml of vapor. This 0.35 ml, released at the bottom of the distilling flask, constitutes a small bubble that escapes without great disturbance.

The experiment is now repeated at 20 torr pressure. The 1-mm^3 portion of liquid evaporates at lower temperature. Simple chemical arithmetic, supplemented by Boyle's law, reveals this time a 10,000-fold expansion from the liquid state. What was a small bubble in the previous experiment increases to about 10 ml. Such great expansion magnifies the nuisance of bumping and excessive splashing.

Published records of boiling points at reduced pressure are often in error, sometimes by 10° or more. Such errors arise in part from the marked change in boiling point with change of pressure at low temperatures. An error of 2 or 3 torr in measurement of pressure in a distillation near atmospheric pressure is of little importance; it is increasingly serious as the pressure of the distillation decreases. For example, the boiling point of benzyl cyanide (§36) changes by 12° (from 62 to 74°) between 1 and 2 torr, and by 3° (from 103 to 106°) between 9 and 10 torr; the change between 90 and 100 torr is also 3° (from 161 to 164°), or no more than a third of a degree per torr.

Error is also caused by the gas introduced through the capillary and by the pressure drop in the tubing between the distilling flask and the manometer. It is especially important to use as fine a capillary as possible, to minimize the error from this source. In careful fractional distillation at reduced pressure, it is now common practice to omit the capillary and prevent bumping by stirring the contents of the distilling flask with a magnetic stirrer (§28).

67. High-Vacuum Distillation. In ordinary distillation, even under reduced pressure, a dynamic equilibrium exists between the liquid and vapor phases. Vapor molecules also encounter molecules of air. As a result a large proportion of the molecules that evaporate from the liquid surface return to it. With very high-boiling compounds, and especially with heat-sensitive materials, more effective purification may often be accomplished by greatly reducing the pressure and placing a condenser close enough to the surface of the liquid being purified so that essentially all the molecules that evaporate reach the condenser and none returns to the starting liquid. This is possible when the distance from liquid surface to condenser is no greater than the "undeflected path"[1] of the vapor molecules.

[1]The term "undeflected path" denotes the distance that most of the molecules of the particular vapor will travel under the very low pressure involved, without serious deflection from a rectilinear path. This is greater than the mean free path of the molecules. See Weissberger, *Technique of Organic Chemistry*, **4** (2nd ed.), 499.

Under these conditions boiling does not occur, and there is no "sharp" temperature at which evaporation begins. Separation of mixtures depends upon differences in rates of evaporation instead of differences in vapor pressure, although these two phenomena are often roughly parallel. The operation has been termed *molecular distillation,* unobstructed-path distillation, and high-vacuum distillation. In practice, the limits of pressure are roughly 0.1–1000 μm (1μm = 0.001 torr); the limits 0.1–10μm have also been specified by other writers.

The technique and apparatus of high-vacuum distillation are varied and often complex. The References should be consulted for a more extended discussion.

68. Aspirator Pumps. At normal temperatures a minimum of about 20 torr is all that can be expected from an aspirator or common filter pump. Theoretically the aspirator lowers pressure in the distillation system to a value equal to the vapor pressure of the water passing through the device. For example, at 14° the vapor pressure of water is 12 torr. Leakage around stoppers, tubes, etc., may readily add 5–10 torr more to the actual pressure in the system. In winter, with water coming directly from contact with ice, a minimum (theoretical) pressure of 4.57 torr would be possible; actually, about 8 torr would be more likely in laboratory apparatus.

69. Mechanical Pumps. Electrically driven mechanical pumps, with working parts immersed in oil of very low vapor pressure, are required for pressures below the limits of aspirators. With inlet plugged for testing purposes, such a pump will lower pressure to 0.01 or even 0.001 torr when new and in perfect adjustment. In actual practice, with used pumps connected to laboratory apparatus, pressure as low as 1 torr is difficult to attain and 2–10 torr is more common.

The oil in a mechanical pump is readily contaminated by volatile organic material, which reduces the efficiency of the pump. For this reason a dry ice–acetone trap or equivalent is placed between the distillation system and the pump to protect the latter. Corrosive vapors, which damage the pump, are often stopped by such a trap, but if acid vapors such as hydrogen chloride are given off during the distillation, it is necessary to include a trap packed with solid alkali such as potassium hydroxide. Ascarite (alkali on asbestos), the material usually used to absorb carbon dioxide in quantitative organic combustion analysis, is especially effective in traps to absorb acid vapors.

70. Manometers. The closed-tube manometer shown in Fig. 4 is now the most common device for measurement of pressures during distillation, but great care must be taken to prevent any gases or vapors from reaching the closed space over the mercury. Access of such vapor is minimized by use of the constricted, ring-sealed insert shown in Fig. 4, which also prevents breakage of the closed end of the tube if the pressure is suddenly released. Even a slight amount of vapor in the closed space causes major error in pressure readings. To prevent errors the manometer should be checked before use by attaching it directly to a good mechanical pump that lowers the pressure to below 0.1 torr.

The old-fashioned open manometer shown in Fig. 6 is now seldom used but is still a more practical, less error-prone, and less expensive device than any other for beginning organic laboratories. In advanced work McLeod gages, often of quite simple construction, are becoming common for pressures below 10 torr (see the References).

71. Manostats. Sometimes it is desirable to regulate pressure at a value greater than the pressure limit of the pump. A manostat, or pressure regulator, is then employed (see the References).

In elementary work a simple hand-regulated needle valve may replace the manostat, but it requires rather constant attention.

72. Effective Condensation. The foregoing admonitions deal largely with the procedures and devices chosen to make sure that the vaporization part of reduced pressure is properly conducted. Of like importance, though more simple, are the warnings to effect complete condensation as well.

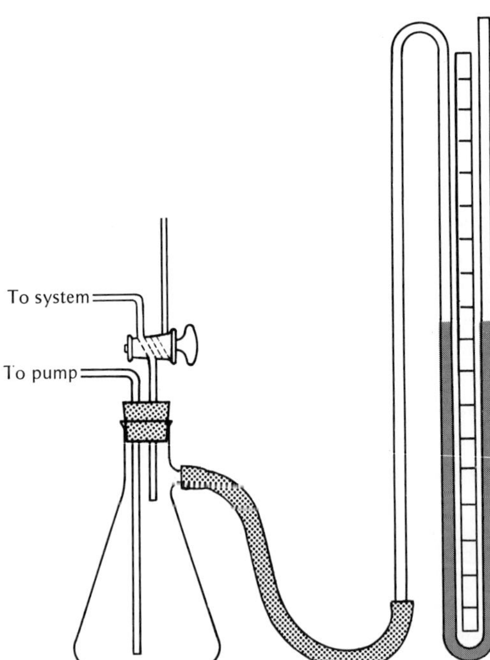

Fig. 6. Manometer.

In the traditional apparatus (Fig. 4) the standard commercial Claisen flask often does not have a long enough sidestem to reach down far enough into the condensing bulb. Hot vapors are thus short-circuited into the pump line and lost. Special adjustment of sidestem by the local glass-blower should be sought. Furthermore, the outer surface of the condensing bulb should be scrubbed carefully with household washing powder to remove grease. Domestic water flow will then spread smoothly over the condensing surface.

With standard-taper equipment there may be loss if the adapter does not have the inset dropping tip shown in Figs. 2 and 5, or if the distillate is too volatile. It sometimes helps to cool the receiver in ice.

73. Changing Receivers. It is awkward to change receivers during distillation under reduced pressure. With most higher-boiling organic compounds it is important that the undistilled product remaining in the distilling flask after a forerun has been distilled should not be exposed to air at the high temperature of the heating bath. One should first lower the heating bath and allow the flask to cool before breaking the vacuum. The receiver can then be exchanged, the reduced pressure re-established, and finally the hot bath raised around the flask. It is sound practice to leave most of the protective system in Fig. 4 under reduced pressure by using the three-way stopcock to break the vacuum near the receiver. If the vacuum is broken simply by disconnecting the rubber tubing at the receiver, air rushes through the trap and carries volatile material into the pump. The manometer shown in Fig. 4 should be protected by keeping the stopcock closed as shown except while a reading is being taken. If a water aspirator is used as a source of reduced pressure, care must be taken to break the vacuum before turning off the water in order to prevent suckback (see §96).

A variety of fraction cutters have been devised for use in fractional distillation under reduced pressure (§78); they are often used for simple reduced-pressure distillation as well.

DISTILLATION WITH A FRACTIONATING COLUMN

74. Simple Distilling Column. It is often desirable to interpose some kind of vertical tube between the boiler and condenser in apparatus for distillation. Under such conditions only part of the vapor reaches the condenser, as discussed in Chapter 5. A residue falls back into the boiler and is called the "condensate" or reflux.

Such a vertical tube, known as a distilling column, has little value unless it contains a series of partial obstructions, circuitous passageways, or "packing" material. Such inserted objects delay downward passage of liquid and expose maximum liquid surface to rising vapor. Much ingenuity has been displayed in column design.

75. Column Packing. Glass beads made by cutting up glass tubing into short sections are used as packing for the Hempel column. Such ready-cut material is available commercially; ornamental beads with small needle holes are not satisfactory. Indentations at the bottom of the column prevent beads from falling through. For easy tasks the Hempel column is fairly efficient but suffers the disadvantage of excessive hold-up of liquid. Figure 7 shows a Hempel assembly as used in experiments described later. The Vigreux column, Fig. 8, has less holdup but is less efficient than the Hempel because there is less opportunity for prolonged contact of liquid and vapor in the column.

It is often difficult to use a noninsulated column as shown in Fig. 7 because air currents cool the outer surface and interrupt the process of equilibration of descending liquid and rising vapor. Performance is improved if the column is wrapped with an insulating material such as as-

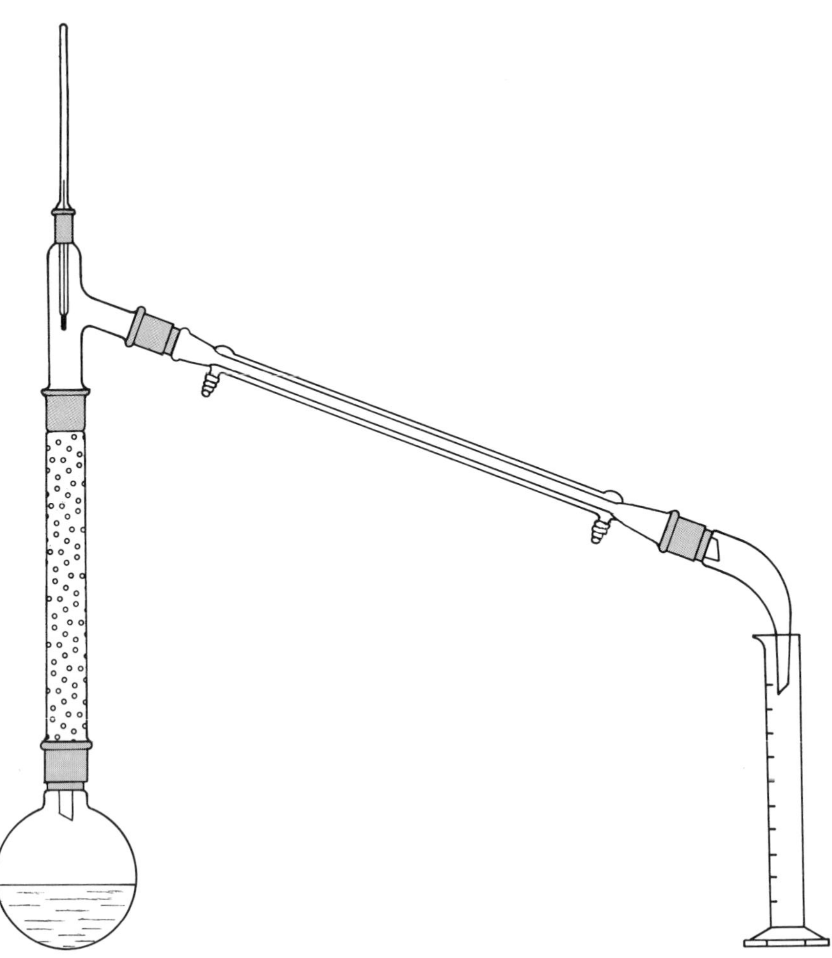

Fig. 7. Apparatus for Fractional Distillation.

bestos paper or glass wool. Simple vacuum-jacketed columns are now quite common. The column must be held as nearly vertical as possible to provide maximum contact between liquid and vapor.

76. Precision Columns. Highly efficient laboratory distilling columns are now available commercially. The best of these is probably the spinning band column, which combines low holdup with low HETP (see §55); it is expensive. Others have packings such as spiral wires, specially shaped metal gauze, glass helices, etc. Heating jackets are often provided for careful control of column temperature or an efficient vacuum jacket is used. The References discuss such columns in detail.

Figure 9 shows a heated column packed with single-turn glass helices.

Fig. 8. Vigreux Column.

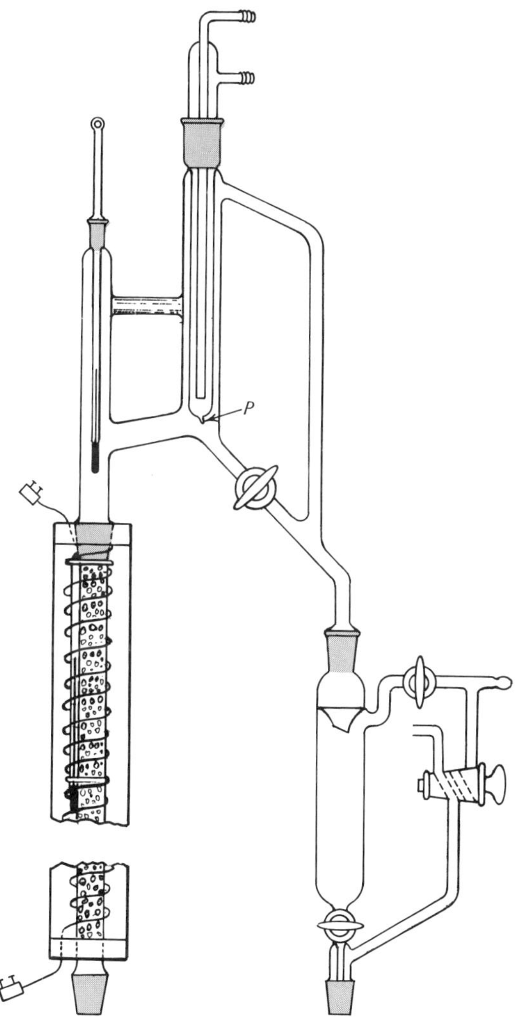

Fig. 9. Precision Fractionating Column.

The distilling head is of the total-reflux, partial-takeoff type now most common. These are much easier to control than the partial-condensation type in which the rising vapors are partially condensed as they rise and everything reaching the top of the column comes over as distillate. The latter are mostly used only if very small amounts of material are available.

77. Operation under Total Reflux. The mixture of vapors from a boiling flask enters the column of Fig. 9 and encounters the glass helices, which cause partial recondensation. Since the column may be of considerable length, the whole current of vapor might be thus condensed and fractionation made impossible were it not for use in the column of an electric heater controlled with high precision at a variable-voltage transformer. Vapor-liquid exchange continues all the way through the efficient helices until at last a small vapor residue reaches first the thermometer and then the reflux condenser at *P*.

After the column has operated for some time under total reflux, the fractionation becomes stable. A small yield may now be tapped off through the stopcock to a receiver. If the substances undergoing fractionation are easily separated, one may safely draw off 1 drop for every 10, or even every 5, that start downward from *P*. A "reflux ratio" of 10 or 5 is thus chosen. With difficult tasks, reflux ratios of 25, 50, or even 100 may be required to prevent breakdown of fractionation. The process is then very slow.

78. Fractional Distillation Under Reduced Pressure. Distillation under reduced pressure has been discussed earlier (§§65, 66). A practical difficulty lies in the greatly increased vapor volumes that must be handled; adequate contact between liquid and vapor is much harder to attain and slower distillations are necessary, especially at greatly reduced pressure. It is also necessary to provide a suitable device to permit exchange of receivers. Figure 9 shows a commonly used fraction cutter. An alternative type, usually called a "cow," is shown in Fig. 10, which also pictures a "modified" Claisen flask with built-in Vigreux column. The disadvantage of the cow is that early fractions are exposed to the vapors of later fractions, so some contamination may occur.

STEAM DISTILLATION

79. Apparatus. Figure 11 pictures the common assembly for steam distillation. The glass tubing used for the exit tube from flask A should be 10 mm or larger in diameter, and the condenser must be efficient because the heat capacity and heat of vaporization of water are high.

Most laboratories are now equipped with piped steam service. An adapter (C in Fig. 11) or a filter bottle serves to prevent condensed steam,

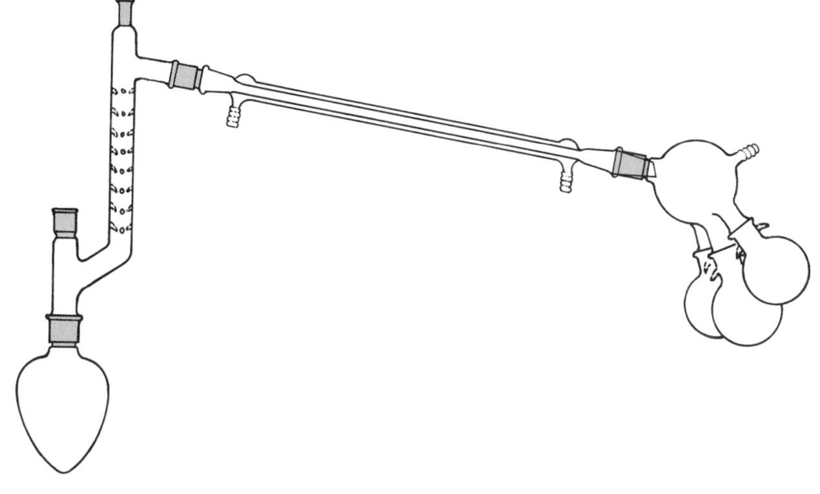

Fig. 10. Apparatus for Distillation Under Reduced Pressure with a Modified Claisen and a "Cow."

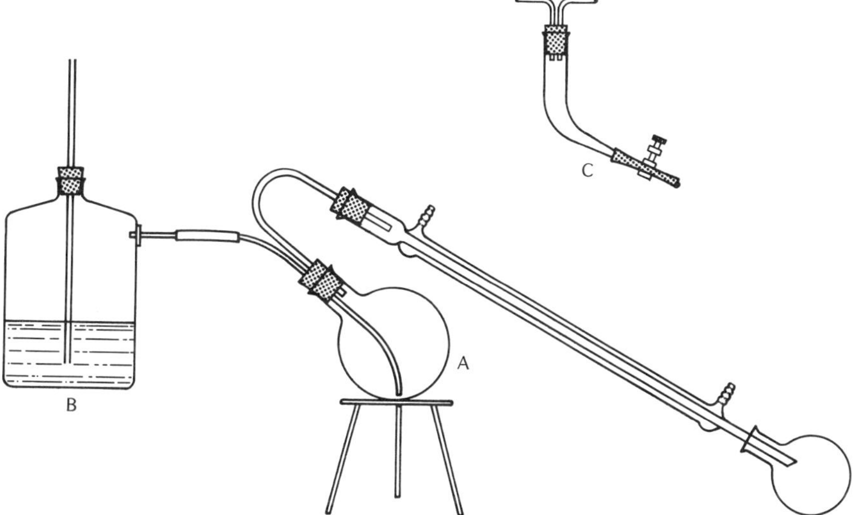

Fig. 11. Apparatus for Steam Distillation.

rust, oil, etc., out of a pipeline from reaching the material being steam-distilled. The drain tube from the adapter is controlled so as to permit constant leakage of water but little or no steam leakage.

When piped steam is unavailable, a steam generator constructed from a gallon can (B in Fig. 11) or a large flask may be used. The safety tube shown in Fig. 11 is essential. It not only affords relief from internal pressure in case the apparatus becomes plugged but also gives warning when the water level in the can drops to the danger line.

Steam distillation may also be carried out in standard-taper equipment. A typical apparatus is shown in Fig. 12.

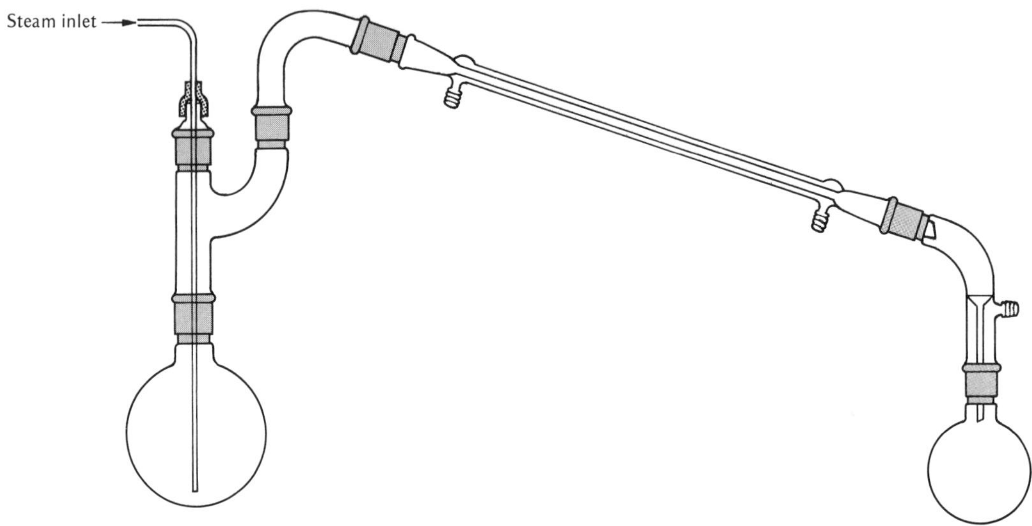

Fig. 12. Standard-Taper Apparatus for Steam Distillation.

80. Conduct of Distillation. Usually a copious current of steam is desirable, particularly with organic compounds of low vapor pressure. The apparatus is therefore designed to permit as rapid flow as possible without blowing liquid over into the receiver. Flask A should be large, and it is often heated with a second burner to retard accumulation of water. Rubber stoppers are often used and they may be wired in. For the theory of steam distillation see §59. Steam distillation is less commonly used today than it was two decades ago.

References

(See Chapter 14 for complete titles.)

In addition to the references at the end of Chapter 5, which give material on technique and apparatus, the following may be consulted: Weissberger, *Technique of Organic Chemistry,* **6,** 57; Müller–Houben–Weyl, **1,** Pt. 1, 777–956, Pt. 2, 423; Bates and Schaefer, 55; Pasto and Johnson, 10; Wiberg, 20; Vogel (1956), 1, 83; (1966), 1, 83.

Chapter 7

Behavior of Solid Substances

81. Melting and Freezing. Normally the isolation, purification, and identification of an organic compound that is solid at ordinary temperatures involves transformations from liquid to solid state and the reverse. One commonly purifies a solid by crystallization from solution; distillation or sublimation may also be used; and zone refining, which involves crystallization from a melt, is of increasing importance. The melting point is usually the first physical property determined for a solid substance; it gives evidence of the purity of the compound and is important for identification. Well-known methods of molecular-weight determination depend upon the depression of the melting point of one compound by known amounts of another. The theory underlying the solid–liquid phase change is based in large measure on the phenomenon of vapor pressure. Both liquid and solid forms of a substance exert vapor pressure; that of the solid rises faster than that of the liquid. Figures 1 and 2 present vapor pressure–temperature diagrams for solid and liquid forms of a compound like camphor, which has a high vapor pressure in both states.

Suppose that a vessel is completely filled with a mixture of solid and liquid camphor, maintained at constant temperature. Molecules of the liquid, attempting to leave as though by evaporation, can escape to no other place than the surface of the solid camphor, where they might adhere. The solid, attempting to evaporate, can go only into the liquid. If the first process were more rapid than the second, the material would eventually all solidify, and one might simply declare that the liquid was *freezing*. If passage from solid to liquid were more rapid, the solid would be *melting*. To decide which of these conflicting processes would prevail, examine Fig. 3, showing vapor-pressure curves of both phases.

Since the two curves are not parallel, they must intersect as shown in Fig. 3. The point of intersection happens to be at the temperature of 179°. One may now propose that solid camphor might exist at tempera-

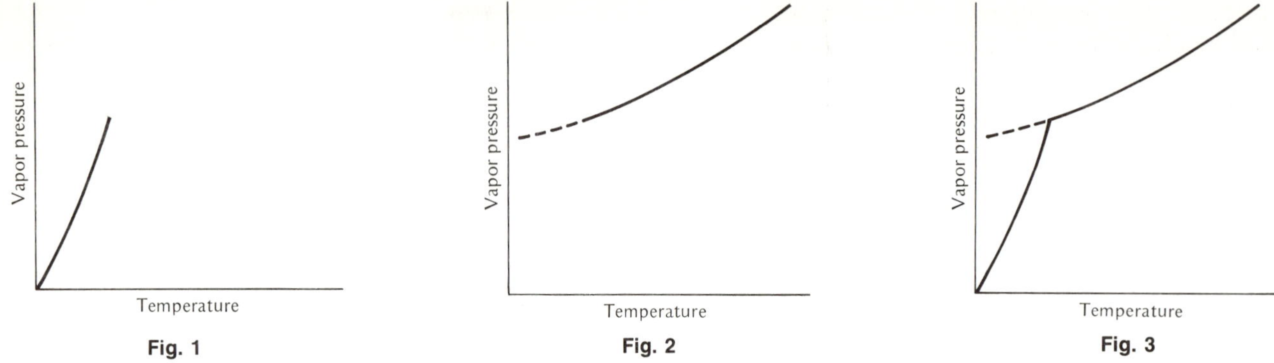

Vapor Pressures of the Solid and Liquid Forms of the Same Substance.

ture T_1, which is any value above 179°. This proposal is at once ruled out, since solid camphor at temperature T_1 would have a greater vapor pressure than liquid camphor and would immediately "distill" into the liquid phase. Similarly, liquid camphor at T_2 would have a greater vapor pressure than the solid and would freeze. Accordingly, at one temperature only, 179°, could both solid and liquid camphor stand together indefinitely. Rates of transfer, solid to liquid and liquid to solid, are alike, and the system is in physical equilibrium. The temperature 179° is known as the melting point[1] of camphor and is characteristic of the substance. It is determined most accurately by an experiment following the foregoing explanation. As long as a mass of finely divided, solid camphor stands unchanged in a bath of liquid camphor, the thermometer will hold strictly to the equilibrium value of 179°. In common laboratory practice, however, a simpler micro method of determining melting point is preferred. See §§107–111.

82. Metastable Condition. From the foregoing explanation it is apparent that a solid cannot exist above the melting point, and therefore the rising vapor-pressure curve of a solid stops sharply at the melting point. Laboratory experience substantiates this conclusion. Similarly, one would surmise that the falling vapor-pressure curve of a liquid would stop sharply at the same point. In practice, however, this is not strictly true. Many liquids may be cooled below their true freezing points, provided

[1]The true melting point is the temperature at which all three phases (vapor, liquid, and solid) coexist. The total pressure of this system is then the vapor pressure of the compound at that temperature. Ordinary melting points are determined at atmospheric pressure and differ slightly from true melting points. The difference is very slight because melting points do not vary rapidly with pressure. However, at very high pressure, startling differences are observed; e.g., at 33,880 atm (35,000 kg per cm²), the melting point of water is 166.6° instead of 0°.

no crystal is at hand upon which the liquid molecules may be deposited in standard geometric pattern. The liquid is then said to be "subcooled" or "undercooled" (often called "supercooled," an unfortunate misnomer patterned after the term "superheated"). Subcooling sometimes deceives a beginner when he tries to locate the equilibrium point by cooling a liquid until it actually solidifies.

For example, pure "glacial" acetic acid is supposed to solidify at 16.67°. In actual practice, however, the falling temperature may reach 15° or even 10° before crystallization starts. When at last crystals appear, the temperature rises, owing to release of heat of fusion. The new crystalline solid melts sharply at 16.67°.

83. Behavior of Impure Preparations. Suppose a small amount of a second substance, naphthalene, is introduced into the equilibrium mixture of pure solid and liquid camphor, all at the temperature of 179°. It is further assumed that no heat is added to the system or is withdrawn.

The naphthalene immediately dissolves in the liquid camphor but not in the solid, which it cannot penetrate. Following Raoult's law, the vapor pressure of the liquid camphor is reduced. The activity, or escaping tendency, of camphor in the liquid state is reduced. Camphor will pass less readily from liquid to solid than from solid to the contaminated liquid. If the temperature were maintained at 179°, the camphor would eventually all melt. But melting uses up heat, and since no heat is being supplied, the temperature must fall.

The phenomenon may be represented as in Fig. 4, which is merely an extension of Fig. 3. The solid camphor at the outset had the vapor pressure and temperature (179°) marked as the point *P*. The impure liquid at

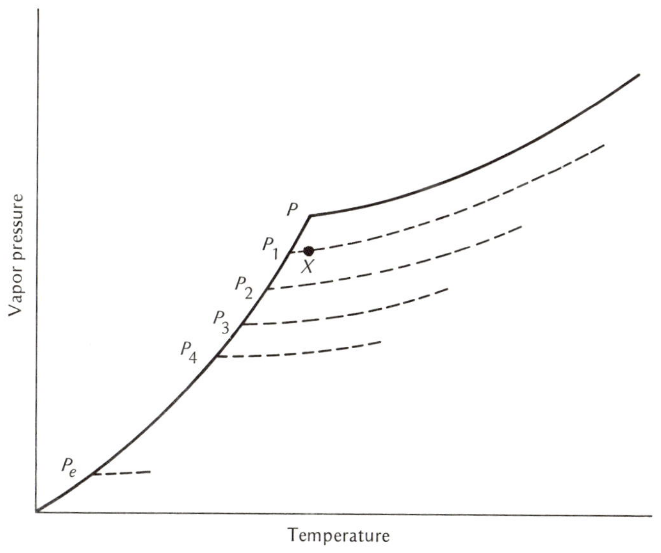

Fig. 4. Effect of Impurity on Melting Point.

[§84] this temperature would have the vapor pressure (i.e., the partial vapor pressure of the camphor component only) marked as the depressed value X. Obviously a phase with pressure P cannot exist in the same vessel side by side with a phase of the same substance in which the pressure is lower than P, as at X. The solid represented at P would melt at once and assume the vapor pressure and temperature indicated by X.

But the melting of a solid requires heat. As none is supplied from the outside, the temperature of the mixture falls. The vapor pressures of the solid, P, and of the impure liquid, X, both fall at the rates marked by their respective curves until they meet at P_1. Here equilibrium ensues, and no further change is seen. Solid camphor and the solution of naphthalene in camphor may remain together indefinitely at the condition P_1, which is below 179°.

Other points still lower in temperature, such as P_2, P_3, and P_4, may be located, showing the equilibrium temperatures of solid camphor in contact with solutions of increasing naphthalene content. Within reasonable limits, Raoult's law now has a new application, namely, that the molecular percentage and thus the molecular weight of a compound may be measured by determining depression of melting point.

If the melting point of camphor can be depressed to P_1, P_2, P_3, P_4, a question naturally arises—How far may it fall? The limit is P_e, the *eutectic point*, which in this case is reached when the fraction of naphthalene

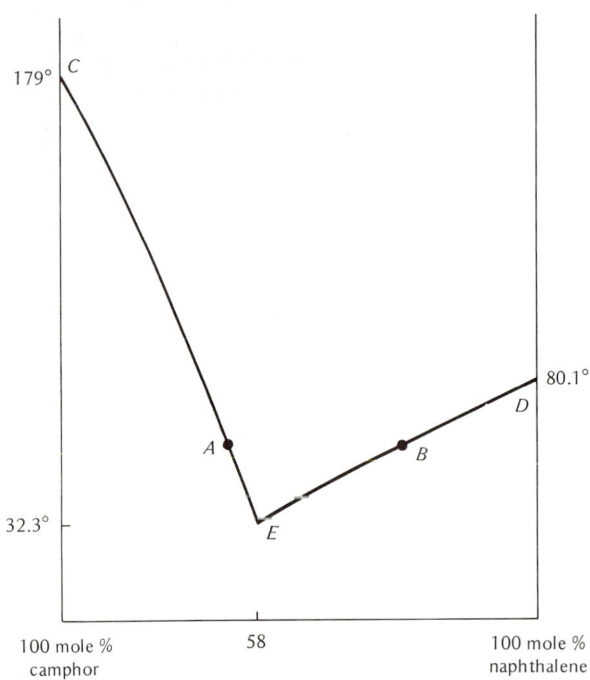

Fig. 5. Melting Point-Composition Diagram.

reaches 42 mole %. It is the temperature below which not even the impure liquid can persist; further refrigeration simply causes both components to solidify completely, and the experiment is done. The eutectic point is better understood from a diagram plotted in a different manner, as seen in Fig. 5.

Melting point is plotted against composition throughout the whole range of possible mixtures of camphor and naphthalene. The curve *CE* indicates the successively lower melting points of camphor as naphthalene is added up to the maximum of 42 mole % impurity. The curve *ED* shows the melting points of naphthalene, with camphor considered as an impurity, up to 58 mole %. Of course, the 42% solution of naphthalene in camphor and the 58% solution of camphor in naphthalene are identical. Therefore the curves must intersect as at *E*, the eutectic point, which happens to be 32.3°.

84. Estimation of Melting Point. A comparison may now be made between conventional melting-point determinations on pure camphor and on camphor contaminated with naphthalene. In the first experiment, pure camphor is unchanged as temperature rises until 179° is reached. At this point, in the *ideal* case, the entire mass melts at this single temperature. The melting point would be described as extremely "sharp." In actual laboratory practice, with slowly rising temperature, a narrow temperature range of a few tenths of a degree would be noted, even with a highly purified specimen. This would still be regarded as "sharp."

Impure solid compounds usually consist of mixtures of crystals. Figure 6 represents in schematic fashion a crystal of otherwise pure camphor with a few tiny crystals of naphthalene scattered over its surface. No change occurs in such a system on heating until the eutectic temperature is just exceeded. Appropriate members of molecules of camphor and naphthalene already in contact at the interface then join (in 58:42 ratio) to form a liquid phase of eutectic composition which contains all the naphthalene. Solid camphor is now in contact with a small coating of eutectic liquid.

Fig. 6. Crystal of Impure Camphor in Melting Point Tube.

Further temperature rise above the eutectic increases the activity (or vapor pressure) of the solid more than that of the liquid. Therefore the solid melts faster than the liquid solidifies, until a new equilibrium is reached in conformity with Raoult's law. More liquid will now be present, but the concentration of naphthalene in it will be lower. As the temperature continues to rise more and more, melting occurs until solid camphor is exhausted and the material is all in the liquid state. Obviously the final melting point—the only really significant melting point of the mixture in question—cannot be as high as 179°, but it may approach that figure if the original fraction of impurity was small.

The melting range thus really extends from the eutectic temperature up to the point of disappearance of the last crystal. In actual practice, with small amounts of impurity, it is usually impossible to detect the initial melting or eutectic condition. The crystals may not be arranged over the surface of a conspicuous crystal as in the ideal illustration proposed above. Thus the temperature may rise many degrees above the eutectic temperature before an appreciable accumulation of liquid occurs. In the case of camphor contaminated with much naphthalene, one might detect initial melting at 80 or 100° and would then expect a protracted range of melting. On the other hand, if the initial melting seemed to occur at 177°, the whole process would be complete within a few tenths of a degree and the preparation adjudged to be nearly pure. If two workers tested the same impure preparation, one might find a higher *initial* melting temperature, but they would both find the same *final* melting point.

On the melting point–composition diagram of naphthalene and camphor are represented two mixtures, A and B, which ostensibly have the same melting point, introducing possible confusion. These are readily distinguished by mixing pure naphthalene with each. In case A the melting point is lowered; in B it is raised.

85. "Mixed Melting Points." This incoherent expression, colloquially used in the laboratory, refers to an important application of the foregoing principles to identification of organic compounds. A concrete example follows.

Example. A research worker has just prepared a compound X which he thinks may be *o*-fluorobenzoic acid (mp 122°). There are, however, several other compounds with approximately the same melting point, including benzoic acid itself. Roughly equal weights of X and of known pure *o*-fluorobenzoic acid from another source are thoroughly mixed, and the melting point of the mixture determined. If X really is the fluoro compound, as suspected, no fall in melting point will be observed. If, however, it happens to be simple benzoic acid or any compound other than *o*-fluorobenzoic acid, a sharp change in melting point will probably be observed, with the following possibilities:

(*a*) A drop of several or many degrees if the two compounds do not react with each other.

(*b*) A rise if the two combine to form a new compound of higher melting point.

Care must be used in the application of this method of identification because one may encounter compounds that form solid solutions (i.e., molecules of one compound fit into the crystal lattice of a second compound without appreciable distortion). This phenomenon is fairly common, and a mixed melting point determination on such compounds gives a melting range (often quite narrow) between the melting points of the two components. This is confusing under two circumstances:

1. If the two compounds have very nearly the same melting point, there will be no depression, and the melting range of the mixture will be the same as that of the pure material or only very slightly broader (if the solution is not ideal). Fortunately this situation is rare.

2. If one is comparing a somewhat impure, unknown compound that melts over a range with a known compound that melts somewhat higher, one then expects the mixture to melt higher than the melting range of the impure substance if the two compounds are identical. There is greater chance to encounter solid solutions under these circumstances because the range of compounds chosen for comparison is greater. Further discussion of solid solutions will be found in the references and in textbooks of physical chemistry.

86. Molecular Weights. The use of freezing-point depression to determine molecular weights is well known. Camphor has a special advantage as solvent because its molal freezing-point depression constant is large (38°) compared with other substances. The foregoing method of molecular-weight determination utilizes camphor and is simple because an ordinary thermometer is used to observe melting-point lowering, and the usual capillary method of melting-point determination is sufficiently accurate. A discussion of this and other cryoscopic methods can be found in the references.

87. Recrystallization. This is a highly selective method of purification because crystals are ordered molecular arrangements such that molecules of one substance can seldom fit into the crystal lattice of another. It is necessary that the substance to be purified and the impurities be sufficiently different in solubility and concentration so that only one kind of crystal forms in the solution. In theory one can then obtain complete purification of a compound, providing a solid solution is not formed with some impurity that is present. In practice, complete removal of the mother liquor (remaining solvent, which contains dissolved impurities) presents problems. See §§102 and 104.

When two or more substances of similar solubility are present in comparable amounts so that mixtures of two kinds of crystals tend to form, one must usually resort to a tedious systematic process of repeated recrystallization and recombination of fractions termed *fractional crystallization*. Frequently one then turns to other separation methods (chromatography, distillation, etc.). Recrystallization is more readily applied to mixtures in which the desired compound is the principal component.

The crystallization process depends both on formation of crystal nuclei and on rate of crystal growth. A discussion of the theory involved will be found in the references.

It is instructive to consider recrystallization in terms of melting point–composition diagrams in which the solvent is one of the components.

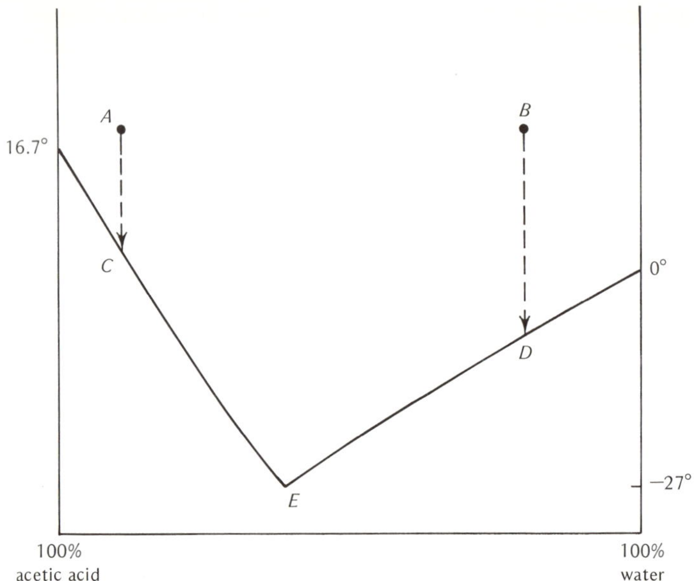

Fig. 7. Melting Points of Acetic Acid–Water Mixture.

Figure 7 for acetic acid–water presents an example which might lead to the blunder of collecting crystals of solvent rather than solute. To obtain pure acetic acid, one must start with a solution of higher acid concentration than eutectic, *E*.

The correct method calls for use of a solution whose concentration (of the desired acetic acid) is in excess of the eutectic composition *E*. Accordingly, a solution as free as possible from water is chosen. For simplicity, suppose 90 g of acetic acid is to be recrystallized from its solution in 10 g of water, and this solution, at room temperature, is represented as point *A* of Fig. 7. One now goes through the process described in §83 but in *reverse*.

Solution *A* is cooled, and the vertical arrow *AC* graphically indicates the first stage of that operation. But *C* (3°) is the so-called freezing point of a 90% mixture. Given time or the aid of a seeding operation, crystals of pure acetic acid will now separate as the temperature continues to fall. Since the solution is being robbed of acetic acid, the graphic path no longer goes directly downward, as with *AC*, but bears to the right, as in *CE*. Before the eutectic point is reached (−27°), the operator filters the mixture, obtaining a substantial yield of the desired pure crystalline acetic acid.

A second experiment shows faulty technique in the use of a mixture whose composition is marked on the right side of Fig. 7. A 20% solution of acetic acid is cooled from room temperature (*B*) to −7° (*D*). At *D* crystals appear, and a large crop of beautiful material is collected along the course *DE*, analogous to the crystallization *CE* above cited. Unfortunately, the crystalline product is ice, not acetic acid, and the principles involved in its appearance are quite like those given above in the AC case.

In the case of acetic acid and water the process of crystallization is simple because the two components are miscible in the liquid state in all proportions, even at low temperatures. When the mutual solubility of the solute and solvent is limited, a new trouble enters.

88. Separation of an "Oil." Attempts to recrystallize certain compounds result in the appearance of so-called oils, which retain impurities and confuse inexperienced workers. Acetanilide and water afford a good example of such troubles, as graphically portrayed in Fig. 8. The simple cooling of a boiling, saturated aqueous solution of acetanilide first causes the separation of the oil; later the oil seems to freeze in a compact mass of questionable purity; finally a terminal yield of relatively pure crystals of acetanilide is obtained from the residual aqueous solution.

In Fig. 8 the melting-point curve starts as usual with the normal melting point of the pure solvent (water) as marked at A, or 0°. As usual it drops to the eutectic E, which in this case happens to be very close to the left vertical axis of the graph. The right branch of the curve rises from E and normally would proceed to B, the melting point of pure acetanilide. This curve is broken, however, into two disconnected sections EC and

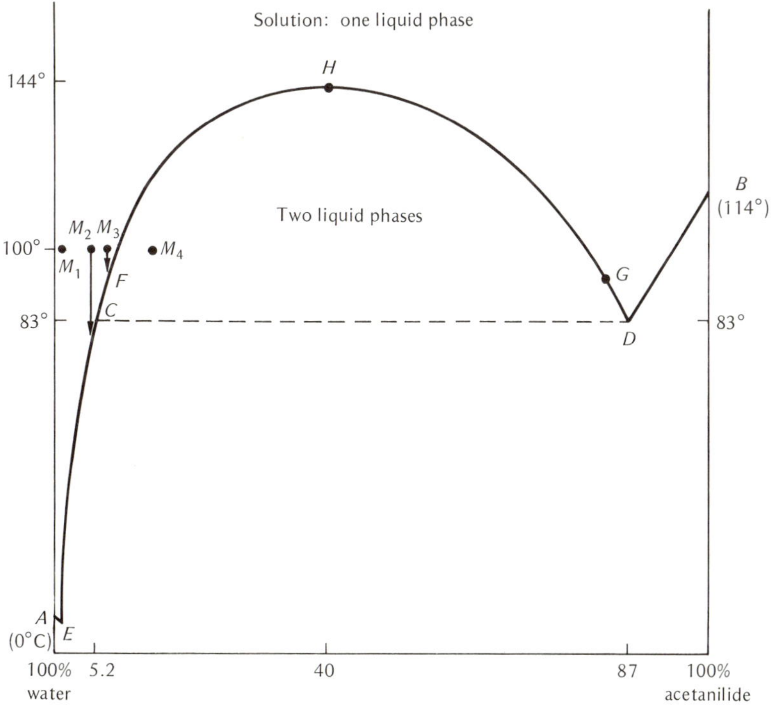

Fig. 8. The System Acetanilide–Water.

DB. The intruding intermediate zone *CDGHF* is a field in which acetanilide and water are not miscible. It is impossible to prepare a solution of these two components with any temperature and corresponding concentration falling within that zone. This fact is likely to disturb the process of recrystallization seriously.

Suppose several mixtures, M_1, M_2, M_3, M_4, be made of solid acetanilide with boiling water, with the purpose of finding a suitable solution from which crystals may be obtained upon cooling. Since the mole % of solute in each of these mixtures is low, their boiling points will all be just slightly above 100°.

Not all the mixtures are satisfactory for production of crystals. M_1, for example, is too dilute. As does the acetic acid mixture, B, in Fig. 7, it will yield ice upon extreme cooling. The very short curve-segment *AE* will be involved in such a procedure.

M_2, a 5% solution, in many respects is ideal for recrystallization. As the solution is cooled, a point just below *C* is soon reached, and from thence to *E* a continuous crystallization of the one substance acetanilide will occur. In actual practice, such a procedure gives beautiful snowlike crystals, especially if the cooling is done rapidly.

M_3, an 8% solution, upon slight cooling graphically encounters the undesirable two-liquid zone at *F*. An oil of composition G separates, and this second liquid phase upon further cooling will collect into a pool of relatively impure material and solidify as a caked mass. After temperature *C* (83°) is reached, only crystals will be produced from the solution. The total recrystallized mass is an unsatisfactory mixture of crude caked material along with much purer white crystals.

M_4, a 15% mixture, is graphically already in the two-liquid field at 100°. It will then behave much as M_3 and give an even more unsatisfactory result than M_3, with increased fraction of the impure cake.

89. Remedies. Three plans are feasible for eliminating the troubles described for cases M_3 and M_4 above.

(*a*) Dilute M_3 or M_4 to concentration M_2, thus avoiding the two-liquid zone. Disadvantage: greater loss of material left behind in the mother liquor.

(*b*) Agitate M_3 or M_4 continuously during the cooling process and particularly at the moment when the temperature reaches 83°. The "oil" is dispersed in fine droplets; when these freeze out in the open water solution, they have an improved chance to squeeze out soluble impurities and yield small crystals of relatively high purity.

(*c*) Add a second and better solvent. The addition of an organic solvent like alcohol or acetone may cause a great reduction in the two-liquid zone *CDGHF* and eliminate the oil. Disadvantage: loss of material in mother liquor, as in (*a*).

Questions

1. Why would one expect the crystals produced from the "oil" (of §88) to be less pure than those produced in the "solution" in the same vessel?

2. Without resorting to any ready-made algebraic formula, outline the steps in a laboratory procedure in estimation of molecular weight of sugar by determination of melting point. Describe both laboratory and arithmetical work.

3. Which would have the lower melting point, a 1% solution of methyl alcohol (mp −98°) in acetic acid (mp 16.67°) or a 1% solution of water (mp 0°) in acetic acid?

4. Suppose a sample of pure (liquid) acetic acid were cooled over a period of 15 min, one calorie per minute being withdrawn. During the first 5 min nothing apparently happened. During the next 5 min the material was subcooled and then suddenly started to crystallize. Crystallization was complete before the final 5 min were completed. Sketch a plausible curve in a plot of temperature versus time elapsed for the whole 15 min period.

5. A mixed-melting-point determination was attempted on a somewhat impure substance, isolated from a reaction mixture, and a known compound melting slightly higher. Although these two compounds were different and gave simple mixtures in the solid state, the melting point–composition diagram was less symmetrical than Fig. 5, and the melting point of the mixture of known and unknown that was taken lay between the melting points of the two starting compounds. Would this be expected to lead to the erroneous conclusion that the compounds were the same? Explain.

90. Purification by Sublimation. Strictly speaking, sublimation is a process in which crystals pass directly into the vapor state. When this phase change is followed by crystallization of the vapor without intervention of the liquid state, a convenient and rapid method of purification may be realized. It is not of great concern to the organic chemist whether the vapor is produced directly from the solid state or whether a liquid is vaporized; for effective purification the important part of the process is direct crystallization from the vapor. These methods are used in the commercial purification of several organic chemicals and are termed *sublimation* by organic chemists.

As does distillation, these methods depend upon differences in vapor pressure to effect separation. However, they have the disadvantage that no procedure has been developed to permit a series of automatic multiple sublimations equivalent to fractionation through a distilling column. When two substances of comparable vapor pressure are to be separated, sublimation is hardly more effective than a single, simple distillation and is seldom used.

The advantage of sublimation over distillation arises from the difficulty with the latter in that volatile impurities tend to dissolve in the newly formed distillate as it appears in the condenser. Sublimation of a solid containing volatile impurities prevents this; as crystals form directly from the vapor, accompanying molecules of impurity, still in the vapor state, find nothing in which they can dissolve and therefore are eliminated. The desired product is then of high purity. Fractional distillation of solids suffers also from the difficulty of preventing solidification in the column and condenser; it is often cumbersome to keep these parts of the apparatus above the melting point of the solid.

Sublimation is limited to solids that display appreciable vapor pressures at their melting points. The table in §91 emphasizes the great variation possible.

91. Extremes of Vapor Pressure. Very great variations of vapor pressure at the melting point occur, as in the following table.

Substance	*Vapor Pressure at Melting Point, torr*	*Melting Point, °C*
Arsenic	34 (atm)	814
Carbon dioxide	5.2 (atm)	−57
Uranium hexafluoride	2 (atm)	69
Perfluorocyclohexane (C_6F_{12})	950	59
Acetylene	891	−82
Hexachloroethane	780	186
Camphor	370	179
Neopentane	275	−16.6
Iodine	90	114
Ammonia	45	−78
Anthracene	41	218
Benzene	36	5.5
Phthalic anhydride	9	131
Napthalene	7	80
Benzoic acid	6	122
Water	4.57	0
Bromine	0.44	−7
Zinc	0.15	419
Sulfur	0.03	119
Acetone	0.02	−95
p-Nitrobenzaldehyde	0.009	106
Toluene	0.001	−95
Mercury	0.000001	−39

92. Simple Sublimation. At atmospheric pressure, sublimation can be carried out only with substances like hexachloroethane, whose vapor pressure reaches 760 torr below its melting point. As the vapor-pressure diagram for this substance shows (Fig. 9) an attempt to heat solid hexachloroethane above 185° (760 torr) in an unstoppered vessel will cause an outpouring of vapor and thus total escape of the material into the outer atmosphere in a manner suggestive of the boiling process. Unless the material were confined within a pressure vessel, phenomena represented by the conventional liquid-vapor-pressure curve *PC* could not even be observed. One may, however, continue to elevate the temperature of the newly produced vapor, but at constant pressure. This procedure may be indicated by the horizontal dotted line *XA*. The heated vapor at condition *A* is now ready for the practical process of transportation to a clean vessel and the completion of sublimation therein.

Vapor A is cooled. Its temperature value retreats to *X* at constant pressure 760 torr. As soon as *X* is reached, further cooling causes the vapor to be converted into a solid, and sublimation occurs. Temperature and pressure continue to fall, and the yield of sublimate grows over the graphic path *XO*.

93. Vacuum Sublimation. Substances such as hexachloroethane are rare, and most solids melt at temperatures below those at which their vapor pressures reach 760 torr. A number of these (e.g., naphthalene,

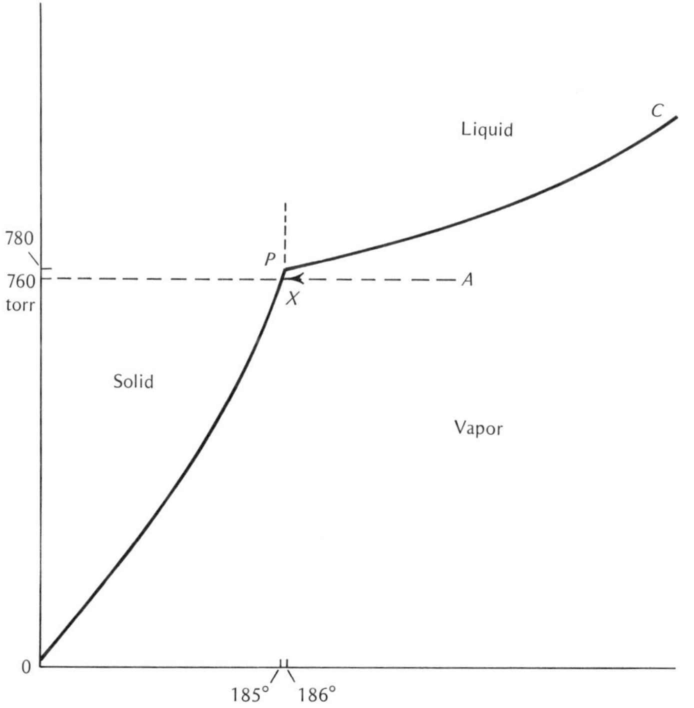

Fig. 9. Sublimation of Hexachloroethane.

camphor, and phthalic anhydride) are readily sublimed under reduced pressure. The vapor-pressure diagram for phthalic anhydride (Fig. 10) shows that at a pressure below 9 mm this compound would behave as hexachloroethane does. By contrast, a compound like acetone melts at an inconveniently low temperature and has so low a vapor pressure at its melting point that sublimation is impractical.

Sublimation under high vacuum is often effective, and like molecular distillation (§67), its use in separation procedures is based on rate of evaporation rather than on vapor pressure.

Apparatus for vacuum sublimation and procedural details are described in §112.

94. Entrainer Sublimation. Solids that must be sublimed under vacuum can sometimes be purified more simply by heating at atmospheric pressure below the melting point and then introducing an inert gas as an entrainer. The vapor is thus swept into a condenser where it crystallizes without first liquefying. A more common procedure, termed *quasisublimation,* is similar but involves further heating of the starting solid above its melting point, to increase the rate of vaporization. Inert gas is passed over the surface of the heated liquid to remove the vapor to a condenser where crystallization occurs, as before, directly from the vapor.

To ensure that crystallization is from the vapor, the partial pressure of the substance being sublimed must be lower than its vapor pressure at its melting point. Otherwise, condensation to liquid will occur, often before contact with condenser walls. A mist of fine droplets will form and may freeze before the walls are reached. Each droplet produces only one or two small crystals, and volatile impurities are literally squeezed out. The product looks like a sublimate and may be of comparable purity.

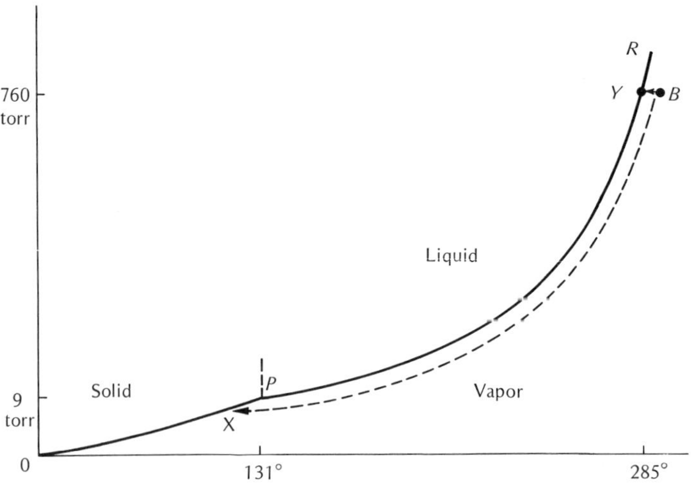

Fig. 10. Sublimation of Phthalic Anhydride.

Probably certain industrial "resublimed" products have been produced in this way.

The process of quasisublimation may be traced for phthalic anhydride on Fig. 10. Suppose heating continues to the boiling point (285°), *Y*, and the vapors are superheated slightly to B. Simple cooling would take the vapors back to the liquid vapor curve *PR* at *Y*, and condensation would occur as in simple distillation. As cooling continued, liquid would continue to condense, following *YP*, and direct transfer of material from vapor to solid would occur only below 131°.

To realize conversion of all the vapor directly to crystals, one must follow *BX*, which is most easily accomplished by diluting the phthalic anhydride vapors with an inert gas that simultaneously reduces temperature to 131° or below. Admixture with cool carbon dioxide or nitrogen is satisfactory and a snowlike mass of sublimate is obtained. Air is less satisfactory as the entrainer because oxygen attacks many organic compounds and a "dust" of finely divided organic material in air may explode like dusty air in flour mills. Sometimes air is used, however, on a small scale with inert compounds (see Experiment 5, §182).

In summary, sublimation is often a convenient and rapid method for purification of solids when impurities are either nonvolatile or considerably more volatile than the desired compound. It is seldom useful when substances of comparable volatility are to be separated. Occasionally a high-boiling liquid can be induced to crystallize by sublimation techniques when other methods fail. Purification of solids is accomplished far oftener by recrystallization or chromatography.

Questions

1. Describe the procedure you would expect to use if you were required to purify the following substances by sublimation: (*a*) camphor, (*b*) sulfur, (*c*) benzene.

2. Suppose the iodobenzene cited in the steam-distillation experiment of §59 were replaced by camphor. Describe the probable behavior of the system when the attempt is made to conduct the steam distillation.

3. What advantages might there be in "steam sublimation" as contrasted with ordinary sublimation in air?

4. On the basis of principles discussed in this chapter, explain the peculiar behavior of liquid carbon dioxide when released in quantity from a cylinder into a cloth bag.

5. Referring to §94, explain in detail why the dilution of pure phthalic anhydride vapor with nitrogen prevents the formation of a (liquid) distillate.

6. When hexachloroethane is stored in a partly filled bottle in a room subject to great changes in temperature from day to night, it may eventually be found that part of the material then adheres to the upper walls of the bottle. Explain.

7. With reference to data in §94 devise laboratory apparatus for preparing "frozen mist" from a sample of crude benzoic acid, and explain how it should be operated.

8. Explain how you would prepare perfluorocyclohexane in the liquid state (cf. §91).

References

(See Chapter 14 for complete titles.)

Melting Points: Weissberger, *Technique of Organic Chemistry,* **1** (3rd ed.), Pt. 1, 287; Müller–Houben–Weyl, **2,** 737; Wiberg, 75; Vogel (1956), 21; Vogel (1966), 23.

Crystallization: Weissberger, *Technique of Organic Chemistry,* **3** (2nd ed.), Pt. 1, 395.

Sublimation: Weissberger, *Technique of Organic Chemistry,* **4** (2nd ed.), 603.

Chapter **8**

Techniques for Purification of Solids and Determination of Melting Points

95. Sequence of Operations. Procedures for isolation and purification of solid organic compounds often start with a mass of crystals that has separated from a liquid reaction mixture and follow the sequence: (1) filtration, (2) washing of the crude crystalline product, (3) recrystallization (redissolving the solid in heated solvent, decolorizing, filtering of the hot solution, and cooling to induce crystallization), (4) isolation of the new crystals, (5) washing, and (6) drying and storage.

96. Filtration. As shown in Fig. 1, a Büchner funnel for large quantities of product or a Hirsch funnel for smaller yields is mounted for suction filtration. The inner test tube is used when both solid product and a limited amount of filtrate are to be preserved with minimum loss in transfer.

The circular piece of filter paper used in the Büchner or Hirsch funnel should be just large enough to cover the whole horizontal surface of the perforated plate without crumpling of the edges of the paper. Where crystals of appreciable size are to be collected, common porous filter paper, such as the creped variety, is desirable.

The flask is now connected to suction equipment with heavy rubber tubing of thickness sufficient to withstand complete evacuation without collapse. A safety bottle should be interposed between a filter pump and the Büchner or Hirsch outfit; otherwise a backwash of water from the pump into the filtrate may easily take place. Such misfortune may readily occur when a pump is drawing vigorously from a flask under a clogged filter. A sudden fall in water pressure, caused by the wide opening of a nearby faucet, allows water to be sucked back from the aspirator pump into the highly evacuated filter flask. Similar backwash occurs when the

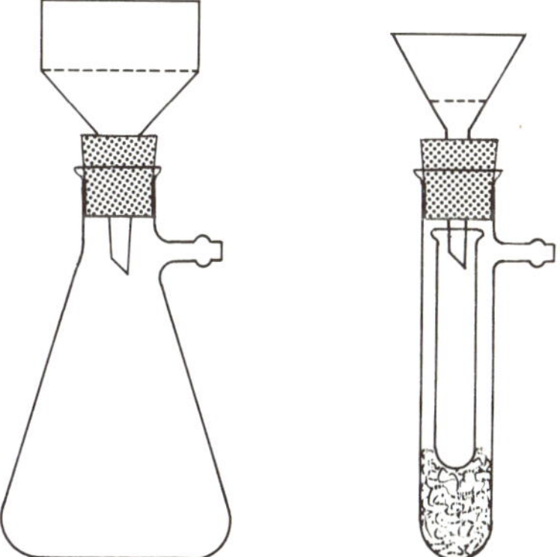

Fig. 1. Büchner and Hirsch Filtration Apparatus.

careless worker fails to disconnect the suction hose *before* he turns off the water faucet serving the filter pump.

Before filtration is attempted, suction is applied and a small amount of solvent, like that from which the crystals are to be separated, is poured over the *entire* surface of the filter paper. Care is now taken that the paper clings tightly all over the plate, without chance for small particles of solid to slip under the paper and reach the filtrate below.

The solvent, now down in the filter flask, is removed if its presence in the coming filtrate would be undesirable. Suction is resumed, and the reaction mixture promptly poured upon the center of the paper. Ordinarily the "mother liquor," or residual solution separated from crystalline matter, passes down rapidly, and, if valuable, is set aside in a separate vessel before the crystals are washed.

Frequently some of the crystals are left behind in this operation. It is then convenient to return part of the filtrate to the original vessel and to use it to flush the residual crystals into the funnel. Such an operation may be repeated as often as necessary.

After all of the crystals have been transferred to the funnel, they are tamped down with a spatula or the top of a glass stopper of "mushroom" type to remove as much of the liquid as possible. In spite of vigorous suction, 5–20% of the wet mass is likely to be liquid; in the case of finely divided, semicolloidal preparations, much more.

97. Washing. With suction off, the crystalline mass in the funnel is moistened with a solvent in which the liquid is soluble and the crystals relatively insoluble. This will usually be the solvent that was used for the reaction. The solvent is usually cold, to reduce the solubility of the crystals; in some instances a different solvent may be used. A spatula is used to mix the crystals with the wash liquid in order to insure thorough contact between them. Care must be taken not to allow crystals to get around the edge of the paper and not to scrape a hole in it. Suction is then reapplied and the crystals pressed down as before.

If the washed crystalline mass is not easily shaken out of the suction funnel, it may be blown out upon a dish from the funnel stem end. In such a procedure a short section of large rubber tubing is interposed, for safety, between funnel and mouth of operator. Still simpler is the plan of grasping the bare end of the funnel with the three lower fingers of the hand and blowing with the mouth pressed firmly against the upper part of the fist.

If the flask is connected to piped vacuum service during suction filtration, great care must be taken to avoid overfilling the vessel with the mother liquor or wash solvent, lest liquid pass into the vacuum line and injure vacuum equipment.

98. Recrystallization. The most common procedure for recrystallization consists of dissolving an impure substance in a hot solvent and allowing the material to crystallize as the solution cools. Impurities are removed during the process mainly because they remain dissolved in the solvent. Coloring matter and certain other impurities are removed by selective adsorption, usually on active carbon (§131). Occasionally an impurity remains undissolved during the initial solution of the compound and is removed by filtration.

99. Choice of Solvent. The success of recrystallization depends first of all on proper choice of solvent. Ideally, the solubility of the desired compound should be high in the hot solvent and very low in the cold; impurities either should be highly insoluble in the hot solvent or should dissolve readily and remain in solution.

Often compounds are encountered for which information on a suitable recrystallization solvent is unavailable. Small-scale trial recrystallizations are then carried out rapidly in 10- by 75-mm test tubes. A few drops of various common solvents (see accompanying table) are used with small portions of the crude, finely divided preparation and the crystals are stirred and crushed under the cold solvent with a stirring rod. If solution occurs at room temperature, the solvent is probably unsuitable. If solution does not occur, the sample is warmed gently on a steam bath or over a small flame (preferably a microburner) with stirring. A few more drops of solvent may be added if only partial solution occurs. (Transfer of solvent is most conveniently done with small, clean dropping tubes drawn out at the end like pipets.) If a homogeneous solution is obtained, it is cooled, and the inside walls of the test tube are scratched if crystallization does not occur readily. If no crystals can be obtained or if solution does not occur, the solvent is probably unsuitable and another should be tried.

Frequently a good recrystallization solvent is one not too like the compound being purified. Unoxidized substances are often better recrystallized from ethyl alcohol than from ether or petroleum solvent, for example. On the other hand, if the solvent and solute are too different, as in the case of water and *p*-dibromobenzene, it is impossible, under any circumstances, to dissolve enough for effective work. Acetone is seldom effective as a recrystallization solvent. Although it dissolves many organic compounds, the variation of solubility with temperature is usually too low. Frequently the mother liquor retains too much of the desired product.

100. Use of Mixed Solvents. Sometimes an organic compound dissolves either too freely or too sparingly in any one available solvent.

[§100] **List of Common Solvents**

Solvent	*Boiling Point, °C*	*Freezing Point, °C*
Diethyl ether	35	−116
Methylene dichloride	40	−97
Carbon disulfide	46	−111
Acetone	56	−95
Chloroform	61	−64
Methyl alcohol	65	−98
Tetrahydrofuran	66	−101
Diisopropyl ether	68	−60
Carbon tetrachloride	76	−23
Ethyl acetate	77	−84
Ethyl alcohol	78	−117
Benzene	80	5.5
Cyclohexane	81.4	6.5
Acetonitrile	82	−44
Isopropyl alcohol	82	−89
Petroleum solvents (variable)	30 to 175	
Water	100	0
Nitromethane	101	−29
Dioxane	102	12
Toluene	111	−95
Pyridine	115	−42
n-Butyl alcohol	118	−89
Acetic acid (glac.)	118	17
m-Xylene	139	−54
N,*N*-Dimethylformamide	154	
Nitrobenzene	211	5.7
Diethylene glycol	245	−10

Mixtures of solvents may then be serviceable. The following solvent pairs are often used:

Alcohol–water.
Benzene–ligroin (petroleum).
Acetic acid–water.
Ether–alcohol.
Ether–petroleum ether.

The solid is usually first dissolved in the solvent of a given solvent pair in which it is more soluble. The second solvent of the solvent pair, in which the solid is less soluble, is now added slowly until crystallization is imminent. During all this procedure the temperature is maintained at a relatively high value, perhaps near the boiling point. Cooling now causes a substantial separation of the desired solid.

Inexperienced workers, anxious for a large yield of recrystallized material, often overdo the performance just described. For example, a compound is dissolved, in quantity, in hot alcohol, and water is added to unwarranted excess. Practically the whole of the original preparation, impurities included, separates from the solution precipitately as an amorphous mass.

A hydrocarbon solvent such as benzene or light ligroin might be paired with ethyl alcohol, provided the latter is absolute or nearly absolute. If the alcohol contains a substantial—and uncertain—quantity of water, such a mixed-solvent plan should be avoided, since the resulting solution is likely to separate into two liquid phases, making trouble in the ultimate process of crystallization.

101. Redissolving the Crude Product. Recrystallization is best carried out in a conical (Erlenmeyer) flask of borosilicate glass (Pyrex, Kimax). The amount of crude product taken should be capable of solution in a volume of solvent that will fill the flask somewhat less than half full. The flask containing finely divided product and less than enough solvent is attached to a reflux condenser, unless a noninflammable solvent is employed, and warmed with swirling to hasten solution. More solvent is added through the condenser, as needed, to just complete solution. Heating is best done with a steam bath or hot plate, but a burner may be used with higher-boiling solvents. Boiling chips are added as needed. Advanced workers often omit the reflux condenser when working with small amounts of material even when using flammable solvents that are moderately volatile, but a flame is then never used and the operation is frequently carried out in a hood. The vapors of many common solvents (notably carbon tetrachloride) are quite toxic.

Once solution is complete (or when all but insoluble impurities have dissolved), it is usually necessary to filter the hot solution. Most crude preparations give slightly off-color, murky solutions that should be clarified with active carbon. See §131 for details. It is important to make certain that the solution is not superheated before active carbon is added; usually one simply removes the flask from the heater briefly and swirls it. Brief heating with the carbon is usually adequate to complete the decolorization.

Hot solutions are most often filtered by gravity through a stemless funnel (Fig. 2). A fairly retentive filter paper is needed to remove active carbon, but more rapid paper is satisfactory in the absence of carbon.[1] The top of the filter paper should fall well below the top of the funnel as shown. A fluted filter paper speeds up the filtering operation and is easily prepared by folding the paper in sixteenths, all in the same direction as

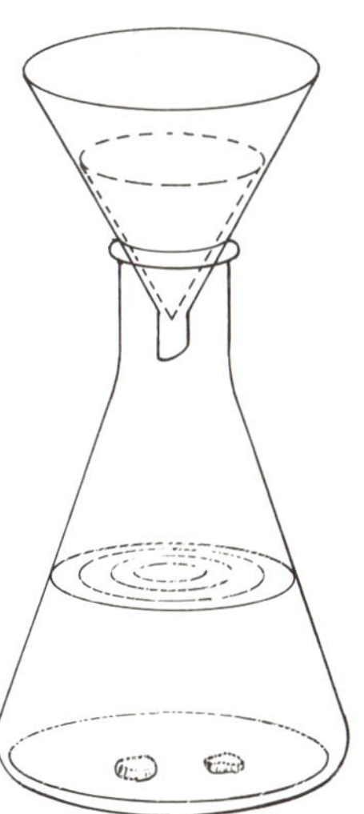

Fig. 2. Use of Stemless Funnel.

[1]Whatman No. 1 paper usually works well for removing active carbon, and the less retentive Schleicher and Schuell No. 410 is excellent for most other organic laboratory applications.

shown in Fig. 3, placing reverse folds between each of these first creases, opening the paper, and dividing the two sections that lie flat against the funnel into smaller folds as shown at *A*. With small papers, only half as many folds may be used. It is best to leave the center of the paper almost uncreased but to reinforce the outer portions of the creases by added pressure with the fingers while the paper is in an accordioned segment, as in Fig. 3(d). Various other methods of folding have been devised. When removing carbon, some investigators prefer to use a paper placed in the funnel in the ordinary way rather than a fluted filter paper because the carbon tends to creep up the folds and may thus get into the filtrate. If the solvent is water, the danger arises that the bottom point of the filter may be softened and burst. In such a case the filter should be pushed down slightly into the funnel, thus providing a constriction sufficient to give support but not enough to plug the funnel stem.

A warm funnel should be used to prevent crystallization on the paper or in the stem, and a slight excess of solvent is added. With small amounts it is convenient to warm the funnel by resting it in the neck of the flask (as shown in Fig. 2) and refluxing solvent around it. This may be done during final warming of the preparation; the funnel also serves as a partial condenser and prevents all but small amounts of solvent from escaping.

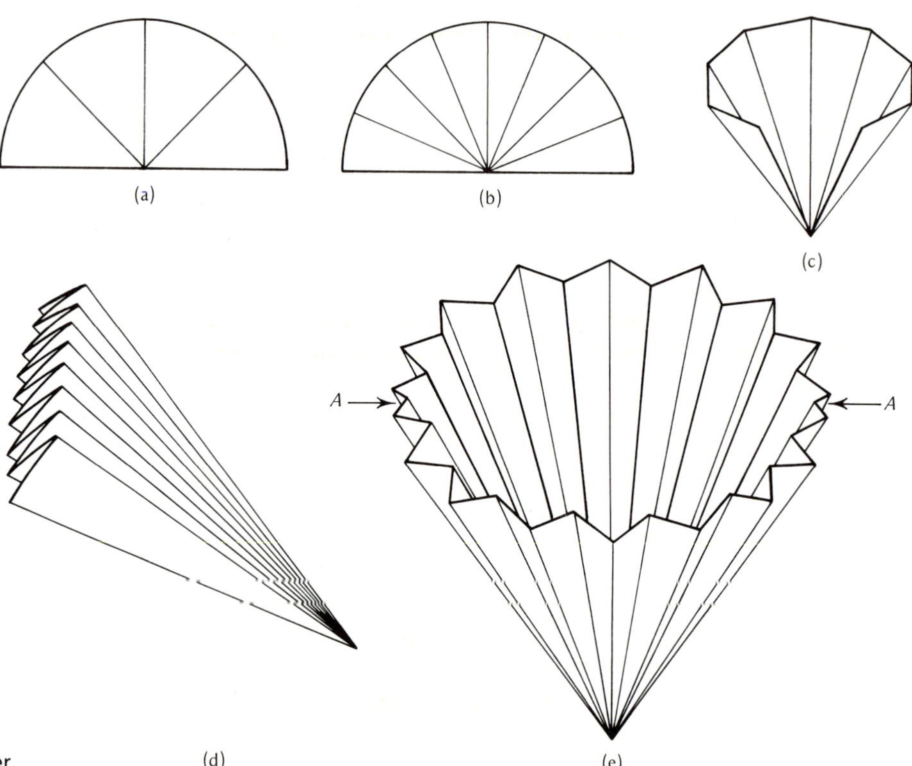

Fig. 3. Folding of a Filter Paper.

Just before filtration is started, the funnel is transferred to the neck of the clean flask that is to receive the filtrate. The solution is then poured quickly into the paper, taking care that the level in the funnel remains well below the top of the paper. As filtration proceeds, the bulk of the solution being filtered is kept warm on the steam bath or hot plate. With a conical flask that is not too full, there is little tendency of the solution's being filtered to run down the outside of the flask. A small crust of crystals may form on the outside of the flask rim, but this is easily washed into the funnel with a little hot solvent at the end of the filtration.

After the solution has run through the paper, a crust of crystals often remains around the tip of the funnel; ill-formed crystals have often formed in the body of the filtrate. It is common practice to rinse the original flask with a little warm solvent and to wash the filter paper with the rinsings as they are filtered. It is convenient to manipulate the solvent with a medicine dropper (best with a 6- or 8-in. glass portion). With practice, all the material may be washed out of the original flask, the crust removed from the flask lip, the paper washed down, and the material gotten into the filtrate with very little loss and without much additional dilution of the filtrate. A boiling chip is then added to the filtrate, which is warmed until all crystals have dissolved and the solvent has refluxed into the funnel long enough to wash down any crystals that remain around the tip. The clear solution is then covered and set aside to crystallize.

With larger preparations it is not effective to warm the funnel as described, and a water or conical steam coil or a funnel-heating device heated by electricity, is employed. Hot solutions are seldom filtered with suction in laboratory operations because the suction causes rapid evaporation and cooling of the solution; premature crystallization, especially in the filter plate of the funnel, is often troublesome.

102. Crystallization. The most desirable crystal size depends upon the particular compound under investigation and upon the difficulty of the purification. Tiny crystals in the form of a fine powder present so much surface that filtration and drying may be difficult. Very large crystals, unless grown under special controlled conditions, usually occlude the mother liquor, which is almost impossible to wash out during filtration. The drying of such crystals is difficult because the mother liquor is held in the interstices of the crystal mass, and evaporation of the solvent is slow; when it has been accomplished, the impurities are left on the surface of the crystals.

For many substances purification is most effective if crystals of medium size (perhaps not more than 3–4 mm long, if needles; or not over 1–2 mm, if the three dimensions are comparable) are obtained. If a difficult separation of two substances of comparable solubility and concentration is being attempted, one takes special pains to allow the dust-free solution to cool slowly and completely undisturbed. Seed crystals of the desired product are introduced as soon as the temperature drops to the point where they will not dissolve. In this way it is sometimes possible to induce the desired component to crystallize in large crystals while the second remains in supersaturated solution. These large crystals can then be recrystallized in smaller size and freed from remaining impurity.

103. Slow Crystal Formation. Occasionally a solution will not deposit crystals on cooling, despite the fact that it is already highly supersaturated. The presence of a protective colloid, perhaps a viscous substance, causes this trouble in many cases. If a so-called seed crystal of the desired substance is introduced, crystallization sometimes starts and continues until equilibrium is reached. Seeding is illustrated in the recrystallization of benzil (§358). In troublesome cases the seeding process may be combined to good advantage with refrigeration in an ice-salt bath at $-10°$ to $-15°$. Cases have been known where a supersaturated solution, contaminated with "tar," which was acting as a protective colloid, required hours of refrigeration with ice and salt for the first crystallization. When the crystalline product was subjected to recrystallization, the time of crystallization was shortened to a few minutes. If crystallization does not start readily, one may assist the process by scratching the inner wall of the vessel with a glass rod. Crystals may then form on the scratch lines.

104. Isolation and Washing. After crystallization is complete, the crystals are collected on the Büchner or Hirsch funnel as described for the original reaction product (§96). Care must be taken not to draw air through the crystals because this will cause evaporation of the mother liquor, and the impurities that were dissolved therein will be deposited on the surface of the crystals. When the mother liquor has drained through, the crystals are pressed down quickly and the suction is interrupted by removing the rubber tubing from the filter flask. A little cold solvent is then added and the filter cake is broken up carefully with a spatula. This allows the solvent to wash mother liquor from the surface of all crystals. Suction is again applied, and the crystals are pressed down to remove as much solvent as possible. The washing procedure should be repeated unless the crystals are so soluble in fresh cold solvent that appreciable losses occur.

An excellent technique, which avoids drawing air through the crystals, involves a rubber dam. This is a piece of thin sheet rubber of the type used by dentists or, better, of neoprene. As soon as the crystals have been transferred to the funnel, quickly smoothed down, and most of the mother liquor drawn through, the rubber dam is stretched over the top of the funnel and fastened with a rubber band. The rubber is drawn down against the crystals by the suction, and forces mother liquor out more effectively than the usual tamping while it also prevents the passage of any air through the crystals. When suction is released, the dam is easily removed, wash solvent applied as described, and the dam used again to press out the solvent.

105. Nonvolatile Solvent. If a crystalline product has been deposited from a nonvolatile solvent, it becomes desirable to wash with a volatile solvent, giving special attention to technique.

Example. A crude product is recrystallized from nitrobenzene (bp 210°), low-boiling solvents having proved unsatisfactory. The residual mother liquor clinging to the new recrystallized product contains impurities. If the crystals were washed first with the volatile solvent benzene, there would be danger of precipitation of certain impurities which, though soluble in nitrobenzene, do not dissolve appreciably in benzene. The crystals are therefore washed first with nitrobenzene to remove the impure mother liquor, and then with benzene to remove the nitrobenzene. The benzene, being volatile, disappears promptly upon drying.

106. Drying. The washed crystals are now placed in an evaporating dish and dried either in the open air or in a vacuum desiccator, as conditions require.

MELTING-POINT TECHNIQUE

107. Determination of Melting Points. High-precision melting points can be determined by obtaining a cooling curve under conditions which assure equilibrium between liquid and solid. This method has been refined so that it can be one of the most precise methods for determining the purity of a solid organic compound. However, it is time consuming and requires a fairly large sample. Details are given in the References.

For most purposes melting points are determined by placing a small amount of solid in a thin-walled capillary tube that is then heated slowly in an air or liquid bath. Melting usually occurs over a temperature range; it is preferable to report this as the "melting range" rather than "melting point," although the latter term is commonly employed. The lower limit of the range is the bath temperature at which melting can first be detected and the upper limit is the bath temperature at which the last crystal disappears. The capillary tube ($\sim$1 mm diameter) must have thin walls for high heat conductivity and heating must be slow enough so that bath and sample are at the same temperature. Thermometer bulb and sample must be in as close contact as possible. Heating may be rapid until a temperature about 10° below the melting range is reached, but must then be no more than a degree per minute if an accurate value is to be obtained. To save time a quick preliminary determination is usually made to find the approximate range, and a fresh sample then used for the final determination.

As explained in Chapter 7 an impure compound usually melts over a temperature range which is lower than the true melting point of the compound; the broader the range the lower the purity. Capillary melting points are seldom completely sharp and a range of 1° is usually acceptable, especially if further recrystallization or other purification does not raise the value. Further recrystallization is necessary if the melting range is broad or the value found is below that reported in the literature. Occasionally sharp-melting mixtures are encountered. These result from formation of solid solutions (see §85) or from chance isolation of a mixture that has the composition of a eutectic. The fact that such sharp-melting material is a mixture is usually discovered when spectra of the supposedly pure compound are taken; however, a few percent of one of the components may escape spectral detection.

108. Melting Point Apparatus. Figure 4 pictures two apparatuses commonly used for determination of melting point. In (*a*) a 100-ml beaker is

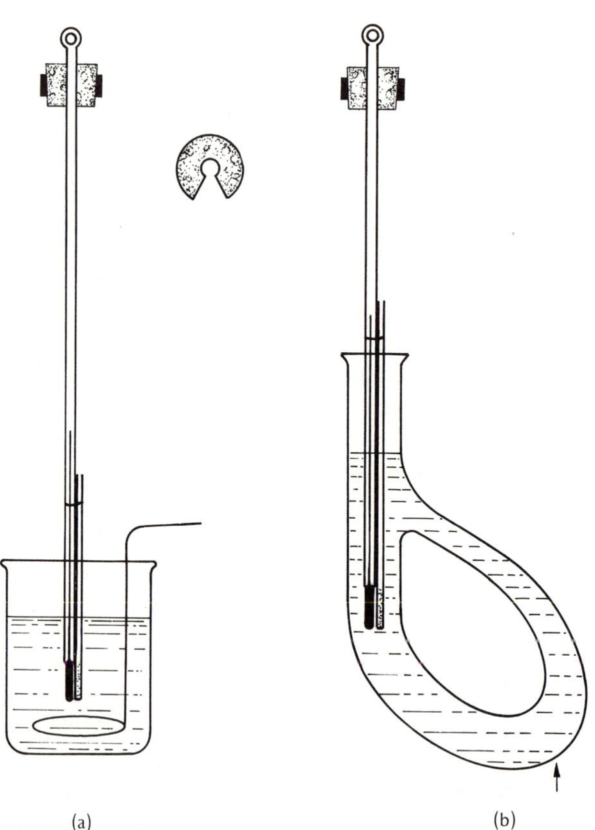

Fig. 4. Apparatus for Determination of Melting Point.

supported on a wire gauze and filled as shown with a suitable liquid. The capillary tube is attached to a thermometer with a small rubber band (usually cut from new, soft $\frac{3}{16}$-in. rubber tubing) so that sample and thermometer bulb are at the same height. The thermometer is then mounted securely in the bath with the bulb about 1 cm from the bottom of the beaker. If the thermometer is fitted with a cork near the top before the capillary is attached, it can easily be held with a clamp. **Warning:** *Take special care that the thermometer does not slip down and break the bottom of the beaker!* A slotted cork is sometimes preferred to permit reading the full length of the thermometer but it must be cut as shown so it still holds the stem securely. The liquid in the beaker must be stirred. The glass stirrer shown, which can be made from a 4- or 5-mm glass rod, is suitable but tedious to use. Mechanical stirring can be provided by a small propeller stirrer of glass or metal attached to a small, inexpensive, constant-speed motor (e.g., a small phonograph motor obtainable from a radio supply house). The beaker is heated by a small Bunsen flame.

Figure 4(*b*) pictures a modified Thiele apparatus. Heat is applied at the point marked by an arrow and stirring occurs by convection. A small flame, usually from a microburner, is the source of heat; effective circulation of the liquid requires a steady source of heat and freedom from drafts around the apparatus.

A 50-ml round-bottom, long-neck flask mounted vertically and carefully heated with a soft, bare flame directly in the center of the bottom of the flask provides equally effective convection mixing and temperature control. The thermometer bulb and sample must be carefully centered in the body of the flask.

More elaborate electrically heated and mechanically stirred baths are usually used in research. The Hershberg apparatus (see References) is one of the best. Such apparatus is available commercially.

The choice of a suitable bath liquid for any melting-point apparatus presents problems. Silicone oils (§20) are probably the best, but the expense may preclude use in elementary laboratories. Vegetable oils and glycerol darken rapidly but are often used. Concentrated sulfuric acid is the most economical, but the hazards involved in its use are so great that it is not recommended. It darkens if contaminated by small amounts of organic matter but may be cleared by addition of a few centigrams (avoid excess!) of powdered sodium nitrate or potassium nitrate to the liquid while hot.

The nuisance of a liquid heating bath may be eliminated by use of a copper or aluminum block in which holes are drilled for the thermometer, the melting-point tubes, a suitable light, and a viewing port. The high heat conductivity of these metals favors uniformity in temperature, and such apparatus is especially convenient for organic compounds melting above 300° where bath liquids darken rapidly or are too volatile. With this apparatus, it is important to heat slowly near the melting point, since heat transfer is less efficient than with liquid baths. See Wiberg, 82, for construction of such a block.

In another type of melting-point apparatus, crystals are placed on the polished surface of a heated metal block in which a thermometer is mounted. The crystals are often placed between cover glasses on such blocks. Electrically heated varieties are commercially available. Special care must be taken to heat slowly near the melting point when such apparatus is used and even then some compounds give quite different values from those observed with the ordinary capillary procedure. It is recommended that any melting points recorded in the literature be taken by the capillary method because it is so universally used; if only the melting point on the block is available, it should be marked as such.

Melting points of very small amounts of material are determined on the hot stage of a microscope. A polarizing microscope is especially convenient for this use (see the References for details).

A simple air-bath device in which crystals are placed directly on a thermometer bulb inside concentric test tubes has been devised [A. May, *Anal. Chem.* **21,** 1427 (1949)]. This apparatus gives quite satisfactory results if heat is applied very slowly near the melting point of the compound.

109. Construction and Use of Melting-Point Capillaries. Thin-walled melting-point-capillary tubes are available commercially and probably cost less than the glass that would be used by students in learning to construct their own tubes. Instructions for preparation of such tubes are included because at least chemistry majors should learn to prepare their own.

The best material from which to prepare such tubes is the thin-walled stock tubing of borosilicate glass from which manufacturers make commercial test tubes. A blast lamp is necessary and 12- to 15-mm tubing is suitable. The tubing should be washed carefully, rinsed with distilled water, and allowed to dry. Inorganic material on the inside surface of the tubing may cause decomposition of the sample at melting temperatures.

In the absence of the stock tubing, a glass rod is fused to the end of a test tube of the desired diameter, as in Fig. 5(*a*). The newly assembled rod-tube must not be allowed to cool, but instead must be immediately converted into miniature tubing. As shown in 5(*b*), either the body of the test tube or the center of a section of stock tubing is heated strongly in a large flame. A little practice enables the beginner to rotate the tube without twisting the softened portion until a length of 3–5 cm of the tube has become very soft but has not collapsed.

At this point the tube is removed from the flame and immediately drawn out *in a straight line* until the external diameter has shrunk to a range of 1–2 mm. The practice of swinging the arms out in a wide curve will yield slightly curved tubes, which may not adhere well to the thermometer in subsequent procedures. The operator should pause for 3 or 4 sec with hands in the outstretched position, to ensure solidification of the slender tube before it can bend.

The melting-point tubing is now cut into 15-cm lengths with a needle-point flame, the sharp edge of a freshly broken piece of porous clay plate, or a carborundum crystal. The flame method forestalls entry of condensable water vapor into the tube. If the clay plate or carborundum crystal is used, both ends of the 15-cm length are immediately sealed before the flame and then the tube may be stored indefinitely. Do not use a file or knife as a cutting tool because these would crush the fragile tubing.

Using clay plate or carborundum, cut a melting-point tube into the desired two single tubes of 7- to 8-cm length. Powder the crystalline material ready for test, and insert the open end of a melting-point tube into the mass. Tap the tube, closed end down, on a wooden table until the particles reach the bottom; or use a file gently to jar the crystals to the bottom of the tube.

110. Errors. In most of the apparatuses described the thermometer is not fully immersed in the bath and the proper correction for emergent stem should be made (§18). In the Hershberg apparatus small Anschutz thermometers are used with total immersion; usually such an apparatus is calibrated with standard compounds (§167), or against other thermometers that have been calibrated. Melting points or melting ranges that have been determined on such an apparatus, or have been corrected for emergent stem, are marked "cor."

If the melting-point tube is heated too rapidly, a positive error in the determination is likely to occur. The mass of finely divided organic compound, perhaps not so compact as it should be, conducts heat poorly and does not reach the true melting temperature as soon as the highly

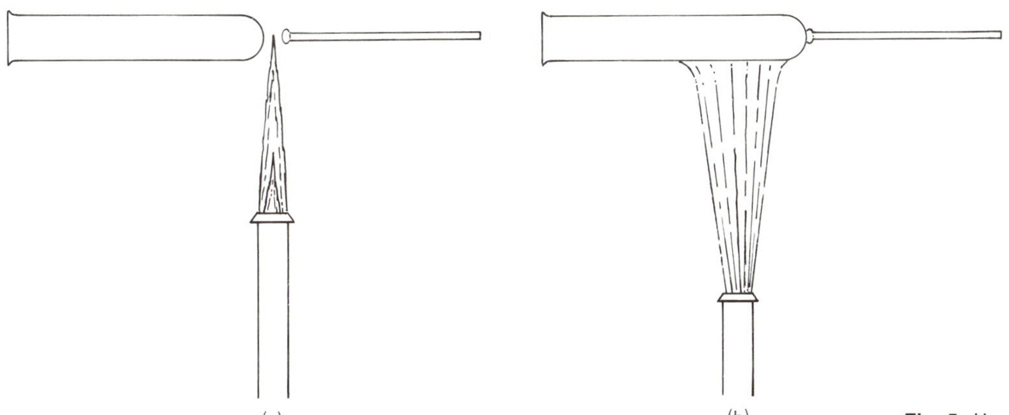

Fig. 5. Use of Blast Lamp.

conductive metallic mercury in the thermometer. Excessively thick melting-point tubes aggravate this difficulty.

It is not advisable to reheat a melting-point sample that has already been melted and resolidified in a previous run. Decomposition often starts immediately upon melting, possibly accelerated by chemical action of the heated glass surface. The used tubes are therefore discarded and new tubes provided if repetitions are necessary. Always compact the material in the tube as much as possible by tapping it firmly against a wooden surface.

111. Elementary Laboratory Practice. Because it is time consuming and a nuisance for beginning students to construct and store a suitable melting point apparatus, it is common practice to provide such an apparatus (or more than one) for use by all of the students. This also provides an opportunity for students to see and use more than one kind of apparatus.

112. Sublimation Apparatus. Apparatus and procedure for simple sublimation and for one form of entrainer sublimation are described in Experiment 5, §180. Many variations exist.

Vacuum sublimation is frequently carried out in apparatus such as shown in Fig. 6. A similar device may be constructed from two sizes of test tubes (the outer one with a side arm) and rubber stoppers. The outer tube serves as the container for the material being sublimed (the sublimand); the inner tube, as condenser. Flat surfaces as shown are preferable because the sublimand should be spread in an even, thin layer at a uniform distance from the condensing surface, to ensure smooth operation. The sublimand should be finely divided and must usually be degassed by gradual heating under partial vacuum before full vacuum is applied. Sometimes the degassing is carried out before the condenser is in place; then it is best to cool the sublimand before breaking the vacuum to insert the condenser.

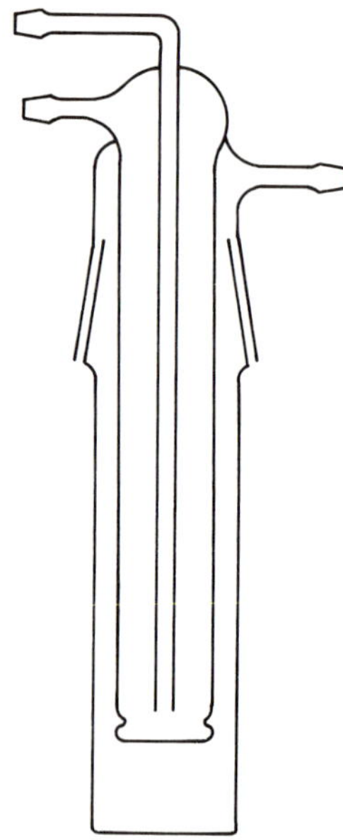

Fig. 6. Sublimation Apparatus.

High-vacuum sublimation must be carried out in apparatus in which the distance from sublimand to condenser surface is less than the "undeflected path" of the vapor molecules. The References should be consulted for details.

References

(See Chapter 14 for complete titles.)

Recrystallization: Weissberger, *Technique of Organic Chemistry,* **2** (3rd ed.), **3** (2nd ed.), Pt. 1, 395, 607, 787, **6,** 13, **7** (2nd ed.); *Techniques of Chemistry,* **2** (3rd ed.); Müller–Houben–Weyl, **1,** Pt. 1, 139, 183, 341, 753; Wiberg, 98; Bates and Shaeffer, 50; Pasto and Johnson, 7.

Melting-Point Determination: Weissberger, *Technique of Organic Chemistry,* **1** (3rd ed.), Pt. 1, 287, **6,** 145; Müller–Houben–Weyl, **2,** 787; Wiberg, 75.

Sublimation: Weissberger, *Technique of Organic Chemistry,* **4** (2nd ed.), 603, **6,** 84; Müller–Houben–Weyl, **1,** Pt. 1, 897; Wiberg, 114.

Chapter **9**

Extraction, Adsorption, and Chromatography

Purification of a compound and separation of a compound from a mixture are often accomplished by introduction of a second phase to which one substance or type will be transferred and thus isolated. In an ideal case the desired compound will be completely transferred to the new phase (or left in the first phase) while all impurities or other components of the mixture will be left behind (or transferred to the second phase). It is more common for the desired compound to be *partitioned* between the two phases (i.e., part of it remains in the original phase and part is transferred to the new one). *Extraction* is one example in which a liquid phase, the extractant, removes a desired substance from a second phase, which may be liquid, solid, or gas. *Washing* is a similar process in which an undesirable impurity is removed, leaving the desired substance in the original phase. *Adsorption* is a selective process involving the adhesion of materials to the active surface of an inert insoluble solid. When a mixture of gases, liquids, or compounds in solution comes in contact with such a solid, separation or purification is effected by partitioning, involving the active surface.

EXTRACTION AND WASHING

113. Theory of Extraction. If a solution is in contact with an immiscible solvent, the solute will distribute itself between the two liquid phases. At equilibrium the ratio of the concentrations of the solute in the two phases will be approximately constant no matter what the total concentration.[1] This ratio, called the *distribution coefficient, distribution*

[1]The precise statement is that the solute will distribute itself so that its activity is the same in each phase. Heterogeneous equilibria are discussed in textbooks on physical chemistry. A. E. Hill has given a clear exposition and summarized the expressions required when association or dissociation complicates the situation. This appeared in *A Treatise on Physical*

ratio, or *partition coefficient,* is approximately equal to the ratio of the solubilities of the solute in the two solvents and is useful as a guide for purification and recovery operations.

Example. A rough calculation, which ignores the mutual solubilities of the two solvents and the changes in volume that occur when a solid is dissolved in a given volume of liquid, may be made as follows:

The distribution coefficient of isobutyric acid between diethyl ether and water at 25° is approximately 3 (between water and diethyl ether it is 0.333). Given 40 g of isobutyric acid in 1000 ml of water this indicates that three fourths of it can be extracted into 1000 ml of ether in a single extraction. Let x equal the weight of isobutyric acid remaining in each milliliter of water solution at equilibrium. Then $3x$ is the weight of acid in each milliliter of ether.

$$\begin{aligned} 1000 \times x &= 1000x \text{ g of isobutyric acid in the water} \\ 1000 \times 3x &= 3000x \text{ g of isobutyric acid in the ether} \\ \hline \text{Total} &= 4000x = 40 \text{ g (the amount of acid taken)} \end{aligned}$$

$$x = 40/4000 = 0.01 \text{ g}$$

and

$$\begin{aligned} 3000x &= 30 \text{ g of acid extracted} \\ 1000x &= 10 \text{ g of acid remaining in the water} \end{aligned}$$

More efficient extraction is attained if the liter of ether is used in smaller portions with several extractions. Thus, if the original solution of 40 g of isobutyric acid in water were extracted with 1000 ml of ether in two portions (500 ml each), one would calculate the amount extracted by the first 500 ml as follows:

$$\begin{aligned} 1000 \times x &= 1000x \text{ g of acid in the water layer} \\ 500 \times 3x &= 1500x \text{ g of acid in the ether layer} \\ \hline \text{Total} &= 2500x \text{ g} = 40 \text{ g} \end{aligned}$$

$$x = 0.016 \text{ g and } 1500x = 24 \text{ g of acid extracted}$$

The aqueous solution, now containing only 16 g of isobutyric acid, is extracted with the second 500-ml portion of ether.

$$\begin{aligned} 2500x &= 16 \text{ g} \\ x &= 0.0064 \\ 1500x &= 9.6 \text{ g of acid extracted} \end{aligned}$$

Chemistry (2nd ed.), H. S. Taylor, ed., Van Nostrand, Princeton, N.J., 1930, p. 467 (a volume long out of print). Chapter 3 of *Physical and Chemical Methods of Separation* by E. W. Berg, McGraw-Hill, New York, 1963, pp. 50–79, is one of the best brief presentations. L. C. Craig and D. Craig give a longer discussion (Weissberger, *Technique of Organic Chemistry,* **3** (2nd ed.), Pt. 1, pp. 149–393).

The total acid extracted is thus 33.6 g, a distinctly better total yield.

Division of the 1000 ml of ether into smaller portions and an increased number of extractions will increase the yield further, but beyond five or six portions the gain is slight. The following equation is useful as a guide to the number of extractions needed to accomplish any desired degree of recovery and may be used to determine the effect of various divisions of a given volume of extractant:

$$C_n = C_o \left(\frac{KV_1}{KV_1 + V_2} \right)^n$$

where C_n is the weight of material that remains in the original solution after n extractions with volume V_2 of extractant, V_1 is the volume of original solution, C_o is the weight of material originally in volume V_1, and K is the distribution coefficient between the liquids making up V_1 and V_2, respectively. To calculate the effect of extracting the isobutyric acid solution three times with 1000 ml of ether in three equal portions, one places $C_o = 40$, $V_1 = 1000$, $V_2 = 333$, $n = 3$, and $K = \frac{1}{3}$.

$$C_n = 40 \left(\frac{333}{333 + 333} \right)^3 = 40 \left(\frac{1}{2} \right)^3 = 5.0 \text{ g}$$

The total acid extracted = 40 − 5 = 35 g. A similar calculation for 1000 ml of ether in four equal portions shows that 4.26 g remains in the water layer.

114. Extraction in Practice. Extraction is seldom the final method of purification of a compound but is often useful to separate the product of an organic synthesis from impurities and by-products that are water soluble. When part or all of a desired compound is obtained in water solution, as is often the case, one may recover it by using a water-insoluble solvent in which it is readily soluble and in which the impurities are insoluble or only slightly soluble. The two liquids are mixed by shaking in a separatory funnel and the layers are separated. The funnel should not be filled to more than about three-fourths capacity. The initial mixing must be done with great care when volatile solvents are used because pressure develops in the funnel: the glass stopper may be forced out, a spray blown into the face of the operator, and the product lost. The hazard is greater with very volatile solvents such as ether, or if the solution is warm. Pressure develops because the two-phase system that is formed when the solvent is poured into a funnel containing an aqueous solution and air is not in equilibrium with the vapor, which is essentially air plus the water vapor above the aqueous solution. When the funnel is closed and shaken, the vapor pressure of the volatile, insoluble solvent is essentially added to the 760 torr already present. This condition is aggravated when the funnel is warmed by the heat of one's hands. It is especially troublesome when a gas is evolved; for example, when an ether solution of an acid is extracted with sodium carbonate solution.

To avoid trouble, one grasps the funnel as shown in Fig. 1 so that both the stopcock and stopper are held in. The funnel is carefully inverted and the pressure is released by slowly opening the stopcock. The stopcock is closed, the funnel is carefully rocked, and the stopcock again is opened (funnel inverted) to relieve pressure. Gentle shaking is next used, and finally the well-stoppered funnel may be shaken vigorously. After the atmosphere in the funnel is saturated with the solvent vapor, further shaking (either with or without addition of new solvent) develops little or no added pressure.

Clean separation of layers does not always occur. Emulsions may be formed. These may sometimes be broken by adding acid or base to change the pH. Addition of common salt may help because its highly ionized solution may aid in electric discharge of offending colloidal particles. It is usually advisable to avoid vigorous shaking of systems that tend to emulsify.

Often the interface presents a fuzzy, emulsified appearance as a result of accumulation of solid particles and trash. If necessary, the whole liquid mass may then be strained through cheesecloth or glass wool, with resulting clarification of the interfacial zone. It may even be advisable to filter through a coarse filter pad. More often the trash may be allowed to pass along with the layer being extracted, and eventually discarded.

After the layers have separated, the lower one is run off through the

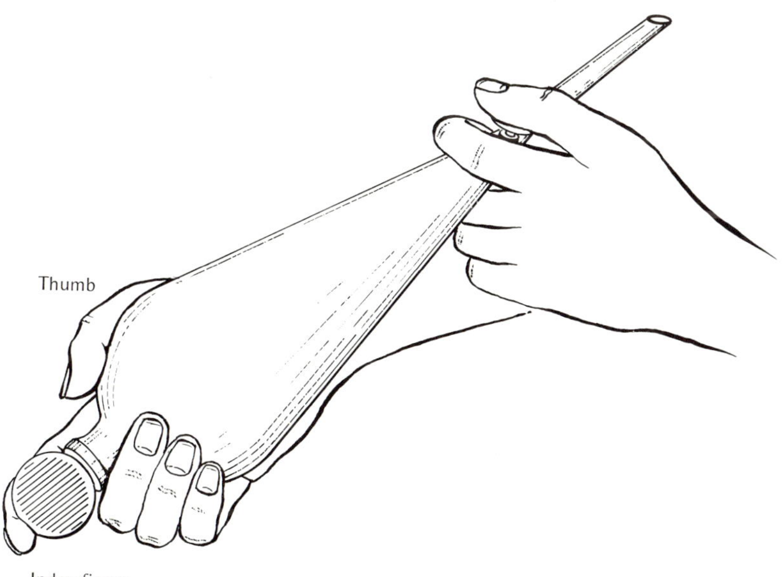

Fig. 1. Manipulation of a Separatory Funnel.

stopcock; if the layer being extracted is the upper one, extraction can now continue by addition of new solvent. Usually the small amount of lower layer remaining in the bore of the stopcock is washed into the receiver by letting a drop or so of fresh solvent wash it through before the funnel is inverted to start the second extraction. If the layer being extracted is the lower one, the upper layer is poured out of the top of the funnel, the lower layer is poured back into the funnel, and extraction continued.

115. Recovery of Extracted Material. The combined extract is dried (see §135), and the solvent is removed by simple distillation if the product boils high enough so that the loss is negligible. Often the volume of extract requires so large a flask that the residual product cannot be distilled from the same flask without appreciable loss. It may be easiest to transfer the residue to a smaller flask with aid of a little fresh solvent. Some chemists prefer to choose a flask of the size appropriate for distillation of the final residual product and to run the dilute solution into this flask gradually from a dropping funnel while the distillation to remove solvent is in progress. Finally, the liquid product, free from solvent, is distilled in the standard manner after a thermometer has been substituted for the dropping funnel.

116. Choice of Extraction Solvent. A good extraction solvent must have low solubility in the second phase (usually water), a low boiling point (to facilitate recovery of the product), and the ability to dissolve the product readily without dissolving the impurities (i.e., the distribution coefficient for the product between solvent and original phase must be high, but for impurities must be low). Of course it is not always possible to find a solvent that has all of these properties and one seeks the best compromise. Common, inexpensive extraction solvents include diethyl ether, diisopropyl ether, lower-boiling paraffin hydrocarbons (petroleum ether, ligroin, pentane, hexane, etc.), benzene, methylene chloride, chloroform, and carbon tetrachloride.

If a distribution coefficient is very large (>100) a single extraction should suffice to transfer a solute from one phase to the other. However, if five extractions have not resulted in adequate separation, it is better to seek a more efficient extractant or to use an apparatus for continuous extraction (§117). One can usually reduce the solubility of an organic product in an aqueous solution and increase the efficiency of the extraction by saturating the solution with a salt; sodium chloride is most often used.

Not many distribution coefficients have been determined accurately or can be found in the literature. The organic chemist usually uses his general knowledge of solvents in choosing a suitable extraction solvent. Occasionally, a rough value of K is obtained by extracting a measured volume of an aqueous solution of known concentration with a measured volume of solvent, separating layers, drying briefly, evaporating the solvent, and weighing the residue.

Diethyl ether is the most common extraction solvent in organic laboratories.

In elementary practice, however, the extreme volatility and flammability of this ether lead to wasteful evaporation and fire hazard. Diisopropyl ether (bp 67.5°) is no more expensive, has much lower fire hazard, and is recommended for beginners, particularly in crowded laboratories.

Both these ethers (and many other ethers as well) form dangerously explosive peroxides when exposed to air for some time. The danger from these unstable compounds is greatest just at the conclusion of a distillation or solvent removal when the relatively nonvolatile peroxide becomes concentrated in the distilling flask. Although ethereal distillations, where a sizable liquid residue remains at the moment of final removal of solvent, have seldom caused trouble, care should be taken to use ether free from peroxides. Beware of ether that has been standing for several months in a partly filled bottle exposed to light and air. Peroxide in ether is detected by liberation of iodine when a sample is shaken with an equal volume of 2% potassium iodide solution containing a few drops of hydrochloric acid. The peroxide may be eliminated by steam distillation of the ether from sodium hydroxide solution, but this operation is not recommended for the elementary laboratory.

If it is desirable to extract an acid from a nonaqueous solvent, aqueous sodium bicarbonate, carbonate, or hydroxide is used. The acid is converted into the salt, which is almost certain to have a large distribution ratio in water and solvent. The acid may thus be extracted almost completely in one operation. Similarly, a base can be withdrawn almost totally in one operation by extraction with an aqueous acid, such as dilute sulfuric acid. If an organic acid or base has been transferred in this way to an aqueous solution, it can be recovered by making the solution acidic or basic with mineral acid or basic reagent, respectively, and re-extracting with organic solvent. In operations of this sort one usually insures clean separations by finally washing the organic layer with a little pure water.

117. Multiple Extraction. Various devices exist to accomplish multiple extraction without the use of large volumes of solvent. Solids may be placed in the porous thimble of a Soxhlet extractor (Fig. 2) and subjected to repeated extraction with a solvent having selective solvent power for either impurities or product. Continuous liquid-liquid extractors depend upon the passage of fine drops of extractant through the liquid being extracted. With both these devices the solvent containing extracted material is collected in a heated flask from which the solvent is continuously distilled, recondensed, and passed again through the material being extracted. Countercurrent extraction, a procedure for repeated, systematic extraction, may be carried out with a series of separa-

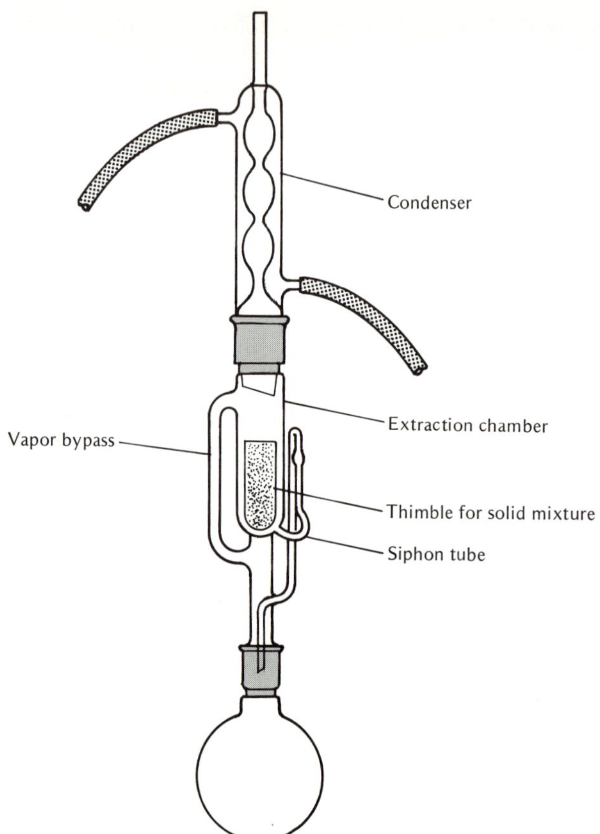

Fig. 2. Soxhlet Apparatus for Extraction.

tory funnels as for simple extraction but is more conveniently performed in special, automatic apparatus. Details of these more advanced techniques are given in the References.

118. Washing of Liquid Preparations. This operation is the same as extraction except that the immiscible solvent used as the wash liquid is chosen to dissolve the impurities rather than the product. Frequently solutions of organic preparations in organic solvents are washed with water to remove small amounts of salts, acids, or bases. When a corrosive liquid such as concentrated sulfuric acid is used to wash out impurities—as in the butyl bromide experiment (§198)—great care must be taken to keep the material away from the skin.

ADSORPTION AND CHROMATOGRAPHY

Solids adsorb materials from solution as a result of van der Waals forces. Such intermolecular forces operate only at very short distances and hold materials on a surface as a monomolecular layer. Competition for positions on the surface exists between the various kinds of molecules in the solution, including solvent molecules. The amount of a solute that will be adsorbed from solution in a solvent will therefore depend on the nature of both solute and solvent as well as on the concentration of solute and the amount of surface available. Adsorbents differ in their relative adsorptive power for various molecules, so that the amount of solute held also depends on the adsorbent used. If a weakly adsorbed compound has saturated a surface, and a solution of a substance that will adhere more strongly is brought into contact with that surface, the more weakly adsorbed compound will be displaced. Similarly, if the original solvent is replaced by one capable of being held more strongly, the amount of solute remaining on the surface will be decreased by a similar displacement. This usually permits easy recovery of adsorbed materials.

Adsorption may serve in various ways as a basis for purification of compounds and separation of mixtures. A discussion of decolorizing procedures will be deferred until chromatography has been described and the factors involved in adsorption have been discussed.

119. Chromatography. The term *chromatography* describes processes in which partitioning between phases is carried out in multiple fashion; one phase, a solid or a liquid adsorbed on a solid, is held stationary, and the second is passed over it. In *adsorption chromatography* a bulk liquid phase (usually a mixture of compounds in solution) comes in contact with an adsorptive, finely divided solid, and selective adsorption of the components of the mixture occurs on the surface of the solid. In *partition chromatography* one phase is a liquid adsorbed on the surface of a solid, and partitioning occurs between this immobile liquid phase and a bulk phase; the process is thus a multiple partitioning between two liquid phases. *Thin-layer chromatography* (*tlc*) is a form of adsorption chromatography in which a thin layer of adsorbent supported on a flat surface is used instead of a column of adsorbent. *Paper chromatography* is an example of partition chromatography in which filter paper serves as the support for the immobile phase. *Gas chromatography* (*gc*) or *vapor phase chromatography* (*vpc*) involves a mobile gas phase and an immobile phase, which is usually a liquid adsorbed on a solid [*gas-liquid phase chromatography* (*glpc* or *glc*)], or less commonly a solid adsorbent. The term *liquid chromatography* is used to designate chromatographic techniques not covered by gas chromatography.

ADSORPTION CHROMATOGRAPHY

120. The Chromatographic Process. Adsorption chromatography is usually carried out by placing the adsorbent in a vertical cylindrical glass tube and passing through it a solution of the mixture to be separated; this is called *column chromatography*. If the amount of solute is small compared with the capacity of the adsorbent, all of the solute will be held in a zone at the top of the column. A suitable solvent or mixture of solvents (§142) is then added, and as it passes down the column, selective desorption and readsorption occur so that the components of the mixture distribute themselves in separate zones on the column, the least tightly adsorbed migrating farthest down, the next adsorbing just above it, etc. The adsorbent with its bands of separated material is called a *chromatogram,* and the process of bringing about the separation is the development of the chromatogram. In most cases the composition of the solvent must be varied during the development by using increasingly more polar solvents in order that smooth separation will occur.

The success of a chromatographic separation depends not only on proper choice of adsorbent and elution solvents but also on suitable column preparation. The ratio of weights of adsorbent to mixture being separated must be high; usually 30:1 is adequate, but sometimes up to 100:1 is necessary. The ratio of column height to diameter is also important; about 7.5:1 is considered optimum. Occasionally a column is packed with dry adsorbent, but wet packing is usually preferred (see §189 for directions for packing the column).

If the components of the mixture are colored or if they fluoresce in ultraviolet light, their positions on the column are easily found. A wide variety of methods has been developed for locating colorless compounds on chromatograms. A few will be mentioned below and many others are cited in the references.

Recovery of the components of the mixture can be accomplished in two ways. One of these involves pushing the column of moist adsorbent out of the tube, separating it into sections each of which contains one component of the mixture, and placing each separately in a quantity of a solvent that will readily dissolve the component and displace it from the adsorbent, which may then be removed by filtration. It is convenient to use a slightly tapered tube when the adsorbent is to be pushed out of the column. This technique is seldom used.

In the second procedure, addition of the development solvent is continued, and the components pass in turn out of the column with the solvent and are collected separately. This is referred to as *elution*.

The zones that contain colorless compounds are more easily identified on extruded columns. Often a thin vertical section of the exposed adsorbent is brushed with a solution of a reagent that gives a color with the compound sought. For example, a dilute solution of potassium permanganate has been used to locate bands containing unsaturated compounds. In the elution technique, tests may be applied to tiny portions of the eluate to identify the desired compound; or continuous measurement of some physical property such as refractive index or conductivity may be useful.

Sometimes marking substances that give a color reaction with the desired compounds are placed on the adsorbent. Thus an indicator may first be adsorbed on the column when acids or bases are to be separated, and bands will be marked by color change. Often an empirical procedure is employed in which succeeding small portions of eluate are collected in systematic fashion and solvent is evaporated rapidly from each as the experiment progresses. A band may then be indicated by the nature of the residue or by portions containing material separated by portions containing only solvent. Chromatography of colorless compounds may also be accomplished by conversion of the mixture to colored derivatives, which may be reconverted to starting materials after separation; for instance, aldehydes and ketones may be separated as their 2,4-dinitrophenylhydrazones.

121. Adsorbents. An effective adsorbent must have a high but selective adsorptive power and high surface area; it must be chemically inert, readily available, and preferably white. Alumina and silica (silicic acid or silica gel) are the most commonly used materials, but activated charcoal, magnesia, calcium carbonate, sucrose, starch, and powdered cellulose have fairly general application; a number of other solids are occasionally used. Except for activated charcoal these materials are listed in order of decreasing adsorption affinity; charcoal is more adsorptive than silica and is sometimes listed as the most active. However, the activity of any adsorbent can be varied over a wide range, and the order is not always the same with different types of solutes. Moreover, different batches of the same adsorbent may vary in adsorptive power owing to the presence of contaminants or to differences in previous treatment. It is usually more satisfactory to control the relative adsorption of the sample by changes in solvent composition rather than by changes in adsorbent. Finely divided material is necessary to give high surface area per unit weight. The color of charcoal prevents visual observation of the progress of a chromatogram, and it is therefore employed mainly for decolorizing (§130).

Water is strongly adsorbed by most adsorbents and deactivates them because it is not easily displaced by other compounds. Heating is usually effective to remove water and activate the adsorbent. The adsorptive power of alumina can be varied over a considerable range by change of the amount of water on the surface. Experiment 7, §188, describes the activation and standardization of alumina to permit effective control of chromatographic separations.

Ordinary alumina often contains sodium carbonate or sodium bicarbonate, which sometimes leads to chemical reactions such as ester hydrolysis or aldol condensation during chromatography. These reactions may be prevented in most instances by use of neutral alumina. Alumina carefully standardized for chromatography is commercially available[2] in three forms: acidic, basic, and neutral. Acidic alumina is an acid-washed alumina giving a suspension in water with a pH of about 4; it is useful for separation of acidic materials (e.g., carboxylic acids). Neutral alumina has a pH of about 7.5 (useful for neutral materials), and basic alumina a pH of about 10 (useful for amines or other basic compounds).

Small particle size is desirable for adsorbents because high surface area gives high adsorption capacity, but liquids often flow very slowly through tightly packed, finely divided solids. To increase flow rate, the adsorbent is sometimes mixed with an inert filter aid such as Filter Cel or Celite. The adsorbent may also be converted to a porous granular form, with little loss of capacity. It is convenient to make a paste of the adsorbent with water, allow it to dry, crush it to suitable size, and reactivate.

122. Solute Structure. The adsorbability of compounds parallels their polarizability; substances with highly polar groups are strongly adsorbed. The following classes of compounds are listed in order of decreasing adsorbability on alumina: acids and bases > alcohols, amines, thiols > esters, aldehydes, ketones > aromatic compounds > halides, ethers > olefins > saturated hydrocarbons.

The effect of the nature of the functional group on adsorbability of a series of aromatic compounds is shown in Table 1. The parent structures, stilbene and azobenzene, are only weakly adsorbed, so that the adsorption affinity of the derivatives is almost entirely dependent on the substitutent. Minor differences between the two series emphasize the fact that some variation can be expected with different solvents and types of parent structure. Variation in the adsorbent may change the order even more. The list given above is thus no more than a rough guide for the planning of experiments on particular mixtures. The effect of a number of functional groups is not generally additive, although adsorption commonly increases with increasing number of like substituents: introduction of groups of a different type may either increase or decrease adsorption. It has been observed that a compound capable of internal hydrogen bonding is adsorbed less strongly than an isomer in which the location of functional groups permits only intermolecular hydrogen bonding.

123. Solvents. Choice of solvent is important for successful chromatography. A mixture to be separated is usually applied to the column in fairly concentrated solution, in the least polar solvent in which it is readily soluble, so that a well-defined zone is formed near the top. Development

[2]Adsorbents Woelm, distributed by Waters Associates, Inc., 61 Fountain St., Framingham, Mass. 01701, and comparable material, Bio-rad Laboratories, 32nd and Griffin Ave., Richmond, Cal. 94804.

Table 1. Functional Groups in Order of Decreasing Adsorption Affinity on Alumina

R is $C_6H_5CH{=}CHC_6H_4$-*p*[a]	*R is* $C_6H_5N{=}NC_6H_4$-*p*[b]
RCO_2H	RCO_2H
$RCONH_2$	ROH
ROH	$RNHCOCH_3$
RNH_2	$ROCOCH_3$
$RNHCOCH_3$	RNH_2
$ROCOCH_3$	$ROCOC_6H_5$
RCO_2CH_3	RCO_2CH_3
$RN(CH_3)_2$	$RN(CH_3)_2$
$ROCOC_6H_5$	RNO_2
RNO_2	$ROCH_3$
$ROCH_3$	RH
RH	

Solvents: (*a*) from carbon tetrachloride, (*b*) from benzene.

of the chromatogram is then started by passage through the column of the least polar solvent that causes movement of the most readily desorbed component of the mixture. Occasionally a single solvent will suffice to give complete separation. More commonly one must gradually vary the solvent composition by employing more and more polar solvents. Table 2 lists the common elution solvents in order of increasing polarity.

Warning: *The presence of even small amounts of water, alcohols, or acids in the less polar solvents will destroy the activity of the adsorbent; pure, dry solvents should be employed.*

If closely related compounds are to be separated, the change in solvent composition may be limited to a few percent every 100 ml, or the composition of the eluting solvent may be changed continuously by dropwise addition of the next solvent in the series to a reservoir of solvent

Table 2. Eluting Power of Solvants (in Increasing Order) Against Alumina[a]

Petroleum ether	Diethyl ether
Cyclohexane	Ethyl acetate
Carbon tetrachloride	Acetone
Trichloroethylene	1, 2-Dichloroethane
Carbon disulfide	Ethanol
Toluene	Methanol
Benzene	Water
Dichloromethane	Pyridine
Chloroform	Organic Acids

[a]This table is a combination of the lists by W. Trappe [*Biochem. Z.*, **305,** 150 (1940)] and H. H. Strain [*Chromatographic Adsorption Analysis*, Wiley-Interscience, New York, 1942] with minor modification.

flowing into the column at a rate to keep the volume in the reservoir constant (*gradient elution*).

Seldom does one have to use all of the solvents listed in any given chromatographic separation. Time is usually saved with new separations by a preliminary run on a small scale with rapid solvent change to determine when the desired product is eluted. Thin-layer chromatography may also be used as a guide for column chromatography.

124. Advantages and Limitations. Adsorption chromatography is especially valuable as a method of separation and purification because it depends on physical properties strikingly different from those governing conventional procedures already described. It thus supplements these techniques. It requires very simple apparatus, can be applied to small quantities, and is advantageous for heat-sensitive materials because it is carried out at room temperature.

The method also has several shortcomings. Because it is highly empirical, a number of trial experiments may be necessary and each application may require considerable time before a satisfactory procedure has been devised. The operation itself may be time-consuming, especially when a series of fractions must be taken and each concentrated; this may involve large volumes of solvents. Finally, the method is usually hard to apply to large-scale preparations.

Mention should be made of the recent development of high-speed, high-performance liquid chromatography techniques that overcome many of the disadvantages referred to above. Special apparatus, which permits operation under pressure, has made rapid separation possible and larger amounts can be handled. The References should be consulted for discussion of these new developments.

125. Thin-Layer Chromatography (TLC). Thin-layer chromatography is adsorption chromatography in which a thin layer of adsorbent supported on a flat surface replaces the column of adsorbent. Silica gel and alumina are the most common adsorbents but a number of others have been used. They must be more finely divided than for column chromatography and usually contain plaster of Paris ($CaSO_4 \cdot \frac{1}{2}H_2O$) as a binder; it is recommended that commercially available adsorbents especially prepared for tlc be used. A suitable layer of adsorbent on a glass plate is easily prepared (§351), or precoated glass, aluminum, or polyester sheets can be purchased.[3] Plastic or aluminum sheets are convenient because they can be cut to appropriate size with scissors, but they are relatively expensive (about $0.75 per 20- by 20-cm sheet).

[3]Polyester sheets from Eastman Kodak Co., Eastman Organic Chemicals, Rochester, N.Y. 14650, or Industrial Chemicals Division, Mallinckrodt Chemical Works, St. Louis, Mo. 63147. See Footnote 2 for glass or aluminum plates; other firms also supply these.

A thin-layer chromatogram is prepared by placing a small spot of mixture near the bottom of the plate and developing by capillary movement of a solvent up the plate in a suitable closed container saturated with the solvent vapor. A single solvent or solvent mixture is used, but additional plates are easily run, or the chromatogram may be dried and then developed further, either in the same direction or at right angles to the direction of first development. Eluents are chosen as for column chromatography and tlc is often used as a guide for larger separations on columns. It is also possible to use larger plates for tlc.

The various techniques used to locate the components on column chromatograms can be used for tlc, but another technique is common. The developed and dried plate is placed in a closed container with a few iodine crystals; iodine vapor is adsorbed into the areas of the plate containing organic compounds and brown spots mark component locations. Another technique is based on the charring of many organic compounds by hot sulfuric acid. The dry, developed chromatogram is sprayed with sulfuric acid, then heated at 100–110° over a hot plate or in a drying oven.

Thin-layer chromatography may be used for tentative identification. A compound will progress up a plate at a reproducible rate relative to the solvent front or an added standard compound if conditions are standardized. R_f or R_x values are obtained, where

$$R_f = \frac{\text{displacement of compound}}{\text{displacement of solvent front}} \qquad R_x = \frac{\text{displacement of compound}}{\text{displacement of standard}}$$

Direct comparison of known and unknown, by spotting both on the same plate, is preferable. R_f or R_x values from the literature are not reproducible enough to be more than general guides. At best the conclusions are no more than tentative.

The special advantage of tlc is the rapidity with which it can be carried out. It is useful for detecting the components of an unknown mixture or crude product, for following the progress of purification, and for monitoring the progress of a reaction. Experiment 38 illustrates the technique.

OTHER CHROMATOGRAPHIC METHODS

126. Partition Chromatography: Paper Chromatography. Partition chromatography resembles a multiple extraction process and may be carried out in a column as for adsorption chromatography. Silica gel is frequently used as a support for the immobile liquid phase.

Paper chromatography is an important example of partition chromatography. Filter paper serves as a support for the stationary phase which is usually adsorbed water. Such paper is nearly pure cellulose and in an atmosphere saturated with water vapor adsorbs about 22% of water. The mobile phase, usually an organic liquid containing some water, travels from a reservoir by capillary action either up a strip or sheet of the paper (ascending chromatography), down the paper (descending chromatography), or radially and horizontally from a central spot on a circular piece of paper.

The techniques of paper chromatography are similar to those of tlc. The mixture to be separated is placed as a spot near the reservoir and is picked up by the mobile phase as it moves past. Each component of the mixture is repeatedly distributed between the two phases and moves in the same direction as the moving solvent front. It is important that this development of the chromatogram occurs in an enclosed vessel and that the atmosphere therein be saturated with the vapor of the solvent.

Under carefully controlled conditions relative rates can be duplicated accurately. They are expressed as R_f values (§125). These values depend not only on the nature of the compound, but also on other factors such as nature of the mobile phase, filter paper, and apparatus. Temperature, atmospheric humidity, and presence of extraneous material also have some effect. The amount of mixture applied to the paper must be small; if the paper is overloaded the spots do not move smoothly, "tailing" occurs, and separation is poor. The variation of R_f values with the structure of the migrating compound has been discussed (consult the References). Because the stationary phase is the more hydrophilic one in conventional paper chromatography (since it must wet and adhere to a hydrophilic support), one would expect the more hydrophilic components of a mixture to be held more tightly in the stationary phase and to move more slowly (lower R_f). This is observed.

Paper chromatography is carried out on very small amounts of material (5–300 μg = .005–3 mg) and is applicable mainly to polyfunctional or highly polar compounds such as sugars or amino acids. When colorless compounds are involved, the developed chromatogram is usually sprayed with a reagent that reacts with the components of the mixture to give colored substances. Experiment 7, §190, gives an example of paper chromatography.

127. Gas Chromatography or Vapor Phase Chromatography. *Gas-liquid chromatography* (*glc*), sometimes referred to as *glpc* (*gas-liquid phase chromatography*), is a form of partition chromatography in which the stationary phase is a high-boiling liquid adsorbed on a finely divided solid or on the walls of capillary tubing, and the mobile phase is a mixture of compounds in the vapor state carried by a permanent gas such as helium. Separation depends on solubility equilibria between components in these phases. This technique has become one of the most versatile and widely used separation methods in chemistry. It requires very small amounts of material, is applicable over a wide range of temperatures, can be adapted for quantitative work, and is more rapid than most other analytical techniques. A brief account of glpc is given below, but the References should be consulted for further information.

A related chromatographic method, *gas-solid chromatography* (*gsc*), involves partial and selective adsorption of the components of the vapor phase on the surface of a finely divided solid rather than selective solution in a liquid phase on a surface. This method is much less commonly used than glpc and will not be discussed further here.

A gas chromatograph consists of the components shown schematically in Fig. 3. Carrier gas, usually from a high-pressure cylinder, is supplied through a pressure gauge and reduction valve at 30–100 psi such that a constant flow of 50–100 ml/min is maintained through the apparatus. The gas flows first through an injector block which can be kept at a temperature (0–350°) such that the least volatile component of the mixture is quickly volatilized into the gas stream. The mixture to be separated is forced into the injector block through a septum by means of a microsyringe. Solid samples are usually introduced in solution in a volatile solvent that passes quickly through the column so that it does not interfere with the determination.

The sample is then carried into a column housed in a heated oven. Packed columns are usually made of narrow metal tubing 6–10 ft long (sometimes shorter or much longer) packed with a finely divided solid support that holds the liquid phase on its surface. After packing, the column is coiled to a spiral or other shape that fits conveniently into the oven. Glass capillary columns, often very long, in which the liquid is held on the interior surface, are also used. The column may be held at constant temperature or its temperature may be increased in a regular, reproducible, "programmed" way as the determination progresses.

From the column the gas stream containing the separated components of the sample flows into the detector where a property that varies with the composition of the stream is measured. This property is often the thermal conductivity of the gas stream, or the gas may be ionized by suitable means and the extent of ionization measured. The result is an electronic signal, which is usually recorded on a strip chart recorder to provide a chromatogram like that shown in Fig. 4. The gas stream passes out of the detector through an exit port where the rate of flow is measured and where a sample collector may be attached.

Individual peaks on the chromatogram represent detection of some

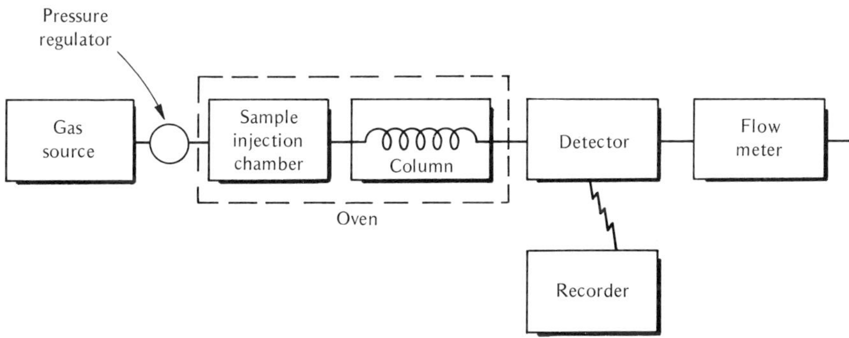

Fig. 3. Essential Features of a Gas Chromatograph.

material being eluted from the column. A peak is characterized by a retention volume or retention time. The *retention volume* is the total volume of gas that has passed through the column to elute the compound that gives that peak. This volume is a function of the volume of gas in the column at any instant, the volume of the adsorbed liquid phase, the flow rate, and the temperature. Retention volumes can be related to the distribution coefficient of the compound between the adsorbed liquid and the gas; it is useful in the theory of glpc and may be used in quantitative determinations. The *retention time* is the time from injection of the sample to the maximum of the elution peak. It is easier to measure than retention volume and is more commly used. It depends on the retention volume.

Under carefully controlled conditions, retention times are useful for tentative identification of an unknown. The retention time of a peak is compared with the retention times of known compounds that might be anticipated as components. Some of each known may then be added to the mixture to see if the peak under examination is in fact increased in area. However, many compounds have closely similar retention times so that this does not assure identification. If the known and unknown are not separated on several columns, the identification is somewhat more certain, but final identification requires isolation from the exit gas of the compound causing the peak and the usual spectral or chemical identification.

The area of the peak in the chromatogram is proportional to the amount

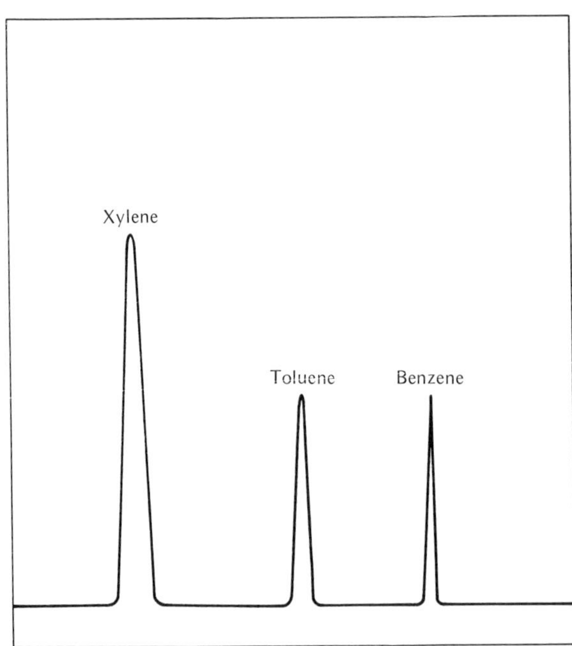

Fig. 4. Gas-Liquid Chromatogram of a Mixture of Benzene, Toluene, and Xylene. Stationary Phase: Carbowax.

of the compound that has been eluted from the column. Occasionally two very similar compounds will give essentially the same response in the detector so that a given amount of either compound will give the same peak area; however, this is rare. For quantitative work a known mixture of the components is prepared, and the response of each component relative to one of them taken as a standard, or to some added standard compound, is determined.

$$\text{Response ratio } \frac{\text{A}}{\text{B}} = \frac{\text{wt. A/wt. B}}{\text{area A/area B}}$$

The ratios obtained in this way are used to calculate the relative amounts of the various components. Yields cannot be calculated without use of the added standard compound unless all products appear as peaks in the chromatogram and there has been no mechanical loss of material. Peak areas are often obtained from electronic integrators installed on the recorder, but may also be determined with a planimeter, estimated by triangulation, or obtained by simply cutting out the peaks with scissors and weighing the paper.

The versatility of glpc lies in the wide choice of liquids available for stationary phases. These liquids are most often adsorbed on 30–60 mesh granules of firebrick or diatomaceous earth, which can be purchased unwashed, acid washed, base washed, or trimethylsilated. The liquid phase is a high-boiling liquid chosen to take advantage of structural differences between the components of the mixture. The volatility of a component in the immobile phase depends on the strength of its interactions with the liquid; these may involve hydrogen bond formation, dipole-dipole interactions, or complex formation. Nonpolar liquids interact weakly with dissolved nonpolar substances, and the retention times of such substances are roughly proportional to their boiling points. The liquid must be thermally stable and essentially nonvolatile (vp < 0.1 torr) at the maximum temperature involved. Liquids commonly used include dimethylsiloxane polymers (silicones or silicone rubbers), polyethylene glycols (Carbowaxes), hydrocarbon greases (Apiezon L or M), and a number of high-boiling, pure compounds such as didecyl phthalate, tricresyl phosphate, and 1, 2, 3-tris (2-cyanoethoxy) propane. Lists of such compounds with types of substances on which they have been used successfully and maximum temperatures for use are given in the References.

Column packings are prepared by dissolving the liquid phase (usually 5–25% of the weight of the solid support) in a volatile solvent (e.g., methylene chloride or methanol), mixing this with the solid support, and removing the solvent by heating gently while agitating the slurry (usually under reduced pressure). The packing material is then poured slowly into a straight piece of metal tubing of the desired length. The tube is tapped or vibrated constantly during the filling operation to insure a uniform tightly packed column. A glass wool plug is placed in one end of the tubing before filling, and another plug fills the other end to hold the packing in place. The column is then coiled to fit the oven, attached by suitable gastight fittings to the injection block and detector, and conditioned. Conditioning consists of passing an inert gas through the column while it is heated to the maximum temperature at which it will be used. This removes any volatile impurities and produces a column that gives a steady base line on the strip chart. After each use the column should be purged at this maximum temperature while gas continues to flow. Packed columns for analytical work are usually 1–50 ft long and $\frac{1}{8}$ or $\frac{1}{4}$ in. in diameter. Copper, aluminum, or stainless steel tubing is commonly used. Packed columns can be obtained commercially.

A thermal conductivity detector measures the thermal conductivity of the eluent gas stream against a reference stream of the carrier gas. A Wheatstone bridge is used in which the resistances are heated filaments, two in the eluent stream and two in the reference stream. When no component is being eluted, the bridge remains in balance; but as a component emerges, the conductivity of the eluent stream changes, causing more or less heat to be transferred from the filament to the gas. The change in temperature of the filament changes the resistance and hence the flow of the electrical current; this change is recorded to produce the chromatogram. Hydrogen or helium are the carrier gases giving highest sensitivity and helium is preferred because it is more inert.

In an ionization detector the eluent stream is partially ionized; the ions migrate to a polarized electronic grid and produce a current that can be recorded to give the chromatogram. In the flame ionization detector the ions are produced by combustion. The eluent gas is mixed with a stream of hydrogen and oxygen, which is burned in the detector. The component being eluted undergoes partial combustion to produce ionic fragments or undergoes ionization induced by the energy produced by combustion of the hydrogen. The extent of ionization depends on the nature of the component. Ionization may also be induced by low-energy β- or α-rays. If the separated components are to be collected from the emerging gas stream, it is necessary to pass only a small amount of the stream through the ionization detector, since the components are destroyed in the detector. When ionization detectors are used, the carrier gas is usually nitrogen or argon. These detectors are somewhat more sensitive than thermal-conductivity devices.

Gas-liquid phase chromatography is limited by the thermal stability of the compounds to be separated and by the small amounts of material that can be used without overloading the column. Sample size is often around 1 mg, and the separation of 50 mg is laborious unless special preparative gas chromatographs are available; these are expensive. A discussion of glpc on a preparative scale is left for more advanced books.

128. Ion-Exchange Chromatography. Ion-exchange chromatography is a form of column chromatography in which the stationary phase is a polymer containing acidic or basic functional groups capable of undergoing exchange of ions; it is used for exchanging anions or cations, for deionizing solutions, and for separating such polar substances as carboxylic acids, amines, and amino acids.

129. Gel Permeation Chromatography. Also called gel filtration, gel permeation chromatography employs a cross-linked polymer as the stationary phase. The polymers contain pores large enough to hold certain size molecules but small enough to exclude others. Separation is based almost entirely on molecular size, and the method is applied mainly to macromolecules.

Molecular sieve chromatography is a term embracing gel chromatography and also including synthetic zeolites (inorganic) as the stationary phase.

These and other kinds of chromatography are described in the References.

CLARIFICATION PROCEDURES AND BATCH ADSORPTION

130. Decolorization. A simple form of batch adsorption has been used traditionally to decolorize organic solids during recrystallization. In this procedure the crude compound is dissolved in the recrystallization solvent, a small amount of decolorizing carbon added to the solution, suspension boiled briefly, the carbon removed by filtration, and the solid allowed to crystallize.

131. Active Carbon. This adsorbent differs from many others because it has few or no polar groups and does not participate in hydrogen bonding. The ease of adsorption on this material therefore depends mainly on the polarizability of the compound. Polarizability increases with degree of unsaturation, and aromatic compounds are usually more polarizable than aliphatic. Coloring matter in crude organic preparations is usually highly unsaturated, or aromatic, and of fairly high molecular weight. Active carbon selectively adsorbs such material. The smallest amount of decolorizing agent that will do the job should be used because the desired compound will also be adsorbed if the surface of the adsorbent is not saturated by the coloring matter. In practice, a few tenths of a percent of active carbon is tried first, and one seldom uses more than 20 mg per gram of dry compound.

An important function of carbon as a decolorizing agent depends upon its ability to adsorb finely dispersed particles from colloidal suspension. Turbidity or off-colors (brown, olive, purplish gray, and other tints not recognized as primary colors) frequently indicate colloidal suspensions. Unless colloidal matter is removed, it will often be adsorbed on the surface of the crystals of product; it frequently interferes with the crystallization process. The ability to flocculate and hence remove colloidal matter is dependent on the nature of the charge on both the colloid and the adsorbent; sometimes change of pH will increase the effectiveness of the decolorization. Decolorizing carbons also differ in their acidity. Chemically activated, or steam-activated and washed carbons are usually weakly acidic (pH 6.0–6.5; e.g., Nuchar); unwashed steam-activated carbons are basic (pH 9–11; e.g., Darco and Norite). Boneblack or animal charcoal, formerly the only available decolorizing carbon, is much inferior to the modern activated carbons mentioned above, which come from vegetable sources. Untreated boneblack may contain up to 80% of calcium phosphate and other calcium salts; this ash is especially troublesome in clarifying acidic solutions of organic compounds.

132. Procedures. Gravity filtration is best during decolorization because active carbon will usually pass through filter paper if vacuum filtration is employed. Sometimes a less porous filter paper must be used. Occasionally the finely divided, active carbon is peptized by adsorption of charged ions or particles, forming a colloidal suspension that passes through the filter paper and even resists centrifugation. This is most often observed in alcoholic solutions. The difficulty may often be overcome if a little filter-paper pulp, a finely divided powder of opposite charge, or a salt containing a polyvalent ion is stirred into the solution; these flocculate the colloid and permit efficient filtration.

Decolorization depends also on the solvent used. Frequently the following order of effectiveness is observed: water > methanol > ethanol > ethyl acetate > acetone > chloroform. Decolorization is sometimes effected in diethyl ether, and when this is not a suitable solvent for recrystallization, it is removed after filtration, and a more suitable solvent is employed.

Decolorization may be carried out by filtration of the solution through a chromatographic column, using any suitable adsorbent. Active carbon is employed in this way when filtration difficulties are encountered in the conventional clarification. It is usually best to use a poorly adsorbed solvent, such as ether or a saturated hydrocarbon, for decolorization by chromatography. With colorless adsorbents, use of a short column results in coloring of all the adsorbent and avoids adsorption of the desired product.

Directions for decolorizing with active carbon sometimes specify long boiling. This should not be necessary if simple adsorption were the only phenomenon involved because this usually occurs rapidly at room temperature. Brief heating should aid the flocculation process when colloids must be removed, but long heating should ordinarily be avoided. It has been suggested that decolorization may sometimes involve a reducing or oxidizing action of the carbon, which may depend on adsorbed carbon monoxide or oxygen, respectively.

133. Batch Adsorption Techniques. These techniques have been used industrially or in the laboratory for a variety of purifications other than simple decolorization. They are described in more advanced texts.

References

(See Chapter 14 for complete titles.)

Extraction and Washing: Weissberger, *Technique of Organic Chemistry,* **3** (2nd ed.), Pt. 1, 149, **6,** 97; Müller–Houben–Weyl, **1,** Pt. 1, 222; Wiberg, 179; Bates and Schaefer, 45; Pasto and Johnson, 19.

Adsorption and Chromatography: Weissberger, *Technique of Organic Chemistry,* **5, 10, 12, 13;** Müller–Houben–Weyl, **1,** Pt. 1, 465, **2,** 869; Wiberg, 149; Bates and Schaefer, 72; Pasto and Johnson, 24.

The literature on chromatography is voluminous and only a few of the monographs will be mentioned: O. Mikeš, *Laboratory Handbook of Chromatographic Methods,* Van Nostrand Reinhold, New York, 1967; L. R. Snyder, *Principles of Adsorption Chromatography,* Marcel Dekker, New York, 1968; J. C. Giddings, *Dynamics of Chromatography,* Marcel Dekker, New York, 1965; J. J. Kirkland, ed., *Modern Practice of Liquid Chromatography,* Wiley–Interscience, New York, 1971; G. Zweig and J. R. Whitaker, *Paper Chromatography and Electrophoresis,* 2 vols., Academic Press, New York, 1967 and 1971; G. Zweig and J. Sherma, eds., *Handbook of Chromatography,* CRC Press, Cleveland, O., 1972; C. W. C. Peereboom, "Paper Chromatography and Thin Layer Chromatography," Ch. 1 in *Electrical Methods, Physical Separation Methods,* Vol. 2c of C. L. Wilson and D. W. Wilson, eds., *Comprehensive Analytical Chemistry,* American Elsevier, New York, 1971; K. Randerath, *Thin-Layer Chromatography* (2nd ed.), Academic Press, New York, 1966; H. Determann, *Gel Chromatography,* Springer-Verlag, New York, 1969; H. Purnell, *Gas Chromatography,* John Wiley & Sons, New York, 1962; A. B. Littlewood, *Gas Chromatography* (2nd ed.), Academic Press, New York, 1970; R. A. Jones, *An Introduction to Gas-Liquid Chromatography,* Academic Press, New York, 1970; I. I. Domsky and J. A. Perry, eds., *Recent Advances in Gas Chromatography,* Marcel Dekker, New York, 1971; D. W. Grant, *Gas-Liquid Chromatography,* Van Nostrand Reinhold, New York, 1971.

In addition a number of serial publications devoted exclusively to chromatography or offering important current information on this subject are published. These include the following: *Advances in Chromatography,* Marcel Dekker, New York (since 1965); *Chromatographia,* Pergamon, New York (since 1968); *Chromatographic Reviews,* American Elsevier, New York (since 1959); *Journal of Chromatographic Science* (formerly *Journal of Gas Chromatography*), S. T. Preston, Evanston, Illinois (since 1963); *Journal of Chromatography,* American Elsevier,

New York (since 1958); *Progress in Separation and Purification,* Wiley–Interscience, New York (since 1968); *Separation Science,* Marcel Dekker, New York (since 1966); *Gas Chromatography Abstracts,* The Institute of Petroleum, London (since 1958); *Progress in Thin-Layer Chromatography,* Ann Arbor Science Publishers, Ann Arbor, Mich. (since 1970).

Chapter 10

Drying of Organic Preparations

It is almost always necessary to dry a reagent, solvent, or product during some stage of the conduct of an organic reaction. A brief introduction to general principles and procedures that permit this operation to be conducted efficiently and with a minimum of loss is given below.

134. Gases. Gases are ordinarily dried by passage over some inorganic compound that will combine with the water. Molecular sieves (§136) are especially effective for this purpose, but alumina, silica gel, and many of the drying agents in Table 1 are also used. Gases are sometimes dried by passage over a surface chilled with solid carbon dioxide or liquid nitrogen.

135. Liquids. The drying operation is especially important with liquids because final purification is often by distillation. If water is present it will usually distill with the liquid either as a result of steam distillation when undissolved (§59) or as an azeotropic mixture when dissolved (§§45–49). Even if the vapor-pressure relations permit separation of water by distillation, this usually requires a careful fractionation that is time consuming and involves appreciable loss. However, if one has an ample supply of a hydrophobic liquid such as benzene, it is conveniently dried by distillation. Any undissolved water steam-distills quickly and the benzene–water azeotrope then distills until all water is gone. The rest of the distillate is anhydrous.

In most cases it is best to dry liquids by placing them in contact with a solid drying agent that forms a hydrate, reacts chemically with water, or adsorbs it efficiently. The drying agent should be insoluble in the liquid, not react chemically with it, and be removed easily and completely. Table 1 gives some of the common drying agents, the products of their reaction with water, and comments on their uses and drying characteristics.

Drying agents vary in their capacity for water (weight of water taken up

Table 1. Common Drying Agents

Drying Agent[a]	*Products with H_2O (Vapor Pressure at 25°*[b]*)*		*Comments*
$CaSO_4$ $2.40	$CaSO_4 \cdot \frac{1}{2}H_2O$ $CaSO_4 \cdot 2H_2O$	(0.005) (9)	Porous granular form (Drierite) very rapid and efficient; capacity low to hemihydrate; dihydrate forms slowly. May be used with most organic compounds. Preliminary drying with high-capacity agent recommended.
$CaCl_2$ $3.86	$CaCl_2 \cdot H_2O$ $CaCl_2 \cdot 2H_2O$ $CaCl_2 \cdot 4H_2O$ $CaCl_2 \cdot 6H_2O$	(0.3) (1) (3) (5)	Moderate capacity, intensity, and speed; porous granular forms available. Good preliminary drying agent. Cannot be used for alcohols or amines (compounds formed); contains some $Ca(OH)_2$ so cannot be used with acids, phenols, or esters.
$CuSO_4$ $5.86	$CuSO_4 \cdot H_2O$ $CuSO_4 \cdot 3H_2O$ $CuSO_4 \cdot 5H_2O$	(0.8) (5) (8)	Moderate capacity, intensity, and speed. Seldom used because more expensive. May form chelates or insoluble salts.
$MgSO_4$ $4.70	$MgSO_4 \cdot H_2O$ $MgSO_4 \cdot 2H_2O$ $MgSO_4 \cdot 4H_2O$ $MgSO_4 \cdot 5H_2O$ $MgSO_4 \cdot 6H_2O$ $MgSO_4 \cdot 7H_2O$	(1) (2) (5) (9) (10) (12)	Excellent general drying agent. Intense to monohydrate, high capacity to heptahydrate. Quite rapid. Neutral. Used as a powder so more careful filtration needed. Often picks up moisture from the air during storage but easily dehydrated by mild heating.
K_2CO_3 $2.94	$K_2CO_3 \cdot 1\frac{1}{2}H_2O$	(1.1)	Good capacity for intensity shown but somewhat slow. Basic so cannot be used with acidic substances.
Na_2SO_4 $2.10	$Na_2SO_4 \cdot 7H_2O$ $Na_2SO_4 \cdot 10H_2O$	 (22)	Poor intensity but very high capacity. Rather slow. Excellent for preliminary drying. Neutral. Used as a powder so more careful filtration needed.
P_2O_5 $3.64	H_3PO_4	(0.000)	Very high intensity. Strongly acidic; use limited to saturated and aromatic hydrocarbons, halides, and ethers. Tends to form a sticky coating over unused reagent which decreases capacity. A granular form (Granusic) is more efficient, but expensive ($7.28/lb).

Table 1. Common Drying Agents *(continued)*

Drying Agent[a]	*Products with H_2O (Vapor Pressure at 25°*[b]*)*		*Comments*
BaO $4.30	$Ba(OH)_2$	(0.0006)	Very high intensity. Strongly basic. Fairly often used for alcohols and amines. Capacity rather low. More expensive if carbide free.
CaO $2.96	$Ca(OH)_2$	(0.003)	Like $Ba(OH)_2$ but lower intensity and higher capacity.
NaOH $2.08	$NaOH{\cdot}H_2O$	(1.5)	Strongly basic. Pellets contain about 2% water. Capacity at high intensity limited by lack of porosity. Appears to form a water solution on the surface when used as a drying agent. Forms a number of higher hydrates at lower temperatures. Used to dry amines.
KOH $2.75	$KOH{\cdot}H_2O$ $KOH{\cdot}H_2O$ $KOH{\cdot}2H_2O$	(0.04)[c] (0.08)[c] (1.2)[c]	Strongly basic. Pellets contain about 1% K_2CO_3 and 14% H_2O. Capacity at high intensity limited by lack of porosity. Appears to form a water solution on the surface when used as a drying agent. To obtain anhydrous KOH requires high-temperature fusion in a metal vessel. Used to dry amines.

[a]Prices (taken from a 1973 catalog) are for 1-lb bottles and are simply to give an idea of the relative cost of the different drying agents. Certain of the compounds (e.g., $CaCl_2$) can be obtained at substantially lower prices in bulk and the quality is often suitable. Educational institutions may receive discount from prices shown.

[b]The figures in parentheses are usually the partial vapor pressures of water in torrs for systems such as anhydrous salt–lowest hydrate–water vapor or two adjacent hydrates and water vapor (e.g., $CaSO{\cdot}H_2O$–$CaSO_4{\cdot}2H_2O$–H_2O vapor), and were taken from *Gmelins Handbuch der Anorganischen Chemie* (8th ed.). Some are milligrams of residual water per liter of dried air [from J. H. Bower, *Bur. Std. J. Res.*, **12**, 241 (1934)]. Since 1 mg of water in 1 liter of air exerts a partial pressure of approximately 1 torr at 25° this introduces no error. The figures are rounded and in some instances one figure was arbitarily chosen when several values were given (e.g., $CaCl_2$–$CaCl_2{\cdot}H_2O$–H_2O). See Weissberger, *Techniques of Chemistry*, **2** (3rd ed.), 566–67, for tables giving similar or related information.

[c]Another report gives 1.54 torr for $KOH{\cdot}H_2O$ and 3.59 for $KOH{\cdot}2H_2O$, both at 27.75°.

per gram of agent) and drying intensity (the percentage of water remaining in the liquid when in equilibrium with the product formed by reaction of the agent with water). The latter is roughly proportional to the partial pressure of water vapor in equilibrium with the product; these vapor pressures are given in Table 1. It is not uncommon for a drying agent to form a series of hydrates containing increasing amounts of water. With these the lowest hydrate may give intense drying but have low capacity, while the higher hydrates give decreased intensity and increased capacity (see Table 1 for examples). Drying agents also vary in the rate of reaction with water. This often depends more on particle size and porosity of the solid than on intrinsic reaction rate. In general the lower the solubility of water in the liquid, the easier it is to dry.

It is ordinary practice to add a small amount of the drying agent to the liquid in a conical flask, which permits maximum contact between liquid and agent. The flask is swirled and much of the water is taken up quickly. The flask is then allowed to stand for a few hours, or overnight, to complete the drying; occasional swirling hastens the process. The amount of drying agent depends on the quantity of water present. It is usually sufficient to cover just the bottom of the flask; after swirling, some of the granules of solid should appear loose and not caked. A large excess results in greater loss of product by adsorption on the solid. It is sometimes advantageous to dry a liquid at lower temperatures because this lowers the solubility of water in the liquid and increases the intensity of the drying agent to some extent. However, if a cold liquid is exposed to moist laboratory air during filtration, more water may be introduced than was removed by the lower temperature. Most alcohols, ethers, and ketones are extremely hygroscopic when very dry, and special precautions must be taken to protect them from moist air during laboratory operations.

It is usually necessary to remove the drying agent before distillation because most hydrates break down at moderate temperatures (30–50°) and the water then distills with product. For example, $MgSO_4 \cdot 7H_2O$, $Na_2SO_4 \cdot 10H_2O$, and $CaCl_2 \cdot 6H_2O$ decompose at 48°, 33°, and 30°, respectively. Even if the boiling point of the liquid is below the temperature at which the hydrate decomposes, the intensity of drying decreases as the temperature is raised. With drying agents such as barium oxide or phosphorus pentoxide these difficulties are not encountered, but compounds are often sensitive to strong bases or acids at distillation temperatures, and it is then necessary to remove these drying agents as well. Some workers prefer to filter through a small plug of cotton or glass wool in an ordinary funnel; others prefer a filter paper, which has less tendency to allow small particles of solid to slip through. If a solution is being dried one can reduce losses by use of a little fresh solvent to wash the spent drying agent and filter paper.

With very wet products it is better to add the drying agent in portions and decant between each addition. When a drying agent of high intensity and low capacity, such as calcium sulfate, is to be used, preliminary drying with a cheap agent of high capacity is recommended. If in the course of workup a solution is manipulated in a separatory funnel just before drying, it is often efficient to do the preliminary drying in the separatory funnel; the partially dried product is easily decanted into a conical flask and the spent drying agent rinsed with fresh solvent. Sometimes the first small portion of agent goes into solution in water to form a concentrated aqueous layer that is easily separated.

136. Drying Agents That Require Special Comment. Magnesium perchlorate, $Mg(ClO_4)_2$ is a very effective drying agent sold under the trade name Anhydrone. It gives a trihydrate and the equilibrium vapor pressure of water in this system is only 0.002 torr. Anhydrone is used for absorption of water in quantitative analysis of organic compounds for carbon and hydrogen.

> **Warning:** *An organic liquid should never be dried by placing this drying agent in it because there is great danger of forming explosive organic perchlorates. Even small amounts of these perchlorates can explode with devastating force. Anhydrone should not be used as a desiccator desiccant for this same reason.*

Molecular sieves. This term is applied to crystalline zeolites (aluminosilicates) that are characterized by a porous structure having channels of fixed size. A number of these occur naturally, but those used as adsorbents in chemistry are synthetic materials that were developed by the Linde Company, a division of Union Carbide Corporation. The channels are normally filled with water and this is removed during manufacture, by heating under reduced pressure, to leave the crystal structure intact. They are called molecular sieves because the channels are of a specific size into which certain small molecules, including water, will fit, but larger molecules will not. Molecular sieves are useful as intense drying agents of moderate capacity. Types 3A, 4A, and 5A are most commonly used. These designations refer to the approximate diameters of the channels in Ångström units. The material is available in cylindrical or spherical pellets $\frac{1}{8}$ in. and $\frac{1}{16}$ in. in diameter, and as powders of varying mesh size; it is relatively expensive ($7.60/lb). With organic molecules too large to fit into the channels drying is very effective. For example, the water content of diethyl ether is reduced to $<0.001\%$ when it stands over a 4A molecular sieve, and the sieve absorbs $>9.5\%$ of its own weight of water. However, smaller molecules of liquids are also adsorbed (e.g., methanol, ethanol, or propanol) and likewise many gases (e.g. carbon dioxide, hydrogen sulfide, ammonia, ethane, ethylene, acetylene, and propylene). Water is usually adsorbed more strongly so molecular sieves can still be used to dry these compounds, but less effectively. The water content of ethanol is reduced to 0.25% when it stands over 4A sieve, and the sieve has taken up 7% of its own weight of water. Molecular sieves are widely used commercially to dry gases or liquids and to maintain a dry atmosphere in certain applications. Special grades for use in gas chromatography are available. See the References for additional information.

Sulfuric acid is an effective drying agent, but it is seldom used to dry organic liquids. The vapor pressure of water over sulfuric acid is about 0.008 torr at 25° (0.3 torr for 95% acid). Only aliphatic hydrocarbons and saturated alkyl halides can be dried by washing with concentrated sulfuric acid, which also serves to remove olefins and oxygen- or nitrogen-containing compounds that may be present as impurities (e.g., Experiment 9, §198, *n*-Butyl Bromide). However, most organic liquids are less stable if distilled in the presence of traces of strong acids, so even with *n*-butyl bromide, residual sulfuric acid is removed with dilute base and

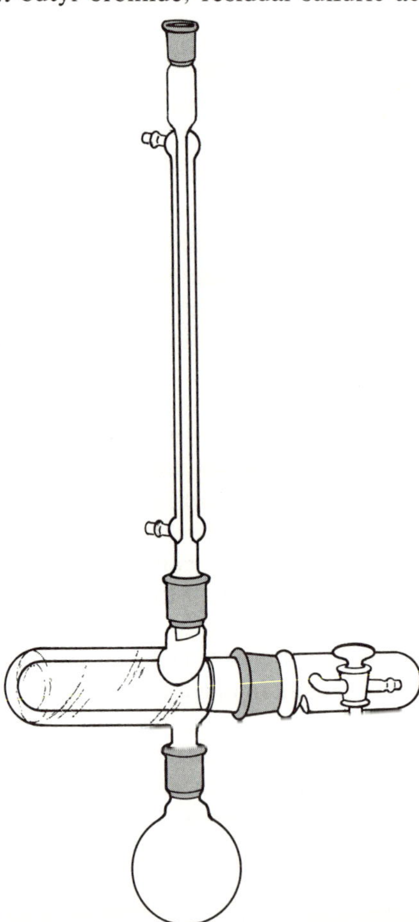

Fig. 1. Modified Abderhalden Drying Apparatus.

water; the preparation is finally dried in the conventional way over calcium chloride.

137. Solids. In elementary practice a freshly prepared wet solid is ordinarily dried for a day or more in the open air. The crystals, lying either on filter paper or directly on a dish, are loosely covered to exclude dust. The compound is sometimes placed in an oven to hasten drying, but only if its melting point is well above oven temperature and the compound is completely stable; it is best to test a small sample in the oven first.

Vacuum desiccators are especially useful to hasten the drying of solids. Material to be dried is placed in a dish or other container on a rack above the desiccant, and the desiccator is evacuated to a residual pressure of 20 torr or below. The ground glass contact flange on the lid of the desiccator and the stopcock must be carefully lubricated so that the vacuum is maintained and the lid can be removed easily. Transfer of water vapor from product to desiccant is more rapid than when a blanket of relatively dense air intervenes.

Sulfuric acid is often used as a desiccant, but care must be taken to avoid contamination of the product by acid. Danger of splashing may be reduced by placing glass beads in the acid. Sulfuric acid should not be used if the product is highly sensitive to acids. Molecular sieves are also effective desiccants and many of the drying agents listed in Table 1 have been used.

When the product is dry, great care must be taken to readmit air slowly, lest the loose, dry particles of organic material be blown out of the dish.

Vacuum-drying ovens are sometimes used for larger amounts of products that have higher melting points.

Drying of solid samples for analysis is usually done by heating under vacuum. Figure 1 pictures a modern version of the Abderhalden drying apparatus that is used for this purpose. The solid to be dried, usually in a small vial, is placed in that portion of the inner tube of the apparatus which can be heated by the hot vapors of a refluxing liquid. The temperature of this chamber is determined by the liquid chosen. The water vapor is taken up by an efficient drying agent such as phorphorus pentoxide, which is outside the heated zone. This apparatus serves as a small vacuum oven for drying of all kinds of small samples.

Questions

1. Explain how benzene (bp 80.6°) can be freed from water by distillation, in view of the fact that water boils nearly 20° higher than the benzene.

2. A certain preparation of 2-heptene, C_7H_{14}, was transparent and clear. Upon distillation, this liquid yielded at first a cloudy distillate. Explain.

3. Choose an appropriate drying agent for each of the following liquids:

(*a*) chlorobenzene
(*b*) formic acid
(*c*) acetonitrile
(*d*) butyraldehyde
(*e*) an ethereal solution of butylamine
(*f*) propyl alcohol
(*g*) methyl formate
(*h*) *m*-xylene
(*i*) methylene dichloride
(*j*) formamide

4. Name two substances either of which might properly be distilled with the drying agent still in the liquid.

5. In which liquid would the partial vapor pressure of water be greater, a 99% solution of butyl alcohol in water or a 99% solution of methyl alcohol? Give two reasons, based on different principles, to support your conclusion.

6. What objections might be offered to use of the particular drying agents in cases cited below? Explain in detail.

(*a*) phosphorus pentoxide for isopropyl alcohol
(*b*) potassium hydroxide for methyl butyrate
(*c*) sodium sulfate for acetone
(*d*) magnesium sulfate for acetic acid
(*e*) calcium chloride for aniline

7. Which would be more easily dried, isobutylamine or *n*-hexylamine?

8. A certain preparation of bromobenzene, dried thoroughly with anhydrous magnesium sulfate, was distilled without removal of the spent drying agent. Predict and explain the resulting difficulties.

9. Why is the ethereal extract of an organic compound dried before instead of after evaporation of the ether?

10. Why should one conduct the final distillation of a liquid preparation after drying rather than before, no ether being involved?

11. With §134 in mind, how would you calculate the thoroughness of desiccation when a moist gas is passed slowly over solid carbon dioxide?

12. Calculate the capacity of each of the drying agents listed in Table 1.

References

(See Chapter 14 for complete titles.)

Weissberger, *Technique of Organic Chemistry,* **3** (2nd ed.), Pt. 1, 787, **7** (2nd ed.), 289; *Techniques of Chemistry,* **2** (3rd ed.), 565; Müller–Houben–Weyl, **1**, Pt. 1, 757, Pt. 2, 213, 770, 869; Wiberg, 20, 109, 147.

Molecular Sieves: C. K. Hersh, *Molecular Sieves,* Van Nostrand Reinhold, New York, 1961; review: V. Ya. Nikolina, I. E. Neimark, and M. A. Piontkovskaya, *Russian Chemical Reviews* (Engl. translation of *Uspekhi Khimii*), **29**, 509–21 (1968).

Chapter 11

Solvents and Solubility

Solvents are of great importance in organic chemistry. Although chemical transformations can be carried out and studied in many environments (gas, liquid, solid, and polymer phases, phase interfaces, etc.), liquid solutions have been more used than any other. It is usually more convenient to carry out reactions in solution, easier to control conditions, and easier to isolate products. The nature of the solvent has a great effect on the rates of many organic reactions and rates are more readily determined in solution.

Solvents are also important because most organic solids are purified by crystallization. Proper choice of solvent is the key to successful and convenient purification.

Liquid-liquid extraction relies on the relative solubilities of solutes between to immiscible liquids. With repeated extractions, even similar compounds can be separated.

Finally, the solubility of an unknown compound in various solvents is widely used to aid in identification. This aspect of the subject is discussed and illustrated in the section on Qualitative Organic Analysis.

An understanding of solubility and of the effect of solvents on reaction rates and equilibria depends on knowledge of chemical bonding, of the polar nature of many covalent bonds as well as of ionic bonds, and of the nature of intermolecular forces. In liquids there is a balance between kinetic energy and the intermolecular attractive forces. This results in an irregular, constantly changing arrangement of the molecules, which are close together so that intermolecular forces become very effective. Textbooks of general chemistry and organic chemistry discuss these topics in an introductory way, and more detailed discussions are presented in physical chemistry; this material will not be reviewed in the present chapter.

138. Solubility. It would be convenient to have rules by which one could calculate the solubility of any organic compound in any solvent. Unfortunately the ability to dissolve is governed by too many factors to permit formulation of rules; hence for practical laboratory work in organic chemistry, a simple, empirical approach is often sufficient.

Solution of an ionic solid in a polar solvent such as water involves separation and solvation of the ions. The crystal structure is lost, and the lattice energy of the crystal is compensated for by the solvation energy. Similarly, solution of a polar nonionic solid in a polar solvent results in intermolecular attractive forces between the molecules of solvent and solute; these may not be as strong as those involved in ion solvation, but they also are electrostatic in nature. Again there is loss of the lattice energy of the solid, which is compensated for by these attractive forces. Since the melting of a solid also involves overcoming the lattice energy of the crystal one might expect some correlation between melting point and solubility. The rough rule is that for two closely related substances, the higher melting will be the less soluble. Table 1 presents several examples. As a consequence of this behavior, solids of low melting point are often relatively difficult to purify by crystallization because they have high solubility even in cold solvents.

Table 1. Solubilities and Melting Points of Closely Related Compounds

Compound	*Melting Point (°C)*	*Solubility*[a]
Fumaric acid	287	0.7 g in 100 ml H_2O, 25° 5.75 g in 100 ml C_2H_5OH, 29.7°
Maleic acid	130.5	74 g in 100 ml H_2O, 25° 69.9 g in 100 ml C_2H_5OH, 29.7°
Anthracene	217	0.03 g in 100 ml C_2H_5OH, 25°
Naphthalene	80.2	11.8 g in 100 ml C_2H_5OH, 25°
o-Nitrobenzaldehyde	44	71% in C_6H_6, 25°
m-Nitrobenzaldehyde	58	47% in C_6H_6, 25°
p-Nitrobenzaldehyde	106	8% in C_6H_6, 25°
o-Dinitrobenzene	118.5	27.1 g in 100 g $CHCl_3$, 17.6°
m-Dinitrobenzene	90	32.4 g in 100 g $CHCl_3$, 17.6°
p-Dinitrobenzene	174	1.82 g in 100 g $CHCl_3$, 17.6°
o-Chlorobenzoic acid	142	2.13 g/liter in H_2O, 25°
m-Chlorobenzoic acid	158	0.385 g/liter in H_2O, 25°
p-Chlorobenzoic acid	243	0.068 g/liter in H_2O, 25°

[a]For several of these compounds the solubility in various other solvents is also given in A. Seidel, *Solubilities of Organic Compounds* (3rd ed.), Van Nostrand Reinhold, New York, 1941.

It is fairly common to classify solvents in terms of polarity, although it is difficult to give a quantitative meaning to this term. Water is the prototype of the polar solvent. When the shape of a molecule and the distribution of electrons that make up the chemical bonds are such that the centers of positive and negative charge do not coincide, the molecule will possess a permanent dipole and can be called polar. The dipole moment or the dielectric constant might then be thought to measure the polarity of the substance. However, solubility is determined by the strength of the intermolecular forces between the unlike molecules relative to the forces between the like molecules. This depends on the shapes of the molecules, their sizes, their polarizability, and how shielded the dipoles are. Intermolecular forces include ion-dipole, dipole-dipole, dipole-induced dipole and London dispersion forces. Hydrogen bonding may make an especially effective contribution. Thus dielectric constant and dipole moment are at best no more than rough measures of solvent polarity.

Alternative terms sometimes applied to solvents are hydrophobic (water-hating), corresponding to non-polar, and hydrophilic (water-loving), corresponding to polar.

It is useful to arrange a few common solvents in order of increasing polarity: hexane $<$ benzene $<$ diethyl ether $<$ ethanol $<$ water. Any two of these in adjacent positions are mutually very soluble; those far apart are mutually almost insoluble. Furthermore, substances related chemically to a given compound in the list are likely to dissolve readily in that compound and to have decreasing solubility in solvents that are progressively farther away in the list. The familiar generalization "like dissolves like" summarizes these facts; it rests on the observation that intermolecular forces of attraction tend to be greater between chemically similar molecules than between dissimilar pairs. However, the generalization is only a rough guide, and it is often difficult to make predictions.

The list of solvents for chromatography arranged in order of increasing eluting power against alumina (§123) is roughly in order of increasing polarity because adsorption on alumina also depends on polarity. It can serve as a guide to solubility.

Rough generalizations can be made about the polarity of various classes of compounds. One observes that unsaturation in a molecule increases polarity slightly. Saturated hydrocarbons are nonpolar, and simple alkyl halides are only slightly more polar. However, the presence of several halogens may make considerable difference; for example, dichloromethane is slightly polar, but chloroform is quite a bit more so with its hydrogen having weak hydrogen-bonding power. On the other hand, carbon tetrachloride is a useful nonpolar solvent.

Oxygen-containing compounds are always somewhat polar, but the nature of the functional group is important. Ethers, esters, and ketones act as acceptors in hydrogen-bond formation but not as donors. Diethyl ether and ethyl acetate are common solvents of intermediate polarity.

Ethers are more inert, which makes diethyl ether an especially widely used reaction solvent (e.g., for Grignard reactions). The slight solubility of water in this ether (1 g/75 g ether) is sometimes an advantage. Formation of an amine salt is far easier in this solvent than in one like pentane. If hydrogen chloride is passed into an amine in pentane, the hydrochloride carries down any traces of water with it, and a sticky precipitate results. In dry ether one obtains a crystalline hydrochloride, and traces of water remain in the solvent. Other useful ethers include dioxane and tetrahydrofuran. Among ketones, acetone is unique for its very powerful solvent action. It is classed among the dipolar aprotic solvents discussed in §141. Butanone is also a widely used solvent of intermediate polarity.

Alcohols are the solvents most like water because they can act both as donors and acceptors in hydrogen-bond formation. Ethanol is of greatest importance because it can dissolve a wide spectrum of both polar and nonpolar substances. It is widely used both as a reaction solvent and for crystallization. Mixtures of alcohol with water also find wide use. Methanol is somewhat more polar than ethanol.

Nitrogen compounds are also polar. Primary and secondary amines can undergo hydrogen bonding like alcohols. Their basicity makes them reactive and hence unsuitable as solvents with many compounds. Tertiary amines can act only as hydrogen-bond acceptors, and they are used as solvents somewhat oftener; their basicity is often a drawback. Pyridine, a weakly basic tertiary amine, does find wide use. The higher amides are solids, but formamide and *N*-methylformamide are useful as polar protic solvents. Liquid amides that have no hydrogen on nitrogen (e.g., *N, N*-dimethylformamide) are highly polar, aprotic solvents and will be discussed in §141. Nitro compounds (nitromethane, nitrobenzene) and nitriles are also polar and classed as polar aprotic solvents. Many of these compounds have quite high boiling points, a feature that makes them less useful as crystallization solvents. Acetonitrile, however, is quite widely used for this purpose.

139. Solubility by Reaction. Sometimes a rapid chemical reaction will occur between solvent and solute. If the product is very soluble, it may appear that the solute has high solubility. Common examples involve acids and bases. Benzoic acid is only moderately soluble in water but dissolves readily in dilute alkali because the ionic salt is formed and is highly water soluble. Similarly, aniline is sparingly soluble in water but dissolves in dilute hydrochloric acid wherein anilinium chloride forms rapidly. Neutralization of the solution with strong base regenerates the slightly soluble aniline.

140. Solvent Effects on Organic Reactions. Reaction rates, equilibria, and product distribution of organic reactions are often sensitive to the solvent used. A qualitative theory to account for solvent effects was advanced by Hughes and Ingold in 1935; it has been summarized by Ingold [C. K. Ingold, *Structure and Mechanism in Organic Chemistry* (2nd ed.), Cornell University Press, Ithaca, N.Y., 1969 (1st ed., 1953)]. Ions or polar molecules placed in a polar solvent are solvated and become more stable. The reactants and transition state for any given reaction will be solvated to different extents. When the solvent is changed from a less polar to a more polar, the heat of activation for a given reaction will decrease if the transition state is more polar than the reactants and increase if the transition state is less polar. Entropy changes will also occur, but the assumption is made that these will be less important than the heat of activation. The effect of solvent change on equilibrium will parallel that on rate: a more polar solvent will increase exothermicity if products are more polar than reactants.

Ingold equates solvent polarity with power to solvate; it will increase with the molecular dipole moment of the solvent but decrease with increased thickness of shielding of dipole charges. The extent of solvation of a charged entity is expected to increase with the charge but decrease with increasing dispersal of charge; the decrease of solvation due to charge dispersal is assumed to be less than that due to destruction of charge.

The theory was applied mainly to nucleophilic aliphatic substitution and elimination; S_N2 reactions of four charge types and S_N1 reactions of two charge types were discussed. The predictions are given in Table 2.

Table 2. Predicted Solvent Effects on Rates of Nucleophilic Substitution

Mechanism	*Initial State*	*Transition State*	*Effect of Activation on Charges*	*Rate Effect for Increased Solvent Polarity*
S_N2	$Y^- + RX$	$Y^{\delta-}\cdots R\cdots X^{\delta-}$	Dispersed	Small decrease
	$Y + RX$	$Y^{\delta+}\cdots R\cdots X^{\delta-}$	Increased	Large increase
	$Y^- + RX^+$	$Y^{\delta-}\cdots R\cdots X^{\delta+}$	Reduced	Large decrease
	$Y + RX^+$	$Y^{\delta+}\cdots R\cdots X^{\delta+}$	Dispersed	Small decrease
S_N1	RX	$R^{\delta+}\cdots\cdots X^{\delta-}$	Increased	Large increase
	RX^+	$R^{\delta+}\cdots\cdots X^{\delta+}$	Dispersed	Small decrease

Several examples in agreement with these predictions were presented. Ingold also presented predictions and experimental verification for E1 and E2 eliminations; effects both on rates and on the relative proportions of substitution and elimination were considered. The reference should be consulted for details. It was clearly pointed out that the theory is no more than qualitative. One must always be certain that the mechanism remains the same as the solvent is changed. Predictions may be invalid if the position of the transition state along the reaction coordinate varies too greatly. Many investigations have been carried out on the effect of solvents on reactions and many more are needed. However, the theory does serve as a practical guide for choice of solvent when one is attempting to increase yields, shorten reaction times, and decrease side reactions in a reaction of the type discussed.

An interesting example of the solvent affecting the product distribution is the acylation of naphthalene. When the reaction is carried out in

$$C_{10}H_8 + CH_3\overset{O}{\overset{\|}{C}}Cl \xrightarrow[C_6H_5NO_2]{AlCl_3,\ 95°} \beta\text{-}C_{10}H_7\overset{O}{\overset{\|}{C}}CH_3$$

$$C_{10}H_8 + CH_3\overset{O}{\overset{\|}{C}}Cl \xrightarrow[CH_2Cl_2]{AlCl_3,\ 0°} \underset{93\%}{\alpha\text{-}C_{10}H_7\overset{O}{\overset{\|}{C}}CH_3} + \underset{\text{trace}}{\beta\text{-}C_{10}H_7\overset{O}{\overset{\|}{C}}CH_3}$$

nitrobenzene, the β-isomer is obtained in 90% yield; however, when the solvent is carbon disulfide or methylene chloride, the α-isomer predominates (93% + a trace of β-isomer). Steric hindrance becomes a factor in nitrobenzene solvent since the bulk of the electrophile (acetyl chloride–aluminum chloride complex) is increased by solvation with nitrobenzene; therefore, the less hindered β-position of naphthalene is attacked.

In contrast to ionic reactions (and predictable from the preceding theory), homolytic reactions, which involve neutral radicals, are not as sensitive to solvent change. The References or textbooks of physical organic chemistry should be consulted for more information.

141. Dipolar Aprotic Solvents. One group of examples of spectacular rate changes with changes of solvent involves the contrast between protic and dipolar aprotic solvents for bimolecular reactions between anions and uncharged species. These solvent effects were extensively studied by A. J. Parker, who has reviewed his work and that of others [*Quart. Rev.*, **16,** 163 (1962); *Advances in Organic Chemistry, Methods and Results,* **5,** 1 (1965); *Advances in Physical Organic Chemistry,* **5,** 173 (1967); *Chem. Rev.*, **69,** 1 (1969)].

Dipolar aprotic solvents are highly polar liquids that do not contain hydrogens attached to oxygen or nitrogen and cannot donate hydrogen to form hydrogen bonds. Only solvents with dielectric constants >15 are included because in solvents of lower dielectric constant, ion aggregation is so extensive that it is difficult to observe the behavior of solvent-separated ions. Common members of the class are *N, N*-dimethylformamide (DMF), *N,N*-dimethylacetamide (DMAC), dimethylsulfoxide (DMSO), hexamethylphosphoramide (HMPT), *N*-methylpyrrolidone (NMePy), acetone, acetonitrile, nitromethane, nitrobenzene, benzonitrile, sulfur dioxide, and propylene carbonate.

The reactions considered are S_N2 substitutions, S_N2 addition-eliminations (at aryl, vinyl, and carbonyl carbons), and E2 eliminations. With few exceptions the rates in dipolar aprotic solvents are faster. Table 3 gives two examples that have been examined in several solvents; the rate changes are very large. In other instances rate changes may be less. These S_N2 reactions are faster in the aprotic solvents because the reactant anion, $Y\!:^-$, is strongly solvated through hydrogen bonds in the protic solvents; this greatly reduces its reactivity. In the first example, an S_N2 displacement in a saturated aliphatic system, solvation of the alkyl halide and the transition state are far less important. In the second example, a bimolecular addition-elimination on an activated aromatic nucleus, the solvation of the transition state plays a larger role, although changes in solvation of the anion still dominate. The rate change with type of solvent is considerably smaller for such addition-eliminations at carbonyl carbon (e.g., ester hydrolysis) because negative charge is much more localized on oxygen in the transition state. This results in increased solvation of the transition state when the solvent is protic owing to hydrogen bonding; the rate is thus increased in protic solvents, and the rate increase on transfer to dipolar aprotic solvents is less.

When a bimolecular reaction does not involve attack by an anion, the effect of change from protic to dipolar aprotic solvent may be very different. For example, molecule-molecule reactions

$$Y\!: + RX \rightleftharpoons \left[Y^{\delta+}\cdots R\cdots X^{\delta-}\right]^{\ddagger} \rightleftharpoons YR^+ + X^-$$

such as the Menschutkin reaction

$$(C_2H_5)_3N\!: + C_2H_5I \longrightarrow (C_2H_5)_4N^+X^-$$

are rather insensitive to protic-dipolar aprotic solvent transfer. Many dissociative reactions, rearrangements, and bimolecular solvolyses of RX are slower in dipolar aprotic than in protic solvents. The references cited above give further information.

142. Practical Considerations. It must be emphasized that many considerations besides rate enter into choice of solvent for a given reaction. It is important that the solvent be inert and not undergo some reaction

Table 3. Rates in Protic and Dipolar Aprotic Solvents (25°, *k* Relative to Methanol)

Reaction A: $CH_3I + Cl^- \longrightarrow CH_3Cl + I^-$

Reaction B: $4\text{-}NO_2C_6H_4F + N_3^- \longrightarrow 4\text{-}NO_2C_6H_4N_3 + F^-$

Solvent	CH_3OH	$HCONH_2$	CH_3NO_2	CH_3CN
Reaction A	1	16	16×10^3	4×10^4
Reaction B	1	6.3	3.2×10^3	7.9×10^3
Solvent	DMSO	DMF	Acetone	
Reaction A		8×10^5	1.6×10^6	
Reaction B	7.9×10^3	3.2×10^4	7.9×10^4	
Solvent	DMAC	NMePy	HMPT	
Reaction A	2.5×10^6	7.9×10^6		
Reaction B	1×10^5	2.0×10^5	7.9×10^7	

with the reactants or products to give by-products. Ease of isolation and purification of the product is often a decisive consideration. Availability and cost of the solvent are important. In many instances recorded in published papers, choice of solvent has been made simply on the basis of the solvent reported in the older literature for the synthesis being carried out or for a similar synthesis. Since many of the dipolar aprotic solvents were not cheap or readily available until fairly recently, they are less often used than their special properties warrant. Some are high boiling and water insoluble so their removal after reaction presents problems. However, DMF, DMAC, DMSO, NMePy, and propylene carbonate are relatively inexpensive and miscible with water; they should be considered along with acetone and acetonitrile for those reactions that are accelerated by dipolar aprotic solvents.

References

(See Chapter 14 for complete titles.)

Weissberger, *Technique of Organic Chemistry,* **7** (2nd ed.), *Techniques of Chemistry,* **2** (3rd ed.); Müller–Houben–Weyl, **1,** Pt. 1, 453, Pt. 2, 765; Wiberg, 240; C. Marsden and S. Mann, *Solvent Guide* (2nd ed.), Wiley–Interscience, New York, 1963; T. H. Durrans, *Solvents* (7th ed.), Chapman and Hall, London, 1957; G. Charlot and B. Tremillion, *Chemical Reactions in Solvents and Melts* (translated by P. T. T. Harvey), Pergamon Press, Elmsford, N.Y., 1969; E. S. Amis, *Solvent Effects on Reaction Rates and Mechanisms,* Academic Press, New York, 1966.

Chapter 12

Organic Bases and Acids

143. Fundamental Definitions. The classical definition for acids (compounds that yield hydrogen ions in water) and bases (compounds that yield hydroxide ions) would be adequate if all reactions were carried out in water. Two useful extensions have been proposed redefining the acid-base concept: the hydrogen ion transfer approach (Brønsted–Lowry) and the electron-pair transfer approach (Lewis).

In the Brønsted–Lowry treatment, an acid is a compound that can donate a proton to another species, while a base is defined as a species that will accept a proton.

Concurrently with development of the protonic treatment of acids (1923), G. N. Lewis defined bases as compounds having unshared electron pairs (or even a shared pair as in a π-bond) that can be donated to form a covalent bond with another compound. A "Lewis acid" is defined as a compound that can accept an electron pair into its valence shell (e.g., BF_3, $AlCl_3$).

Much of organic chemistry relies on acid-base relationships. The solubility of an amine in dilute hydrochloric acid illustrates this relationship.

$$C_6H_5NH_2 + HCl \rightarrow C_6H_5\overset{+}{N}H_3Cl^-$$

A further example will be demonstrated in Experiment 30 when a "Lewis acid" is used to form an ionic complex to promote both the acylation and alkylation of aromatic compounds.

144. Basic Strength. The term "strength" as applied to a base refers to the ability of that substance to withdraw, in a reversible equilibrium reaction, protons from some very weak acid chosen as a test standard. With the precaution to exclude extremely weak and extremely strong bases, water may be taken as this test-standard reagent. The base under test, according to its strength, takes a proton from each reacting water molecule, releasing in each case a hydroxyl ion. The equilibrium concentration of such accumulation of OH^- is, of course, readily measured with a pH meter. The numerical result calculated from this simple experiment is known as the basic ionization constant, or dissociation constant, recorded in handbooks and illustrated by examples in the accompanying table. *n*-Propylamine is chosen here as an example for actual calculation.

Example. The *n*-propylamine is merely added to water, and the solution diluted to 1 *M* concentration for arithmetical convenience. The following reaction proceeds at once to equilibrium:

$$C_3H_7NH_2 + HOH \rightleftharpoons C_3H_7NH_3^+ + OH^-$$

The pH-meter test now shows that the concentration of hydroxyl ion is approximately 2×10^{-2} *M*; i.e., about 2% converted into the ionic state. Without significant error, one may now ignore the concentration of water in the foregoing equation and write a mass-action equation as follows:

$$\frac{[C_3H_7NH_3^+] \times [OH^-]}{[C_3H_7NH_2]} = K_b$$

where K_b is the ionization constant of the base being tested. With substitution of numerical values already suggested,

$$\frac{(2 \times 10^{-2})(2 \times 10^{-2})}{(\sim 1)} = K_b = 4 \times 10^{-4}$$

which roughly is in agreement with the table.

This base would be regarded as rather weak in common inorganic chemistry, in contrast with hydroxyl ion, a very strong base. In organic chemistry, however, it is considered fairly strong. Aniline, with a K_b of 4.6×10^{-10}, is typical of a weak base. In a saturated aqueous solution of aniline only a very small fraction of 1% of the aniline molecules are found to have accepted protons from water.

145. Acids. Water also serves as the reference compound for acids, but here it acts as a base. The equilibrium is shown below for acetic acid.

$$H_2O + CH_3CO_2H \rightarrow H_3O^+ + CH_3CO_2^-$$

The concentration of water remains constant and is again omitted; thus

$$\frac{[H_3O^+] \times [CH_3CO_2^-]}{[CH_3CO_2H]} = K_a$$

where K_a is the ionization constant of the acid.

In this case the constant is recorded (see Table 1) as 1.8×10^{-5}. This is a median value in organic chemistry and indicates that acetic acid is an acid of moderate strength. Hundreds of carboxylic acids are found in this class, with constants from 10^{-4} to 10^{-5}. Those acids whose K_a value is in the range of 10^{-1} to 10^{-3} are rated as strong. This statement means that a large fraction of the molecules of such strong acid are found to have reacted with water in their equilibrium mixture with that solvent. Those acids with constants of 10^{-10}, or thereabouts, exist largely in their original molecular form in the water equilibrium and are called weak. Constants of extremely low values, such as 10^{-20} to 10^{-43}, apply to substances scarcely recognized as acids in common chemical operations. They are not derived from precise measurements but instead are the result of semiquantitative experiments.

Hydrogen chloride is so strong an acid that it yields its protons to water completely in ordinary dilute solutions:

$$H_2O + HCl \rightarrow H_3O^+ + Cl^-$$

$$\mathrm{H{:}\overset{H}{\ddot{\underset{..}{O}}}{:}} + \mathrm{H{:}\ddot{\underset{..}{Cl}}{:}} \rightarrow \left[\mathrm{H{:}\overset{H}{\ddot{\underset{..}{O}}}{:}H}\right]^+ + \mathrm{{:}\ddot{\underset{..}{Cl}}{:}}^-$$

The resulting ionic pair is known as an *oxonium salt,* analogous to ammonium salts, although the laboratory label probably reads "Acid, Hydrochloric, HCl, Dilute." The positive portion of this particular

Table 1. Ionization Constants

Acids	K_a	*Bases*	K_b
Trifluoroacetic acid	(too strong)[a]	Methoxide (OCH_3^-)	(too strong)
Trichloroacetic acid	2×10^{-1}	Guanidine	3×10^{-1}
Dichloroacetic acid	5×10^{-2}	Dipropylamine	1.02×10^{-3}
Chloroacetic acid	1.6×10^{-3}	Tripropylamine	5×10^{-4}
Lactic acid	1.4×10^{-4}	*n*-Propylamine	3.9×10^{-4}
Benzoic acid	6.6×10^{-5}	Ammonia	1.8×10^{-5}
Acetic acid	1.8×10^{-5}	Strychnine	1×10^{-7}
Carbonic acid (first H^+)	3×10^{-7}	Pyridine	2.3×10^{-9}
Phenol	1.3×10^{-10}	Aniline	4.6×10^{-10}
Sucrose	1.8×10^{-13}	Dimethylaniline	2.4×10^{-10}
Water	10^{-14}	α-Naphthylamine	9.9×10^{-11}
Ethyl alcohol	10^{-18}	Urea	1.5×10^{-13}
Phenylacetylene	10^{-21}	*o*-Chloroaniline	9.2×10^{-14}
Triphenylmethane	$\sim 10^{-33}$	Water	10^{-14}
Diphenylmethane	$\sim 10^{-35}$	Acetamide	(too weak)
Phenylmethane (toluene)	$\sim 10^{-35}$	Triphenylamine	(too weak)
Methane	$\sim 10^{-43}$	Acetone	(too weak)
		Diphenyl ether	(too weak)

[a]Observations "too strong" or "too weak" mean that the constant is outside the range in which water is acceptable as the standard test reagent.

oxonium salt is called hydronium ion. Similar conditions probably exist in the case of trifluoroacetic acid shown in the table. In other words, a precise name for dilute trifluoroacetic acid might be hydronium trifluoroacetate.

146. Extremely Weak Bases and Acids. If a base or an acid is weaker than water itself, it cannot be detected as a base (or acid) in water solution. For example, the attempt to neutralize triphenylamine (a base weaker than water) with aqueous hydrochloric acid fails since any available acidic hydrogen has already been taken out of the field of reaction by the stronger base HOH to yield H_3O^+.

Similarly, the very weak base ether does not react with dilute aqueous hydrochloric acid but reacts readily with anhydrous HCl to form a salt $(C_2H_5)_2OH^+Cl^-$. Some of these "ether oxonium" salts may even be crystallized.

By converse reasoning, the acidic character of methanol cannot be detected in aqueous solution by titration even with the strong base Na^+OH^-. The attempt is doomed to failure from the start, since the compound that would supposedly be produced (H—OH) is a stronger acid than the methanol that one is proposing to use as a reagent. The reverse reaction actually occurs.

$$CH_3OH + Na^+OH^- \xrightarrow[]{\text{(cannot produce)}}\!\!\!\!\!\!\!/\;\; HOH + Na^+OCH_3^-$$

147. Practical Applications. From the above numerical data and theory a number of practical applications are derived, as follows:

(*a*) Ordinary carboxylic acids, containing carbon, hydrogen, and oxygen only, are so much weaker than the common strong inorganic acids such as hydrochloric and sulfuric that they may readily be produced from their salts by mere admixture of the latter with the strong acid. Water does not appreciably interfere with this process, since the oxonium ion, H_3O^+, which holds the acidic protons in aqueous solutions of strong acids, is itself a stronger acid than the carboxylic acids that are to be formed. If the desired carboxylic acid is to be isolated by distillation, the nonvolatile reagent sulfuric acid is employed so that any excess of the reagent will not be distilled. On the other hand, if the organic acid is to be isolated by crystallization, hydrochloric acid is employed so that any excess will easily be dissipated by simple evaporation.

(*b*) By analogy with (*a*), it becomes evident that a weak organic base may be expelled from its salt by admixture of a strong base. For example, aniline hydrochloride (§318) is converted into free aniline base by the addition of hydroxyl ion in the form of sodium hydroxide:

$$C_6H_5NH_3^+ + OH^- \rightarrow C_6H_5NH_2 + H_2O$$

(*c*) Phenols are so much weaker than carboxylic acids that separations of the former from the latter are occasionally based upon this fact. The mixture of phenol and carboxylic acid is treated with sodium bicarbonate. The carboxylic acid yields its proton to the bicarbonate ion, producing carbonic acid, a weaker acid, while the phenol does not react because it is weaker than carbonic acid. The unreacted phenol is then isolated by ether extraction, distillation, or other process to which the carboxylic acid, now in the form of its salt, does not respond.

(*d*) The fact that acids weaker than water may not be neutralized with aqueous bases is turned to advantage in the purification of ethyl alcohol contaminated with acids. The addition of sodium hydroxide converts into nonvolatile salts all carboxylic acids, all phenols, and of course all strong inorganic acids. Barring the presence of neutral or basic volatile substances, a distillate containing only alcohol and water may easily be obtained from the mixture.

(*e*) Closely connected with the foregoing principle is the rule that salts of acids weaker than water must be carefully protected from moisture. Water, the stronger acid, would of course react with such compounds as sodium ethoxide, calcium carbide (acetylide), and diethylzinc. Even salts of acids slightly stronger than water, such as sodium acetoacetic ester, react with water so extensively as to upset synthetic procedures involving those salts. Naturally, all such salts are prepared with complete avoidance of water.

(*f*) Many organic compounds not commonly recognized as bases will form salts with anhydrous acids. Such compounds are too weak to react even with the relatively strong H_3O^+ acid. Given a strong anhydrous acid like concentrated sulfuric acid, a reaction takes place and the base dissolves. Advantage is taken of this situation in qualitative organic analysis. See §374. Here cold concentrated sulfuric acid is used to identify oxygen and nitrogen derivatives by observing its power to yield protons to these compounds, thus forming salts that dissolve in the acid. It is probable that concentrated sulfuric acid plays the same role in the well-known acid treatment of petroleum widely employed in refineries. To be sure, such concentrated sulfuric acid is not 100% pure. It may be reported to contain 5–8% of water. Such water is, of course, not free (basic) H_2O but is in the form of H_3O^+ or similar coordinated hydrate. The residual acid therefore serves virtually as a nonaqueous reagent and can form salts of ethers, aldehydes, ketones, esters, nitriles, etc. "Fuming" sulfuric acid, containing no water (or H_3O^+), is often used to increase the effectiveness of such acid treatment.

Structural formulas of these salts offer no fundamental novelty in arrangement. For example,

$$\mathrm{C_3H_7{-}O{-}\overset{+OH}{\overset{\|}{C}}{-}C_3H_7\ \ HSO_4^-} \qquad \mathrm{C_3H_7{-}\underset{H}{\overset{+}{O}}{-}\overset{O}{\overset{\|}{C}}{-}C_3H_7\ HSO_4^-}$$

Propyl butyrate acid sulfate (optional structure)

$$\left[\mathrm{C_5H_{11}{-}\underset{H}{S}{-}C_5H_{11}}\right]^+ \mathrm{HSO_4^-} \qquad \mathrm{C_6H_5{-}C{\equiv}\overset{+}{N}H\ HSO_4^-}$$

Diamyl sulfide (a "sulfonium" salt) acid sulfate — Benzonitrile acid sulfate

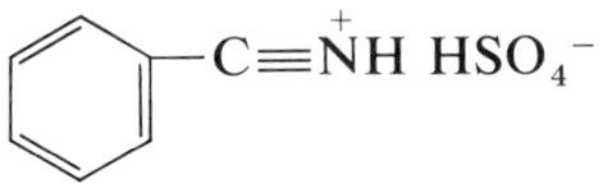

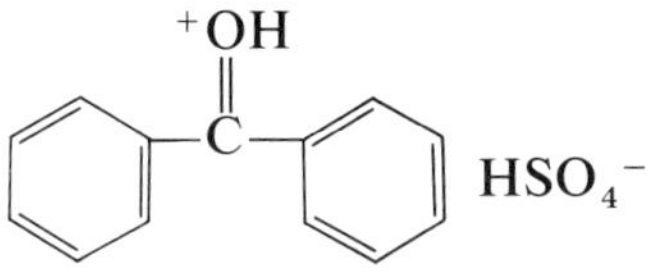

$$\mathrm{C_6H_5{-}\overset{+OH}{\overset{\|}{C}}{-}C_6H_5\ \ HSO_4^-} \qquad \mathrm{C_6H_5{-}\overset{H}{C}{=}\overset{+}{O}H\ \ HSO_4^-}$$

Benzophenone acid sulfate 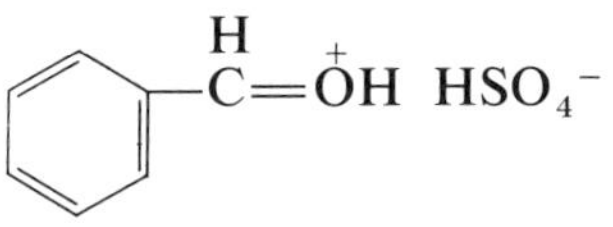 — Benzaldehyde acid sulfate

When a compound such as one of the above is poured upon ice (thus avoiding high temperatures in dilution), the aqueous base H_2O furnished by the ice immediately reacts with the oxonium or sulfonium salt, regenerating the weak organic base which then separates in liquid or solid form. For example,

$$\mathrm{C_3H_7{-}O{-}\overset{+OH}{\overset{\|}{C}}{-}C_3H_7\ HSO_4^- + H_2O \rightarrow C_3H_7{-}O{-}\underset{O}{\underset{\|}{C}}{-}C_3H_7 + H_3O^+ + HSO_4^-}$$

Concentrated sulfuric acid, particularly the anhydrous and fuming grades of the reagent, is known to be excellent for dissolving organic matter that adheres to soiled glassware. It is likely that the largest part of this powerful solvent effect is due to the ability of the anhydrous reagent to convert oxidized matter in the organic stains or tars into salts that are, of course, very soluble in the sulfuric acid.

(*g*) Salts of extremely weak amine bases may be prepared with the aid of like principles, using a convenient diluent. For example, in the molecule of a dinitroaminobenzoic acid the basic character of the amine group is suppressed by the accumulated force of three radicals of somewhat acidic character. Instead of water, anhydrous alcohol, ether, or mixtures of the two, are then used as solvents. Gaseous hydrogen chloride then causes the precipitation of the desired hydrochloride in an effective manner.

(*h*) Compounds of the toluene and ethane types are so low in acidic strength that their sodium salts are prepared only with elaborate precautions. Furthermore, these salts are so reactive that admixture of moisture or even ordinary air causes violent decomposition and spontaneous combustion. Accordingly they are rarely used in common organic synthesis.

Questions on Acid-Base Theory

1. How would you recover the principal compound cited, without concern as to loss of the contaminant or contaminants?

(*a*) cyclohexane (containing cyclohexene)
(*b*) *n*-hexyl alcohol (containing caproic acid)
(*c*) ethyl alcohol (containing aniline)
(*d*) phenol (containing naphthalene)
(*e*) *tert*-butyl alcohol (containing acetic acid and butylamine)
(*f*) aniline (containing nitrobenzene)

2. How would you isolate, in purified state, each of the compounds in each of the following mixtures?

(*a*) tri-*n*-butylamine and triphenylamine
(*b*) *p*-aminobenzoic acid and aniline
(*c*) *n*-octane, aniline, and 2-hexene
(*d*) *o*-toluidine, benzoic acid, and *m*-xylene
(*e*) *o*-cresol, phenylacetic acid, and toluene
(*f*) dibutyl ether and bromobenzene

Chapter 13

Spectroscopic Techniques

148. General Introduction. Determination of the structure of organic compounds is a primary concern of the organic chemist. Earlier methods involved finding the composition of the compound, identifying the functional groups present by characteristic chemical reactions, and establishing the final structure by degradation reaction, by synthesis, or by conversion to a compound or a derivative of known structure. The process was often difficult and sometimes gave incorrect results because one or more of the reactions employed led to rearrangement or other unexpected results.

The organic chemist has long been aware that spectroscopy gave vital information on structure, but early instruments were difficult to use and the information obtained was hard to interpret. Gradually the techniques were improved, the instruments made more sensitive, more reliable, and more convenient, and the body of knowledge greatly increased. Nuclear magnetic resonance (nmr) spectroscopy and mass spectroscopy were added to infrared (ir) and ultraviolet (uv) spectroscopy. Today the organic chemist turns first to spectroscopic methods to solve problems of structure and routinely determines infrared and nuclear magnetic resonance spectra. Mass spectra are becoming easier to obtain, and ultraviolet spectra are reported for most conjugated unsaturated compounds. Spectroscopic methods are also used to follow the courses of reactions both qualitatively and quantitatively.

A brief introduction to ir and nmr spectroscopy is given in this chapter. Spectra for a number of starting materials or products are included in the experiments in Part 2, and references are given in §§572–575 for experiments that involve spectroscopic techniques and are suitable for undergraduate courses. The section of this manual on qualitative organic analysis (§§370–403) presents classical methods; however, a combination of chemical methods and spectroscopic techniques is usually required for complete qualitative analysis. The instructor will determine the extent to which spectra will be used in the experiments on qualitative organic analysis. In some laboratories copies of the spectra of the unknown are supplied, and students use chemical transformations to confirm their spectroscopic interpretations.

The material presented here is necessarily brief. Most textbooks of organic chemistry contain a chapter on spectroscopy. The literature on spectroscopy and its application to organic chemistry is voluminous, but several excellent short texts are available (for example, the books by Dyer and by Pasto and Johnson). Each student should read additional material. A few of the most useful books are listed in §§151–152.

149. Infrared Spectroscopy. Infrared radiation involves that part of the electromagnetic spectrum of longer wavelength than the visible region and shorter than the microwave region. Of greatest practical use in organic chemistry is the very limited portion of the infrared region between about 2.5 and 15 μ (4000 and 660 cm^{-1}). Band positions are expressed either in wavelength [the common unit of which is the micron (μ), equal to 10^{-4} cm] or in frequency [usually expressed in terms of wavenumbers (ν), whose unit is the reciprocal centimeter (cm^{-1})]. These are related by the expression:

$$\nu = \frac{1}{\mu} \times 10^4$$

Radiation in the infrared region can be absorbed by an organic molecule and converted into energy of molecular vibration. An organic molecule exists not as a group of atoms but as a highly mobile collection of vibrating atoms on a rigid framework. It can be likened to a group of balls connected by springs. There are two types of molecular vibrations: stretching and bending. Stretching is a vibration along the bond axis such that the distance between the two atoms is increased or decreased. Bending or deformation involves a change in atomic position with respect to the bond axis. Irradiation of a molecule with a frequency equivalent to a particular mode of molecular vibration will cause the molecule to absorb that radiation, providing the vibration is accompanied by a net change in dipole moment of the molecule. By varying the frequency of the incident radiation and determining how much is absorbed at various wavelengths, as it passes through the sample, one obtains an absorption pattern. Such a plot of intensities (percent of incident radiation that is transmitted or absorbed) versus frequency of radiation or wavelength is the infrared spectrum of the compound.

A very simple molecule can give rise to a highly complex spectrum. This enables the chemist to match the spectrum of an unknown against that of an authentic sample to establish identity. In addition, certain groups give rise to vibrational bands at or near the same frequency regardless of the structure of the rest of the molecule, thus yielding useful structural information by simple inspection of the spectrum.

The positions of the peaks in the spectrum depend on the frequencies of the natural molecular vibrations in the molecule. These molecular vibrations can be compared to a simple harmonic oscillator and an application of Hooke's law readily allows one to predict fairly accurately the frequency (ν) of a given vibration. Thus

$$\nu \text{ (in cm}^{-1}) \propto \sqrt{\frac{k(m_1 + m_2)}{m_1 m_2}}$$

where k = force constant (strength or rigidity of bond), and m_1 and m_2 = effective masses of the two atoms involved in the vibration. A brief inspection of this equation allows one to recognize a number of general correlations. Increasing atomic weight should cause a decrease in the absorption band frequency while factors producing a change in the force constant should cause a corresponding change in the observed frequency; for example, increased bond order (as by proceeding from a single to a double to a triple bond) should increase the frequency while increased bond length should decrease the frequency.

The degree of absorption or the intensity of a given peak is determined for the most part by the overall net change in dipole accompanying that particular molecular motion. If no net change occurs, there will be no absorption. In some instances this change in dipole may be very small, as in the relatively nonpolarized $>C=C<$ group, resulting in weak absorption; a large change, as in the highly polar carbonyl group, gives rise to very strong infrared absorption.

The number of infrared absorption bands is increased by vibrations due to combinations and harmonic oscillations. The harmonic or overtone vibrations occur at frequencies, which are integral multiples of the fundamental frequency; a harmonic or overtone band usually is about .01 as intense as the fundamental band. The combination bands, on the other hand, occur at frequencies resulting from the sums of or differences between two or more fundamental or harmonic vibrations. Much of the uniqueness of an infrared spectrum arises from these bands.

A given functional group usually absorbs in the same general region, more or less independent of the rest of the molecule. This is a consequence of the fact that the energy of absorption depends on two factors: the masses of the atoms, which do not change, and the force constant between them, which should vary only slightly with different environments. This behavior has been substantiated by a large number of examples, and correlation tables have been developed as an aid in interpreting these seemingly complex spectra, making this a very powerful analytical tool. In general the spectra can be divided into two main regions: 2.5–7.7 μ (4000–1300 cm^{-1}), which is referred to as the functional group region, and 7.7–15 μ (1300–660 cm^{-1}), which is called the "fingerprint" region. The former region usually consists of vibrations between two atoms; such vibrations depend on the functionality of those atoms and are largely independent of the overall structure. For example, the carbonyl absorption of acetone is the same as that of di-*n*-butyl ketone. However, structural influences do manifest themselves by causing slight variations in the position of the absorption band. For example, the carbonyl absorption band of acetamide (5.97 μ, 1670 cm^{-1}) is different from the carbonyl absorption band of cyclohexanone (5.83 μ, 1715 cm^{-1}) owing to electronic and steric effects of the overall structure. The principal motions in the fingerprint region are single-bond stretching and bending vibrations of polyatomic systems involving motions of bonds linking substituents to the remainder of the molecule (in other words, complex vibrational-rotational excitations of the entire molecule) and combination and harmonic frequency bands. There are a few functional group absorptions in this region (e.g., C—O stretches), but they are the exception rather than the rule.

The physical state of the sample to be analyzed is of limited concern, for spectra of solids, liquids, and gases can readily be determined. The sample must be dry, however, as water absorbs strongly in the region 2.7 μ (~3710 cm^{-1}) and near 6.15 μ (1630 cm^{-1}). These absorptions would obscure an important region of the spectrum or cause erroneous structural assignments (e.g., a nonexistent OH group). Gases are normally placed in glass-enclosed cells equipped with sodium chloride windows as these alkali halides are transparent in this region of radiation. Liquids can be placed between two NaCl plates neat or placed in a NaCl cavity cell neat or in solution. When samples are run in solution, a solvent sample is normally employed in a double beam spectrometer in order to cancel out absorptions due to solvent. Solids can be handled as solutions, as suspensions in Nujol or Fluorolub spread in a thin film between two salt plates, or as a solid solution, for example, a potassium bromide pellet containing approximately 1–2% of the solid to be analyzed. Because of the large difference in instrumentation design, the actual physical manipulations will be left to the instructor to demonstrate.

Warning: *Do not use any water in the salt plates as this will pit and ruin them.*

To aid you in the interpretation of your spectra, some characteristic group frequencies are listed in Table 1. Also, several typical spectra are shown in Figures 1–7, on which characteristic bands have been identified.

Table 1. Representative Infrared Absorptions

Infrared Functional Group Region	*Wavelength, μ*	*Frequency, cm^{-1}*
2.5–3.1 μ Region		
O—H stretch		
Alcohols and Phenols		
free O—H	2.75–2.8 (sharp)	3636–3571
hydrogen-bonded		
dimeric	2.80–2.90 (sharp)	3571–3448
polymeric	2.94–3.10 (broad)	3401–3226
Carboxylic acids	3.0–4.0 (v. broad)	3333–2500
N—H stretch		
Amines and Amides		
free N—H		
primary	2.8–2.9 and 2.9–3.0	3571–3448 and 3448–3333
secondary	2.85–2.95	3509–3390
hydrogen-bonded	~2.99–3.25 (broad)	3344–3077
Acetylenic C—H	3.0–3.1 (v. sharp)	3333–3226
3.2–4.0 μ Region		
C—H stretch		
Alkenes	3.23–3.32	3096–3012
Aromatic	~3.30	~3030
Alkane	3.37–3.52	2967–2841
Aldehyde $\left(-\overset{\overset{\displaystyle O}{\|}}{C}-H\right)$	3.45–3.55 and 3.6–3.7	2899–2817 and 2778–2703
O-Methyl, *N*-Methyl	3.54–3.62	2825–2762
4.2–5.0 μ Region		
—C≡Y stretch		
Nitrile (—C≡N)	4.42–4.51	2262–2217
Acetylene (—C≡C—)	4.43–4.76	2257–2101
5.5–6.0 μ Region		
>C=O stretch		
Anhydride $\left(-\overset{\overset{\displaystyle O}{\|}}{C}-O-\overset{\overset{\displaystyle O}{\|}}{C}-\right)$	~5.5 and ~5.7	~1818 and ~1754
Acid chlorides		
aliphatic $\left(R\overset{\overset{\displaystyle O}{\|}}{C}Cl\right)$	~5.55	~1802

Table 1. Representative Infrared Absorptions (*continued*)

Infrared Functional Group Region	*Wavelength, μ*	*Frequency, cm⁻¹*
Acid chlorides (*cont.*)		
aromatic $\left(\mathrm{Ar\overset{O}{\overset{\Vert}{C}}Cl}\right)$	~5.68	~1761
Ester		
saturated $\left(\mathrm{R\overset{O}{\overset{\Vert}{C}}OR}\right)$	5.71–5.78	1751–1730
α,β-unsaturated and aromatic	5.78–5.82	1730–1718
Acid		
saturated $\left(\mathrm{R\overset{O}{\overset{\Vert}{C}}OH}\right)$	5.80–5.88	1724–1701
α,β-unsaturated and aromatic	5.88–5.94	1701–1684
Aldehydes		
saturated $\left(\mathrm{R\overset{O}{\overset{\Vert}{C}}H}\right)$	5.75–5.81	1739–1721
α,β-unsaturated $\left(\mathrm{{>}C{=}\overset{\vert}{C}{-}\overset{O}{\overset{\Vert}{C}}{-}H}\right)$	5.78–5.93	1730–1686
aromatic $\left(\mathrm{Ar\overset{O}{\overset{\Vert}{C}}H}\right)$	5.83–5.89	1715–1698
Ketone		
cyclopentanone 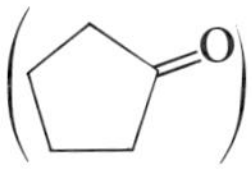	~5.75	~1739
saturated $\left(\mathrm{R\overset{O}{\overset{\Vert}{C}}R}\right)$	~5.83	~1715
aromatic $\left(\mathrm{R\overset{O}{\overset{\Vert}{C}}Ar}\right)$	5.88–5.95	1701–1681
α,β-unsaturated $\left(\mathrm{{-}\overset{\vert}{C}{=}\overset{\vert}{C}{-}\overset{O}{\overset{\Vert}{C}}{-}R}\right)$	5.94–6.01	1684–1664
Amide $\left(\mathrm{{-}\overset{O}{\overset{\Vert}{C}}{-}N{<}}\right)$	5.88–6.14	1701–1629

Table 1. Representative Infrared Absorptions (*continued*)

Infrared Functional Group Region	*Wavelength, μ*	*Frequency, cm^{-1}*
6.0–7.4 μ Region		
>C=C< stretch		
Monoene	5.95–6.17	1681–1621
Diene (C=C—C=C)	~6.06 and ~6.25	~1650 and ~1600
Arene	6.15–6.3	1626–1587
	6.25–6.35	1600–1575
	6.6–6.7 and 6.9	1515–1493 and 1450
NO_2 stretch		
Nitroalkane (RNO_2)	~6.4 and ~7.4	~1563 and ~1351
Nitroarene ($ArNO_2$)	~6.6 and ~7.4	~1515 and ~1351
—NH_2 bend	6.1–6.3	1639–1587
—CH_2—deformation	6.74–6.92	1485–1445
C—CH_3 deformation	6.8–7.0 and 7.25–7.32	1471–1429 and 1379–1366
Fingerprint Region 7.4–11 μ		
C—O stretch		
Phenols (Ar—OH)	~8.33	~1200
Alcohols		
primary (RCH_2—OH)	~9.5	~1050
secondary (R_2CH—OH)	~9.1	~1100
tertiary (R_3C—OH)	~8.7	~1150
Ethers (R—O—R)	8.7–9.35	1150–1070
C—N stretch		
Amines		
aliphatic (R—N<)	8.2–9.8	1220–1020
aromatic (Ar—N<)	7.4–8.0	1350–1250

Fig. 1. Infrared Spectrum of Isobutylamine (thin film). (© Sadtler Research Laboratories, Inc., 1962.)

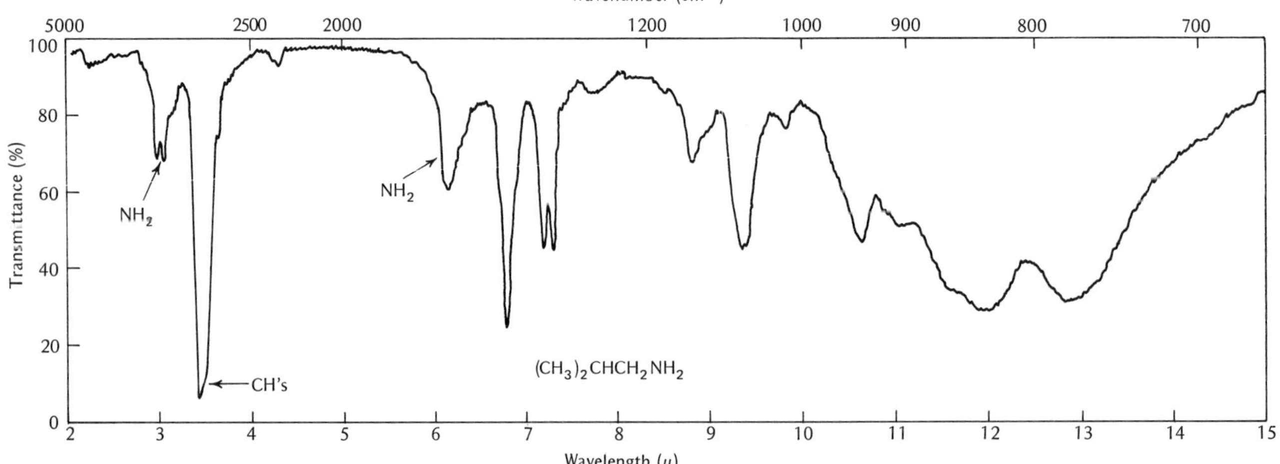

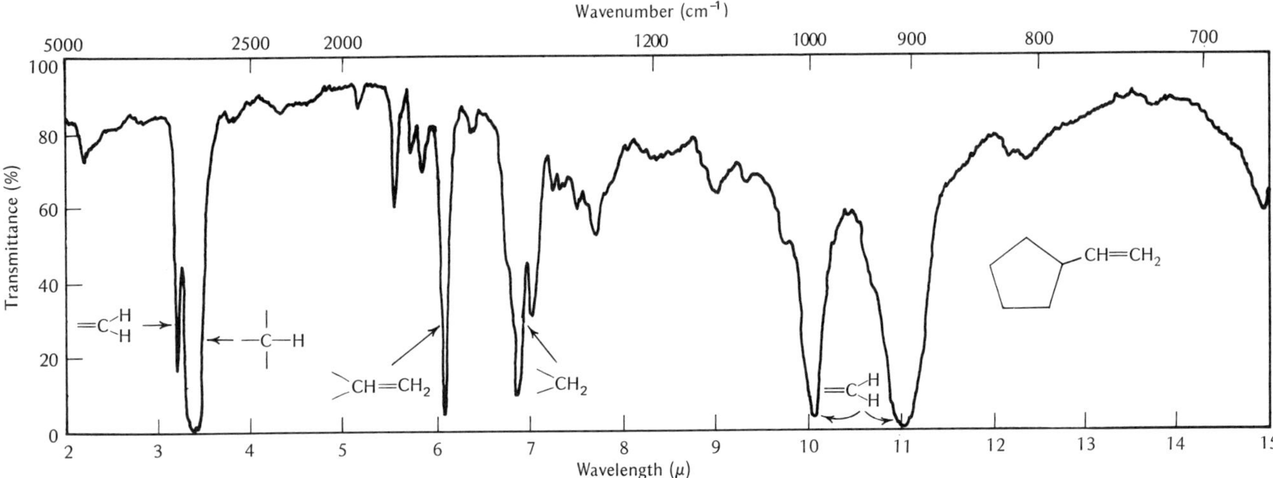

Fig. 2. Infrared Spectrum of Vinylcyclopentane (thin film). (© Sadtler Research Laboratories, Inc., 1962.)

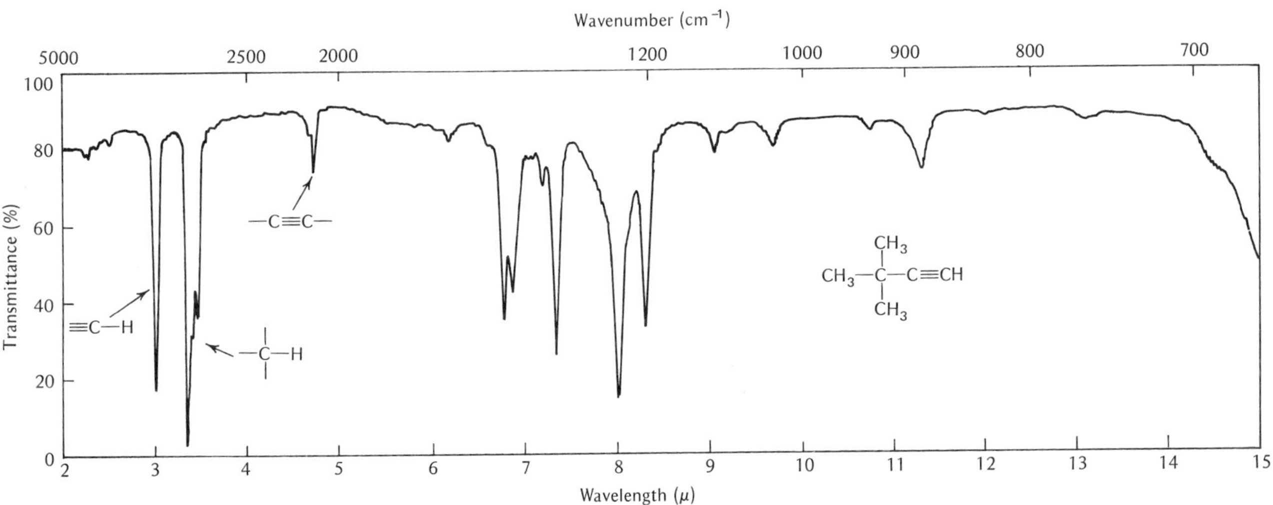

Fig. 3. Infrared Spectrum of 3,3-Dimethyl-1-butyne (thin film). (© Sadtler Research Laboratories, Inc., 1972.)

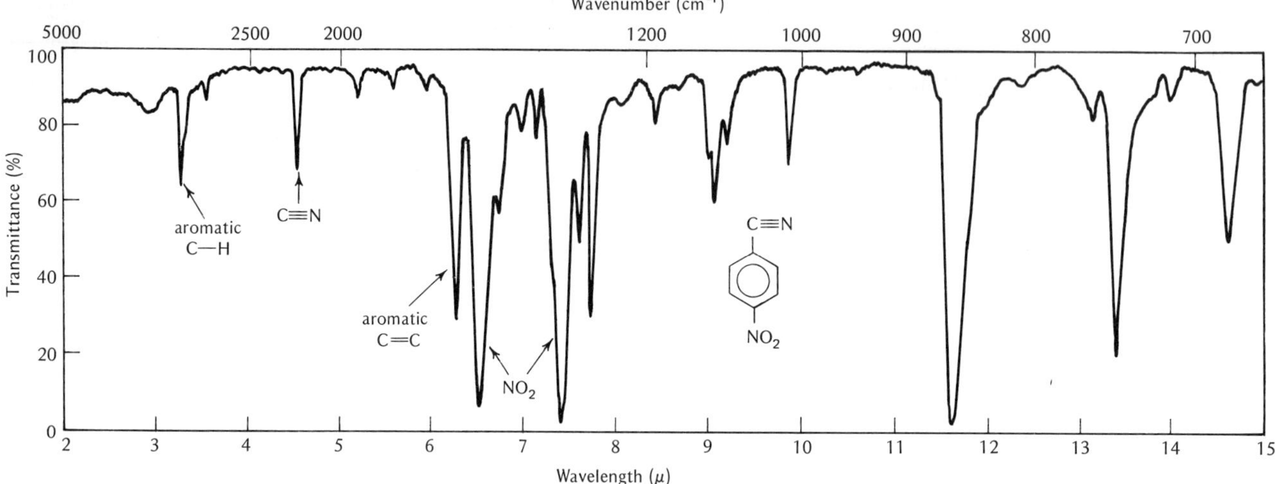

Fig. 4. Infrared Spectrum of *p*-Nitrobenzonitrile (KBr wafer). (© Sadtler Research Laboratories, Inc., 1964.)

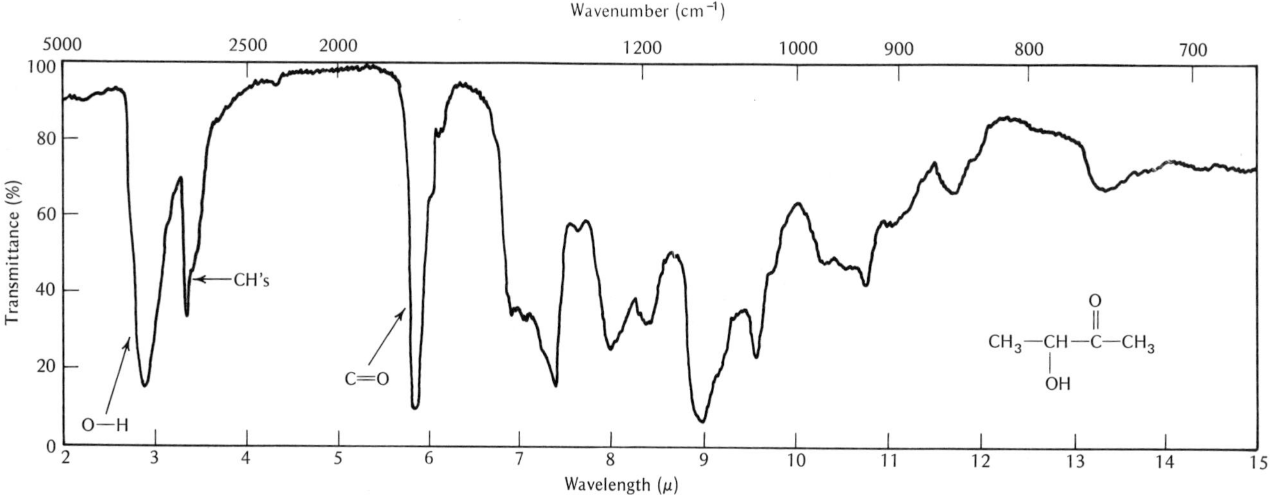

Fig. 5. Infrared Spectrum of 3-Hydroxy-2-butanone (thin film). (© Sadtler Research Laboratories, Inc., 1962.)

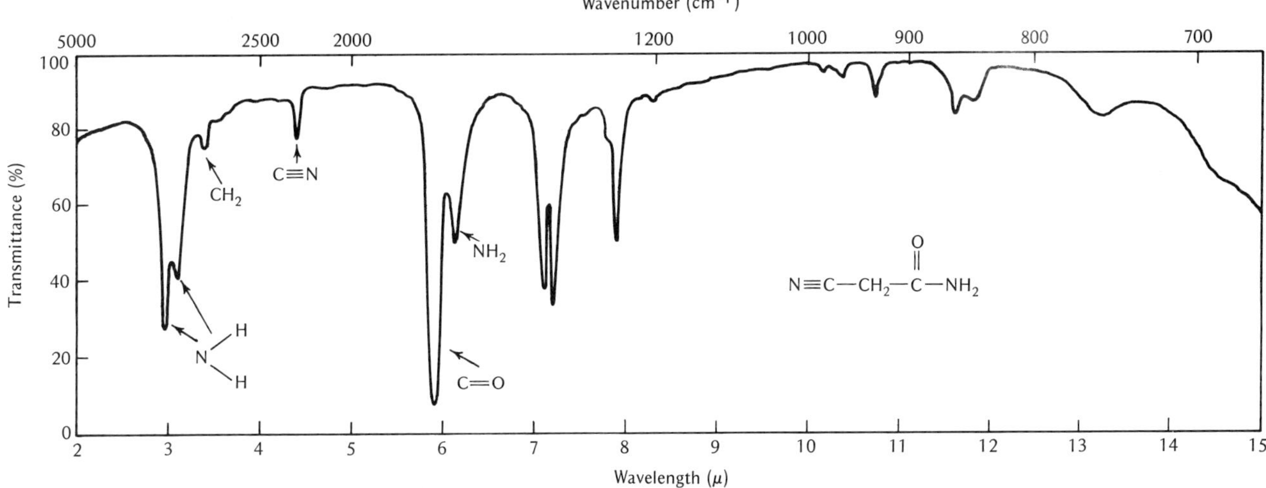

Fig. 6. Infrared Spectrum of Cyanoacetamide (KBr wafer). (© Sadtler Research Laboratories, Inc., 1970.)

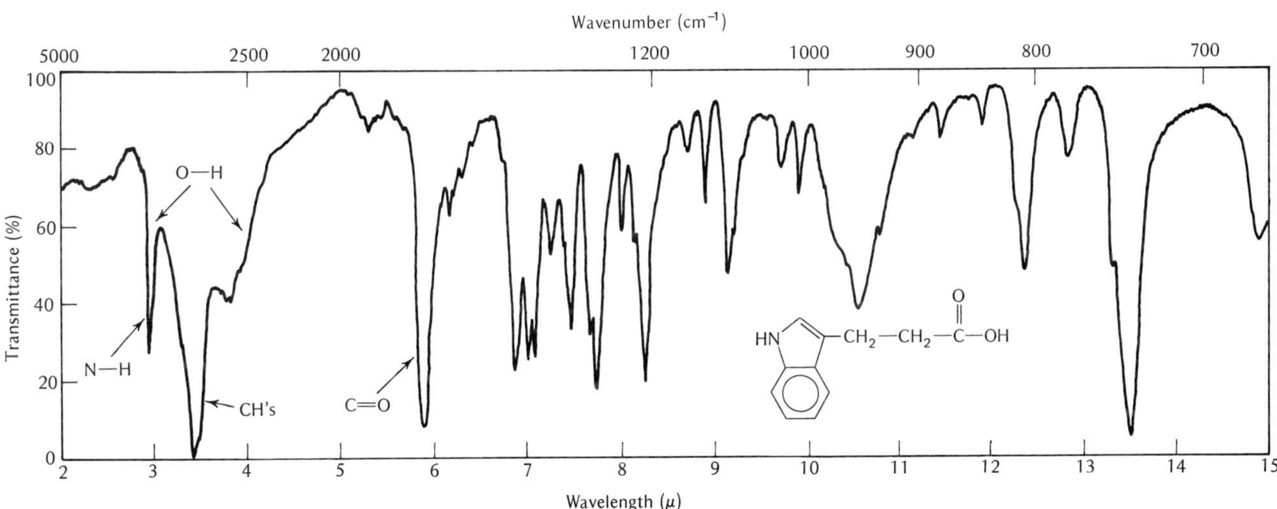

Fig. 7. Infrared Spectrum of Indole-3-propionic Acid (mineral oil mull). (© Sadtler Research Laboratories, Inc., 1970.)

150. Nuclear Magnetic Resonance (NMR) Spectroscopy. The hydrogen nucleus (and the nuclei of certain other atoms) possesses the property of nuclear spin, that is, it has properties such that the nucleus can be likened to a tiny bar magnet. When a hydrogen nucleus is placed in a magnetic field, it lines up with (more stable) or against (less stable) the applied field, corresponding to two energetically different spin quantum states. If the hydrogen nucleus is then subjected to electromagnetic radiation of the proper frequency so that the energy supplied by the radiation will be equal to the energy difference between the two spin states, that is, sufficient to convert protons from the more stable to less stable states, the radiation will be absorbed as the hydrogen nucleus passes from one spin state to the other. It is this absorption of electromagnetic radiation that is measured by the nmr spectrometer. In practice, the sample of an organic compound is normally dissolved in an appropriate solvent and subjected to a magnetic field of about 14,000 gauss. The frequency of electromagnetic radiation required for absorption by a hydrogen nucleus is then about 60 megahertz (MHz).

CHEMICAL SHIFT. The nmr spectrum, a plot of frequency against the intensity of absorption, gives information about the molecular environment of each proton in the molecule. Such a spectrum can be obtained either by sweeping the radiation frequency over a small range at constant magnetic field or, more commonly and more easily, by sweeping the magnetic field at constant frequency. The absorption position in the spectrum is called the chemical shift of the proton in question; chemical shift depends on the atom or group to which it is bonded. Because the strength of the applied magnetic field cannot be determined to a high degree of accuracy (about 1 part in 10^8 would be required), the exact absolute chemical shift cannot be measured easily. However, the chemical shift relative to an arbitrarily chosen standard can be accurately determined. Tetramethylsilane (TMS) has been chosen as that standard. A small amount of TMS is usually added to the sample as an internal standard, and the chemical shift values are recorded as values relative to TMS.

The units most often used to express chemical shift values are parts per million (δ). The hydrogens of TMS are arbitarily assigned the value of 0.00 δ, and the chemical shifts of all other hydrogens are expressed as parts per million downfield (to the left of the TMS absorption) or upfield (to the right of the TMS absorption) from the TMS signal. The chemical shift is sometimes given in two other units. One is tau, τ, which is equal to $10 - \delta$. The other is cycles per second (cps, or Hz). For a 60-MHz instrument, the chemical shift in cps is given by $\delta \times 60$.

The chemical shifts of the various hydrogens of an organic compound depend upon their electronic environment. If a hydrogen nucleus is shielded from the externally applied magnetic field by electron density around it, its signal appears at high field (the right portion of the spectrum). However, if adjacent atoms withdraw electron density from the hydrogen nucleus, it is deshielded and its absorption occurs at lower field (the left side of the spectrum). For example, the methylene hydrogens of benzyl acetate absorb at lower field (4.99 δ) than the methyl hydrogens (1.96 δ) because both the electronegative oxygen and the benzene ring withdraw electrons from the methylene group (Fig. 8).

A number of common chemical shifts are listed in Table 2. As the alkyl chloride, bromide, and iodide data indicate, the chemical shift values of hydrogens bonded to carbons bearing electronegative substituents increase with increasing electronegativity of the electron-withdrawing group. In addition, the deshielding effect of a substituent decreases as the number of intervening bonds to a given hydrogen increases. For example, hydrogens of R—O—C$\underline{H}_3$ absorb at about 3.3 δ, but the methyl hydrogens of R—O—CH$_2$—C$\underline{H}_3$ absorb at about 1.3 δ.

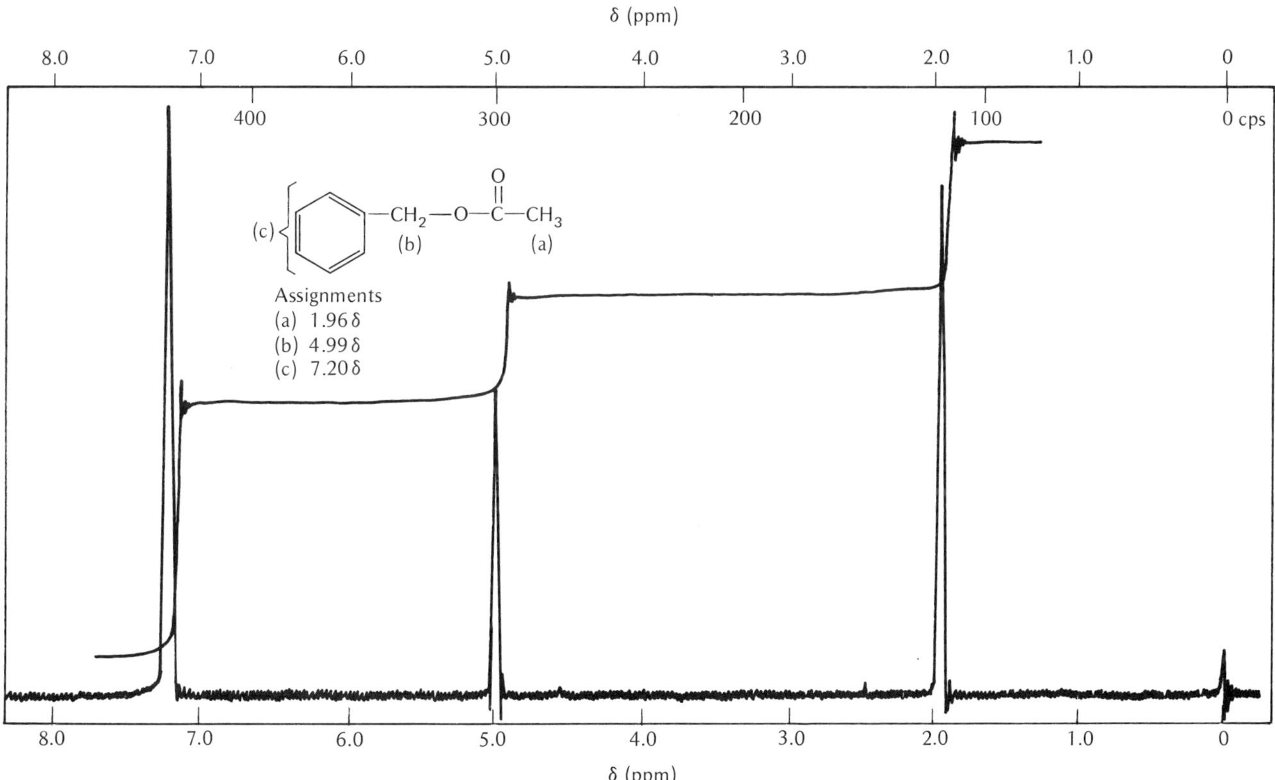

Fig. 8. NMR Spectrum of Benzyl Acetate. (© Sadtler Research Laboratories, Inc., 1971.)

[§150]

Table 2. Typical Hydrogen Chemical Shifts

Compound	*Chemical Shift* (δ)
$RC\underline{H}_3$	0.9
$R_2C\underline{H}_2$	1.3
$R_3C\underline{H}$	1.5
$RC\underline{H}_2Cl$	3.5–4.0
$RC\underline{H}_2Br$	3.0–3.7
$RC\underline{H}_2I$	2.0–3.5
$R{-}O{-}C\underline{H}_3$	3.2–3.5
$R{-}O{-}CH_2C\underline{H}_3$	1.2–1.4
$R{-}O{-}(CH_2)_2C\underline{H}_3$	0.9–1.1
$C{=}C{-}\underline{H}$	5.0–5.3
$C{\equiv}C{-}\underline{H}$	2.5
$Ar{-}\underline{H}$	6.5–8.0
$C{=}C{-}C\underline{H}_3$	1.7
$C{\equiv}C{-}C\underline{H}_3$	1.8
$Ar{-}C\underline{H}_3$	2.3
$R{-}\overset{\overset{O}{\|}}{C}{-}C\underline{H}_3$	2.2
$R{-}\overset{\overset{O}{\|}}{C}{-}O{-}C\underline{H}_3$	3.6
$R{-}O{-}\underline{H}$	3.0–6.0
$Ar{-}O{-}\underline{H}$	6.0–8.0
$R{-}\overset{\overset{O}{\|}}{C}{-}\underline{H}$	9.0–10.0
$R{-}\overset{\overset{O}{\|}}{C}{-}O{-}\underline{H}$	10.5–11.5
$R{-}N\underline{H}_2$	1.0–4.0
$Ar{-}N\underline{H}_2$	3.0–4.5
$R_2N{-}C\underline{H}_3$	2.2

PEAK AREAS. The area under a peak in an nmr spectrum is directly proportional to the number of hydrogens causing the absorption. Modern nmr spectrometers electronically integrate the area under each peak and record an integration line on the spectrum. The number of hydrogens absorbing at each peak is directly proportional to the change in height of the integration line as it passes over the peak. For example, in the spectrum of benzyl acetate (Fig. 8), the ratio of the hydrogens is 5:2:3 (aromatic hydrogens/methylene hydrogens/methyl hydrogens).

SPIN-SPIN COUPLING. Another aspect of the nmr spectrum which is valuable in structure determination is spin-spin coupling. To illustrate this feature of the spectrum, consider the case of ethyl bromide (Fig. 9). Instead of two peaks with an area ratio of 2:3, the spectrum consists of two groups of peaks, a quartet and a triplet, in a 2:3 ratio. The spin-spin splitting phenomena result because there is magnetic interaction, called coupling, between hydrogen nuclei which are within a few bonds of each other. Generally, only nonequivalent hydrogens on the same carbon or on immediately adjacent carbons will cause spin-spin splitting with a

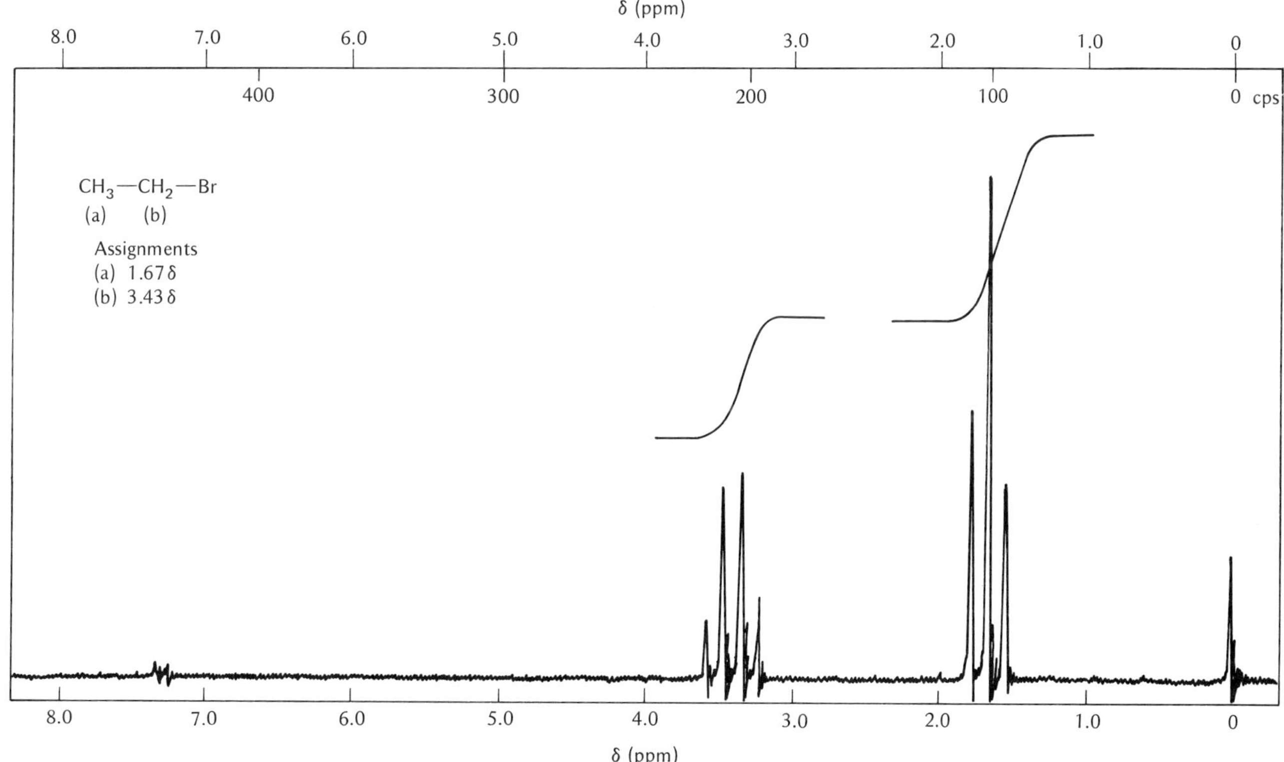

Fig. 9. NMR Spectrum of Ethyl Bromide. (Courtesy Varian Instrument Division.)

[§150] given hydrogen. In ethyl bromide, the methyl hydrogens are coupled with the set of methylene hydrogens, which are in turn split by the methyl hydrogens. In general, if there are n equivalent hydrogens coupling with a particular hydrogen (or set of equivalent hydrogens), the absorption of that hydrogen will be split into $n + 1$ lines. For example, the methylene hydrogens of ethyl bromide are coupled with the set of three equivalent methyl hydrogens ($n = 3$), resulting in a quartet. The methyl hydrogens are split by two equivalent methylene hydrogens ($n = 2$), resulting in a triplet absorption pattern.

INTERPRETATION OF NMR SPECTRA. Analysis of the nmr spectrum of an organic compound can yield a great deal of information concerning the type of functional groups to which the absorbing nuclei are attached. Analysis of the integration curve provides data about the relative numbers of each type of hydrogen in the compound. Interpretation of the spin-spin splitting patterns yields valuable information concerning the number of hydrogens adjacent to the absorption hydrogen. By analyzing each of the above features of the spectrum, it is often possible to draw one

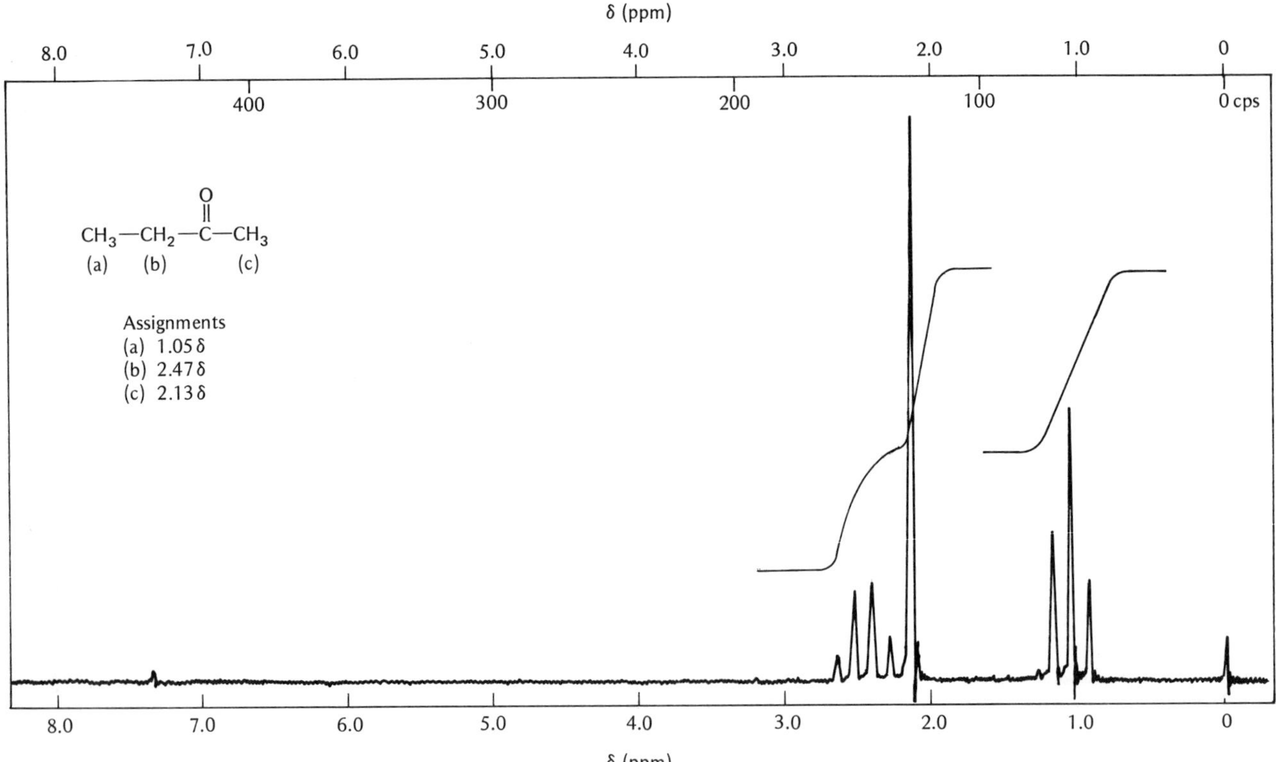

Fig. 10. NMR Spectrum of 2-Butanone. (Courtesy Varian Instrument Division.)

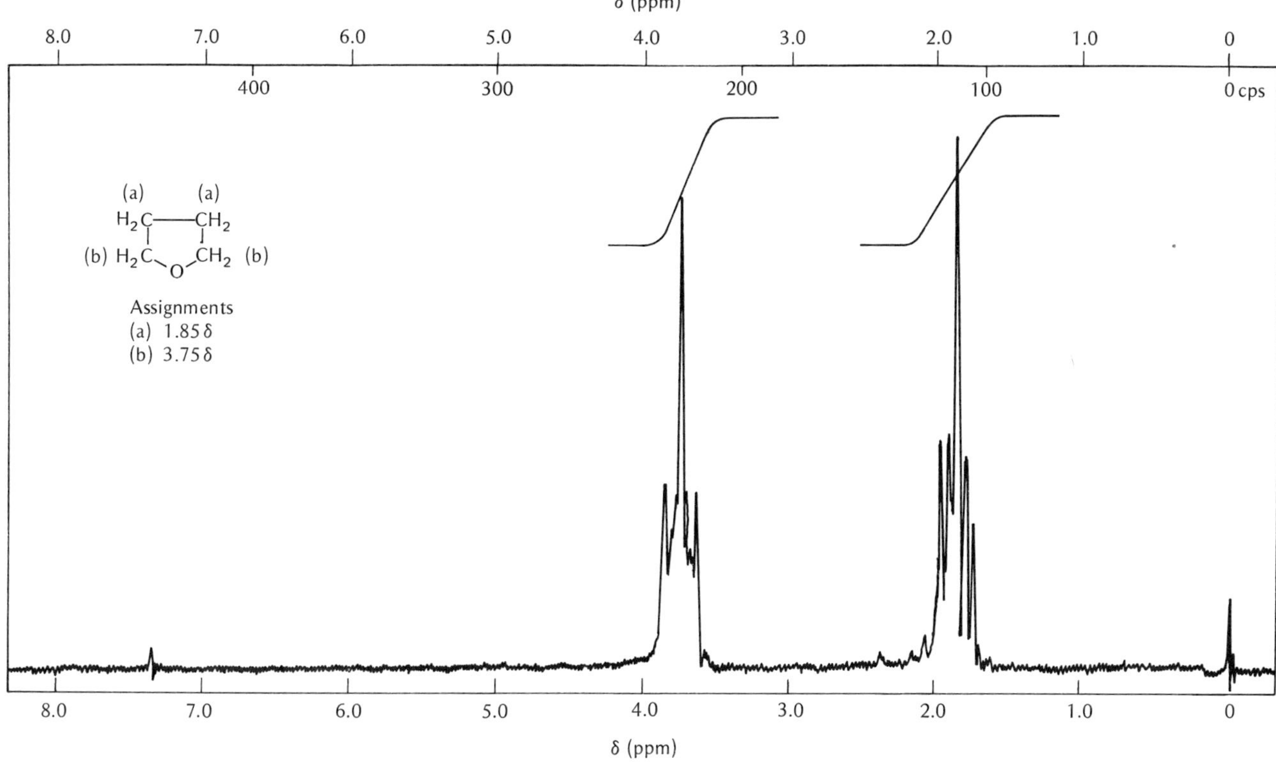

Fig. 11. NMR Spectrum of Tetrahydrofuran. (Courtesy Varian Instrument Division.)

or more structures that fit all the data. The exact identity of the compound may then be determined by using all of the available information, including physical constants, chemical behavior, and other spectral data.

The spectra in Figs. 10–12 demonstrate how nmr spectroscopy may be used to distinguish among three isomers of the molecular formula C_4H_8O.

151. General Books Covering Several Spectroscopic Methods

1. *Applications of Absorption Spectroscopy of Organic Compounds,* J. R. Dyer, Prentice-Hall, Englewood Cliffs, N.J., 1965.
2. *Spectroscopic Methods in Organic Chemistry,* I. Fleming and D. H. Williams, McGraw-Hill, New York, 1966; *Spectroscopic Problems in Organic Chemistry,* McGraw-Hill, New York, 1967.
3. *Organic Spectroscopy,* P. Laszlo and P. Stang, Harper & Row, New York, 1971.
4. *Spectrometric Identification of Organic Compounds* (2nd ed.), R. M. Silverstein and G. C. Bassler, John Wiley & Sons, New York, 1967.

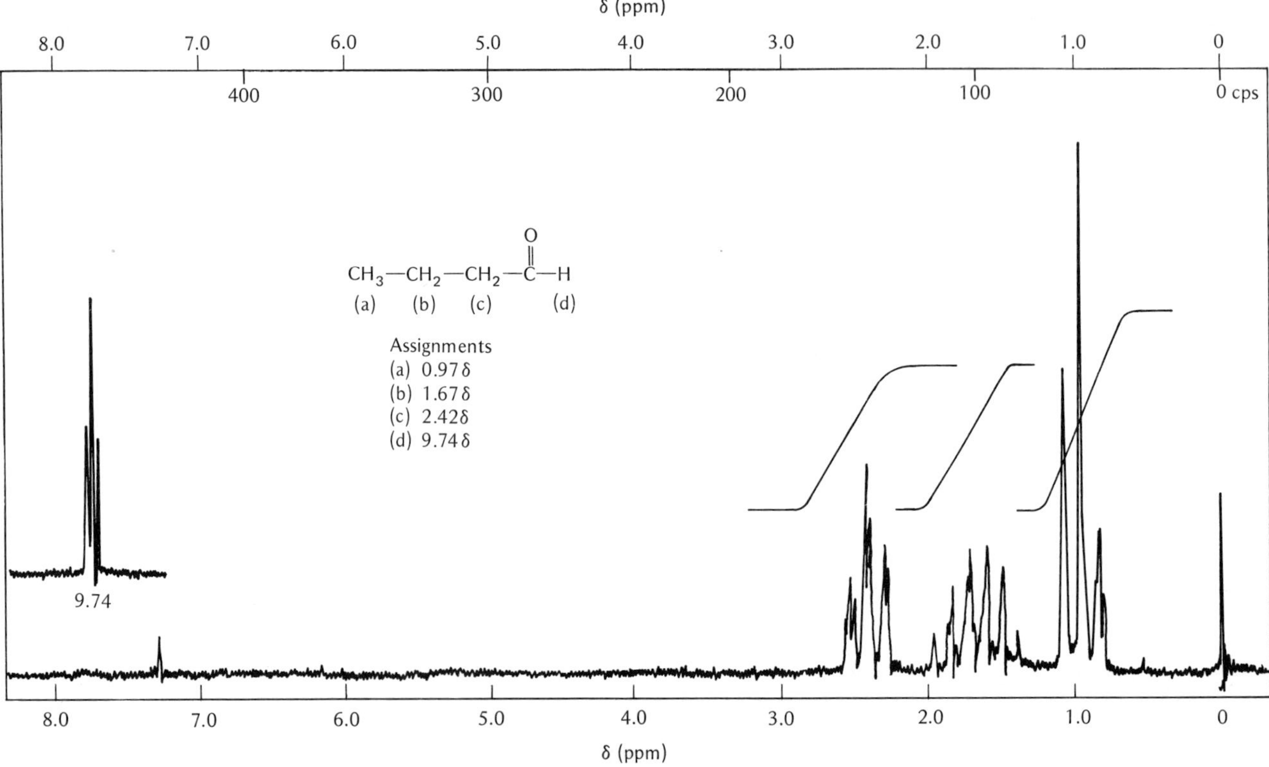

Fig. 12. NMR Spectrum of Butanal. (Courtesy Varian Instrument Division.)

5. *Organic Structure Determination,* D. J. Pasto and C. R. Johnson, Prentice-Hall, Englewood Cliffs, N.J., 1969, pp. 83–294.
6. *Technique of Organic Chemistry,* A. Weissberger, ed.: **9** (2nd ed.), Pt. 1, *Chemical Applications of Spectroscopy,* W. West, ed., 1968; **11,** *Elucidation of Structures by Physical and Chemical Methods,* Pt. 1, K. W. Bentley, ed., 1963. See Chapter 14.
7. *Physical Methods of Chemistry,* A. Weissberger and B. W. Rossiter, eds., (Weissberger, *Techniques of Chemistry,* **1**). See Chapter 14.

152. Books Covering a Single Spectroscopic Method

Infrared Spectroscopy

1. *The Infrared Spectra of Complex Molecules,* L. J. Bellamy, John Wiley & Sons, New York, 1958.
2. *Infrared Spectroscopy* (2nd ed.), R. T. Conley, Allyn and Bacon, Boston, 1972.
3. *Laboratory Methods in Infrared Spectroscopy,* R. G. J. Miller, ed., Heyden and Son, London, 1965.

4. *Infrared Absorption Spectroscopy,* K. Nakanishi, Holden-Day, San Francisco, 1962.
5. *Infrared Band Handbook* (2nd ed.), H. A. Szymanski and R. E. Erickson, Plenum, New York, 1970.
6. *The Aldrich Library of Infrared Spectra,* C. J. Pouchert, Aldrich Chemical Co., Milwaukee, 1970.

ULTRAVIOLET SPECTROSCOPY

1. *An Introduction to Electronic Absorption Spectroscopy in Organic Chemistry,* A. E. Gillam and E. S. Stern, Edward Arnold Publishers, London, 1957.
2. *Ultraviolet and Visible Spectroscopy* (2nd ed.), C. N. R. Rao, Butterworths, London, 1967.
3. *Theory and Applications of Ultraviolet Spectroscopy,* H. H. Jaffé and M. Orchin, John Wiley & Sons, New York, 1962.

NUCLEAR MAGNETIC RESONANCE

1. *High Resolution Nuclear Magnetic Resonance,* J. A. Pople, W. G. Schneider, and H. J. Bernstein, McGraw-Hill, New York, 1959.
2. *Nuclear Magnetic Resonance for Organic Chemists,* D. W. Mathieson, Academic Press, New York, 1967.
3. *Interpretation of NMR Spectra; An Empirical Approach,* R. H. Bible, Plenum Press, New York, 1965.
4. *Nuclear Magnetic Resonance,* W. W. Paudler, Allyn and Bacon, Boston, 1971.
5. *High Resolution Nuclear Magnetic Resonance Spectroscopy,* J. W. Emsley, J. Feeney and L. H. Sutcliffe, Pergamon Press, Oxford, Vol. 1, 1965, Vol. 2, 1966.
6. *Nuclear Magnetic Resonance Spectroscopy,* F. Bovey, Academic Press, New York, 1969.
7. *High Resolution NMR; Theory and Chemical Applications,* E. D. Becker, Academic Press, New York, 1969.
8. *Applications of Nuclear Magnetic Resonance Spectroscopy in Organic Chemistry* (2nd ed.), L. M. Jackman and S. Sternhell, Pergamon Press, Elmsford, N.Y., 1969.
9. *An Introduction to the Analysis of Spin-Spin Splitting in High Resolution Nuclear Magnetic Resonance Spectra,* J. D. Roberts, Benjamin, New York, 1961.
10. *Progress in Nuclear Magnetic Resonance Spectroscopy,* J. W. Emsley, J. Feeney, and L. H. Sutcliffe, eds., Vols. 1–9, Pergamon Press, Elmsford, N.Y., Vol 1, 1966.
11. *Annual Reports on NMR Spectroscopy,* E. F. Mooney, ed., Academic Press, New York, Vol. 1, 1968 (first two volumes entitled *Annual Review of NMR Spectroscopy*).

12. *High Resolution NMR Spectra Catalog,* Varian Associates, Palo Alto, Cal., Vol. 1, 1962, Vol. 2, 1963.
13. *Nuclear Magnetic Resonance Spectra and Chemical Structures,* W. Brügel, Academic Press, New York, 1967.

MASS SPECTROSCOPY

1. *Introductory Mass Spectrometry,* S. R. Shrader, Allyn and Bacon, Boston, 1971.
2. *Mass Spectrometry of Organic Compounds,* H. Budzikiewicz, C. Djerassi, and D. H. Williams, Holden-Day, San Francisco, 1967.
3. *Interpretation of Mass Spectra,* F. W. McLafferty, Benjamin, New York, 1966.
4. *Introduction to Mass Spectrometry,* H. C. Hill, Sadtler Research Laboratories, Philadelphia, 1966.
5. *The Mass Spectra of Organic Molecules,* J. H. Beynon, R. A. Saunders, and A. E. Williams, American Elsevier, New York, 1968.
6. *Ion-Molecule Reactions,* E. W. McDaniel et al., Wiley–Interscience, New York, 1970.

Chapter 14

The Literature of Organic Chemistry

153. Abstract Journals. The information in reports of original investigations in organic chemistry is readily available because it has been so beautifully classified, indexed, and abstracted. At present the leading abstract journal is *Chemical Abstracts* (started in 1907), which is published twice a month and prints abstracts of all the articles on every phase of chemistry that appear in over 10,000 scholarly and technical journals. Starting with 1962 there have been two volumes a year. Subject, author, formula, patent, and several specialized indexes appear promptly for each volume (the subject index, hardest to prepare, is received about one year after the last issue of each volume). Decennial indexes have been prepared for the first fifty years of *Chemical Abstracts* and quinquennial indexes since then. The *Seventh Collective Index* covering 1962–66 is now complete and publication of the eighth is underway.

The German chemical abstract journal, *Chemisches Zentralblatt,* was started much earlier than *Chemical Abstracts* (1830) and is important for its coverage of the early period of organic chemistry. It was superior to *Chemical Abstracts* before 1935; earlier *Chemical Abstracts* formula indexes are somewhat unreliable. Political and economic troubles made *Chemisches Zentralblatt* less complete from about 1935 to 1950. It ceased publication after the last 1969 issue, and its staff undertook abstracting of chemical publications in German for *Chemical Abstracts.*

A comprehensive Russian chemical abstract journal, *Referativnyi Zhurnal Khimiya,* is now published. There is little need to consult it because Russian papers are carefully abstracted in *Chemical Abstracts.* On the other hand, it is becoming increasingly important to have available Russian publications that report original work. A number of these now appear in English translation.

154. The Literature Search. Any comprehensive search of the original literature for all the information on a given subject is a difficult job requiring patience and ingenuity. A thorough understanding of the indexing system used in the subject index of the abstract journal is necessary, and every advantage must be taken of other secondary sources and of references to earlier work that appear in more recent original papers. *Chemical Abstracts* publishes with each annual index a discussion of the system used. Every serious student of organic chemistry must be familiar with this system.

It is far harder to cover information that has appeared more recently than the last subject index. Some chemists regularly scan each current issue of *Chemical Abstracts* for material on topics vital to their work. This imposing task is made somewhat easier by the division of *Chemical Abstracts* into eighty sections dealing with various divisions of chemistry. Most of the papers of interest to organic chemists appear in sections 21 through 34; an individual subscription to these *Organic Chemistry Sections* cost thirty-five dollars a year in 1973 to members of the American Chemical Society (an annual subscription to the complete *Chemical Abstracts* with indexes cost $2400 in 1974). Nevertheless, the job is very time consuming and still leaves a time lag between appearance of the original paper and appearance of the abstract. Other chemists make use of other auxiliary publications such as *Chemical Titles, Current Contents,* and *Current Abstracts of Chemistry and Index Chemicus.* Many others regularly scan a group of important journals as they appear, reading papers of special interest to them. Often an important new paper will contain references to all the recent original work so that a subject outside one's regular interests can be brought up to date with references that have appeared since the most recent subject index.

An additional aid in literature search is *Science Citation Index,* an international interdisciplinary index to the literature of science, medicine, agriculture, technology, and the behavioral and social sciences, published by Institute for Scientific Information, Inc. It was started in 1961 and is becoming more comprehensive and useful each year. The first volume described this type of index as follows: "A citation index is a directory of cited references where each reference is accompanied by a list of source documents which cite it. The most characteristic feature of the citation index is that the user begins a search with a specific known paper (target reference). From this starting point one is brought forward in time to subsequent papers related to the earlier paper." Indexes appear quarterly and these are combined into an annual index. The 1972 index appeared in ten large volumes covering 2425 source journals. A five-year cumulative index covering 1965–69 is also available. Although coverage of chemistry is limited to only the important journals (around 200–300), the different approach to literature coverage and the interdisciplinary nature of *Science Citation Index* make it very useful. The *Guide* published as part of the annual source index should be consulted for more information.

An important recent trend in abstracting and indexing the chemical literature involves arrangements for machine searching. *Chemical Abstracts* can now be obtained on magnetic tapes. A computer question on a given compound, class of compounds, or chemical subject can be formulated in detailed terms so that a printout of all references on the topic can be obtained in so far as it is covered in *Chemical Abstracts*. Attempts are being made by the American Chemical Society to realize a truly international service by cooperative agreements with foreign chemical societies, and the problem of coverage of all branches of science, with a minimum of duplication, by cooperation between Chemical Abstracts Service, Bioscience Information Service, and Engineering Index is also being explored. Machine searching of the literature is clearly the trend of the future; it is required by the greatly increased volume of science publication and is made possible by developments in computer science. The exact form of search is in a state of rapid change.

155. Beilstein's *Handbuch der organischen Chemie*. A search for all information on a given organic compound is a less taxing problem than a search for a general subject, but it may be very time-consuming if the compound is a common one. The most useful and convenient source of references to the older literature is Beilstein's *Handbuch der organischen Chemie,* written in German. The fourth (current) edition of this great encyclopedia consists of five sets of volumes. The main series covers the literature through 1909, the first supplement (erstes Ergänzungswerk) from 1910 through 1919, and the second supplement from 1920 through 1929. A third supplement covering the literature from 1930 through 1949 was started in 1958 but was still no more than half completed in 1973; a fourth supplement to cover 1950 through 1959 was started in 1972. Each supplement contains the same number of volumes as the main series, and the arrangement of material is parallel. Each page of the supplement receives a double number: a conventional page number and, in the center of the top of the page, the number of the page in the main series on which information on the same compounds is found. As soon as a compound is located in one series, it can be quickly found in the others.

Three procedures are available to locate a given compound in "Beilstein." One usually uses the cumulative formula index that lists compounds in order of increasing complexity of their molecular formulas. One must understand German nomenclature to select the desired compound from among its isomers. A second cumulative index lists compounds alphabetically by name. It is much less convenient unless one has an excellent understanding of German nomenclature. The third method depends upon a clear understanding of the logical arrangement of the compounds in the *Handbuch*. One then merely looks where the compound should be.

"Beilstein" is divided into the following major divisions:

Acyclic Compounds	Volumes I–IV
Isocyclic Compounds	Volumes V–XVI
Heterocyclic Compounds	Volumes XVII–XXVII
Natural Products Not Found Earlier[1]	Volumes XXX–XXXI[1]

Within each of the first two major divisions, compounds are listed by basic classes:

1. Hydrocarbons
2. Hydroxy (oxy) compounds
3. Carbonyl (oxo) compounds
4. Carboxylic acids
5. Sulfinic acids
6. Sulfonic acids
7. Selenium compounds
8. Amines
9. Hydroxylamines
10. Hydrazines
11. Azo compounds
12. Other groups (often rare)

Polyfunctional molecules are found under the class that appears last in the list. For example, hydroxyketones are under "Carbonyl Compounds," but aminoketones are under "Amines." Within each class, compounds are listed in order of decreasing saturation, and after this, by increasing molecular weight. Compounds that contain functional groups not listed as basic classes are considered as derivatives and are listed immediately after the parent compounds. Thus ethyl butyl ether is a derivative of butyl alcohol and nitrobenzene, a derivative of benzene. Three classes of derivatives are distinguished and appear in the following order:

1. Functional derivatives or compounds that can be hydrolyzed to the parent compound (the order of these is based on the principle of latest position as for the main classes);
2. Substitution derivatives (compounds containing halogen, nitroso, nitro, and azido [N_3] groups, in that order);
3. Sulfur compounds that are listed as derived from the corresponding oxygen compounds.

Heterocyclic compounds include all compounds with one or more atoms other than carbon in a ring. Even lactones and cyclic anhydrides

[1]Through 1973 only rubber and related polymers, carotenoids, and carbohydrates were covered, and these only through the first supplement. Such natural products as steroids, alkaloids, and many others have not yet been covered. Volumes XXVIII and XXIX contain the indexes.

are found here. Heterocycles appear in order of a *hetero number,* as indicated in the "Beilstein" guide or in the index, and within each parent heterocyclic system in the same order as described for the first two major divisions.

A more detailed account of this system is found in *A Brief Introduction to the Use of Beilstein's Handbuch der organischen Chemie* by E. H. Huntress, John Wiley & Sons, New York, 1938, and a still more complete exposition was published in Vol. I of "Beilstein" (in German). For one familiar with the system it is easier to locate a compound by the third method than by the first.

It is important that a compound is not listed in "Beilstein" until its composition has been confirmed by analysis. For this reason the first reference to a given compound in the *Handbuch* may not be the first reference on the compound.

After the literature search for a given compound has been carried as far as is possible in "Beilstein," it is continued in the abstract journals. These secondary sources serve mainly as a means of locating the original references. No serious chemist is satisfied until he has consulted original papers, and any report he makes on a compound should be based on reading of the original references as far as this is possible.

Chapter 15 outlines an introductory problem in literature searching and report writing on the synthesis of specific organic compounds.

156. Organic Synthesis. The problem of choosing the best method for synthesis of a given compound or of devising a satisfactory method of synthesis for a new compound is facilitated by the existence of a number of special publications. The most important of these are listed below with brief descriptions; the special abbreviations that will be used in this manual are also given (e.g., OS for *Organic Syntheses*).

1. **Organic Syntheses** (OS), John Wiley & Sons, New York. America's most famous publication in organic chemistry. This series of annual volumes originated under the inspiration of Roger Adams and has been extended by numerous associate editors and contributors. It is virtually a huge laboratory manual of organic preparations, often on rather large scale, from well-tested directions; annual volumes since 1922. Collective volumes (OS–CV), covering ten-year periods, with corrections, are available. The latter compilations should be consulted, where possible, in preference to the original single volumes of *Organic Syntheses,* since they record changes in laboratory directions based on several years of trial by various readers of the original volumes. If the preparation of a particular compound is not included, a similar compound may be found; the "type of reaction" and "type of compound" indexes in the collective volumes are useful for this purpose.

2. **Methoden der organischen Chemie,** revised fourth edition edited by E. Müller, Georg Thieme Verlag, Stuttgart. The third edition (1922–25) is universally known as Houben–Weyl, and this new fourth edition will be designated here as Müller–Houben–Weyl. This work is planned for sixteen volumes. Volume I, Parts 1 and 2 (about 2000 pages), covers general laboratory practice and appeared in 1958 and 1959. Volume II, "Analytical Methods," and Volume III, "Physical methods" (two parts), have also appeared. Only three parts of the eight-part Volume IV on general chemical methods had been published by 1973. However, many of the remaining volumes on special chemical methods, organized by classes of compounds, were available by this date. Volume XVI will be the collective index. This great comprehensive series is invaluable as a guide to methods of synthesis of organic compounds and as a source of relevant references for the details of methods to accomplish specific kinds of transformations.

3. **Organic Reactions** (OR), John Wiley & Sons, New York. Also initiated by Roger Adams. A series of treatises, each on a single type of synthetic reaction, by a specialist in the field. In each treatise a few examples of actual preparative directions are given for specific compounds. Volumes appear about every 18 months; Vol. 20 was published in 1973.

4. **Synthetic Methods of Organic Chemistry,** W. Theilheimer, ed., S. Karger, New York. An annual series that covers the current literature on organic synthesis. Typical preparations are given briefly. Careful classification, based on the type of transformation, makes this series easy to use. Volume 26 appeared in 1972.

5. **Reagents for Organic Synthesis** (ROS), by L. F. Fieser and M. Fieser, John Wiley & Sons, New York, Vol. 1, 1967, Vol. 2, 1969, Vol. 3, 1972. These volumes are of great value to organic chemists because they survey a large number of reagents used in organic chemistry, detail their usefulness, set out their limitations, give the commercial source if there is one, and present leading references on their synthesis and use.

6. **Advances in Organic Chemistry, Methods and Results,** E. C. Taylor, ed., Wiley–Interscience, New York. A series aimed at providing critical appraisal and evaluation of new aspects of organic chemistry ripe for further development and of novel extensions of well-established methods. Volumes 1 and 2 appeared in 1960, Volume 8 in 1972.

7. **Newer Methods of Preparative Organic Chemistry,** 6 vols., W. Foerst, ed., Academic Press, New York (German edition,

VERLAG CHEMIE, WEINHEIM, 1963–70). Reviews of important organic reactions with brief procedures. In 1948 Interscience Publishers published a translation, with some revision, of the first German edition of Vol. 1 (1943), later reproduced (1963, in German) as the first volume of the continuing series. The German edition of Vol. 6 appeared in 1970 and the English edition in 1971. The individual reviews have usually been published earlier in *Angewande Chemie* (also in the International Edition in English) but frequently without tables or procedures. No further volumes are planned in this series.

A number of shorter books on synthesis of organic compounds have also appeared. Several of these are listed below. Some of them deal with rather special topics.

1. *Art in Organic Synthesis,* N. Anand, J. S. Bindra, and S. Ranganathan, Holden-Day, San Francisco, 1970.
2. *Reduction Techniques and Applications in Organic Synthesis,* R. L. Augustine, Marcel Dekker, New York, 1968.
3. *Survey of Organic Syntheses,* C. A. Buehler and D. E. Pearson, Wiley–Interscience, New York, 1970.
4. *Systematic Organic Chemistry* (4th ed.), W. M. Cumming, I. V. Hopper, and T. S. Wheeler, Constable and Co., London, 1950.
5. *Compendium of Organic Synthetic Methods,* I. T. Harrison and S. Harrison, Wiley–Interscience, New York, 1971.
6. *Organic Synthesis,* 2 vols., V. Migrdichian, Van Nostrand Reinhold, New York, 1957.
7. *Advanced Organic Synthesis: Methods and Techniques,* R. S. Monson, Academic Press, New York, 1971.
8. *Organic Syntheses with Isotopes,* 2 vols., A. Murray III and D. L. Williams, Wiley–Interscience, New York, 1958.
9. *An Advanced Organic Laboratory Course,* M. S. Newman, Macmillan Publishing Co., Inc., New York, 1972.
10. *Principles of Organic Synthesis,* R. O. C. Norman, Methuen and Co., London, 1968.
11. *Purification of Laboratory Chemicals,* D. D. Perrin, W. L. F. Armarego, and D. R. Perrin, Pergamon Press, Elmsford, N.Y., 1966.
12. *Organic Functional Group Preparations,* S. R. Sandler and W. Karo, Academic Press, New York, Vol. 1, 1968. Vol. 2, 1971, Vol. 3, 1972.
13. *Preparation of Organic Intermediates,* D. A. Shirley, John Wiley & Sons, New York, 1951.
14. *Free Radical Reactions in Preparative Organic Chemistry,* G. Sosnovsky, Macmillan Publishing Co., Inc., New York, 1964.
15. *Organic Photochemical Syntheses,* R. Srinivasan, ed., Wiley–Interscience, New York, 1971 (marked Volume 1 but no further volumes had appeared by 1974).

16. *Deuterium Labeling in Organic Chemistry,* A. F. Thomas, Appleton-Century-Crofts, New York, 1971.
17. *Modern Reactions in Organic Synthesis,* C. J. Timmons, ed., Van Nostrand Reinhold, New York, 1970.
18. *A Textbook of Practical Organic Chemistry* (3rd ed.), A. I. Vogel, John Wiley & Sons, New York, 1956.
19. *Elementary Practical Organic Chemistry, Part 1, Small Scale Preparations* (2nd ed.), by A. I. Vogel, John Wiley & Sons, New York, 1966.
20. *Synthetic Organic Chemistry,* R. B. Wagner and H. D. Zook, John Wiley & Sons, New York, 1953.
21. *Semi-Micro Organic Preparations* (2nd ed.), J. H. Wilkinson, Oliver and Boyd, Edinburgh, 1958.

Finally, mention must be made of three serial publications devoted entirely to organic synthesis:

1. *Synthesis, International Journal of Methods in Synthetic Organic Chemistry,* monthly, Georg Thieme Verlag, Stuttgart, and Academic Press, New York (since 1969).
2. *Synthetic Communications,* Marcel Dekker, New York, Vol. 1, 1971 (successor to *Organic Preparations and Procedures,* Vol. 1, 1969).
3. *Annual Reports in Organic Synthesis,* Academic Press, New York, Vol. 1, 1970.

157. Sources of Chemicals. To many students in organic chemistry courses an "available" chemical is one that can be signed out from the storeroom or perhaps one that is listed in the catalog of a chemical supply firm. A practicing chemist needs a wider knowledge of the sources of chemicals. Many chemical supply houses synthesize in their own laboratories relatively few of the chemicals listed in their catalogs. Thousands of chemicals have commercial uses which involve their production on a large scale by a company synthesizing chemicals for a particular industry (e.g., agriculture, cosmetics, dyes, etc.) or by a company based on a given type of raw material (petroleum, coal tar, naval stores). These companies seldom find it profitable to package their products for sale in small amounts. Chemical supply firms do a great service to college laboratories and research organizations by undertaking a rebottling operation to make such chemicals available in convenient amounts. Such an operation is usually much less expensive than synthesis on a small or medium scale. Most supply houses also have facilities for synthesizing compounds on a moderate scale. Information is not publicly available on the relative proportion of chemicals merely rebottled compared with those made by the firm. There also exist a number of companies in the business of carrying out syntheses to order. Often they also offer for sale a number of chemicals for which they have developed special synthetic know-how.

Some firms of this sort specialize in a given type of chemical (hydrocarbons, organofluorine compounds, organometalics).

It is now possible to find a source for a needed chemical without searching through an extensive collection of catalogs. *Reagents for Organic Synthesis* by Fieser and Fieser (§156) gives information on the sources of the reagents discussed. Even more complete is *Chem Sources* – USA, a volume published annually by Directory Publishing Co., Inc., Flemington, New Jersey. The 1973 edition gives information on 664 companies that supply chemicals and lists alphabetically over 54,000 organic and inorganic chemicals. The cost ($48.00) does not seem exorbitant for the service rendered.

158. Organic Laboratory Technique. The most important modern treatise in English on laboratory practice in organic chemistry is *Technique of Organic Chemistry*, Arnold Weissberger, ed., Wiley–Interscience, New York. It appeared in fourteen volumes starting in 1948 and several volumes have gone through two or three editions. In 1971 this series was combined with *Technique of Inorganic Chemistry* and the title changed to *Techniques of Chemistry*. Titles of the fourteen volumes of *Technique of Organic Chemistry* with the most recent edition indicated are:

I. *Physical Methods of Organic Chemistry* (3rd ed.), 4 parts
II. *Catalytic, Photochemical, and Electrolytic Reactions* (2nd ed.)
III. Part I. *Separation and Purification;* Part II. *Laboratory Engineering* (2nd ed.)
IV. *Distillation* (2nd ed.)
V. *Adsorption and Chromatography*
VI. *Micro and Semimicro Methods*
VII. *Organic Solvents* (2nd ed.)
VIII. *Investigation of Rates and Mechanisms of Reactions* (2nd ed.), 2 parts
IX. *Chemical Applications of Spectroscopy* (2nd ed.), 2 parts
X. *Fundamentals of Chromatography*
XI. *Elucidation of Structures by Physical and Chemical Methods*, 2 parts
XII. *Thin-Layer Chromatography*
XIII. *Gas Chromatography*
XIV. *Energy Transfer and Organic Photochemistry*

In the series *Technique of Chemistry* the volumes through 1973 are:

I. *Physical Methods of Chemistry*, 5 parts, incorporating 4th completely revised and augmented edition of *Physical Methods of Organic Chemistry*
II. *Organic Solvents* (3rd ed.)
III. *Photochromism*

IV. *Elucidation of Organic Structures by Physical and Chemical Methods* (2nd ed.), 3 parts

This series has become the standard reference work for organic chemists seeking a broad knowledge of organic laboratory techniques or searching for special methods to overcome unfamiliar difficulties.

A second important source of information on laboratory practice is the Müller–Houben–Weyl *Methoden der organischen Chemie* already discussed (§156). This series is organized quite differently from the Weissberger series, it is based on German apparatus and practice, and the fact that it is in German makes many chemists pass it by; nevertheless, it is an important reference work that deserves more extensive use in the United States.

Several shorter volumes on organic laboratory practice are available. Three of the best are *Research Techniques in Organic Chemistry* by R. B. Bates and J. P. Schaefer, Prentice-Hall, Englewood Cliffs, N.J., 1971; *Organic Structure Determination* by D. J. Pasto and C. R. Johnson, Prentice-Hall, Englewood Cliffs, N.J., 1969; and *Laboratory Technique in Organic Chemistry* by K. B. Wiberg, McGraw-Hill, New York, 1960. The textbooks by Vogel (Items 18 and 19, §156) are also useful.

References in this manual to these books on technique will simply give the author's or editor's name; for example, Wiberg, 179 indicates page 179 in *Laboratory Technique in Organic Chemistry* by K. B. Wiberg. Because both *Technique of Organic Chemistry* and *Techniques of Chemistry* are edited by Weissberger, the titles are also indicated for these series; Weissberger, *Technique of Organic Chemistry,* **6,** 84, indicates page 84 in Vol. 6 of *Technique of Organic Chemistry,* A. Weissberger, ed. Items 18 and 19, §156, both by A. I. Vogel, are designated Vogel (1956) and Vogel (1966), respectively.

Chapter 15

Writing a Research Report

The results of scientific research are of little value unless they are known to others who can make use of them. Written reports are more important than oral communications. These take the form of papers published in scientific journals, of patents, or of research reports to those who may be directing the work.

Primary considerations in writing such reports are clarity and accuracy. The style should be as concise as these considerations permit. Scientific publication of all kinds is now so voluminous that a scientist must write well if his work is to receive attention. The present chapter suggests several types of projects that can be undertaken by students in an organic chemistry course for practice in report writing.

159. Report on Methods of Synthesis of an Organic Compound. One project involves a report of methods used to synthesize a given organic compound. A detailed search is made in the library for all of the papers that contain accounts of synthesis of the compound. A report is then written that should include

1. A brief description of the first synthesis of the compound including method, yield, date, and reference.
2. A summary of all different methods used to synthesize the compound (with references).
3. Critical appraisal of these methods including availability and cost of starting materials and reagents, relative duration and ease of laboratory operations, and yields.
4. Precise directions for the synthesis judged best.
5. The physical constants of the compound.

If the compound has specific uses, shows unusual reactions, is of interest for other investigators, or is of theoretical interest, these features should be mentioned (perhaps in an introductory or a concluding paragraph) and references to the sources of such information should be included.

References cited in the report should be those that give experimental details for the synthesis or provide information judged useful to anyone attempting the synthesis in the laboratory. However, if very few papers even mention the compound, all should be cited. Several styles exist for citing references, and the instructor usually prefers to specify the style to be used (see §161).

The compound must be carefully chosen if the project is to be stimulating and rewarding. It should be a compound for which a moderate number, but not an excessive number, of references will be found by the student in the course of the literature search. At least the key papers should be in English unless the student undertaking the project has a reasonable knowledge of the foreign language most heavily involved or help is available (from the instructor or a friend) to translate the more essential parts of such papers. Abstracts in *Chemical Abstracts* may be used for less important papers if the original is unavailable or is in a language that the student does not understand. If only an abstract was consulted, this should be indicated in the reference. More interest will be generated if the compound is one that has been synthesized by several methods so that these can be evaluated on the basis of the material in the literature. It is especially interesting if certain standard synthetic methods have not been applied to the compound so that the student can consider whether or not a better synthesis might be accomplished by a method that has not been reported previously. Interest is also aroused if the literature suggests that the compound was prepared for some practical end (e.g., as a potential drug that might have a certain type of physiological action) or was of interest because its structure had special features or it could be used to give information about the mechanism of a reaction.

A laboratory course in organic chemistry often includes an individual project that may take the form of synthesis of a compound by a method requiring several steps. Directions for a few synthetic sequences are found in laboratory manuals, but more often the student is expected to look up synthetic methods for the compound, choose the best one (with the assistance of the laboratory instructor), and carry out the practical work. Such a project should require a written report comprising a concise description of the laboratory work actually done, the results obtained, and the literature search. Scientific papers in current journals such as *The Journal of the American Chemical Society* and *The Journal of Organic Chemistry* can serve as models for the report on the practical work. Usually a more detailed report on the literature search will be required than can be included in papers in these journals.

160. Other Written Reports. Laboratory projects of several kinds may be undertaken by organic chemistry students. For example, a general synthetic method may be applied to preparation of a new compound or to one never before prepared by that method. Projects in physical organic chemistry may involve investigation of a reaction mechanism. A special laboratory technique (for example, glc) may be applied to the study of a reaction to learn more about the side products produced or to discover the effect of variation of conditions on the yield of a desired product.

A written report should be prepared on any individual project undertaken. Necessary library work may be more difficult for such a project than for synthesis of a given compound. Often the literature may have too many articles or the relevant references may be hard to find. Selection of those papers which should be cited often requires care. These individual problems are best resolved by discussion with the laboratory instructor. Preparation of a written report on such a laboratory project is often the best way to learn how to write a report.

161. Style and Organization. A written report should present as clearly and directly as possible the objectives of the work, the results, and the conclusions. Clarity in writing rests on thorough understanding of the problem undertaken and of the significance of the results. One must be able to convey this understanding to the reader in a coherent, organized, and concise fashion.

Discussion of writing techniques is beyond the scope of this book. Students are urged to read one of the books on report writing that are listed in the References; the first two are brief and were written especially for chemists. Class discussion of writing techniques is highly desirable.

Special mention must be made of the varied conventions used in citing references. The style used in books is often different from that used in articles published in scientific journals. In the latter, references are usually numbered consecutively throughout the article and designated in the text by number (superscript or in parentheses). The references are then most often placed at the bottom of the page or column in the numerical order in which each appears for the first time. A reference is not reprinted if it appears again later in the article; only its number is inserted in the appropriate place in the text. In some journals references are collected at the end of each article. Usually the style for designating authors, journal or book, volume, and so on is that shown in the examples below.

In books, references are often numbered consecutively throughout a chapter and collected at the end of each chapter (sometimes at the end of the book). It is not uncommon to write authors names with last name first followed by initials (e.g., Jacobs, T. L. and Macomber, R. S.) and to alphabetize the references by names of the first authors. The conventions used are readily apparent for any book or journal when one starts to read the particular item. Scientific reports frequently follow the style described for articles in scientific journals.

Laboratory manuals are somewhat different from other books because so few references are used. When this is true it is more convenient for the reader to have the reference placed in the text in parentheses. General references to subject matter in any chapter are commonly placed at the end of that chapter without numbering. These conventions have been followed in the present laboratory manual.

Individual instructors usually wish to specify the format and conventions to be followed in the report; editors of scientific publications and research directors in industrial organizations do the same. The following suggestions may be followed in the absence of other directions:

1. A typewritten report should be double-spaced. Margins should be adequate, and the left-hand margin should be wider if the report is bound in a folder. For dissertations the left-hand margin is commonly $1\frac{1}{2}$ in., the others $1\frac{1}{4}$ in. Pages should be numbered.
2. The title of the report, the author, the course designation, and the date should appear on the outside of the folder or on a title sheet.
3. References should be numbered consecutively and placed at the end of the report. In the body of the paper they should be designated by a number in parentheses. The following examples illustrate a common style for citing references to journal articles, patents and books:

   T. L. Jacobs and R. S. Macomber, *J. Amer. Chem. Soc.*, **91,** 4824 (1969)

   L. H. Shepherd, Jr. (Ethyl Corporation). U.S. patent 3,597,488, August 3, 1971 [*Chem. Abstr.*, **75,** 88751 (1971)].

   L. F. Fieser and M. Fieser, *Style Guide for Chemists,* Van Nostrand Reinhold, New York, 1960, pp. 65–66.

   Abbreviations of journal titles should conform to these specified in *Chemical Abstracts Service Source Index.*
4. The following abbreviations for common expressions should be used: mm, cm, g, ml, sec, min, hr, bp, mp, fp, nmr, glc. Periods should not be placed after these, but in. for inch and no. for number require periods.

References

W. G. Crouch and R. L. Zetter, *A Guide to Technical Writing* (3rd ed.), Ronald Press, New York, 1964.

D. H. Menzel, H. M. Jones, and L. G. Boyd, *Writing a Technical Paper,* McGraw-Hill, New York, 1961 (paperback).

J. H. Mitchell, *Writing for Professional and Technical Journals,* John Wiley & Sons, New York, 1968.

F. H. Rhodes, *Technical Report Writing* (2nd ed.), McGraw-Hill, New York, 1961.

R. R. Ward, *Practical Technical Writing,* Knopf, New York, 1968.

Part 2

Laboratory Experiments

Experiment 1

Preliminary Laboratory Work

Note. Of the following introductory operations, perform only those specified by the instructor, according to the arrangements of your laboratory.

162. Inspection of Locker Equipment. First make sure that the laboratoy locker contains all the apparatus specified on the official list. A copy of the list may be obtained from the storeroom if one is not found in the locker. Before making complaint about possible deficiencies in equipment, examine the entire outfit in detail and bring all troubles to the storeroom at one time. Return defective pieces and obtain any missing articles on the first day of laboratory work. If any piece of apparatus is slightly damaged, but still fully serviceable, you may be instructed to keep it; but you should write a brief memorandum of such defects on a slip of paper and have this signed by the storekeeper.

Note particularly the possibility of inconspicuous defects in those of the following pieces which may be on your list:

Water-cooled West condenser: Watch for concealed cracks in the inner tube at points where the outer tube is attached to the inner.

Thermometer: If a thermometer has been abused by careless cooling in faucet water, it may have inconspicuous cracks near the mercury well.

Gas hose: If used tubing is furnished, beware of rubber that is cracked or has weak spots due to fire damage.

Separatory or dropping funnels: See that the glass stoppers are the ones that belong to the funnels and thus fit properly.

163. Broken Glassware. Do not discard broken glassware that may be repaired to advantage by a glassblower. Consult your instructor who will advise you of arrangements for repair service or give credit in exchange

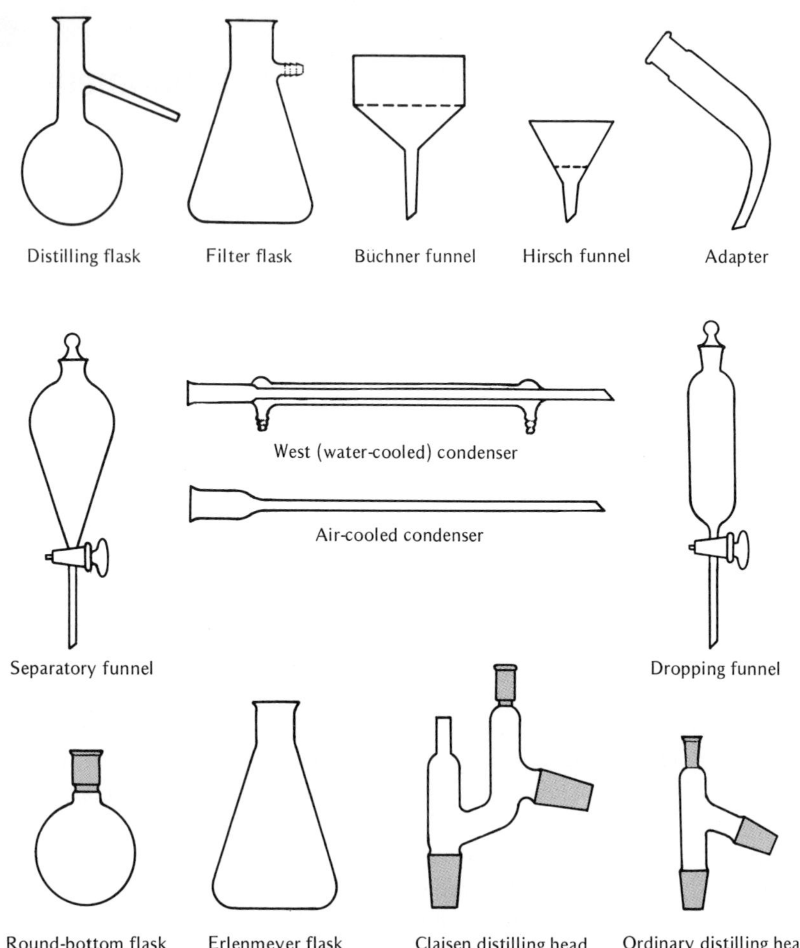

Apparatus Used in Organic Chemistry

for new equipment. For example, distilling flasks and dropping funnels with broken delivery stems are often readily mended.

164. Preparation of Apparatus for the First Experiments. Examine the directions for the first two experiments that have been assigned. If standard-taper equipment is not being used prepare the necessary glass, rubber, and cork parts for assembly as needed later. Be sure all needed equipment is on hand for these experiments.

165. Interruptions of Experiments. The limitations of fixed laboratory periods raise frequent questions by students as to possibility of interrupting an experiment. Such questions have been foreseen in several cases, particularly during the earlier assignments, by the insertion of a star in the margin at the end of the paragraph where the interruption is permis-

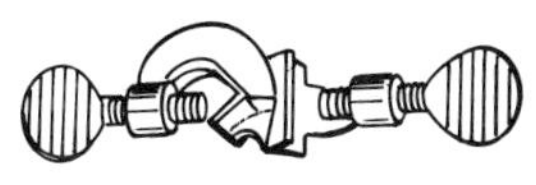

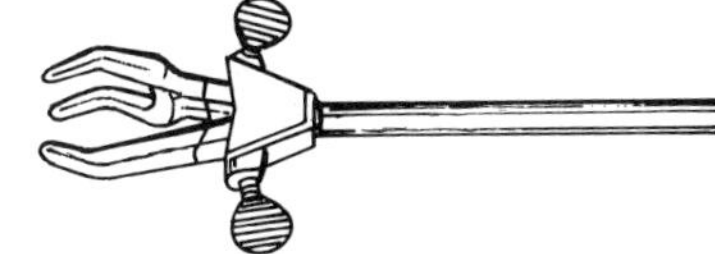

Condenser Clamp and Holder

sible. The asterisk means that it would be allowable to interrupt this experiment at this point, without necessity of resuming operations described in the next paragraph, for a week thereafter.

166. Thermometers. Ordinary chemical thermometers are made of soft glass and break if cooled suddenly from a high temperature (e.g., in a stream of cold water). Graduation marks are points of weakness. To insert the thermometer in a stopper, grasp the portion of the stem near the stopper, work slowly, and use a turning motion. Avoid prying leverage applied at a distance. For rubber stoppers lubricate the thermometer with water or glycerine.

Most thermometers are calibrated by the manufacturer for "total immersion," i.e., with the entire thermometer at the temperature being recorded. An error of a degree or more is common. It is recommended that each thermometer be calibrated at 0° and 100° (and at 218°, the boiling point of naphthalene, if the range extends that high). A calibration record should be placed with the thermometer as well as in the notebook. Succeeding students need not repeat this routine operation unless they doubt the recorded values, have to calibrate a new thermometer, or are directed to do so by the instructor.

Calibration at 0° is done in clean, finely crushed ice to which just enough distilled water has been added to produce a slushy mass without air spaces. Ice must extend to the bottom of the vessel since the density of water increases between 0° and 4° and a lower water layer free from ice will be at a temperature above 0°. Stir the slush carefully but thoroughly with the thermometer and record the correction after the entire mercury column has been immersed for 5 min or more.

Calibration at 100° is done with ordinary distilling apparatus (Fig. 1, §62, 250- or 500-ml flask). In place of the wire gauze use a piece of asbestos board in the center of which a hole about 3 in. in diameter has been cut. Place 100 ml of distilled water in the flask and distill steadily over a small Bunsen flame at the rate of about one drop of distillate per second. If the distillation is too slow, cool air surges back from the condenser; if too fast, masses of superheated steam will reach the thermometer. In a well-conducted experiment there should be a steady drip of water from recondensation on the thermometer bulb. A superheating error is especially common with substances of high heat capacity, such as water. Omission of the asbestos board permits the flame to heat the upper walls of the flask above the liquid level; this results in superheating and high

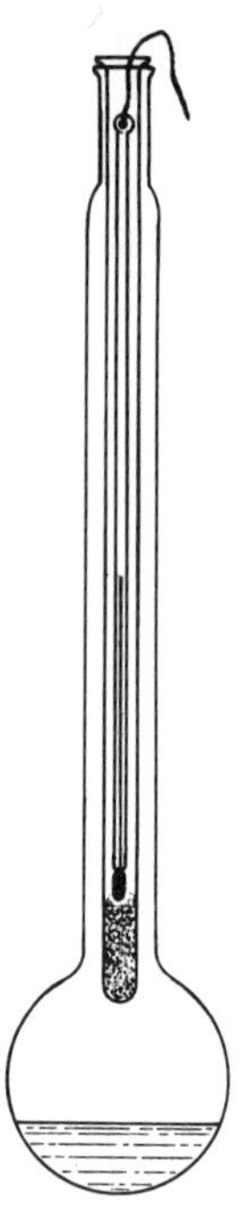

readings. A stem correction and pressure correction (if the barometer reads other than 760 torr) must be applied to the reading (see §§18, 35, and 570).

Calibration at 218° is done in the apparatus shown. A special flask of borosilicate glass with neck about 50 cm long is clamped in a vertical position. A slender test tube of thin borosilicate glass (flanged at the end so that it can be centered without other support, long enough to contain the full length of the thermometer, and provided with a small wad of glass wool as a cushion) is suspended in the flask. Naphthalene is boiled in the flask so that condensing vapors surround this tube to above the 218° calibration but do not escape at the top. In about 10 min the thermometric reading should be constant, and may be compared with the known boiling point of naphthalene (see §570 taking account of the barometric pressure). This calibration unit may constitute a permanent laboratory installation. Alternatively a simple distillation set up with air condenser (§62, Fig. 3) may be used. A stem correction must then be applied, and it may be necessary to warm the condenser to prevent obstruction with crystals.

167. Calibration by Melting Point. A thermometer may be tested by performance of the standard melting-point determination on certain stable, highly purified organic compounds known to give good results. The following compounds are especially recommended by S. C. Bunce (1953):

p-Dichlorobenzene	53.2°
Triphenylmethane	93.2°
Urea	132.7°
Succinic acid	182.8°
p-Nitrobenzoic acid	238.9°
Anthraquinone	284.8°

Experiment 2

Purification by Recrystallization

168. Optional Procedures. If time is available, the procedure of §171 for purification of an unknown crude preparation offers practice on the type of problem most frequently encountered. The shorter experiments of §169 illustrating recrystallization from water and §170 on recrystallization from an inflammable solvent offer an adequate introduction to the technique of recrystallization as it is involved in most experiments in this manual. Chapters 7 and 8 should be studied with care before any of these experiments is undertaken.

For the longer experiment *the student should hand in, some days in advance, as directed by the laboratory instructor,* a clean, dry, corked 6-in. test tube labeled with experiment number and student's name. A sample of a crude organic preparation will then be issued. This material is nominally a single substance, but may be either a commercial product of technical grade or a crude laboratory preparation.

169. Acetanilide. Place 5 g of crude acetanilide in a conical flask of 300- to 500-ml capacity. Measure out 100 ml of water, add about 30 ml of this to the crude solid, and boil gently over wire gauze. A discolored "oil" appears; this is a saturated solution of water in acetanilide that is able to exist as a separate phase in contact with the more watery phase, which itself is a saturated solution of acetanilide in water. See §88 for discussion of such a system of two liquid phases.

Continue adding small portions of the original supply of water, keeping temperature of the solution at boiling, until the last residue of oil disappears. Then add 5 ml more to ensure complete solution.

Now allow the solution to cool slightly, swirl the flask, and add 0.5 to 1 g of decolorizing carbon. The first small portion of carbon should be added cautiously to avoid excessive foaming and loss of material by overflow. Finally boil the mixture for a minute or two.

Filter the hot solution as described in §101 and conduct the processes of crystallization, isolation, washing, and final drying as described in §§102–106. The product should be placed in a dry vial, suitably labeled, and preserved for determination of melting point.

170. Recrystallization from Alcohol. Obtain a 5-g sample of an organic compound which may be recrystallized to advantage from alcohol. One of the following would be suitable: benzil, 2,4-dinitrotoluene, naphthalene, *p*-nitrobromobenzene.

Attach a water-cooled reflux condenser to a 200- or 300-ml conical flask and mount vertically as shown in Fig. 1, §19. Place the sample of crude organic compound in the flask, see that the condenser is firmly attached both to flask and hose bringing in water service, and add about 15 ml of ethyl or isopropyl alcohol through the top of the condenser. A less safe alternative procedure omits the condenser as in §169. Read §63, which describes the fire hazard of flammable solvents.

Heat the flask over a steam cone or water bath. If the compound does not completely dissolve in the boiling alcohol, continue to add solvent until the crystals disappear; then add 5 ml more of the alcohol.

Remove the conical flask from the condenser and heating bath, and proceed with the operations of filtration, cooling, isolation of newly formed crystals, washing, and final drying of the purified product, as described in §§101–106. Preserve the product for determination of melting point. The product should be placed in a dry vial, suitably labeled, and preserved for determination of melting point.

171. Unknown Crude Preparation. This experiment illustrates a situation encountered frequently in the practice of organic chemistry. When solids of unknown or incompletely known character are obtained from a reaction mixture, they are usually most easily purified by recrystallization. The general procedures for choice of suitable solvent and conduct of recrystallization are given in Chapter 8. Study this chapter carefully and follow the directions given there.

172. Selection of Solvent. It will be found that the unknown crude preparation issued can be recrystallized from one of the common solvents available in the laboratory or from a mixed solvent. See §100. The list of common solvents (§99) contains some that are more costly and may not be available. Carry out the preliminary tests for selecting a suitable solvent, as detailed in §99. Do not expect to dissolve particles of dust, cork, or other trash that may happen to accompany the preparation and which are obviously not the desired organic compound. It is not necessary to remove such insoluble impurities or to decolorize with activated carbon during these preliminary tests. If the small test sample, completely soluble when hot, sets, on cooling, to a solid cake that appears unworkable on a filter, it may become fluid enough to pour when stirred with a glass rod; if not, a more dilute solution should give better results.

Sometimes a compound forms a supersaturated solution in a given solvent, especially when colloidal material is left in suspension as may occur in these preliminary tests. If the finely divided solid is insoluble in the cold solvent even after stirring and crushing under the solvent, yet is soluble in hot solvent but fails to crystallize when cooled, do not conclude that the solvent is unsuitable until you have cooled the sample thoroughly, scratched the inside wall of the test tube with a glass rod, introduced a minute seed crystal with more scratching, and finally allowed the test tube to remain in ice for several minutes, with intermittent scratching.

173. Recrystallization Procedure. On the bases of the preliminary experiments decide the size of flask, quantity of solvent, fire precautions, etc., that are necessary to permit purification of about half of the sample issued to you. Follow the directions in §§101–106. If more than about 50 ml of solvent is necessary to dissolve this amount, take less sample so that it will be convenient to work in a 125-ml conical flask. Purity of product is of greater importance than yield in this experiment; so, more dilute solutions may be desirable. However, good technique is marked by recovery of material of high purity in reasonably good yield.

The material supplied will doubtless require decolorization with activated carbon. Read §101 and §§130–132 and proceed as directed. Sometimes decolorization must be carried out in a solvent that is unsuitable for recrystallization. The decolorized solution is then warmed to remove the first solvent, the recrystallization solvent is added, the material is brought into solution, and crystallization is allowed to occur without another filtration. If a mixed solvent is used, it is often advantageous to carry out the decolorization in the solvent component that more readily dissolves the solid and to add the second component only after filtration is complete. The second component should be added to the hot solution and crystallization allowed to occur slowly as cooling proceeds. If difficulty is encountered in the crystallization, owing to formation of an oil, read §§88–89.

When the final yield of crystalline material has separated from solution, collect it on a filter and wash it cautiously with a few cubic centimeters of cold solvent of the kind you have been using. Now set aside a small sample (5–10 mg, estimated) to dry between filter papers and eventually to serve for the first test of melting point. Immediately recrystallize the remainder from fresh solvent, and dry a small sample of the product as before. If this product melts as does the small sample from the first recrystallization, it may be assumed to be pure; if not, repeat the recrystallization until a constant melting point is obtained. Sometimes a compound will crystallize with solvent of crystallization. The solvate generally

[§173] melts at lower temperature than does the unsolvated compound and sometimes can be detected because the sample melts, resolidifies, and melts again at a higher temperature. If a solvate is suspected, the compound should be recrystallized from a different solvent and the two products compared.

The main portion of product is finally dried and stored in a dry vial neatly labeled with your name, the date, and the observed melting point. See Experiment 3.

Experiment 3

Determination of Melting Point

The theory underlying the determination of melting points is given in Chapter 7, §§81–85, and procedures are described in Chapter 8, §§107–111. Read this material carefully before carrying out this experiment.

174. Substances for Test. Obtain samples of materials as assigned, possibly including one or more of the following suggestions. If a purified compound is issued to you, as in (*a*), (*b*), or (*c*) below, a supply equal in volume to no more than a good-sized table pea seed is adequate, since this is a micro method.

(*a*) Purified benzoic acid. This compound is stable and commonly available in high purity. It is excellent material for first practice in technique of determination; mp 122.4°.

(*b*) Low-melting compounds. In very brief courses where only a single determination is made, it is economical to use water as bath liquid. Choice of material is limited to compounds melting considerably below 100°, including (1) *p*-dichlorobenzene, (2) 2,4-dinitrotoluene, and (3) naphthalene.

(*c*) Pure samples of the compounds listed in §167 (Experiment 1) may be issued and will serve to check both the thermometer calibration and the skill of the student.

(*d*) "Unknown" organic compound. The student should hand in a clean, dry, corked 6-in. test tube, with label bearing his name, in advance of the day of the experiment. A sample of a high-grade (already purified) organic compound will then be provided.

(*e*) Purified compound from Experiment 2, §§169, 170, or 173.

175. Choice of Method and Procedure. Provide the necessary equipment from the particular method designated by the instructor for this experiment. Apparatus for determination of melting points is described in §108. If a capillary tube is to be used in the determination, it may be supplied from a commercial source or prepared by the student as described in §109. The procedures are described in §§107 and 108.

176. Melting Points of Mixtures. If time is limited, several students may cooperate in this experiment, in which the melting points of several different mixtures of the same two substances are determined, and the resulting data consolidated.

Weigh out to the accuracy of about 0.01 g approximately 0.50 g of benzoic or cinnamic acid, and 0.50 g of urea. Mix the powdered solids thoroughly and determine the melting point of the mixture. Since a mixture normally does not melt sharply, it should be understood for the purpose of this experiment that the melting "point" shall be the temperature when the last crystal in a sample disappears.

Now prepare as many mixtures as may be convenient over a range of 10, 20, 30, to 90% of urea. (Intervals of 5% will be still better if sufficient help is available.) Determine melting point of each mixture, and plot the entire group of data in the form of a melting point–composition diagram, as described in §83.

Questions

1. Why should one record the temperature of the final melting rather than that of a mixture just starting to melt?

2. Why does the melting point–composition curve follow such an irregular course?

3. In what way would you expect the curve to differ if the mixture used had been benzoic acid with cinnamic acid instead of acid with urea?

Experiment 4

Fractional Distillation

Practice in fractional distillation can be obtained during purification of the product of a reaction, or in an experiment designed to emphasize the improvement in separation obtainable with a fractionating column instead of a simple distilling flask. Procedures for both are given below. Chapters 5 and 6, which cover the theory and technique of distillation, should be studied before carrying out this experiment. The reaction chosen is oxidation of isopropyl alcohol with sodium dichromate in sulfuric acid to produce acetone. Related oxidations are described in Experiment 8. Chapter 3 on calculation of quantities of materials and on balancing oxidation-reduction equations should be studied if this reaction is to be carried out. The balanced oxidation-reduction equation and a table of amounts of reactants should be placed in the notebook before the reaction is started.

177. Oxidation of Isopropyl Alcohol. At or near room temperature an oxidizing agent such as a dichromate in the presence of strong acid will convert a simple secondary alcohol into the ketone. Careful control of temperature is essential; the temperature must be high enough to obtain a reasonable rate of reaction, yet low enough to prevent further oxidation and cleavage of the ketone to carboxylic acids. After oxidation is complete the only volatile substances left in solution are the ketone and water. Simple distillation affords a mixture of ketone and water. The resulting distillate is used to demonstrate fractional distillation (§178)

Do not interrupt this experiment until the first simple distillation has been completed (at the end of §177 as marked by a star). For laboratory courses where time is limited, the oxidation may be omitted and a mixture of acetone and water provided for §178.

The oxidation is conducted most conveniently in a 700- or 1000-ml flat-bottom flask (Florence flask), but may be carried out in a 500-ml distilling flask if care is taken so that no liquids accidentally run out of the sidearm during additions, and if the dichromate is added somewhat more cautiously (heat is dissipated more slowly in the smaller flask). Use of the distilling flask avoids the transfer (last paragraph, this section).

In the flask place 25 ml of water, and add carefully, with shaking, 50 ml of concentrated sulfuric acid. Cool the flask under the tap; then add 100 ml of water and 0.5 mole of isopropyl alcohol. Be careful to note whether anhydrous or aqueous alcohol is furnished. Commercial isopropyl alcohol varies from nearly pure anhydrous material to the constant-boiling aqueous mixture containing approximately 12.5% of water. Make whatever allowance is necessary in your calculations.

Provide an ice bath of size large enough to permit the complete immersion of the body of the flask. This bath should contain at least 500 g of finely crushed ice together with enough water to form a thin slush that will readily allow the entrance and removal of the flask from time to time in the later procedures. In a small beaker containing 25 ml of warm water, dissolve that quantity of sodium dichromate (anhydrous or dihydrate) which will be about 3% in excess of the amount required by theory to convert the alcohol into the ketone. For method of calculation see §§14–16.

A thermometer is now mounted in the flask in such a manner that it is held rigidly, with its bulb in the reaction mixture, but with some arrangement for adding a liquid reagent from time to time. An old common cork, bored somewhat off-center to receive the thermometer and with one side cut off, will be satisfactory.

Immerse the flask in the ice bath, and add the dichromate solution in from 10 to 15 portions. After each addition shake the vessel for a moment in the open with a gentle rotary motion, and immediately plunge it into the ice bath. If the temperature threatens to rise above 40°, remove the flask from the bath, shake again, and return promptly to the ice bath. As soon as the temperature falls to 25°, add a new portion of dichromate. Do not try to lower the temperature below 25°, lest the oxidation be held up and a large amount of unreacted material accumulate, ready to undergo too vigorous a reaction when the temperature is allowed to rise again. At the end of this operation the color of the mixture should be green (like chromic salts), possibly with a slight olive or yellowish tint, indicating indirectly that no more alcohol is left to react.

When the temperature of the reaction mixture shows no further tendency to rise spontaneously, transfer the material to a 500-ml distilling flask and add a boiling stone. Distill through a water-cooled condenser into a 100- or 125-ml distilling flask used as receiver. After the temperature of distillation reaches the boiling point of water, continue distilling until about 10 ml more of liquid comes over at this (constant) temperature. Now discard the dark-colored aqueous residue. If the aqueous solution is to be stored until the next laboratory period, stopper tightly to avoid evaporation. ★

178. Fractionation. The distillate obtained is a mixture of the desired ketone with water. The *entire quantity* of this product is now to be distilled twice. The first distillation gives data for a distillation curve (see §54) for simple distillation; the second produces the product to be handed in and gives data for a second distillation curve showing the improvement realized with the particular fractionating column used.

The distillate from §177 is transferred to a distilling flask that it fills no more than half full (usually a 125-ml flask). A boiling stone is added and the distilling flask is attached to a condenser and curved adapter. A 50- or 100-ml graduated cylinder is used as the receiver and the adapter must project into the receiver to minimize loss by evaporation. The receiver may be cooled in ice if the laboratory is especially warm, and the space between adapter and receiver may be lightly sealed with cotton to reduce evaporation. The apparatus of Fig. 1, §62, is suitable if a cylinder replaces the conical flask as the receiver.

Now distill at a steady, uniform rate the entire contents of the flask and record in the notebook the *boiling point* and *yield so far collected* at each 2- or 3-ml interval. In spite of all care, a few drops of less volatile liquid will always be found as a residue in the distilling flask after the apparatus has become cool. This material is nearly pure water; add it to the distillate because the object in this experiment is to distill an identical mixture by two different methods. Transfer the distillate in the graduated cylinder to a 200-ml round-bottom flask to which is fitted a fractionating column, as shown in Fig. 7, §75. From this flask and column redistill the aqueous ketone solution slowly (1 or 2 drops per sec), making a record of temperature and yields as in the previous distillation. There must be no interruption of the distillation while these data are being taken. Watch out particularly for drafts in the laboratory which cause irregular cooling of the column. Any fall in temperature is prima facie evidence that the apparatus is not being manipulated properly. In other words, distillation is probably not continuous. The first fraction should distill close to the boiling point of acetone (56.5°) and the boiling range should be recorded on the label when the product is handed in.

When the acetone is nearly all distilled, the rate of distillation will fall and it will be necessary to apply more heat (gradually increase the heat). Soon the temperature will rise suddenly. When it reaches 65°, change receivers. A test tube or small flask will serve as receiver until the graduated cylinder is again available. Preserve the main distillate as your acetone preparation. Continue the distillation with its accompanying periodic records until you can get no more liquid to go over into the receiver. ★

Draw on one graph the two curves representing the two distillations—with and without the fractionating column. On this graph plot the boiling points as ordinates and yield in milliliters as abscissas. Either draw this diagram in the notebook or prepare it on a small piece of cross-sectional paper and paste it in the notebook. Yield of acetone is 32 ml.

For questions on the chemistry of this experiment see Experiment 8. The acetone preparation may be saved for examination of ketone reactions as outlined in Experiment 8, §192.

179. Separation of Methanol from Water. A student may carry out both a simple distillation into multiple receivers and a fractional distillation. It is recommended, however, that the experiment be conducted as a cooperative effort with a few students finding the result of simple distillation and others using different distilling columns to determine the relative efficiencies.

Aqueous methanol (100 ml) is subjected to fractional distillation into multiple receivers as described in §50. An apparatus like that shown in Fig. 1, §62, is used for simple distillation, and an apparatus like that in Fig. 11, §75, or a more elaborate distilling column if available, to demonstrate fractional distillation through a column. Provide as receivers five conical flasks, 50- to 125-ml size, and mark these A, B, C, D, and E. Provide also 20 or more clay boiling stones, size about 2 mm in diameter, or the equivalent in carborundum crystals.

Before starting the experiment, prepare in the notebook a blank form similar to that of §50, but specially arranged for the distillation of aqueous methanol. The following entries are arbitrarily chosen as somewhere nearly suitable for the first two runs of the simple distillation through an inefficient column.

Distillation Fractions		*A*	*B*	*C*	*D*	*E*
No. 1	Boiling Range	65–70°	70–80°	80–90°	90–95°	95–100°
	Yield, ml	______	______	______	______	______
No. 2	Boiling Range	65–68°	68–80°	80–90°	90–97°	97–100°
	Yield, ml	______	______	______	______	______

No. 3, etc. (Leave the remainder of the page blank, with room for three or more runs, for which temperature ranges are to be chosen by you as the experiment proceeds.)

Those students using more efficient columns may find narrower initial and final ranges more suitable, and fewer fractions may be better. The following directions apply for the less efficient equipment.

Place 100 ml of the aqueous methanol in the 200-ml boiler flask. Add one of the small boiling stones, complete the connection of apparatus, and distill into each receiver the distillate that comes over at the boiling range indicated. After fraction D has been collected, the residue in the distilling flask may simply be poured, instead of distilled, into receiver E, to save time. Record *directly in the notebook* the volume of each fraction collected throughout the experiment.

Return fraction A to the boiler, add a new boiling stone, and redistill into receiver A until the range for new fraction A (65–68°) has been covered. Discontinue distillation, pour contents of B into the boiler, add a boiling stone, and collect two new fractions at the new temperature ranges: into A, 65–68°; and into B, 68–80°. At 80°, stop distillation and add the contents of C. Distill three fractions—again at new ranges—into A, 65–68°; into B, 68–80°; and into C, 80–90°. Now add contents of D to the boiler and collect four fractions: into A, 65–68°; B, 68–80°; C, 80–90°; and D, 90–97°. The residue (97–100°) is poured into E.

Depending on the efficiency of the column used, more or less progress in the task of fractionation will be seen in comparison of the two yield records as suggested above. On the basis of this experience, enter in the notebook a new set of boiling ranges for distillation No. 3, and carry out the program so established. If the third run of distillations does not finish the job satisfactorily, choose new boiling ranges for a fourth and possibly even a fifth fractionation.

Students carrying out the experiment with an efficient fractionating column will probably need to run a single distillation or at most only one additional distillation in the manner described above. Usually the distillation must be carried out more slowly. Smaller columns might be used with corresponding smaller amounts of starting mixture. After the boiling point of pure water is reached there is no need to complete the distillation; simply allow the column to drain and record the volume of boiler residue.

Prepare distillation curves as described in §54, placing the results for several distillations involving columns of varying efficiency on a single graph.

Questions

1. Draw ideal distillation curves for the distillations in §§178 and 179.
2. How does fractional distillation on a laboratory scale as in this experiment differ from distillations of similar mixtures in industrial practice?

Experiment 5

Sublimation

180. Hexachloroethane. In this experiment a 2-g sample of crude crystalline hexachloroethane, is placed in an evaporating dish of about 4-in. size. A piece of 3-in. filter paper is placed over the substance in the dish, and the whole is covered with a watch glass placed convex side up. The evaporating dish is now placed in a sand bath and heated to a temperature not exceeding 190° (preferably about 185°), the temperature at which the vapor pressure of hexachloroethane reaches a value near 1 atm. It should be noted that too much heating of the sand bath will drive the volatile substance entirely out of the apparatus only too easily, so that close attention should be paid to the temperature.

The vapors of the hexachloroethane gradually work their way upward and finally cause an accumulation of sublimate upon the watch glass. After the transfer from dish to glass seems to be complete, allow the sand bath to cool to about 120° or less. Now remove the dish and lift the watch glass over to a paper or dry open dish, and scrape off the crystalline sublimate.

Since the melting point of hexachloroethane is above the true sublimation point, it is necessary to seal a sample in a completely closed tube, for a determination of the melting point. If you wish to try this experiment, use only a small quantity, and do not heat beyond the melting point. For the theory of sublimation, with discussion of the behavior of hexachloroethane, see §§90–93 and 112.

181. Hexachloroethane. Optional Brief Test. Place about 0.5 g of hexachloroethane in a 6-in. test tube, insert a slotted cork (not airtight), and clamp the test tube over a ringstand at about a 45° angle. Heat the closed end gently with a burner, and note the appearance of sublimed crystals on the upper walls of the test tube.

182. Anthracene. About 3–4 g of this hydrocarbon is placed in a 400-ml beaker that rests upon wire gauze over a stand and burner. The beaker is covered with a piece of stiff card or asbestos paper, about 15 cm square, which has a circular opening about 2–3 cm in diameter cut through the middle. A piece of filter paper, slightly oversize, is wedged tightly against the perforated plate of a 3-in. Büchner funnel. The funnel is inverted over the beaker and card and is connected by rubber hose to the suction service or aspirator. It should be held in position securely with a clamp.

Apply gentle suction, thus drawing air from the beaker through the paper. Heat the anthracene so that vapors, mixed with air, may pass freely into the funnel, which serves as subliming chamber. If the heating is too rapid, vapors pass entirely through the paper and sublimate gets out into the rubber tube. When little or no more product seems to be accumulating, allow the apparatus to cool, collect the anthracene, and wash out the residue of tarry anthracene with a practical grade of benzene, toluene, or xylenes mixed with abrasive washing powder.

Experiment 6

Extraction from Solution

Chapter 9, §§113–116, and Chapter 11, §138, should be studied before this experiment is carried out.

183. Adipic Acid. In this experiment adipic acid, $HO_2C(CH_2)_4CO_2H$, a well-known reagent used in the manufacture of nylon, is extracted from aqueous solution. Since this acid is intermediate, from the solubility standpoint, between the hydrophilic and hydrophobic types (§138), the choice of extraction solvent may have marked influence on the question as to which extraction is best. Suggested solvents for comparison are ethyl and isopropyl ethers. According to time available in the course, comparison is made of the behavior of two or more extraction solvents in the hands of one or several students. The mechanical details of laboratory procedure in either case would be similar.

Briefly, the plan calls first for provision of an adipic acid solution, made without high precision of measurement but later titrated carefully by ordinary acidimetry. A sample or samples of the solution are extracted with a given solvent. The residual solution in each case, after extraction, is titrated with standard sodium hydroxide. Simple calculations then reveal the amount of adipic acid extracted.

184. Procedure. Weigh to the accuracy of about 0.1 g not less than 1.5 nor more than 2 g of crystalline adipic acid, and place in a flask of 500- to 1000-ml capacity. Add about 125 ml of water for each gram of adipic acid taken. Warm the mixture gently, with shaking, until the acid is completely dissolved. Mix thoroughly and cool to room temperature. Now place 30 ml of this solution, measured as accurately as possible with a graduated cylinder, and 30 ml of diethyl ether, similarly measured, in a separatory funnel of 125- to 250-ml capacity. Read carefully the discussion in §114 of the technique in shaking a separatory funnel containing ether; also make certain that there is no flame in the vicinity.

Following the precautions suggested above, shake the funnel gently, relieve internal pressure, then shake thoroughly to establish equilibrium.

Allow the aqueous layer to settle, and separate it sharply from the ether, allowing the entire aqueous portion to run into a clean dry flask labeled "Aqueous Residue No. 1." Pour the residual ether extract into a bottle labeled "Ethyl Ether Extracts."

185. Second Solvent. Titration. Repeat the experiment with a new 30-ml portion of adipic acid solution, but use diisopropyl ether instead of diethyl ether. Mark the new aqueous residue as No. 2, and pour the ethereal extract into a bottle labeled "Diisopropyl Ether Extracts."

Fill a buret with standard sodium hydroxide solution whose concentration is in the vicinity of 0.2 *N*. Now titrate each of the 30-ml aqueous residues, in separate experiments, using phenolphthalein as indicator. Finally titrate a sample of 30-ml volume, measured with the same graduate, of the original unextracted adipic solution, marked as "No. 3."

From the three titration values calculate:

(*a*) The weight of adipic acid in each of the three titrated samples.
(*b*) The weight of adipic acid, per liter, in the aqueous solution before and after extraction in each case.
(*c*) The weight of adipic acid per liter in each ethereal extract.
(*d*) The distribution coefficient of adipic acid in ether and water for each type of ether.

186. Optional Experiment. A number of other compounds may replace adipic acid in the procedure of §§184–185. Acetic acid and propionic acid are suitable. It is suggested that different students use different acids and compare their results.

Questions

1. Why is one of the two distribution coefficients greater than the other?

2. Suppose di-*n*-amyl ether were used in a third extraction experiment similar in plan to §184. In what way would you expect the distribution coefficient to differ from that obtained with di-isopropyl ether?

3. Suppose you were the instructor in this course and had become tired of running this experiment each year exactly as described above. A substitute for adipic acid was to be considered. Predict any changes or difficulties involved in the directions for the experiment if the adipic acid were replaced with (*a*) trifluoroacetic acid, (*b*) palmitic acid, (*c*) bromobenzene.

4. Many physicians, concerned about adequate supplies of vitamin A in diet of patients, advise against taking "mineral oil" (intestinal lubricant) for digestive disorders. If vitamin A is assumed to be a mixture of $C_{20}H_{29}OH$ and $C_{23}H_{31}OH$, explain the physicians' opinion.

Experiment 7

Chromatography

The theory and practice of chromatography are described in Chapter 9. In this experiment the standardization and use of activated alumina are illustrated. An experiment employing paper chromatography is also given. Thin-layer chromatography is illustrated in Experiment 38.

187. Preparation of Alumina. Commercial, powdered activated alumina such as supplied by Harshaw Scientific under the designation "Alumina, Activated, Chromatographic Powdered, 90% Al_2O_3, Catalyst Grade," Al-0109P or Al-0101P may be moderately active (around III) as received, but an opened bottle that has been standing in the storeroom is usually of low activity (V). It is dehydrated by heating to 360° for 5 hr, which gives material of activity I. Less adsorptive grades are obtained by adding water as follows: II, 3% water; III, 6%; IV, 10%; and V, 15%. The instructor may prefer to have alumina of several different grades of activity prepared and placed on the reagent shelf by laboratory assistants.

188. Standardization. The adsorptive power of adsorbents varies with many factors. A simple, empirical standardization of alumina makes use of the behavior of a series of azo dyes on a chromatographic column of fixed dimensions. These dyes, in order of increasing adsorbability on alumina, are the following: (A) azobenzene, (B) *p*-methoxyazobenzene, (C) 1-phenylazo-2-naphthol, (D) 1-[4-(*o*-tolylazo)-2-methylphenylazo]-2-naphthol, (E) *p*-aminoazobenzene, (F) *p*-hydroxyazobenzene.[1] The order of adsorbability is different for different adsorbents and even for different solvents.

Prepare a chromatographic column in a glass tube (10 cm long and 1.5 cm inside diameter) attached to a short outlet tube (4-5 mm in di-

[1] H. Brockmann and H. Schodder, *Ber.*, **74,** 73 (1941); H. Brockmann, *Discuss. Faraday Soc.*, **7,** 58 (1949). 1-Phenylazo-2-naphthol (C) is called Sudan Yellow by Brockmann. It has Colour Index number 12055 and is listed as Sudan Yellow, Biological Stain (National) #NA-620, by Fisher Scientific Co. D is called Sudan Red by Brockmann and Sudan IV

ameter) by first inserting a small cotton plug and then introducing the alumina as a slurry in a mixture of benzene and petroleum ether (1:4 by volume) until a column 5 cm high is in place. The column outlet may be closed with a short length of rubber tubing and a screw clamp. The chromatographic tube may be assembled with corks, or a short length of 4-mm glass tubing may be sealed on the bottom of an 18 by 150 mm culture tube or test tube. The tube should be tapped while the slurry is poured in, to ensure even packing. Care must be taken that solvent covers the adsorbent at all times. Place a piece of filter paper (cut with a cork borer to fit the column precisely and to lie flat on the alumina) or a very small cotton plug on top of the alumina to keep the surface from being disturbed when the dye solution is poured in. Let solvent flow out until the solvent level in the column stands just above the surface of the alumina. Then pour 10 ml of one of the dye solutions[2] onto the column, and allow the solvent to flow out at a rate of 20–30 drops per min. Just before the last of the solution enters the alumina, the development solvent (a mixture of 1 volume of benzene and 4 volumes of petroleum ether) is introduced. A total of just 20 ml of this solvent is used to develop the chromatogram.

The activity of the alumina is indicated by Roman numerals; activity I indicates that with the A/B test solution, (A) is found at the bottom of the column, (B) at the top, and there is no dye in the filtrate. With alumina of activity II, test solution A/B gives (A) in the filtrate and (B) at the bottom of the column; with B/C (B) will again be at the bottom of the column and (C) at the top. An alumina of intermediate activity between I and II may be encountered and should be designated I/II. The procedure with the other dye solutions is similar. The lowest number corresponds to the most active alumina. Run enough tests to establish the activity of the alumina you are using.

by Fisher Scientific Co. It has Colour Index number 26105. These dyes have the following structures:

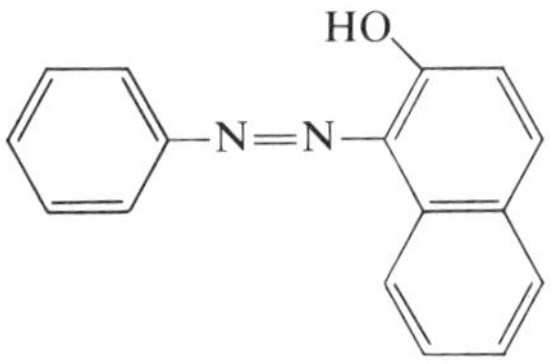

HO, N=N, (C)

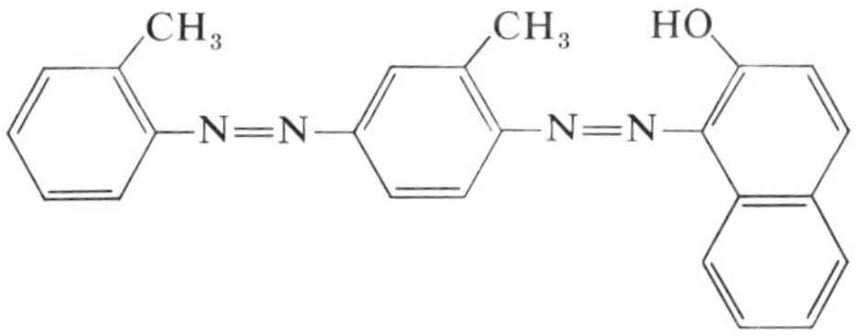

CH_3, CH_3, HO, N=N, N=N, (D)

All the compounds needed for this standardization are available, but Aldrich Chemical Co. seems to be the only source for *p*-methoxyazobenzene (B).

[2]The dye solutions are made up by dissolving 20 mg of each of two dyes (adjacent to each other on the list) in 10 ml of pure benzene (distilled over potassium hydroxide) and diluting to 50 ml with petroleum ether. A hexane fraction such as Skellysolve B, bp 60–70°, is satisfactory.

189. Separation of a Mixture of *o*- and *p*-Nitroanilines

Warning: *This experiment will fail if stopped before both compounds are off the column.*

A commercial chromatography tube that has an inside diameter of 20 mm is very satisfactory for this separation, although the dimensions are not critical and a suitable tube may be constructed from glass tubing and a cork. Commercial tubes usually have sealed-in fritted discs (coarse porosity). The diameter of the glass outlet tube should permit a short piece of rubber tubing to slip over it and fit tightly enough so that it does not slip off. The tubing should be closed with a screw clamp.

In a beaker prepare a slurry from dry benzene and 50 g of alumina. The adsorbent for this experiment should be of activity IV. Activity III alumina effects a good separation, but much solvent is required for elution and the time required is longer. Alumina of low activity such as V does not accomplish a clean separation. Unless the chromatography tube has a fritted disc, put a small plug of cotton or glass wool in the bottom and pour the slurry into the tube rapidly with tapping to ensure even packing. Drain benzene from the tube until the solvent stands just above the surface of the alumina. Collect the benzene in a clean conical flask or beaker and use it to wash down any alumina that adheres to the inside of the tube. Tap the tube until the top alumina surface is level, and cover this with a piece of filter paper cut to size with a cork borer; a small plug of glass wool will also serve. Allow benzene to flow out of the column until the solvent level is just above the surface of the alumina. *Do not allow the liquid level to drop below this surface; otherwise the adsorbent will become dry and give poor separation.* Save the benzene for use in the development of the chromatogram. Carefully place 3 ml of a solution of *o*- and *p*-nitroanilines on the column with a dropping tube or a small calibrated pipet. This solution should contain 0.55 g of the para compound and 0.7 g of the ortho in 100 ml of benzene; it is approximately saturated with *p*-nitroaniline.

Allow benzene to flow slowly from the tube, and when the solution level is just at the alumina surface, use 10 ml of benzene to wash down any of the nitroaniline mixture that has adhered to the column wall. Allow benzene to flow as before, and repeat the washing with 10 ml more benzene. When this has just about run into the alumina, fill about half of the space above the alumina in the chromatographic tube with benzene. Allow a regular, dropwise flow of solvent to continue, and observe the formation and separation of bands. When the first band reaches the bottom of the column, change to a clean receiver and collect all this band in one flask. Add more benzene as needed, but try to regulate the amount of solvent so that it is near the level of the alumina when the last of the band leaves the column. At this point change the eluant from benzene to benzene–ether (1:1 by volume) and continue elution. When the second band reaches the bottom of the column, change to a clean receiver and collect all this band in a single fraction.

Evaporate both fractions nearly to dryness on a steam bath and recrystallize each. It will be necessary to transfer the fractions to small vessels (e.g., 10-ml conical flasks) and to work with care because complete recovery will give only 16.5 mg of para and 21 mg of ortho. The melting point of *o*-nitroaniline is 71.5°; of *p*-nitroaniline, 147.5°.

Questions

1. Which compound elutes first? Why?

2. If *m*-nitroaniline were put in the mixture, where would you expect it to move with respect to the ortho and para isomers? Why?

190. Paper Chromatography. Paper chromatography is described in §126. Dyes are particularly convenient for chromatographic separation by this technique. A careful study of the identification of synthetic coloring used in foods, drugs, and cosmetics has been published [D. H. Tilden, *J. Assoc. Offic. Agr. Chemists,* **35,** 423 (1952), **36,** 802 (1953)], and a simplified general procedure for separation of about 20 of the common F, D, and C dyes has appeared [F. J. Bandelin and J. V. Tuschkoff, *J. Am. Pharm. Assoc., Sci. Ed.,* **49,** 302 (1960)]. The present experiment uses colored inks, most of which contain mixtures of dyes. The developer solution is that used by Bandelin and Tuschkoff. The directions given below are suitable for examination of food, drug, or cosmetic dyes.[3]

MATERIALS NEEDED. *Whatman No. 1 filter paper*[4]: strips 19.2 cm long and 1.25 cm ($\frac{1}{2}$ in.) wide marked with lines 6 cm and 16 cm from one end, and provided with parallel slits near that end (see accompanying illus-

[3]The standard reference work for dyes is *The Colour Index,* published by The Society of Dyers and Colourists. Bradford, Yorkshire, England, 2nd ed., 1956, 3rd ed., 1971. A list of F, D, and C dyes appeared as an appendix in Vol. 1, pp 1800–1802, of the second edition of *The Colour Index.* In the third edition there is a more extensive section describing the public concern about and legislative control over dyes used in food and drugs. Included here is the list of dyes permitted for use in foods in the United States, along with lists of dyes permitted for use in drugs, in drugs and cosmetics, and in cosmetics. Certification is by the Food and Drug Administration of the U.S. Department of Health, Education, and Welfare under the Federal Food, Drug, and Cosmetic Act, Part 8, Title 21, Code of Federal Regulations. Lists for England and for other countries are also given (see Vol. 2, pp. 2773–2788, of the third edition).

Dyes are designated in *The Colour Index* and in many other places by CI followed by a number, indicating the Colour Index number. To find the structure of a dye or its commercial names (often several for one given dye), one looks under the CI number in the proper section of *The Colour Index.*

[4]Available in 600-ft rolls, $\frac{1}{2}$ in. wide, or in sheets that can be cut into strips on a paper cutter.

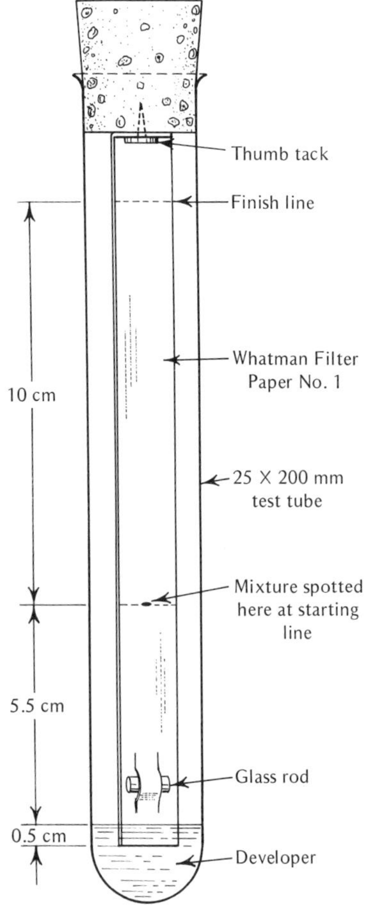

tration). These are cut conveniently with a sharp razor blade. Handle the filter paper as little as possible.

Test tubes: 25 by 200 mm with corks to fit.

Glass rod: Short pieces about $1\frac{1}{2}$ cm long and 4 mm in diameter.

Thumb tacks.

Developer solution: Dissolve 2 ml of 28% ammonium hydroxide and 2 ml of isobutyl alcohol in enough distilled water to make 100 ml of solution.

Dye solutions: Colored inks such as washable blue, washable black, permanent emerald green.

PROCEDURE. Carefully place 2 ml of the developer solution in each test tube; use a pipet and be sure none of the solution gets on the wall of the tube. Stopper the tube. Place a spot, about 3 mm in diameter, of one of the dyes on the center of the starting line of a strip of the paper (see the illustration). This can be done conveniently with a wooden medical applicator stick or other small round piece of wood. A capillary tube can also be used. Allow the spot to dry (best in a current of warm air). Fold about 1 cm of the top end of the strip at right angles, place a short length of glass rod through the slits to weight the bottom of the paper, and fasten the paper to the underside of the cork with a thumb tack. Lower the paper into the tube so that it hangs vertically and does not quite touch the bottom. The paper stretches slightly when it is wet. The distance from the surface of the developer to the starting line should be 5.5 cm.

Allow the chromatogram to develop undisturbed. When the solvent front reaches the finish line, stop the development by removing the paper, and mark the position of the spot. Suspend the paper where it can dry without touching any place except that above the solvent line. Measure the distance the dye has traveled, and record the R_f value.

The dye spot should travel as a compact spot. Some "tailing" or "streaking" often occurs if too much of the dye has been applied to the paper. On the other hand, with violet, green, or blue dyes it is often necessary to apply the dye more than once to get a concentration that can be seen on the chromatogram.

If the procedure is to be used for identification, it is necessary to run a number of known dyes and to compare R_f values obtained with values for the dyes in the mixtures. There is no reason to expect the dyes in inks to be confined to F, D, and C dyes.

Naming of dyes is an interesting exercise in chemical nomenclature.

Experiment 8

Carbinol-to-Carbonyl Oxidations and Reactions of Aldehydes and Ketones

191. General Comment on Carbinol-to-Carbonyl Oxidations. Oxidation of a secondary alcohol to a ketone

$$-\overset{|}{\underset{\underset{H}{|}}{C}}-OH + Cr_2O_7^{--} \longrightarrow -\overset{|}{C}{=}O + Cr^{3+}$$

is easily carried out in the laboratory with sodium dichromate and sulfuric acid. Careful control of temperature is essential; the temperature must be high enough to obtain a reasonable rate of reaction, yet low enough to prevent further oxidation and cleavage of the ketone to carboxylic acids. A procedure for oxidation of cyclohexanol to cyclohexanone is given in §193. Oxidation of cyclohexanol with cleavage is described in §241. A procedure for oxidation of isopropyl alcohol to acetone was given in Experiment 4, §177.

Preparation of an aldehyde by oxidation of a primary alcohol is more difficult in practice because the aldehyde is readily oxidized further to a carboxylic acid. A more important side reaction is rapid formation of hemiacetal from aldehyde and starting alcohol in the acid solution, and rapid oxidation of the hemiacetal to an ester.

$$RCH_2OH \xrightarrow{[O]} RC\!\begin{smallmatrix}\nearrow O\\ \searrow H\end{smallmatrix} \underset{}{\overset{RCH_2OH}{\rightleftharpoons}} R\overset{OH}{\underset{H}{C}}-OCH_2R \xrightarrow{[O]} RC\!\begin{smallmatrix}\nearrow O\\ \searrow OCH_2R\end{smallmatrix}$$

The usual procedure involves removal of the aldehyde from contact with oxidizing agent and starting alcohol by distillation as rapidly as it is formed. The aldehyde is more volatile than the corresponding alcohol, carboxylic acid, or hemiacetal so that much of it can be isolated in this way, but some reacts before it can be volatilized and yields are only moderate.

192. Oxidation of a Primary Alcohol. Synthesis of Propionaldehyde. To 20 ml of water in a small flask add cautiously 10 ml of concentrated sulfuric acid, and cool the solution to 40° or below. In this solution dissolve 0.1 mole of *n*-propyl alcohol. Transfer the solution to a simple distillation apparatus (Fig. 1, §62, 125-ml flask) with care not to allow liquid to run into the sidestem of the flask. The distilling flask should be equipped with a dropping funnel instead of the thermometer shown. Add a boiling stone.

In the dropping funnel of the apparatus place in a 10-ml water solution that quantity of sodium dichromate (anhydrous or hydrated) theoretically required to convert the alcohol into aldehyde. Heat the liquid in the flask to boiling. When alcohol vapor threatens to go over into the condenser, remove the burner and run in the dichromate solution as fast as may be necessary to keep reaction and distillation going with little or no aid from the flame.

When the oxidation reaction is complete, discard the green residue in the flask, wash out the flask, replace the dropping funnel with a thermometer, and redistill the crude aldehyde–alcohol distillate, collecting all the new distillate coming over up to about 90°. The product contains considerable alcohol and water which has been carried over during the rapid distillations; it is suitable for the tests described in §194.

If an individual student wishes to collect from fellow workers 50 ml or more of the mixed distillate, he may then obtain by fractional distillation a preparation of the propionaldehyde alone bp 48.8°. In case equipment with motor stirrer (§26) is available, much better yields of purified propionaldehyde may be obtained by the method of Hurd and Meinert (OS–CV **2**).

193. Cyclohexanone. Prepare a solution of oxidizing agent by dissolving 21 g (0.07 mole) of sodium dichromate dihydrate in 120 ml of water and *to this dichromate solution* cautiously adding 17 ml of concentrated sulfuric acid. Cool this solution to room temperature. Place 21 ml (20 g, 0.2 mole) of cyclohexanol and 60 ml of water in a 500-ml round-bottom flask and to this mixture add the dichromate solution in 10 to 15 portions, adding at a rate sufficient to keep the temperature at 40–50°. Swirl the flask during the addition and if necessary cool in an ice bath. If cooling becomes necessary, reduce the rate of addition of dichromate solution. Do not maintain a very low temperature with an ice bath, as the oxidation may be retarded enough to allow a large amount of unreacted material to accumulate, resulting in a dangerously vigorous, exothermic reaction when external cooling is finally removed. When the addition is complete, allow the flask to stand with occasional swirling for one hour to complete the oxidation. At the end of this operation, the color of the mixture should be green (like chromic salts), possibly with a slight olive or yellowish tint, indicating indirectly that no more alcohol is left to react. Add an additional 100 ml of water, fit the flask with a short fractionating column and water-cooled condenser, and distill into a graduated cylinder, used as a receiver, until 100 ml of distillate has been collected. The cyclohexanone will form a layer on top. Salt out the product by saturating the aqueous layer with NaCl and separate the organic layer. Extract the aqueous layer with two 15-ml portions of chloroform. Combine the chloroform extracts with the organic layer already separated, and dry over magnesium sulfate. Filter the solution into an appropriate size distilling flask, distill off the solvent from a water bath, and finally distill the cyclohexanone, collecting the fraction boiling at 151–156°.

194. Tests for Aldehydes and Ketones. Acetone from Experiment 4 and propionaldehyde from §192, or samples of these compounds from the reagent shelf, should be used for the following tests.

2,4-DINITROPHENYLHYDRAZONE. This is a general reagent for preparation of a solid derivative of an aldehyde or ketone. Follow the directions of §393. Propionaldehyde 2,4-dinitrophenylhydrazone melts at 155°, acetone 2,4-dinitrophenylhydrazone at 128°.

OXIME. This derivative is less commonly used than a 2,4-dinitrophenylhydrazone. Follow the directions of §393. Acetone oxime melts at 59°. An oxime may also be prepared from propionaldehyde but it may be hard to obtain crystals because propionaldoxime melts low (40°).

TOLLENS' TEST. See §381.

FEHLING'S TEST. Prepare 10 ml of Fehling's solution just before use by mixing 5 ml of "No. 1" (a 3% solution of cupric sulfate, cryst.) and 5 ml of "No. 2" (a 15% solution of Rochelle salt in 5% sodium hydroxide). The mixture should be dark blue and clear. Heat to boiling, and gradually add 2 ml of the propionaldehyde solution dropwise. Reddish brown cuprous oxide is precipitated.

BISULFITE ADDITION COMPOUND. Shake a mixture of 2 ml of acetone and 4 ml of a freshly prepared, saturated solution of sodium bisulfite. Collect some of the crystalline product and test its solubility in (*a*) ether and (*b*) water. What practical use is found for this reaction? The bulsulfite addition compound of propionaldehyde may also be prepared.

IODOFORM. The fact that acetone is a *methyl* ketone is indicated by the iodoform reaction; proceed as in §381.

SCHIFF'S TEST. To 5 ml of "Schiff's reagent" add 1 drop of the propionaldehyde preparation. The resulting bright color, characteristic of ordinary aliphatic aldehydes, is not obtainable from (pure) ketones. Schiff's reagent is made by dissolving 0.1 g of fuchsin (rosaniline hydrochloride dye) in 100 ml of water, saturating the solution with sulfur dioxide, allowing the colorless mixture to stand overnight, and then filtering if necessary.

195. Chromic Anhydride Test for Primary and Secondary Alcohols. Tertiary alcohols are not readily oxidized by hexavalent chromium compounds. This has been made the basis of a test to distinguish them from primary and secondary alcohols. Carry out this test on *tert*-butyl alcohol, isopropyl alcohol, and *n*-propyl alcohol, or other appropriate alcohols as described in §380.

Questions

1. The three reagents used in preparation of propionaldehyde might be introduced in several orders of addition. For example, criticize the proposal to put the sodium dichromate and sulfuric acid in the flask, and then to drop in the alcohol.

2. What is the purpose of the sulfuric acid?

3. Suppose *n*-valeraldehyde were desired instead of propionaldehyde. Suggest changes in technique that you think might be necessary.

4. How is oxidation of isopropyl alcohol to acetone carried out in industrial practice?

5. Suppose you had carelessly added an excess of dichromate and allowed the temperature to rise. What products in addition to expected ketone would be obtained from (*a*) isopropyl alcohol, (*b*) 2-butanol, (*c*) 3-pentanol, (*d*) 1-phenylethanol?

6. Using any one of the methods for balancing redox equations, write balanced equations for oxidation of (*a*) isopropyl alcohol to acetone, (*b*) *n*-propyl alcohol to propionic acid. Use dichromate as the oxidizing agent.

7. The nmr and ir spectra for cyclohexanol and cyclohexanone are shown in Figs. 1–4. Identify the characteristic ir bands for each compound. Identify the signals for the various kinds of protons in the nmr spectra. An nmr spectrum is also given (Fig. 5) for a mixture of cyclohexanol and cyclohexanone, which was obtained in the oxidation, §193. Determine the proportions of the two compounds.

Wavenumber (cm⁻¹)
5000
2500
2000
1200
1000
900
800
700
100
80
60
40
20
0
Transmittance (%)
2
3
4
5
6
7
8
9
10
11
12
13
14
15
Wavelength (μ)

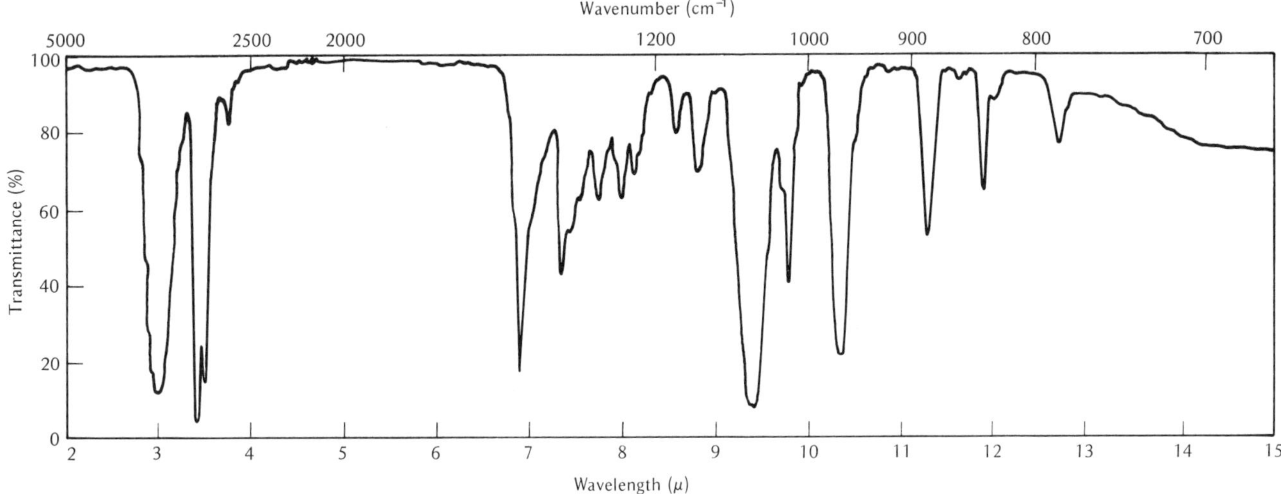

Fig. 1. Infrared Spectrum of Cyclohexanol (© Sadtler Research Laboratories, Inc., 1962.)

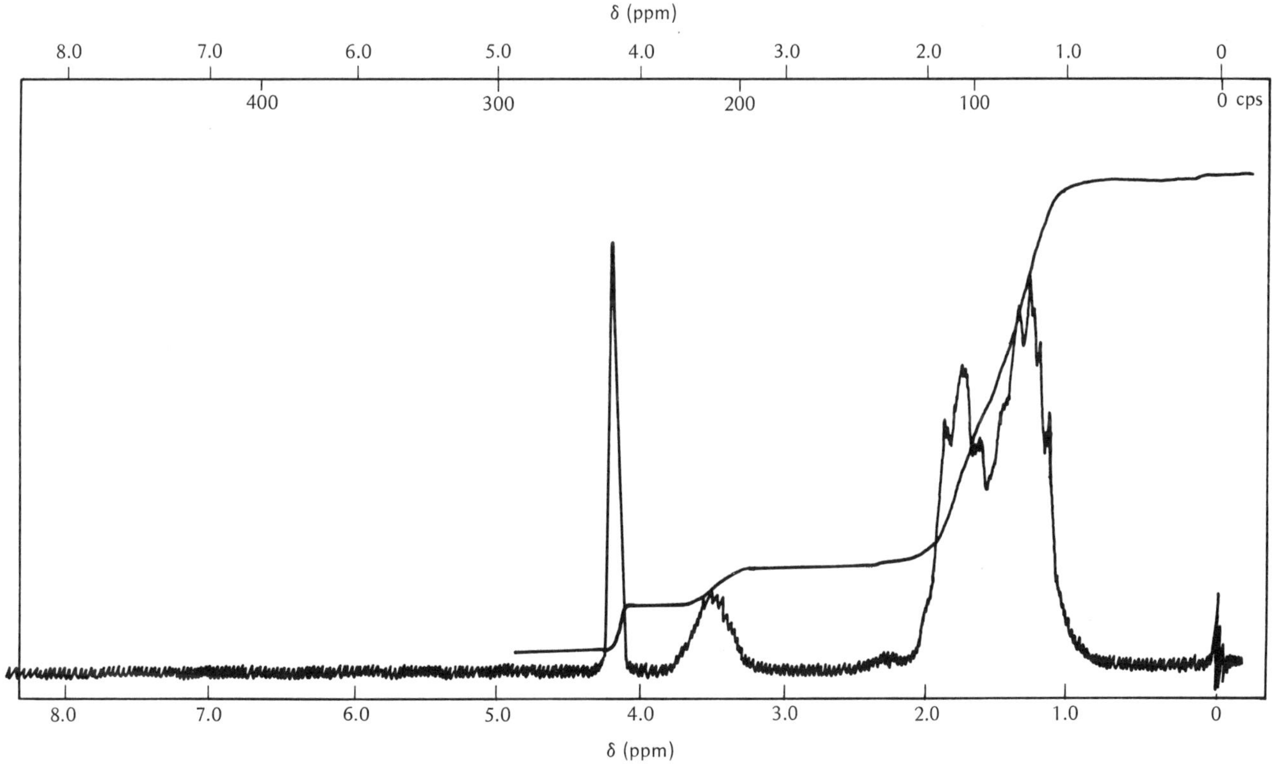

δ (ppm)
8.0
7.0
6.0
5.0
4.0
3.0
2.0
1.0
0
400
300
200
100
0 cps
8.0
7.0
6.0
5.0
4.0
3.0
2.0
1.0
0
δ (ppm)

Fig. 2. NMR Spectrum of Cyclohexanol (© Sadtler Research Laboratories, Inc., 1967.)

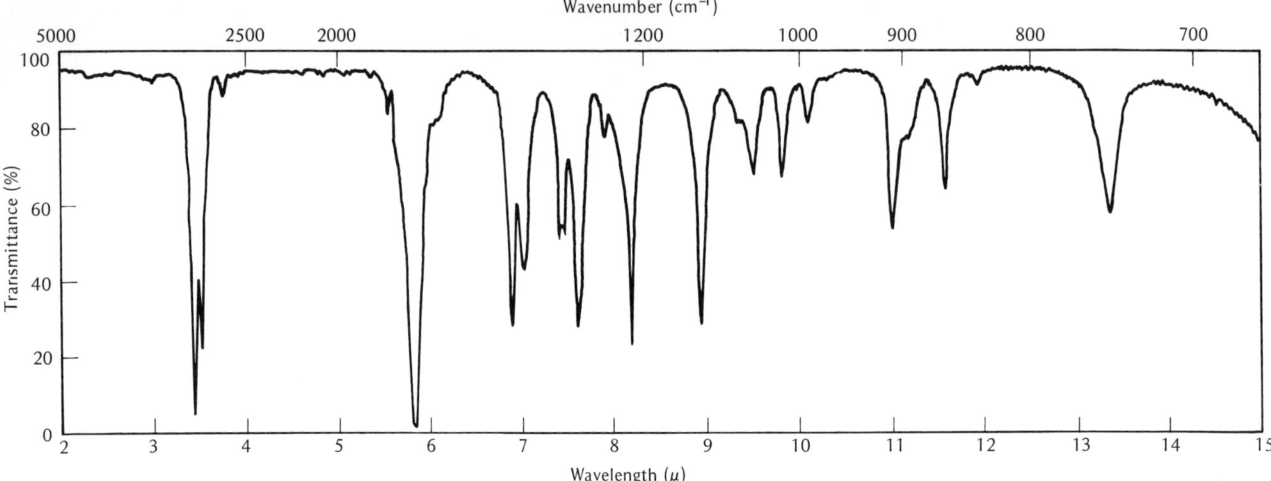

Fig. 3 Infrared Spectrum of Cyclohexanone (© Sadtler Research Laboratories, Inc., 1962.)

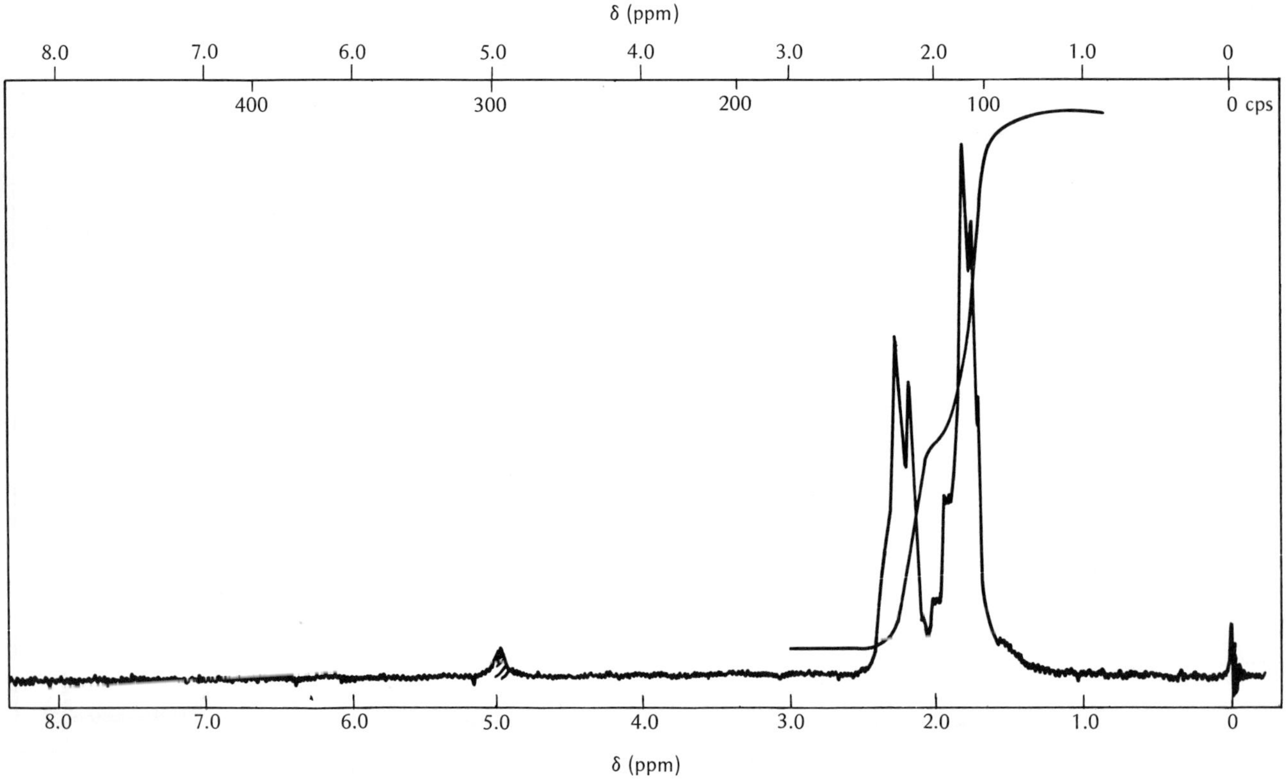

δ (ppm)

8.0 7.0 6.0 5.0 4.0 3.0 2.0 1.0 0

400 300 200 100 0 cps

8.0 7.0 6.0 5.0 4.0 3.0 2.0 1.0 0

δ (ppm)

Fig. 4. NMR Spectrum of Cyclohexanone (© Sadtler Research Laboratories, Inc., 1971.)

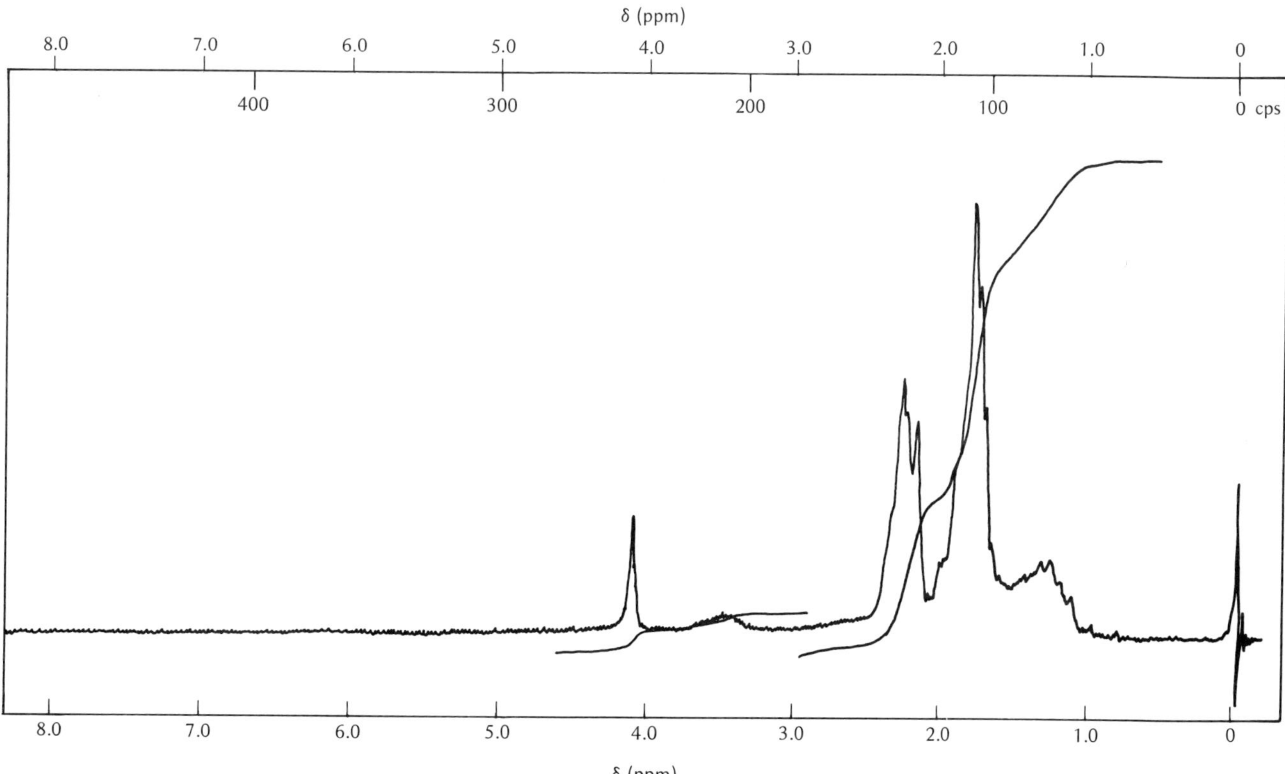

Fig. 5. NMR Spectrum of the Mixture Obtained by Oxidation of Cyclohexanol, §193.

Experiment 9

Nucleophilic Substitution Reactions at Saturated Carbon

This experiment illustrates this type of nucleophilic substitution by procedures for replacement of the hydroxyl of an alcohol by halogen. Experiments 10, 11, 12, and 20 describe other such substitution reactions.

196. Mechanisms of Substitution and Elimination Reactions. Reactions in which a functional group bonded to a saturated carbon is replaced by another group are among the most frequently encountered and important in organic chemistry. Many of these involve attack on the carbon by a nucleophilic reagent which presents an unshared electron pair to the carbon and so displaces the group already present. The group leaves with the electron pair originally involved in its bond with the carbon. The mechanism is represented by the equation below:

$$Nu:^{-} \quad \gt C{-}X \rightleftharpoons \overset{\delta-}{Nu}:\cdots\cdots C \cdots\cdots \overset{\delta-}{:X} \rightleftharpoons Nu{-}C\lt \quad :X^{-}$$

Nucleophile — Substrate — Transition state — Products

The rate of the reaction is first order in nucleophile and in substrate or second order overall; it is designated S_N2. Instead of the charge-type shown where the nucleophile is a negative ion and the substrate neutral, the nucleophile can be neutral (e.g., an amine) and the leaving group can be positively charged so that it ends up neutral:

$$\gt C{-}\overset{+}{O}H_2 \longrightarrow H_2O$$

Substitution at saturated carbon can also occur by ionization of the substrate which produces a carbonium ion; the leaving group becomes a neutral molecule or anion. The carbonium ion reacts with a nucleophile in a rapid second step. The rate of this reaction depends only on the concentration of the substrate and is independent of the concentration of the nucleophile or nucleophiles that eventually end up in products. This mechanism is designated S_N1 and is represented as

$$\text{C–X} \rightleftharpoons \text{–}\overset{+}{\text{C}} + :X^- \xrightleftharpoons{\text{Nu:}} \text{Nu–C} + \text{C–Nu}$$

Which mechanism is involved in any given substitution depends on the structure of the substrate, the nature of nucleophile and leaving group, the solvent and the conditions (temperature, etc.). If the carbon to which X is attached is primary and not too hindered by adjacent substituents, an S_N2 mechanism is common; if the carbon is tertiary, the S_N1 mechanism usually predominates. Secondary substrates often react by both mechanisms.

Substitution reactions by either mechanism are usually accompanied by elimination reactions in which hydrogen (or sometimes another group) attached to the adjacent carbon (β-carbon) leaves along with X; a double bond forms between the α- and β-carbons. These are called β-eliminations and also involve more than one mechanism.

The carbonium ion shown in the S_N1 mechanism may lose a β-hydrogen rapidly to a base (nucleophile) in the solution and so yield an olefin instead of a substitution product. This is the E1 mechanism.

β-Elimination reactions may also involve rate-determining attack by a base on a β-hydrogen of the substrate:

$$\text{B: H–}\overset{\beta}{\text{C}}\text{–}\overset{\alpha}{\text{C}}\text{–X} \longrightarrow \text{B---H---C}{=\!=}\text{C---X} \longrightarrow \text{C=C} \quad \text{BH}$$

Transition state

The rates of such reactions are second order and are designated E2. Another second order process involves either a slow removal of the β-hydrogen by base to form a carbanion, which then loses X to form the olefin, or a rapid removal of the β-hydrogen in an equilibrium step followed by rate-controlling loss of X to give olefin. The latter is termed E1cb (first order elimination on the conjugate base). These second order eliminations compete with substitution reactions.

A clear understanding of the mechanisms of substitution and elimination reactions is necessary for the chemist who is preparing organic compounds by such reactions. The present experiment and Experiments 10–12 and 20 give examples of practical syntheses based on these reactions. Experiment 13 emphasizes that many products may be produced as a result of the competition between various reaction paths. The description given above is very brief; students should study the longer discussion of substitution and elimination reactions in a standard textbook.

197. Conversion of Alcohols to Alkyl Halides. Conversion of alcohols to alkyl halides is often accomplished with hydrohalogen acids. The synthesis of *n*-butyl bromide, §198, illustrates a procedure in which hydrobromic acid, formed from an inorganic bromide and sulfuric acid, is used to convert a primary alcohol to an alkyl bromide. With higher molecular weight primary alcohols, dissolved inorganic salts decrease the solubility of the alcohol and this procedure gives lower yields. Hydrobromic acid containing sulfuric acid but no dissolved salts is effective with these (OS–CV 1). The sulfuric acid increases the rate of the reaction and the yield but is seldom used for secondary or tertiary alcohols because dehydration to olefins and rearrangements occur more readily. Isobutyl alcohol also gives a lower yield of bromide by this procedure. With such alcohols, hydrobromic acid that does not contain sulfuric acid is better, or phosphorus tribromide is used. Hydrogen bromide gas has also been used. Secondary alcohols give some rearranged bromide with any of these reagents. Thionyl chloride is the most effective reagent for converting alcohols to alkyl chlorides. Primary alcohols yield chlorides in fair yield by reaction with concentrated hydrochloric acid and zinc chloride. Tertiary alcohols and such reactive alcohols as benzylic react with cold concentrated hydrochloric acid to give good yields of corresponding chlorides. Phosphorus trichloride is seldom an effective reagent for the preparation of alkyl chlorides.

References for detailed directions for preparation of various alkyl halides are given in §200. An excellent discussion of methods to prepare alkyl halides including methods to obtain iodides and descriptions of other reagents is given in *Survey of Organic Syntheses* by C. A. Buehler and D. E. Pearson, Wiley–Interscience, New York, 1970.

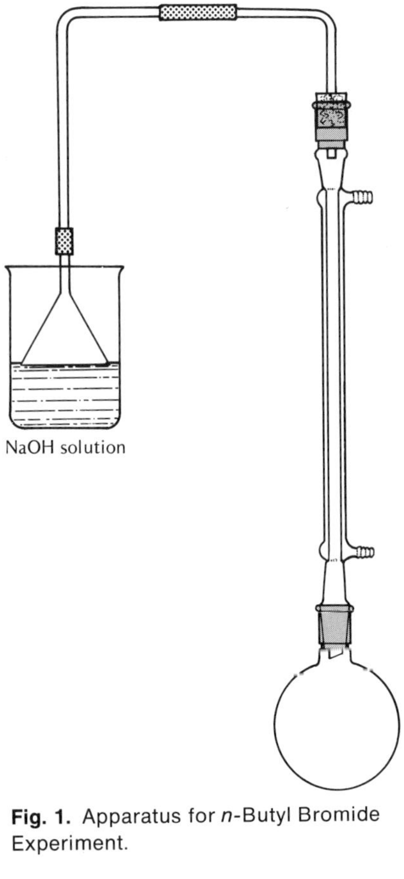

NaOH solution

Fig. 1. Apparatus for *n*-Butyl Bromide Experiment.

198. *n*-Butyl Bromide. In a 500-ml round-bottom flask fitted with reflux condenser and accessories, as shown in Fig. 1, place 0.4 mole of *n*-butyl alcohol and 0.5 mole of potassium or sodium bromide. If sodium bromide is used, note whether or not it is a hydrate; take this into account calculating molar quantity of the salt and providing the correct amount of water later. Mount the flask over wire gauze and burner.

Under the assumption that the apparatus is not to be operated in a hood, some form of "trap" is needed to absorb the offensive hydrogen bromide vapors. The funnel and beaker arrangement shown in Fig. 1 is appropriate, provided care is taken that the funnel does not dip down into the sodium hydroxide solution more than a millimeter or two; there is then no danger of the alkali being sucked back into the condenser. Some workers prefer the "adapter" device illustrated in Experiment 17.

Prepare a solution of sulfuric acid as follows. To 30 ml of cold water in a 200- to 500-ml flask add 55 ml of concentrated sulfuric acid in a fine stream, keeping the solution well mixed by an occasional swirling motion of the vessel. Cool the final mixture by running cold water over the body of the flask, and pass the cool diluted acid through a funnel into the top of the reflux condenser. Loosen the clamps holding flask and condenser and shake the flask until its contents are well mixed. Now readjust the clamps and start water flowing through the condenser jacket.

Using a small gas flame, heat the solution to boiling and adjust the rate of heating so that the liquid drips back slowly from the end of the condenser. Continue the reflux operation for 30 min, after which about 95% of the possible product will have been produced. An additional half-hour of refluxing might yield 1 or 2 g more. Do not expend time with such prolongation of the experiment unless other work is conveniently at hand. ★

Now set up apparatus for ordinary distillation, as shown in Fig. 1, §62, and distill the reaction mixture until only a water-soluble phase seems to be coming over, that is, until no further globules of "oil" are being formed in the condenser. This result is attained usually by the time the temperature has reached 110°. Note the peculiar behavior of the thermometer and compare its reading with the boiling points of the components of the mixture being distilled. The resulting crude distillate is now to be separated from (worthless) accompanying aqueous matter and washed. See §118 for discussion of technique in washing.

With the aid of a common iron ring, mount a separatory funnel on a ringstand. After seeing that any paper slips are removed from ground-glass parts, treat the dry stopcock with a small amount of stopcock lubricant. With the aid of the funnel, separate the crude reaction product, or oil, from the watery portion of the distillate. Which is the butyl bromide layer, upper or lower? Wash the crude butyl bromide with 25 ml of cold concentrated sulfuric acid (preferably chilled in an ice bath to reduce fume nuisance). Run off the sulfuric acid, decant the butyl bromide into a small flask, and wash the funnel and stem with faucet water. Replace the butyl bromide in the funnel, wash with about 25 ml of 1 *N* sodium hydroxide, and finally with water.

With careful separation of liquids, run the bromide layer into a small dry conical flask. Add about 3 g of granular anhydrous calcium chloride or magnesium sulfate, and allow to stand for a day or more. ★

If it is necessary to complete the experiment more promptly, shake the mixture of drying agent and *n*-butyl bromide until the mixture seems to be clear and free from emulsified water. Filter the mixture directly into the distilling flask with care that the product does not flow out of the sidestem during filtration. A loose wad of glass wool or cotton about the size of a pea may absorb less product than a filter paper, but it is sometimes difficult to keep tiny particles of drying agent out of the filtrate; a small filter paper is more often used. Retain the drying flask with residual drying agent until it is certain from the next process that it is no longer needed.

Attach the distilling flask to a water-cooled condenser and distill. If the first part of the distillate is turbid, the drying process was not complete; return all of the liquid to the drying flask, add a little fresh drying agent, and shake again. Then repeat the filtration and distill. The liquid boiling between 99° and 103° is nearly pure *n*-butyl bromide, which has a recorded boiling point of 101.6°. The yield is about 45 g.

199. *n*-Hexyl Chloride. This preparation must be carried out in a good hood or care must be taken to trap the acid vapors that are given off (see §198). Place 71 g (43 ml, 0.6 mole) of thionyl chloride in a 250-ml flask equipped with a reflux condenser and add 2 drops of pyridine. Mount a dropping funnel at the top of the condenser by means of a slotted cork. Add 51 g (62.5 ml, 0.5 mole) of *n*-hexyl alcohol to the thionyl chloride dropwise with swirling during 20 min. Soon hydrogen chloride begins to evolve and only occasional swirling is then needed. The reaction mixture remains colorless and cools slightly. After addition is complete gradually warm the flask. Remove the source of heat if evolution of gas (hydrogen chloride and sulfur dioxide) becomes so vigorous that it threatens to carry the solution into the condenser. The reaction mixture becomes cloudy and slightly yellow. Steady refluxing should be reached after 15–20 min and should be maintained about 45 min longer. Cool the flask slightly, replace the reflux condenser by a Claisen head and condenser set for distillation, add a boiling chip, and distill. A small forerun containing excess thionyl chloride comes over and is discarded (poured into running water in the hood); the product is collected at 132–134°. Pour the product into 300 ml of water, separate, and wash the *n*-hexyl chloride with water, twice with 10% sodium carbonate solution and again with ★ water. Dry over anhydrous calcium chloride.

Filter to remove drying agent and distill through a short Vigreux column. *n*-Hexyl chloride is collected at 132–135° after a very small forerun. The yield is 40–45 g. Pure *n*-hexyl chloride boils at 135°.

200. Procedures for Other Alkyl Halides. In a laboratory course for chemistry majors, a variety of procedures with different reagents and compounds are often tried in one class by different students so that all members of the class get a feel for the various ways a given type of transformation is effected. The following references offer suitable procedures (compounds and reagents indicated).

n-Amyl bromide and *n*-propyl bromide ($NaBr + H_2SO_4$): L. Joseph, M. K. Ross, and W. G. Vullut, *J. Chem. Ed.*, **26,** 329 (1949).

Isoamyl bromide, *n*-dodecyl bromide, ethyl bromide, *n*-octyl bromide, and 1,3-dibromopropane ($HBr + H_2SO_4$): OS–CV **1,** 25.

Cyclohexyl bromide, 1,10-dibromodecane, octadecyl bromide, etc. (dry HBr): OS–CV **2,** 246; OS–CV **3,** 227.

Isobutyl bromide, *sec*-butyl bromide, *n*-propyl bromide, isopropyl bromide (PBr_3): OS–CV **2,** 358.

Cyclohexyl bromide [$(C_6H_5)_3PBr_2$ in dimethylformamide]: G. A. Wiley et al., *J. Amer. Chem. Soc.*, **86,** 964 (1964). This reagent is also useful to convert phenols to aryl bromides, but higher temperatures are required.

n-Butyl chloride, *sec*-butyl chloride, *n*-propyl chloride (conc. $HCl + ZnCl_2$): OS–CV **1,** 142.

tert-Butyl chloride (conc. HCl): OS–CV **1,** 144. This method is useful with many tertiary alcohols: H. C. Brown and R. S. Fletcher, *J. Amer. Chem. Soc.*, **71,** 1845 (1949). Hydrogen chloride gas may also be used: H. C. Brown and N. H. Rei, *J. Org. Chem.*, **31,** 1090 (1964).

Neopentyl chloride and several simpler chlorides [$(C_6H_5)_3P + CCl_4$]: I. M. Downie, J. B. Holmes, and J. B. Lee, *Chem. and Ind.*, 900 (1966); J. Hooz and S. S. H. Gilani, *Canadian J. Chem.*, **46,** 86 (1968).

Cyclohexyl iodide, isobutyl iodide, *tert*-butyl iodide, 1,6-diiodohexane, *n*-propyl iodide ($KI + 95\%\ H_3PO_4$): H. Stone and H. Shechter, *J. Org. Chem.*, **15,** 491 (1950); OS–CV **4,** 323.

Primary alkyl iodides ($I_2 + P$): OS–CV **2,** 399.

n-Butyl iodide, *sec*-butyl iodide, *tert*-butyl iodide, etc. [$(C_6H_5O)_3P\ CH_3I$]: S. R. Landauer and H. N. Rydon, *J. Chem. Soc.*, 2224 (1953).

Questions

1. Why is it desirable to have the strong acid, required in this experiment, fairly *concentrated,* when a greater amount of hydrogen ion would be available by greater dilution with water?

2. In view of question 1, why not eliminate all water from the main reaction and use straight concentrated sulfuric acid?

3. Write equations showing how an ether might appear as a by-product in this experiment.

4. Write equations showing possible reactions of the sulfuric acid wash liquid in removing (*a*) residual alcohol and (*b*) by-product ether from the *n*-butyl bromide.

5. What difference in technique would you think advisable if ethyl bromide were to be made instead of the butyl derivative? If methyl bromide were required?

6. How could you reverse the main reaction of this experiment; i.e., recover *n*-butyl alcohol from the bromide? Devise apparatus for accomplishing this, keeping in mind the physical and chemical properties of the substances involved.

7. What is the purpose of the sodium hydroxide as wash reagent?

8. Why is the boiling point of the mixture of butyl bromide and aqueous residue below 100°, although water (bp 100°) is the lowest-boiling component in the mixture?

Experiment 10

Ethers

201. Symmetrical Ethers. Synthesis of simple aliphatic ethers ROR, in which the alkyl groups are alike and primary, is often carried out by allowing fairly concentrated sulfuric acid to react with the corresponding alcohol. This displacement reaction is accompanied by a competing dehydration, to form an olefin. Higher temperatures favor this side reaction. With tertiary alkyl groups olefin formation is the principal reaction and almost no ether is formed. Secondary alkyl groups give more olefin than primary, but ethers like diisopropyl ether are sometimes synthesized by the method. Clearly this synthetic method is worthless for forming unsymmetrical ethers because a mixture of alcohols ROH and R′OH will give both symmetrical ethers ROR and R′OR′ along with ROR′.

The reaction of sulfuric acid with alcohols is a good example of the way in which the course of an organic reaction can be modified by conditions. At room temperature the chief reaction is formation of the oxonium salt:

$$\mathrm{ROH + H_2SO_4 \rightleftharpoons [ROH_2]^+[HSO_4]^-}$$

If the alcohol is warmed with an excess of sulfuric acid, an alkyl hydrogen sulfate is formed:

$$\mathrm{[ROH_2]^+[HSO_4]^- \rightleftharpoons ROSO_3H + H_2O}$$

Ether formation occurs at higher temperatures and is favored by excess of the alcohol. It may occur by way of the alkyl hydrogen sulfate or through the oxonium salt. These reactions may occur simultaneously under laboratory conditions.

$$\mathrm{ROSO_3H + ROH \rightleftharpoons \begin{bmatrix} ROR \\ H \end{bmatrix}^+ [HSO_4]^- \rightleftharpoons ROR + H_2SO_4}$$

$$\mathrm{[ROH_2]^+ + ROH \rightleftharpoons \begin{bmatrix} ROR \\ H \end{bmatrix}^+ + H_2O \rightleftharpoons ROR + H_3O^+}$$

Olefin formation, an elimination reaction involving a β-hydrogen of the alcohol, accompanies ether formation. The rate of this reaction increases more rapidly with temperature than does the rate of ether formation.

$$CH_3CH_2CH_2CH_2OSO_3H \rightleftharpoons CH_3CH_2CH{=}CH_2 + H_2SO_4$$
$$[CH_3CH_2CH_2CH_2OH_2]^+ \rightleftharpoons CH_3CH_2CH{=}CH_2 + H_2O$$

The problem of the synthesis of any one of the products shown involves choice of conditions that will give the maximum yield of the desired compound. The facts given above indicate that ether formation is favored by excess of alcohol and by a temperature high enough to bring about ether formation but which is otherwise as low as possible in order to avoid excessive olefin production.

202. Di-*n*-butyl Ether. It has been found that temperatures of 130–140° are optimum for converting *n*-butyl alcohol to di-*n*-butyl ether. This temperature is not readily attained by refluxing a mixture of butyl alcohol and sulfuric acid because water is one of the reaction products, and this forms azeotropic mixtures with both the alcohol and the ether. These azeotropic mixtures boil close to 92°. Water also prevents complete formation of the ether because the reaction is reversible and an equilibrium is attained.

$$2C_4H_9OH \rightleftharpoons C_4H_9OC_4H_9 + H_2O$$

It is therefore necessary to remove water as the reaction proceeds. This is most easily accomplished by a water-separator device shown in Fig. 1.[1]

Determine the volume of the reservoir A on the separator by filling the vessel with water and measuring the volume of the water in a graduated cylinder. Assemble the apparatus shown in Fig. 1 and place enough water in A so that it will overflow when an additional 9 ml has been collected. A wide-neck flask of 200- to 250-ml capacity should be used.[2] The bulb of the thermometer should be about 1 cm above the bottom of the flask. Place 50 g of *n*-butyl alcohol and 7 ml of concentrated sulfuric acid in the flask. Reflux the reaction mixture vigorously over a burner. Water and butyl alcohol will collect in A, which is soon filled, and the upper, nonaqueous layer then flows back into the reaction flask. Heating is continued until the temperature rises to 135° (90–110 min). At this point 6 to 9 ml more water will have collected in A. Further heating yields a vapor that is principally butylene, and the reaction mixture darkens rapidly.

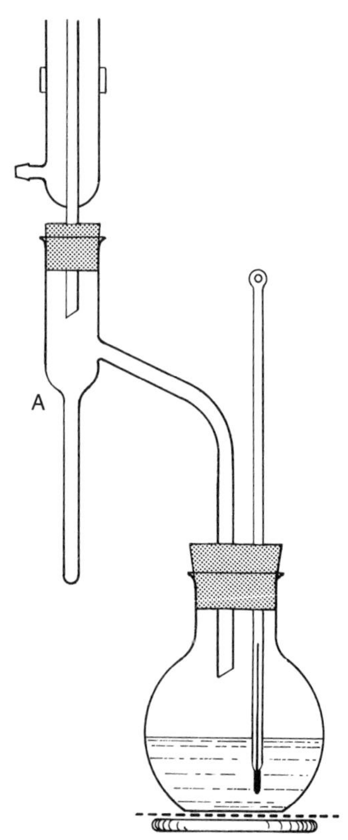

Fig. 1. Water-Separator Device.

[1]This device is an uncalibrated Dean and Stark moisture-determination tube, apparently not available commercially but readily obtained from glass fabricators at less expense than the standard calibrated item. The latter may, of course, be used if available. It is convenient to have a separator that has a 10-ml reservoir.

[2]Ordinary 24/40 standard-taper flasks or 200-ml ring-neck flasks have too narrow necks to accommodate both the separator and the thermometer. A flat-bottom, wide-neck extractor flask such as Corning 5160 is satisfactory.

Allow the reaction mixture to cool, transfer the contents of flask and separator to a separatory funnel containing about 100 ml of water, shake, and separate the layers. Wash the upper layer with 50 ml of water and then with 30 ml of 3 *N* sodium hydroxide. Do not shake the funnel vigorously during the alkali washing or the emulsion that forms may be difficult to break. Wash the upper layer with an additional 30 ml of water and finally with 30 ml of a saturated aqueous solution of calcium chloride. Dry over 3 or 4 g of calcium chloride. ★

Set up a distilling column as shown in §75. The distilling flask should have a capacity of 100 ml or less. Filter the product into this flask and distill slowly until the temperature reaches 135°. Then replace the column by an ordinary distilling head or transfer the flask contents (after cooling) to a small distilling flask and continue the distillation. Collect the di-*n*-butyl ether over the boiling range 140–145°. A yield of 12–15 g is obtained.

203. Di-*n*-butyl Ether—Alternative Procedure. If the water separator described above is not available, the preparation can be carried out by slow distillation of the butyl alcohol–sulfuric acid mixture from a 200-ml flask through a fractionating column. The lower-boiling material distills as a two-phase system and is collected in a graduated cylinder. As each 10 ml of distillate comes over, it is transferred to a separatory funnel; the water layer is returned to the graduated cylinder and the upper layer to the reaction flask. Distillation is continued until 8–9 ml of water has been collected (about $2\frac{1}{2}$ hr). The flask contents are then cooled and the reaction mixture worked up as described. The yield is better by this method (around 25 g), but more time is consumed.

204. Williamson Ether Synthesis. This synthesis involves nucleophilic substitution of the halogen in an alkyl halide by an alkoxide or aryl oxide. It is illustrated here by the synthesis of the important herbicide 2,4-dichlorophenoxyacetic acid, which is generally called 2,4-D.

> **Warning:** *The reagent chloroacetic acid, which serves as the halide in this experiment, is injurious to the skin. If any of this substance touches the skin, wash immediately before further work is attempted.*

In a 400-ml beaker place equimolecular quantities of 2,4-dichlorophenol (8 g) and chloroacetic acid (4.6 g). To these reagents add 4.5 g of solid sodium hydroxide and 25 ml of water. Considerable heat is developed, both from the immediate neutralization of the two organic compounds and the slower reaction to form the ether.

Heat the reaction mixture carefully over wire gauze until the solution approaches dryness. Treat the residue with 160 ml of water, cool the mixture, filter if necessary, and acidify the filtrate with hydrochloric acid. A dense oil, which is the crude 2,4-D, settles to the bottom. Extract the compound with diethyl ether, wash the ethereal solution with water, dry

[§204] over anhydrous magnesium sulfate, filter to remove the spent drying agent, and evaporate the ether. Recrystallize the residue from the evaporation, using benzene as solvent (mp 138°).

Questions

1. What objection would there be to the plan of preparing methyl *n*-butyl ether from methyl alcohol and sulfuric acid, followed by addition of *n*-butyl alcohol?

2. Explain how ether may act as a base.

3. Write an equation for the synthesis of an ether from dimethyl sulfate.

4. In the synthesis of ethyl isobutyl ether, would it be preferable to treat ethyl bromide with sodium isobutoxide or to treat isobutyl bromide with sodium ethoxide? Explain.

5. What side reactions would you expect in the reaction of 2-bromobutane with sodium ethylate in ethanol?

6. Write an equation for the principal reaction of 2-chloro-2-methylbutane with sodium butoxide in 1-butanol. Write equations for the mechanism or mechanisms of this reaction. Write equations for any side reactions.

7. Give a brief description of the use of 2,4-D and related herbicides.

Experiment 11

Nitriles

205. Replacement of Halogen by Cyanide. Cyanide ion is one of the more powerful nucleophilic reagents. Reaction of primary alkyl halides with cyanide ion yields nitriles in excellent yield; secondary alkyl halides react more slowly and give lower yields because elimination to form olefins is more important. Tertiary alkyl halides give mainly elimination. Choice of solvent is important. Alcohol or aqueous alcohol is often used, but dimethyl sulfoxide, a dipolar aprotic solvent (§141), allows more rapid reaction and is often better.[1] Sodium cyanide or potassium cyanide is most often used as the source of cyanide ion, and the solvent must be chosen to give reasonable solubility of these inorganic salts. Cyanide is an ambident anion, capable of attacking either at the carbon end to give nitrile or at the nitrogen end to give isonitrile (R—N=C:). Only small amounts of isonitriles are formed in the usual procedures for preparing nitriles, but appreciable yields are obtained when silver cyanide or cuprous cyanide is employed.

Nitriles are useful for a variety of reactions. Hydrolysis to a carboxylic acid is the most common and is illustrated in §207. An amide is an intermediate in the hydrolysis and may often be isolated because it hydrolyzes less readily than the nitrile (§209). Nitriles yield primary amines by reduction.

206. *n*-Hexyl Cyanide (Heptanonitrile). Place 16 g (0.326 mole) of powdered sodium cyanide and 75 ml of dimethyl sulfoxide in a three-neck flask equipped with a reflux condenser, dropping funnel, and thermometer that dips into the liquid.

> **Warning:** *Sodium cyanide is very toxic; keep it away from even slight cuts or wounds. If cyanide comes in contact with the skin, wash at once.*

[1]This procedure has been applied to a number of alkyl chlorides and bromides; it appears to be quite general. See, for example, L. Friedman and H. S. Schechter, *J. Org. Chem.*, **25**, 877 (1960); R. A. Smiley and C. Arnold, *ibid.*, **25**, 257 (1960); D. A. Argabright and D. W. Hall, *Chem. Ind.*, 1365 (1964).

It is preferable to use magnetic stirring for this experiment, but not necessary; mixing may be accomplished by swirling. Warm the flask to 80° with stirring or swirling (part of the sodium cyanide dissolves) and add 36 g (0.3 mole) of 1-chlorohexane (product from §199) dropwise during 10 min. The temperature usually rises rapidly during the addition and is kept at 140 ± 5° by cooling with water when necessary. If the temperature has not reached 140° by the time half of the 1-chlorohexane is added, warm the solution and maintain it at 140 ± 5°. During the reaction the mixture becomes more fluid and the insoluble salts more crystalline. The temperature drops rapidly after all of the alkyl chloride has been added. Cool the brown reaction mixture, dilute with water to about 300 ml, and extract 3 times with 50-ml portions of ether. Wash[2] the combined ether extracts with 6 *N* hydrochloric acid to hydrolyze the small amounts of noxious isocyanide produced, then with water, and dry over calcium chloride. Filter, remove the ether on a steam bath, taking the usual precautions (§63), and distill the product under reduced pressure, bp 96–97°/50 torr, 74–75°/16 torr. The boiling point at atmospheric pressure is 187°. Yield 20–22 g.

207. Heptanoic Acid. Place 100 ml of 75% sulfuric acid (see §579) in a 250-ml three-neck flask equipped with stirrer, reflux condenser, and dropping funnel. Heat the flask in an oil bath to 150–160° and add 22.2 g (0.2 mole) of *n*-hexyl cyanide with stirring during 15 min. If the amount of *n*-hexyl cyanide available is different, adjust the amount of sulfuric acid accordingly. Continue to heat and stir the reaction mixture for 15 min longer at 160°, then for an hour at 190°. The reaction mixture becomes quite brown. Cool, pour the reaction mixture onto 400 g of cracked ice, and transfer to a separatory funnel. Separate the oil and wash the aqueous layer with 50 ml of ether, which should be added to the oil. This ether solution now contains your product, heptanoic acid, along with small amounts of amide and unhydrolyzed nitrile. Extract the ether solution with 150 ml of 10% sodium carbonate solution (**Warning:** *carbon dioxide evolution*) in two portions. Wash the sodium carbonate extract with a little ether, then acidify with 6 *N* sulfuric acid in the separatory funnel, and separate the oil. *Be very careful to acidify cautiously and swirl thoroughly so that all of the carbon dioxide is evolved before the separatory funnel is closed and inverted.* Wash the sulfuric acid layer with a little ether and combine this with the oil. Wash the ether solution once with water and dry it over anhydrous magnesium sulfate. Finally, filter the solution, remove the ether on a steam bath, and distill the product, best under reduced pressure. The boiling point of heptanoic acid is 108–110°/8 torr; 135°/32 torr; 223.5°/760 torr.

[2]Small amounts of hydrogen cyanide may be liberated during the washing with hydrochloric acid. It is both toxic and volatile, so the washing operation should be carried out in the hood.

208. Benzyl Cyanide (Phenylacetonitrile). Dissolve 30 g of sodium cyanide in 35 ml of water in a 500-ml flask.

> **Warning:** *Sodium cyanide is very toxic; keep it away from even slight cuts or wounds. If cyanide comes in contact with the skin, wash at once.*

Add 50 ml of alcohol and attach a water-cooled condenser to the flask. Add 63 g (0.5 mole) of purified benzyl chloride in several portions through the condenser during 10–15 min. Benzyl chloride is lachrymatory; weigh it in a stoppered conical flask and handle it so a minimum of its vapors contaminate the laboratory. If the laboratory supply is dark colored owing to the presence of undesirable polymers, it may be redistilled in the hood using an air-cooled condenser (bp of benzyl chloride, 179°). Boil the mixture for 3 hr, cool, filter to eliminate crystalline sodium chloride, and distill the alcohol from the filtrate. Dry the residual benzyl cyanide for a few minutes over anhydrous magnesium sulfate and distill the final product. If distillation is conducted at ordinary pressure (bp 234°/760 torr), a fairly good product is obtained, but darker yellow than pure material. A purer product is obtained by distillation at reduced pressure (bp 128°/30 torr). Yield about 40 g.

209. Phenylacetamide. A 500-ml three-neck flask is fitted with a thermometer, reflux condenser, and efficient stirrer. The benzyl cyanide from §208 and 35% hydrochloric acid (4 ml for each gram of nitrile) are placed in the flask, the stirrer started, and a water bath heated to slightly above 40° is placed around the flask. The hydrochloric acid must be at least 30% strength. In 20–40 min the benzyl cyanide dissolves and the temperature rises to about 50°. Higher temperatures are undesirable because the hydrochloric acid is too volatile. Let the homogeneous solution stand in the water bath at 40–50° for an additional 20 min. Then cool the flask in a bath at 15–20°, replace the thermometer by a dropping funnel, and add a volume of cold distilled water equal to the volume of hydrochloric acid used. The rate of water addition is not critical, but the water should not be poured in all at once. Crystals of phenylacetamide soon separate while the water is being added. Cool the reaction mixture in ice for about 20 min and filter with suction. Wash the product on the filter with two 25-ml portions of cold water. The crude product contains a small amount of phenylacetic acid, which is removed by stirring the solid for 15 min with 125 ml of a 10% solution of sodium carbonate. The suspension is then filtered with suction, washed twice with cold water, and recrystallized from ethanol or benzene. The yield before the final recrystallization is 75–80%, mp 154–155°. Recrystallization raises the melting point to 156°.

Reference: OS–CV **3**.

Question

Figures 1–6 represent ir and nmr spectra for phenylacetic acid, phenylacetamide,[3] and phenylacetonitrile. Identify the characteristic ir bands for each compound and the signals for the various kinds of protons in the nmr spectra. If you were attempting to devise a procedure to obtain a maximum yield of the amide from hydrolysis of the nitrile, how might you determine spectroscopically the relative amounts of each of these compounds in the reaction mixture? Would you expect the determination of the amide to be as accurate as of the acid? Explain. How could you tell by ir spectroscopy when all of the nitrile had disappeared?

[3]The nmr spectrum of phenylacetamide shown in Fig. 4 was determined in deuterated dimethyl sulfoxide. $(CD_3)_2SO$, because the compound is quite insoluble in more common nmr solvents. The integration is inaccurate because the solvent is quite hygroscopic and the signal of the H_2O protons falls very near that of the CH_2 protons. The two amide hydrogens are magnetically nonequivalent owing to restricted rotation about the C—N bond; one of the bands is at least partly under the band for the phenyl protons.

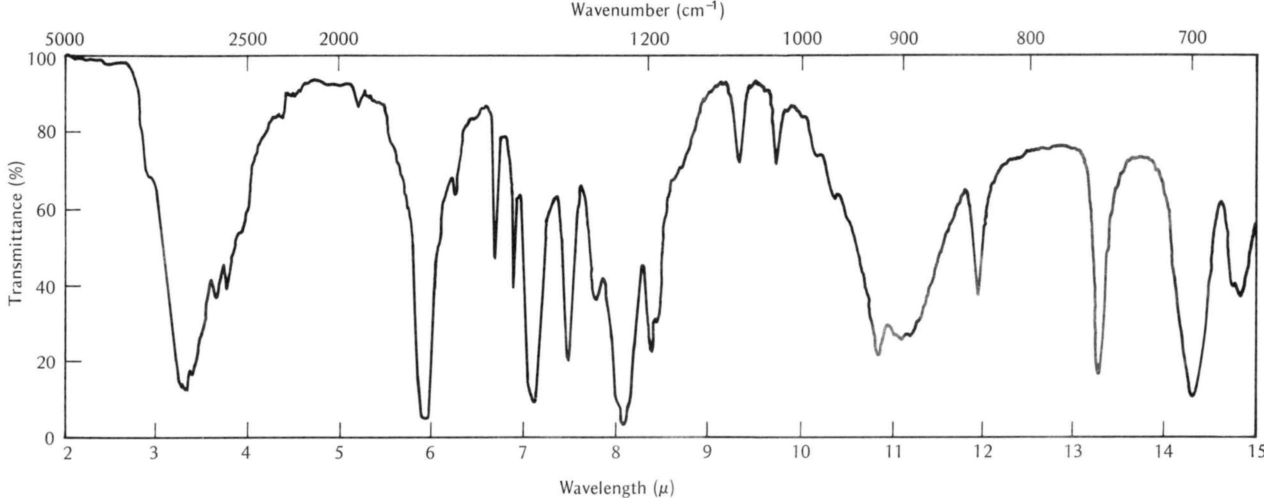

Fig. 1. Infrared Spectrum of Phenylacetic Acid. (© Sadtler Research Laboratories, Inc., 1959.)

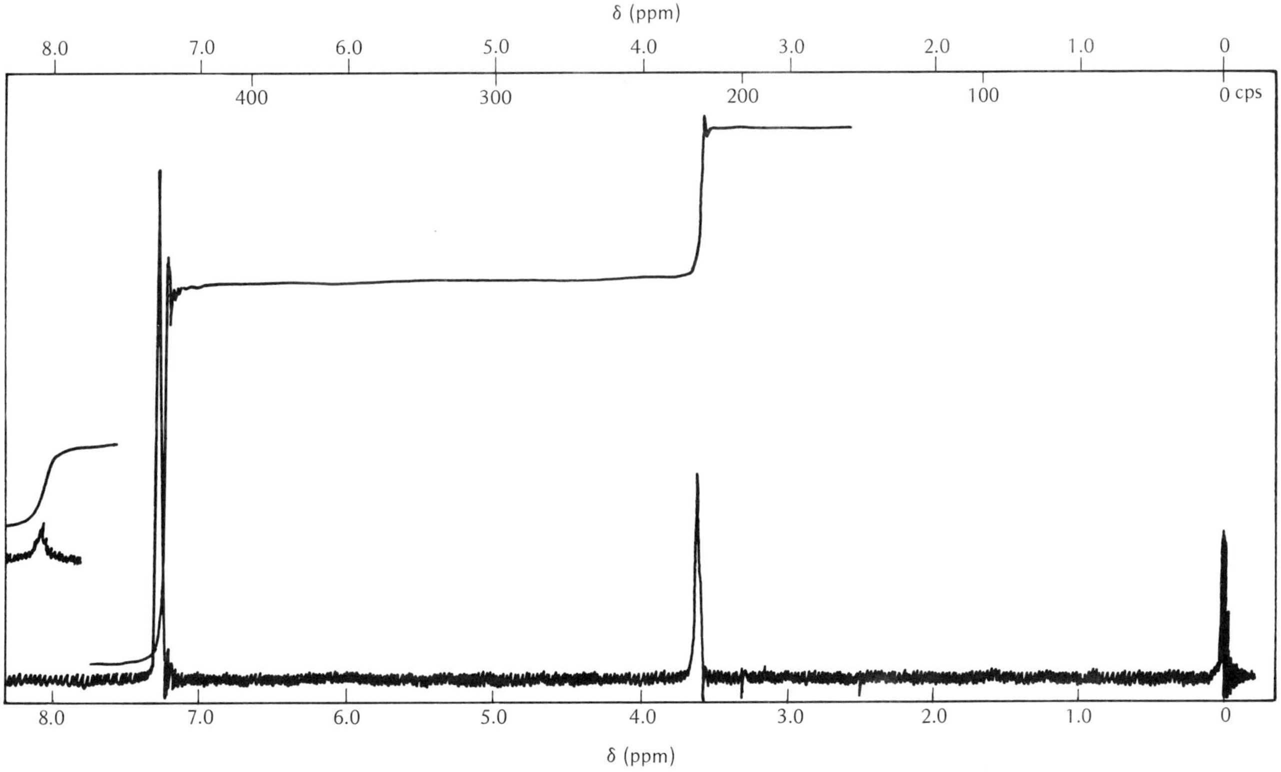

Fig. 2. NMR Spectrum of Phenylacetic Acid. The band shown on the left edge and above the base line is offset 168 cps to the right; its true position is 653 cps (10.88 δ). (© Sadtler Research Laboratories, Inc., 1966.)

Wavenumber (cm⁻¹)
4000 3000 2000 1500 1200 1000 900 800 700
100 80 60 40 20 0
Transmittance (%)
3 4 5 6 7 8 9 10 11 12 13 14 15
Wavelength (μ)

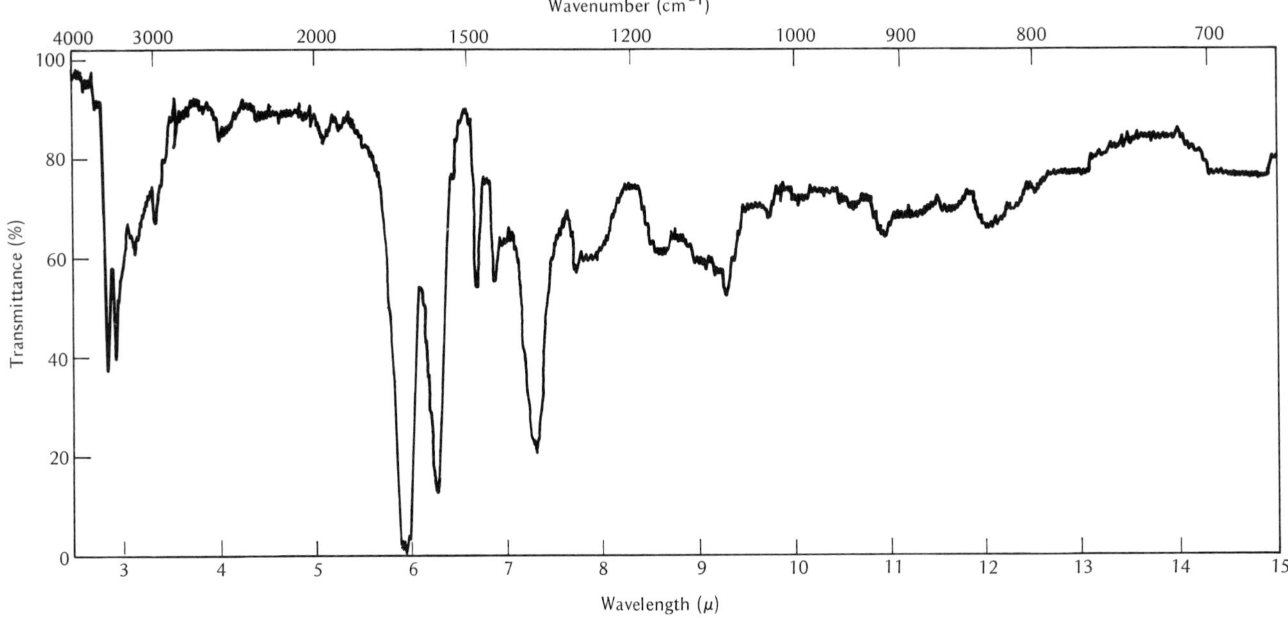

Fig. 3. Infrared Spectrum of Phenylacetamide (in CH_3Cl).

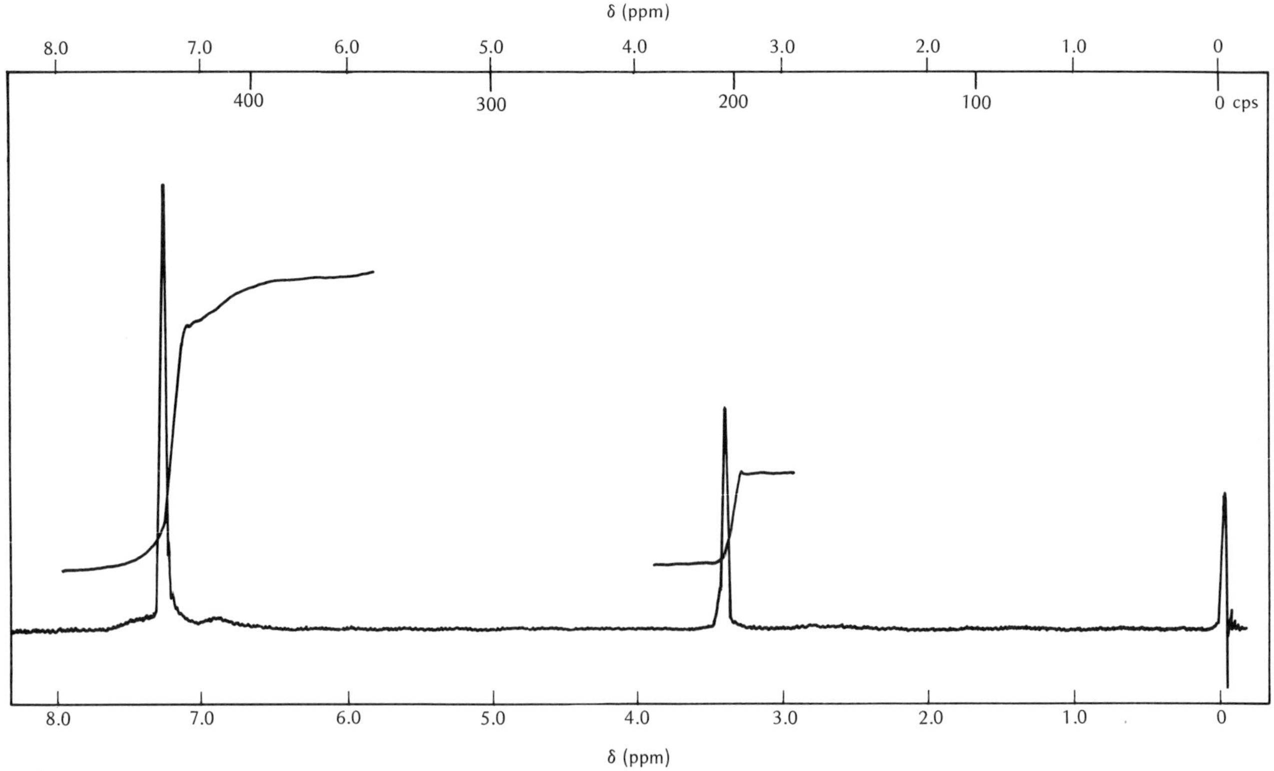

δ (ppm)
8.0 7.0 6.0 5.0 4.0 3.0 2.0 1.0 0
400 300 200 100 0 cps
8.0 7.0 6.0 5.0 4.0 3.0 2.0 1.0 0
δ (ppm)

Fig. 4. NMR Spectrum of Phenylacetamide (see footnote 3).

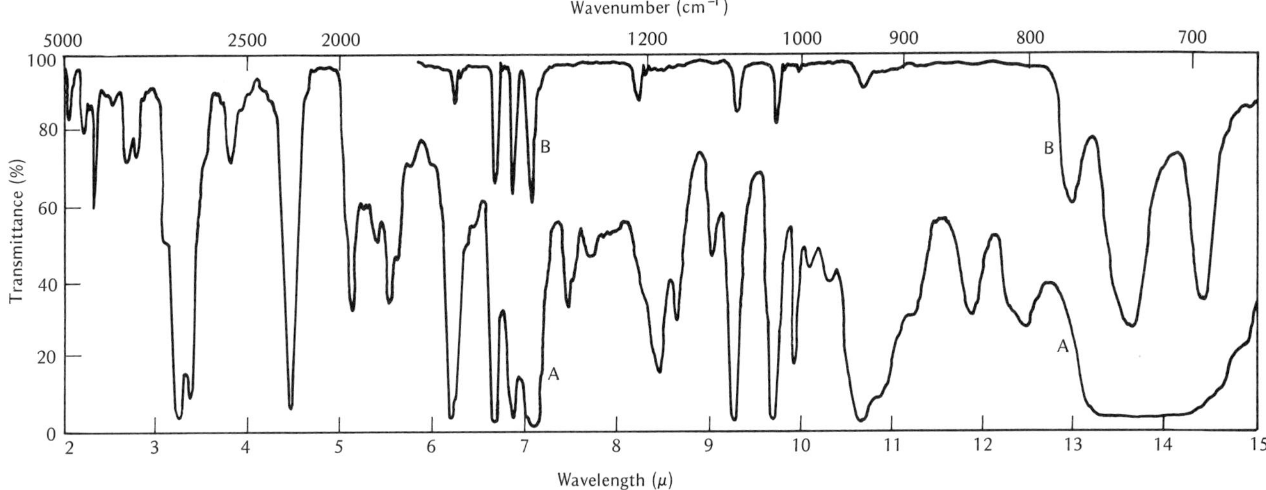

Fig. 5. Infrared Spectrum of Phenylacetonitrile. A, Cell .01 mm; B, between salts. (© Sadtler Research Laboratories, Inc., 1967.)

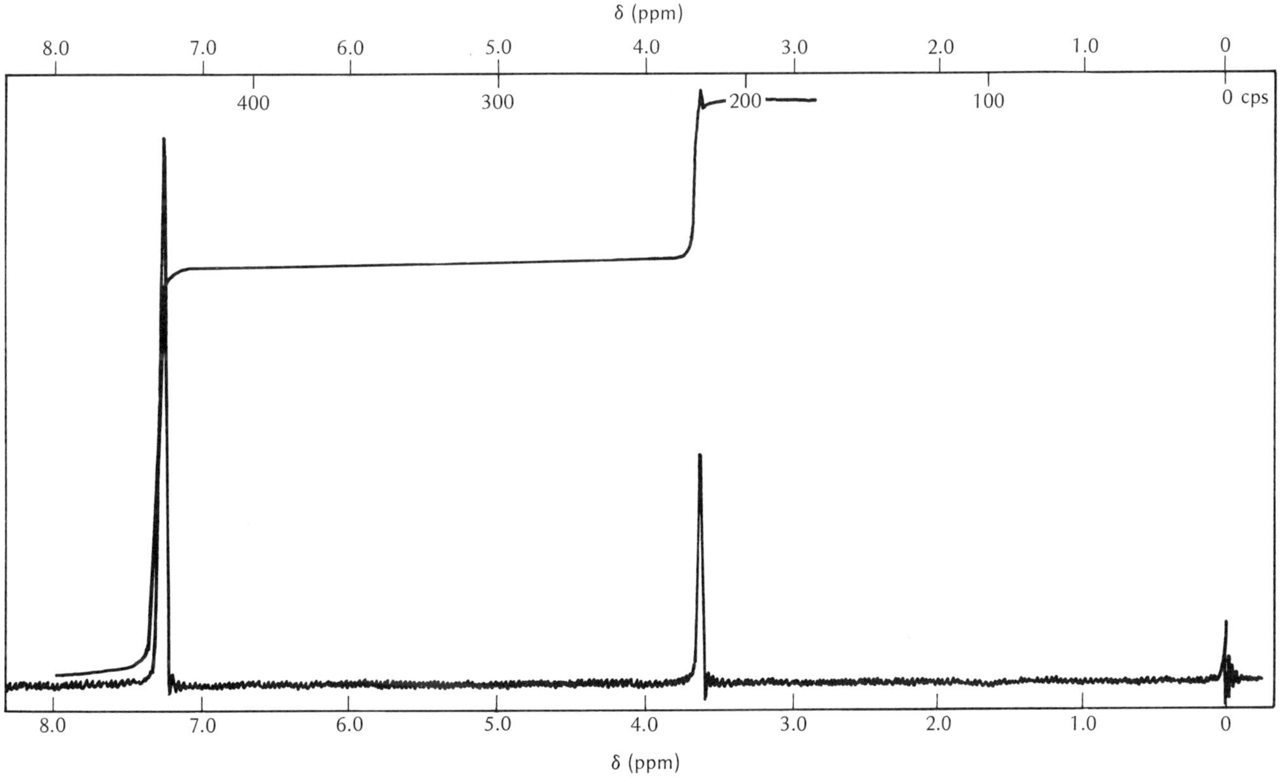

Fig. 6. NMR Spectrum of Phenylacetonitrile. (© Sadtler Research Laboratories, Inc., 1967.)

Experiment 12

Alkylation of Enolate Ions

210. Enolate Ions. Removal of a proton by base from a carbon adjacent to a carbonyl group forms a carbanion that is stabilized by delocalization of the charge to the oxygen of the carbonyl. These carbanions are called enolate ions and are useful nucleophiles in a variety of substitution and addition reactions. When a compound possesses a hydrogen-bearing carbon that is adjacent to two carbonyl groups, the hydrogen is more acidic and more easily removed to form an enolate ion. These "active methylene" compounds are very useful in synthesis. Two of the most common are malonic ester (p$K_a \cong 13$) and ethyl acetoacetate (p$K_a \cong 11$). In this experiment nucleophilic substitution at a saturated carbon is illustrated with enolate ions from these two compounds. Equations are given in the sections in which the procedures are described. Experiments 24 and 25 present nucleophilic additions and nucleophilic substitutions at unsaturated carbon with other enolate ions. All of these reactions illustrate condensations that are among the most important synthetic reactions in organic chemistry.

Many compounds having electron-withdrawing substituents other than carbonyl adjacent to hydrogen-bearing carbons yield useful resonance-stabilized carbanions that react as nucleophiles to form new C—C bonds. Such groups include $-NO_2$, $-SO_2R$, $-SOR$, $-\overset{+}{P}R_3$, $-P(O)(OR)_2$, $-CN$, and $-C_6H_5$ among others. The phenyl group is the least effective of those shown.

Enolate ions are ambident, and the new bond may form either at carbon or at oxygen. Attack on saturated primary carbon (e.g., *n*-butyl bromide) or on active alkyl halides such as benzyl chloride, allyl bromide, or ethyl chloroacetate gives mainly C-alkylation, but secondary halides give lower yields owing to O-alkylation and to elimination to form olefins. Tertiary halides give almost entirely elimination, and aromatic halides do not react.

The material on condensation reactions in a textbook should be studied before laboratory work on these reactions is undertaken.

211. Malonic Ester Synthesis. Diethyl malonate, a relatively inexpensive active methylene compound, is useful to extend an aliphatic carbon chain by two carbons. Its use is illustrated here by synthesis of hydrocinnamic acid. The enolate ion from malonic ester is treated with benzyl chloride, which yields the substituted malonic ester shown. Basic hydrolysis gives the salt of the malonic acid, which is converted to the free acid and warmed to decarboxylate. Compounds having two carboxyl groups attached to one carbon readily lose carbon dioxide by the mechanism shown.

$$H_2C(CO_2C_2H_5)_2 + NaOC_2H_5 \longrightarrow [H\ddot{C}(CO_2C_2H_5)_2]^- Na^+ + C_2H_5OH$$

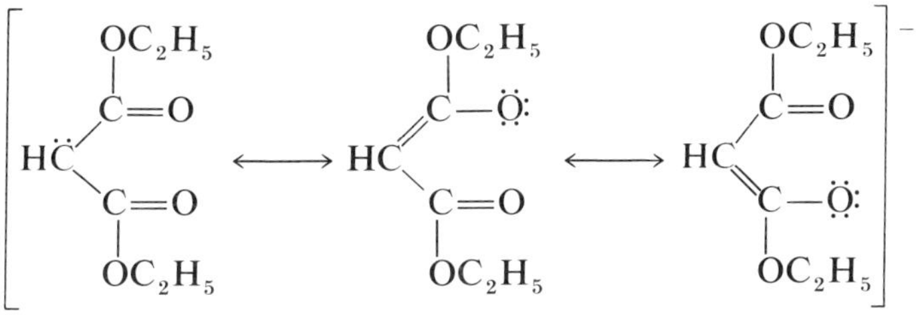

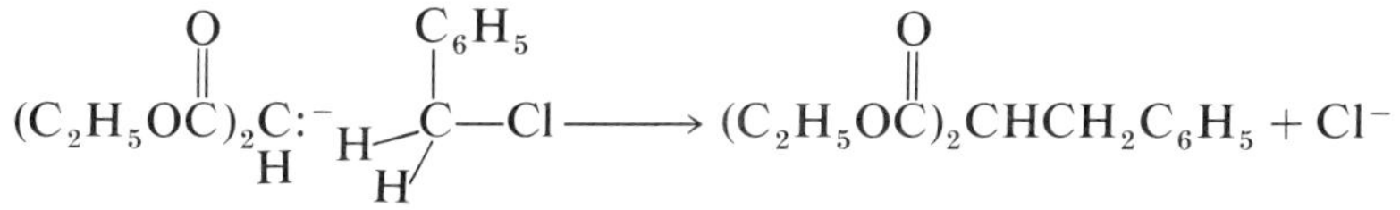

$$\left[H\ddot{C}{<}^{C(=O)OC_2H_5}_{C(=O)OC_2H_5} \longleftrightarrow HC{<}^{\,=C(-\ddot{O}{:}^-)OC_2H_5}_{C(=O)OC_2H_5} \longleftrightarrow HC{<}^{C(=O)OC_2H_5}_{\,=C(-\ddot{O}{:}^-)OC_2H_5} \right]^-$$

$$(C_2H_5O\overset{O}{\overset{\|}{C}})_2\underset{H}{C}{:}^- \quad H{-}\overset{C_6H_5}{\overset{|}{C}}{-}Cl \longrightarrow (C_2H_5O\overset{O}{\overset{\|}{C}})_2CHCH_2C_6H_5 + Cl^-$$
$$\quad\ \ H$$

$$C_6H_5CH_2CH(CO_2C_2H_5)_2 \xrightarrow{KOH} C_6H_5CH_2CH(CO_2^-)_2 \xrightarrow{H_2SO_4} C_6H_5CH_2CH(CO_2H)_2$$

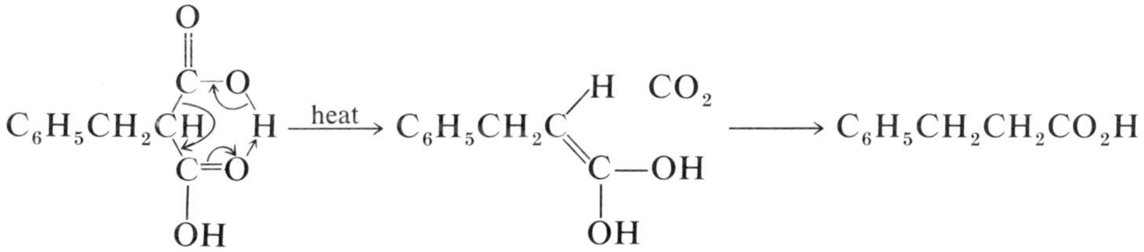

$$C_6H_5CH_2CH{<}^{C(=O)-O}_{C(=O)(OH)}\!\!\cdots H \xrightarrow{heat} C_6H_5CH_2C{<}^{H}_{=C(OH)-OH} + CO_2 \longrightarrow C_6H_5CH_2CH_2CO_2H$$

In this experiment it is necessary to use a sealed stirrer to prevent escape of condensable vapors during stirring. If only ordinary apparatus (with rubber stoppers) is available, use the apparatus shown in Fig. 4, §24. The rubber-sleeve seal shown in Fig. 7, §29, may be used as a substitute for the mercury seal of Fig. 4. If ground-glass apparatus is available, use the outfit of Fig. 5, §24.

In the main flask of the reflux apparatus place 250 ml of absolute alcohol, which must be as nearly anhydrous as possible, preferably not below 99.9%, and under no circumstances below 99.8%. See §213. If apparatus equipped with a mechanical stirrer is not available, a simple flask with reflux condenser may be used, with considerable disadvantage on account of necessity of shaking by hand.

Gradually introduce through the condenser 11.5 g of clean metallic sodium that has been cut into 15 or 20 fragments. After the sodium has all reacted, add with stirring 78 ml of purified diethyl malonate (same as "ethyl malonate"), which has been distilled under reduced pressure if necessary. Now run in very slowly, with stirring, 58 ml of freshly distilled benzyl chloride. Reflux the mixture, with stirring, for 4 hr, during which time a quantity of sodium chloride is precipitated.

Rearrange the condenser for ordinary distillation and distill as much of the remaining ethyl alcohol as possible from a steam bath. To the residue add 200 ml of water and separate the oily layer of crude benzylmalonic ester. Now place this ester in a 500-ml round-bottom flask containing a solution of 75 g of potassium hydroxide in 75 ml of water. Reflux this mixture until hydrolysis is complete (about 2 hr), and then remove residual alcohol by distillation. Wash the resulting aqueous solution of potassium benzylmalonate with a small amount of ether to remove impure, unhydrolyzed oily matter, adding a little water if necessary to dissolve any solid salts present.

Heat the clear potassium benzylmalonate solution to about 80° and slowly add through the reflux condenser 175 ml of 10 *N* sulfuric or hydrochloric acid. After foaming has ceased, reflux the mixture for 3 hr. Cool, add water to dissolve any precipitated salt, then extract the oil with ether. Dry the ethereal solution over anhydrous magnesium sulfate, remove the ether by distillation, and distill the residue under reduced pressure, preserving the fraction boiling at 164–172°/25 torr.

The product will slowly crystallize at room temperature. Recrystallization from petroleum ether, with final cooling to 0°, is more rapid. Yield is 20 g. Melting point of pure hydrocinnamic acid is 48.5°.

212. Preparation of Anhydrous Ethyl Alcohol from 95% Alcohol. If commercial absolute alcohol (ethanol) is not available, provide 250 ml of ordinary aqueous alcohol, density about 0.8, which contains about 0.065 g of water per ml but is commercially designated as 95% by volume, or "190-proof." Place this alcohol together with 75 g of quicklime in a 1-liter round-bottom flask fitted with water-cooled reflux condenser. Allow to stand, well stoppered, overnight or longer, and then reflux over the steam bath for an hour or more. Distill the residual, dehydrated alcohol into a flask with sidestem protected with a calcium chloride tube. In weather not excessively humid this will yield a distillate running from 99.3 to 99.9% alcohol. Percentage of water may be estimated as described in §214.

213. Extremely Dry Ethyl Alcohol. When very thorough dehydration is essential, as in malonic ester syntheses, commercial "absolute" ethanol, or the product of §212 is reworked by the highly efficient method of drying by alkaline hydrolysis. Alcohol of any grade above 99% is suitable as starting material [F. Adickes, *Ber*, **63B,** 2753 (1930)].

Ethyl formate and sodium ethoxide are allowed to react with the water in the alcohol. The ester is used in about 25% excess to ensure completeness of dehydration, and the sodium ethoxide in about 50% excess to ensure final destruction of the volatile ester. The following is an outline indicating substances used and formed in the three stages of the process.

$$C_2H_5OH + H_2O + Na \rightarrow NaOC_2H_5 + NaOH + H_2 \ (+ \text{residual } H_2O)$$

$$\text{Residual } H_2O + NaOC_2H_5 + HCO_2C_2H_5 \rightarrow HCO_2Na + 2C_2H_5OH$$

$$\text{Excess } HCO_2C_2H_5 \text{ (in presence of } NaOC_2H_5) \rightarrow CO + C_2H_5OH$$

The resulting alcohol, distilled in dry apparatus, will be so close to 100% that the figure for water content will be below the limit of experimental error in the bicyclohexyl test outlined below. Workers wishing to test this method of dehydration to the limit of efficiency should arrange apparatus permitting a change from reflux to distillation without breaking connection between condenser and flask.

In a 1-liter round-bottom flask fitted with a reflux condenser place 300 ml of alcohol (99% or better), and add 2 g of metallic sodium, in slices, for each 1 g of water present in the alcohol. When the sodium has completely reacted, add 3 ml of ethyl formate for each 1 g of sodium used. Within a few seconds sodium formate, almost insoluble in anhydrous alcohol, is precipitated.

Reflux the mixture on the water bath for 30 min. Dehydration is complete for all practical purposes in less than 5 min; but the catalyzed disintegration of residual ethyl formate and expulsion of carbon monoxide are somewhat slower.

Distill the reaction mixture with precautions to exclude moisture. Set aside the first 20 ml and collect the next 200 or 250 ml for immediate use in the special experiment requiring the dry solvent (such as §211 or 215). Finally, wash the residue of sodium ethoxide and hydroxide out of the flask as promptly as possible, since this strongly alkaline material attacks glass including the (thermally) resistant Pyrex and Kimax glasses.

214. Estimation of Water in Alcohol. In this test the hydrocarbon bicyclohexyl, formerly commercially known as dicyclohexyl, is used [G. R. Robertson, *Ind. Eng. Chem., Anal. Ed.,* **15,** 451 (1943)].

To 2.0 ml of the alcohol being tested, in a dry 15-mm test tube, add 4.0 ml of bicyclohexyl and stir with a dry thermometer. Heat until the mixture becomes a clear solution, and then allow to cool slowly, with continued stirring. As the test point nears, the liquid becomes opalescent, suggestive of very dilute soap solution. It is still clear enough so that the mercury thread in the immersed section of the thermometer is readily distinguished. Suddenly (within 0.2° temperature range) the liquid becomes completely turbid, and the mercury thread is no longer discernible even through as little as 5-mm thickness of liquid. The temperature at this stage (approximately the critical solution temperature) is noted, and the corresponding percentage of alcohol read from the table given below.

Temperature, °C	23.4	25.4	27.3	29.2	31.0	32.8	41	48	54
Percent alcohol	100	99.9	99.8	99.7	99.6	99.5	99.0	98.5	98.0

215. Alkylation of β-Ketoesters: Ethyl Acetoacetate. This active methylene compound can be used to prepare either methyl alkyl ketones or carboxylic acids. Alkylation of the enolate ion with *n*-butyl bromide gives ethyl 2-acetohexanoate, which in the present experiment is hydrolyzed by dilute alkali and the resulting acid decarboxylated. The alternative cleavage to remove the aceto group occurs in concentrated alkali.

$$CH_3COCH_2CO_2C_2H_5 + OC_2H_5^- \longrightarrow [CH_3COCHCO_2C_2H_5]^- + C_2H_5OH$$

$$[CH_3COCHCO_2C_2H_5]^- + n\text{-}C_4H_9Br \longrightarrow CH_3CO\underset{\displaystyle C_4H_9}{\underset{|}{C}}HCO_2C_2H_5 + Br^-$$

$$CH_3CO\text{—}\overset{\displaystyle H}{\overset{|}{\underset{\displaystyle C_4H_9}{\underset{|}{C}}}}\text{—}CO_2C_2H_5 \xrightarrow{\text{conc. NaOH}} CH_3CO_2^- + C_4H_9CH_2CO_2^- + C_2H_5OH$$

$$CH_3CO\text{—}\overset{\displaystyle H}{\overset{|}{\underset{\displaystyle C_4H_9}{\underset{|}{C}}}}\text{—}CO_2C_2H_5 \xrightarrow{\text{5\% NaOH}} CH_3CO\underset{\displaystyle C_4H_9}{\underset{|}{C}}HCO_2^- \xrightarrow[\text{warm}]{H_3O^+} CH_3COCH_2C_4H_9 + CO_2$$

The malonic ester synthesis usually gives better yields of acids so the acetoacetic ester synthesis is mainly useful to synthesize methyl ketones. Other β-ketoesters and β-diketones can be alkylated similarly. Either a simple round-bottom flask with reflux condenser, preferably the outfit of Fig. 1, §19, or the three neck flask apparatus of Fig. 4, §24, or of Fig. 5, §24, may be used. The condenser should be protected with a calcium chloride tube. Read these directions through and decide on the basis of available time which of the alternative methods and arrangements of apparatus is preferable. Do not interrupt this experiment until the *n*-butyl bromide has been added.

Place 200 ml of extremely dry absolute alcohol (not under 99.8%; see §213) in the flask under reflux condenser, and add 9 g of clean metallic sodium, cut into small fragments, as fast as possible without causing excessive ebullition. To the resulting warm solution of sodium ethoxide, add gradually with shaking or stirring (according to apparatus used), 48 g of purified ethyl acetoacetate. Now add rapidly 55 g of purified *n*-butyl bromide, with shaking or stirring, and reflux the mixture for 5 hr or more. If time is not available to complete the refluxing, stopper the flask tightly and let it stand until the next period. The reaction will go to completion during this period so further refluxing is unnecessary.

Rearrange the apparatus for ordinary distillation and distill out the alcohol. Cool the residual liquid to room temperature and wash with a solution of 3–5 ml of concentrated hydrochloric acid in about 300 ml of water, to dissolve and remove sodium bromide. Mix the residual oil with 350 ml of 5% sodium hydroxide solution. The following alternative procedures are now appropriate: either shake in a stoppered bottle for 30 min and let stand for not over 2 days (preferably only 1 day) or stir vigorously under reflux with a mechanical stirrer for 4 hr. At the end of the chosen period of reaction, separate any unhydrolyzed oily material (separatory funnel). Place the aqueous residue in a 1-liter distilling flask equipped with a dropping funnel instead of a thermometer. Connect the flask to an efficient water-cooled condenser and adapter. Add slowly a solution of 20 ml of concentrated sulfuric acid in 40 ml of water. Carbon dioxide is given off vigorously as the free carboxylic acid is liberated and care must be taken to avoid excessive foaming. After all of the sulfuric acid has been added and most of the carbon dioxide has escaped, distill until about half of the liquid has passed over. Make the distillate alkaline with solid sodium hydroxide, and redistill until 80–90% of the distillate has again passed over.

[§215] Separate the ketone layer and redistill the aqueous layer until about half has passed over. Separate the small additional yield of ketone and combine it with the main product. Extract the water layer of the final distillate with 25–30 ml of ether, and add the ether extract to the ketone product. Wash the combined ketone extract twice with half its volume of saturated calcium chloride solution, dry over magnesium sulfate (5 g), distill. Collect the fraction boiling at 145–152°. Yield is 15–20 g.

Questions

1. Using the same reagents, represent structurally the procedure by which you would prepare a substituted acetic acid instead of methyl amyl ketone.

2. Explain how a sharp deviation from Raoult's law permits the peculiar technique of distillation described in the last paragraph, whereby a relatively less volatile product comes over rapidly.

3. Why do we not use metallic sodium, directly, to prepare the sodium salt of malonic ester, instead of expending time and material going through the sodium ethoxide step? (Consider the mechanical problem.)

4. Why not use sodium hydroxide to form the salt cited in question 3?

5. Why must the alcohol be so nearly water-free?

6. Write equations for the synthesis of 2-benzyl-3-phenylpropionic acid from malonic ester.

Experiment 13

Elimination Reactions and Electrophilic Addition Reactions

216. *β*-Elimination Reactions. *β*-Elimination reactions are discussed in Experiment 9 along with nucleophilic substitution reactions. The present experiment illustrates formation of olefins by dehydration of alcohols in aqueous acidic solutions. The mechanism of dehydration of tertiary alcohols appear to involve formation of a carbonium ion from the conjugate acid of an alcohol:

$$HA + ROH \rightleftharpoons A^- + R\overset{+}{O}H_2 \rightarrow H_2O + R^+ \rightarrow \text{Olefin} + H^+$$

Textbooks sometimes state that this mechanism is involved for primary alcohols but this is unlikely. Either a concerted mechanism or an E2 elimination involving attack on a *β*-hydrogen of the conjugate acid by a base such as bisulfate ion is more probable. Secondary alcohols probably dehydrate by both mechanisms or perhaps by a pathway intermediate between these extremes (see *The Chemistry of the Hydroxyl Group,* S. Patai, ed., Ch. 12 by H. Knözinger, Wiley-Interscience, 1971). The nucleophilic substitution reaction that accompanies these dehydrations is ether formation. Higher temperatures favor the elimination.

Rearrangement and polymerization are other side reactions encountered during dehydration of alcohols. Rearrangement involves carbonium ions as intermediates and occurs at a rate which is competitive with loss of a proton to form an olefin. With hydrocarbon carbonium ions either hydrogen or an alkyl group can migrate with its pair of electrons if an ion of comparable or greater stability can result. The new ion can lose a proton to form an olefin or can rearrange further. Even primary alcohols often yield rearranged olefins. These do not arise during the initial formation of olefin in the dehydration step. Olefins form carbonium ions rapidly by protonation and a terminal olefin forms a secondary or tertiary carbonium ion in this way:

$$(CH_3)_2CHCH_2CH_2OH \longrightarrow (CH_3)_2CHCH{=}CH_2 \ \text{(First product)}$$

$$(CH_3)_2CHCH{=}CH_2 \xrightleftharpoons{H^+} CH_3\overset{CH_3}{\underset{H}{C}}{-}\overset{+}{C}HCH_3 \longrightarrow CH_3\overset{CH_3}{\underset{+}{C}}{-}CH_2CH_3$$

$$(CH_3)_2C{=}CHCH_3 \rightleftharpoons CH_3\overset{CH_3}{\underset{H}{C}}{-}\overset{+}{C}HCH_3 \qquad (CH_3)_2C{=}CHCH_3 \rightleftharpoons CH_3\overset{CH_3}{\underset{+}{C}}{-}CH_2CH_3 \qquad CH_2{=}\overset{CH_3}{C}CH_2CH_3 \rightleftharpoons CH_3\overset{CH_3}{\underset{+}{C}}{-}CH_2CH_3$$

In this system migration of the methyl group would give a carbonium ion of identical structure so that no new products would be observed.

Polymerization is the result of attack by a carbonium ion on an olefin. This yields a new carbonium ion that can undergo all of the reactions of the original ion. Very complicated mixtures result. With tertiary butyl alcohol the process can be formulated as follows:

$$CH_3\overset{CH_3}{\underset{CH_3}{C^+}} \xrightarrow{CH_2=C(CH_3)_2} CH_3\overset{CH_3}{\underset{CH_3}{C}}{-}CH_2{-}\overset{CH_3}{\underset{CH_3}{C^+}} \longrightarrow CH_3\overset{CH_3}{\underset{CH_3}{C}}{-}CH_2{-}\overset{CH_3}{\underset{CH_3}{C}}{-}CH_2{-}\overset{CH_3}{\underset{CH_3}{C^+}} \longrightarrow \cdots$$

$$CH_3\overset{CH_3}{\underset{CH_3}{C}}{-}CH_2{-}\overset{CH_3}{\underset{CH_3}{C^+}} \rightleftharpoons (CH_3)_3CCH{=}C(CH_3)_2 \qquad CH_3\overset{CH_3}{\underset{CH_3}{C}}{-}CH_2{-}\overset{CH_3}{\underset{CH_3}{C^+}} \rightleftharpoons (CH_3)_3CCH_2\underset{CH_3}{C}{=}CH_2 \qquad \cdots{-}\overset{CH_3}{\underset{CH_3}{C^+}} \rightleftharpoons \text{Olefins}$$

In practice, conditions for olefin synthesis vary with the alcohol. Higher temperatures are generally required for primary than for secondary than for tertiary alcohols, but the optimum temperature has to be determined in each instance. Different acids may be used and concentrations varied. It is common practice with volatile olefins to remove them by distillation from the acid medium as rapidly as they are formed, to minimize rearrangement and polymerization. The varying composition of the olefin mixture formed by a given dehydration depends on the particular apparatus as well as on reaction conditions because these determine how rapidly the olefin is removed and hence the extent of the side reactions.

The present experiment gives a procedure for dehydration of isopropyl alcohol, which offers a minimum of trouble with side reactions owing to its simple structure. Propylene is a gas, so skill is required in its manipulation. In a second part of the experiment the student is required to devise his own procedure for dehydration of other alcohols.

217. Electrophilic Addition Reactions. It is convenient to study electrophilic addition reactions as part of the same experiment in which the olefins are produced, particularly for propylene, which is not readily stored. Addition of bromine is typical of these reactions, but others are included. The description of electrophilic addition reactions of olefins should be carefully studied in a textbook before this experiment is undertaken.

218. Propylene and Propylene Dibromide. Set up apparatus as in the accompanying figure, using a 500-ml round-bottom flask, an 8-in. test tube, and an ordinary curved adapter, into which any possible stray bromine vapors may be admitted and absorbed in water. Tygon tubing (substitute for rubber) is recommended as resistant to bromine but is not indispensable. If rubber is used, as much glass tubing as possible is desirable in the line from test tube to adapter, where contact with bromine is more serious. The propylene is passed through the reflux condenser,

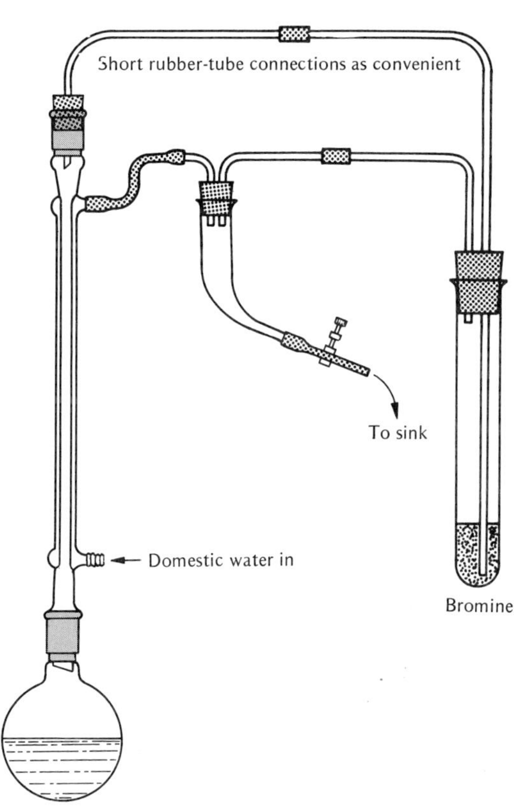

which serves to recondense the isopropyl alcohol that is vaporized during the reaction. Any alcohol vapors that reach the bromine are converted to acetone, which is brominated to bromoacetone, an unpleasant lachrymator.

Carefully pour 30 ml of concentrated sulfuric acid into 15 ml of water in a flask, and mix well. Cool the resulting solution and add it slowly, with thorough mixing, to 45 ml of isopropyl alcohol and a boiling stone contained in the round-bottom flask of the illustration. Now test the assembled apparatus with air or gas to see that there are no leaks.

> **Warning:** *Liquid bromine is now required. Do not take this reagent from stock until you are prepared to go ahead with the experiment. Furthermore, bromine should not be left in a laboratory locker, lest there be evaporation, corrosion of equipment, and an odor nuisance in the laboratory. If the bromine is chilled in ice, or taken just when needed from the refrigerator, it is handled with minimum discomfort. Bromine is a dangerous reagent, attacking the skin vigorously. If even the slightest amount is spilled upon the skin, wash at once under the faucet, and treat the wound promptly with sodium thiosulfate; see information on the inside of the front cover.*

Go to a hood that is in effective operation, place just 8 ml of bromine in the test tube, and complete connection of apparatus. Temporarily disconnect the short Tygon tube at point *A* and attach a piece of ordinary rubber tubing to the outlet tube of the reflux condenser. This rubber tube should be long enough to lead the propylene vapor over the side of the laboratory bench and to a distance from all flames.

Now heat the generator flask until a lively current of propylene is running and until most of the air in the flask and condenser has been expelled. This diversion of the air minimizes subsequent loss of bromine vapors which otherwise are likely to be blown through to the trap with loss of yield. Now reconnect the tube at *A* and pass propylene through the bromine until the latter is decolorized. One must be careful, however, to maintain effective cooling of the condenser so that any escaping isopropyl alcohol will be returned in spite of the fairly rapid rate at which the propylene should be sent over.

The rapid rate may cause slight loss of bromine, as shown by brown fumes in the adapter. In such case the liquid bromine should be cooled with aid of an ice bath. No salt should be placed with the ice, however, lest the bromine freeze. It is also well to avoid excessive aspirator action in the adapter, lest bromine vapors be unduly drawn out into the stream of waste water.

After sufficient propylene has passed to decolorize the bromine, or to reduce the color to pale yellow, immediately disconnect the apparatus at *A* to prevent suck-back of product, and transfer the crude dibromide to a separatory funnel that has a properly lubricated stopcock. Shake the stoppered funnel *very gently*, and relieve gas pressure as described in §114. Wash the product first with water. If it is thought that any bromine persists, wash with a few cubic centimeters of dilute sodium bisulfite solution. Next wash with sodium hydroxide (about 0.5 *N*) and finally with water again. Separate the product carefully, transfer to a small conical flask, dry over granular anhydrous calcium chloride, and distill over wire gauze (yield, 25 g). Boiling point of pure propylene dibromide is 141.6°.

219. Liquid Propylene. If a low-range thermometer and solid carbon dioxide are available, an interesting extension of this experiment may be arranged by condensing a small quantity of the propylene (bp −47°). A small (4- or 5-in.) test tube, fitted like the 8-in. test tube shown in the illustration for §218, is placed upright in a mixture of coarsely crushed solid carbon dioxide (dry ice) and acetone contained in a 250-ml beaker nested in a 400-ml beaker.

After a quantity of liquid propylene has been condensed, the test tube is removed from the alcohol bath, the thermometer is inserted, and the propylene is allowed to boil spontaneously (no flames!) for determination of boiling point.

220. Baeyer's Test. Either the original supply of propylene gas, obtained after completion of §218, or the vapor produced by spontaneous boiling of liquid propylene, as in §219, may be passed through a solution made by adding a few drops of dilute potassium permanganate to about 10 ml of dilute sulfuric acid. A color change indicates oxidation of the alkene at the double bond.

221. Dehydration of Five- and Six-Carbon Alcohols. Several C_5 and C_6 alcohols are suitable for dehydration studies because they are inexpensive and yield olefins that boil at convenient temperatures. The pentenes present greater challenge because they are very volatile (boiling points range from 20° to 39°). Except for 3,3-dimethyl-1-butene, which boils at 41.2° (but is not encountered in these experiments), the acyclic six-carbon olefins boil between 51.1° and 73.2°. Cyclohexene boils at 83°.

Choose one of the following alcohols and devise a procedure for its dehydration. Determine the products of the reaction by vapor-phase chromatography.

2-Methyl-1-butanol	1-Pentanol	3-Methyl-3-pentanol
2-Methyl-2-butanol	2-Pentanol	4-Methyl-2-pentanol
3-Methyl-1-butanol	1-Hexanol	Cyclohexanol

If a five-carbon alcohol is chosen, special care must be taken to prevent major losses of material by evaporation. One simple precaution is to collect the olefin fractions in receivers cooled in ice.

Sulfuric acid is often used for the dehydration, but it also functions as an oxidizing agent to produce sulfur dioxide and products of oxidation of the alcohol or alkene. Phosphoric acid (usually 85%) is also used, but the dehydration is sometimes rather slow; it often causes less rearrangement and less polymerization. Potassium acid sulfate may give superior results with secondary alcohols. The concentration of the acid and the temperature are important. For example, dehydration of *tert*-butyl alcohol takes place at a convenient rate in 40–50% sulfuric acid at 85°, *sec*-butyl alcohol requires 60–65% acid at 100°, and *n*-butyl alcohol requires 75–80% acid at 135–140°.

The "Type of Reaction" index of collective volumes of *Organic Syntheses* is a useful reference for sample procedures for this experiment. Good directions for laboratory experiments on dehydration of specific alcohols and analysis of the olefin mixtures have been published:

3-Hexanol: J. F. McConnell, *J. Chem. Ed.*, **48,** 552 (1971).
2-Methylcyclohexanol: R. L. Taber and W. C. Champion, *J. Chem. Ed.*, **44,** 620 (1967).
3,3-Dimethyl-2-butanol: R. L. Taber, G. D. Grantham, and W. C. Champion, *J. Chem. Ed.*, **46,** 849 (1969).

The alcohols are considerably more expensive than the ones suggested, but the procedures can serve as models for your experiment.

Dehydration of 4-methyl-2-pentanol is an experiment frequently encountered in laboratory manuals. A lecture demonstration version was published by E. J. Nienhouse [*J. Chem. Ed.*, **46,** 765 (1969)], who obtained the following mixture of olefins (given in order of retention times in vpc): 4-methyl-1-pentene (5.2%), *cis* and *trans* 4-methyl-2-pentene (69.4%, not separated), 2-methyl-1-pentene (3.8%), 2-methyl-2-pentene (19.7%), *cis* and *trans* 3-methyl-2-pentene (1.9%, not separated). Compositions appreciably different from this are also reported; variations probably arise from differences in apparatus and conditions.

222. Synthetic Sequence. The more expensive C_5, C_6, or C_7 alcohols can be synthesized by Grignard reactions and their dehydration studied as described above. Such synthetic sequences are excellent projects for longer laboratory courses.

223. Tests for Unsaturation. The usual tests for unsaturation in qualitative organic analysis involve addition of bromine and oxidation with permanganate solution. Both are simple test tube experiments easily observed as decolorization of the bromine or permanganate solution.

The first step in the permanganate oxidation is an electrophilic attack on the double bond to form a cyclic ester of manganese. At low temperatures this can be hydrolyzed to a glycol, but further oxidation and cleavage of the oxidized C—C bond occurs readily.

Bromine adds to olefins by an electrophilic attack forming an intermediate bromonium ion; this is then attacked by bromide ion (in an inert solvent where no other nucleophile is present) to yield a dibromide. The addition is *trans*. If other nucleophiles are present, these compete with bromide ion. Addition of bromine in water solution to indene forming a bromohydrin is an example (§224).

Test the olefin mixtures prepared in §221 for unsaturation by addition of bromine in carbon tetrachloride and with aqueous permanganate solution; procedures are given in §383.

224. 2-Bromo-1-indanol. Suspend 11.6 g (0.1 mole) of indene in 300 ml of water and add enough detergent to completely emulsify the indene when it is shaken. Add a solution of 15 g of bromine in 150 ml of 10% aqueous potassium bromide. The solution must be stirred or shaken and the bromine added slowly in order to obtain the bromohydrin as a solid. If an oil is obtained reflux it with water for a short time to obtain a solid. Filter and recrystallize from benzene. The yield should be 50–80%, mp 130–131°.

Discuss the orientation in this addition reaction. Further experiments with the bromohydrin and with indene are described by J. A. Garrison [*J. Chem. Ed.*, **47**, 300 (1970)].

225. Addition of 2,4-Dinitrobenzenesulfenyl Chloride to Olefins. It would be convenient to find a reagent that adds cleanly to olefins to yield solid derivatives suitable for characterization of the olefin. 2,4-Dinitrobenzenesulfenyl chloride reacts with most olefins by a stereospecific *trans* addition to yield adducts that are solids. With symmetrical olefins a single, sharp-melting compound is produced. Unfortunately unsymmetrical olefins often yield both possible adducts as shown below for propylene.

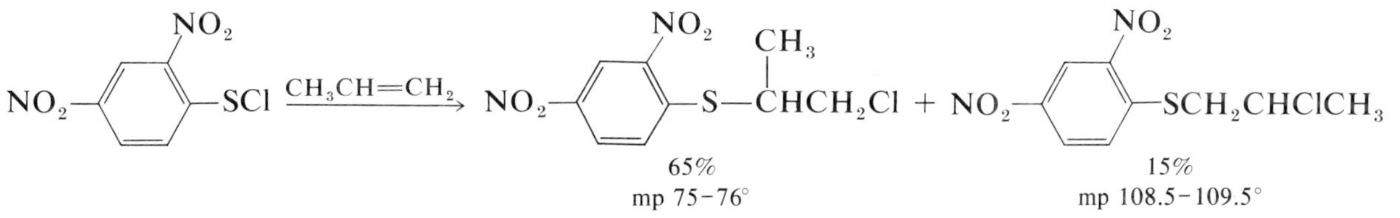

$$2,4\text{-}(NO_2)_2C_6H_3\text{-}SCl \xrightarrow{CH_3CH=CH_2} 2,4\text{-}(NO_2)_2C_6H_3\text{-}S\text{-}CH(CH_3)CH_2Cl + 2,4\text{-}(NO_2)_2C_6H_3\text{-}SCH_2CHClCH_3$$

65%, mp 75–76° (first product); 15%, mp 108.5–109.5° (second product)

Tetrasubstituted olefins do not react.

A procedure suitable for many common olefins is given below. The olefin should be used in 10–100% excess. Glacial acetic acid is the best general solvent for the reaction, but benzene or carbon tetrachloride may sometimes be better. Aqueous potassium iodide is useful to determine completion of the addition; in the presence of unreacted sulfenyl chloride it liberates iodine at once.

$$2\,NO_2\text{-}C_6H_3(NO_2)\text{-}SCl + 2I^- \longrightarrow NO_2\text{-}C_6H_3(NO_2)\text{-}SS\text{-}C_6H_3(NO_2)\text{-}NO_2 + I_2 + 2Cl^-$$

Heat a solution of 0.2 g of 2,4-dinitrobenzenesulfenyl chloride (mol. wt. 234.6) and 0.2–0.3 g of olefin in 5 ml of glacial acetic acid on a steam bath for 15 min or until the potassium iodide test shows that addition is complete. Cool the reaction mixture and collect any product that precipitates. If the cooling is in an ice bath, glacial acetic acid may crystallize (mp 16.6°), and these crystals should not be mistaken for product. If an inadequate amount of product (or no product) is obtained by cooling, pour the reaction mixture (or filtrate) onto 5 or 10 g of crushed ice. An oil or solid separates. If an oil is obtained, it will sometimes crystallize when scratched, or it may be possible to dissolve it in a suitable crystallization solvent and obtain crystals by cooling. Recrystallize the solid and determine the melting point of the dried crystals. Alcohol is often a suitable solvent. Cyclohexene is a suitable inexpensive olefin for this experiment; the adduct melts at 61–62°. Melting points for other adducts are given by Kharasch and Buess, *J. Amer. Chem. Soc.*, **71,** 2724 (1949), the original article from which the procedure was taken.

Questions

1. How can you tell in the laboratory that the reaction of propylene with bromine is not one of substitution rather than addition?

2. What occurs when methanol is submitted to the conditions for dehydration of a primary alcohol?

3. In the preparation of propylene dibromide the crude product is washed with sodium bisulfite solution and then with sodium hydroxide solution. Write equations for the reactions that occur in each instance.

4. Why does propylene dissolve in bromine more rapidly after the reaction of §218 has proceeded for some time?

5. Addition of bromine to *trans*-2-pentene gives a compound with two asymmetric centers. Are all four optically active isomers formed and are the product or products optically active? Explain.

6. Dehydration of 3-hexanol yields a mixture of four olefins (*cis*- and *trans*-2- and 3-hexenes). Complete separation of these by glc has not been accomplished, but a column packed with saturated silver nitrate in benzyl cyanide as the liquid phase and 60–80 mesh Chromosorb W as the support separated the mixture into two peaks: a pair of *cis* isomers (28%) and a pair of *trans* isomers (72%). Each peak was collected and examined by nmr, giving the spectra shown in Figs. 1 and 2. Determine the percent of each olefin in the mixture from the integrations shown on

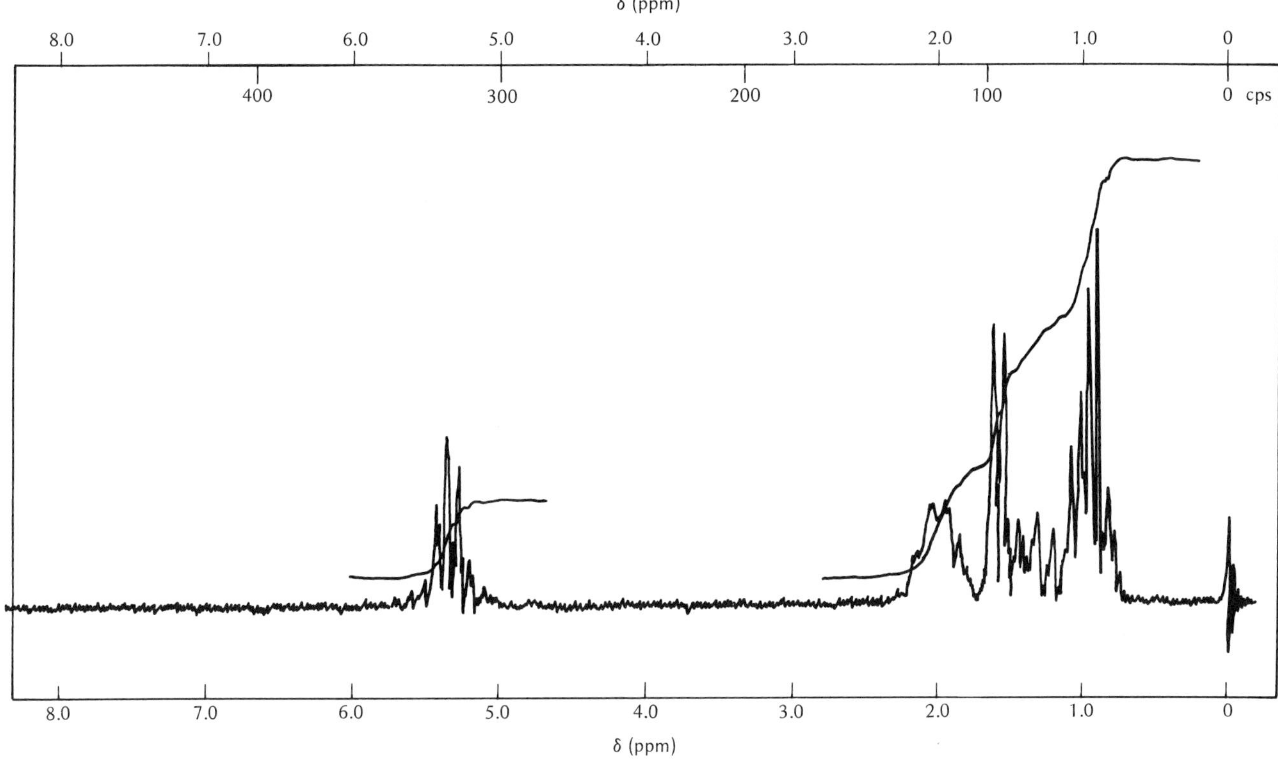

Fig. 1. NMR Spectrum of the Mixture of *cis*-2-Hexene and *cis*-3-Hexene Obtained as Described in Question 6.

the spectra. Figures 3–12 show ir and nmr spectra for the four olefins and for 3-hexanol. Assign as many peaks as you can in these spectra.

7. Dehydration of 1-hexanol gave a mixture of hexenes with the ir and nmr spectra shown in Figs. 13 and 14. Spectra for 1-hexanol and 1-hexene are also given (Figs. 15–18). Assign as many peaks as you can in the spectra. What are the main components of the mixture?

8. Addition of 2,4-dinitrobenzenesulfenyl chloride to propylene gives two adducts, A and B, as explained in §225. The nmr spectra for these adducts are shown in Figs. 19 and 20. Which spectrum corresponds to which adduct? Note that protons α and β to chlorine are deshielded relative to those α or β to sulfur (§159).

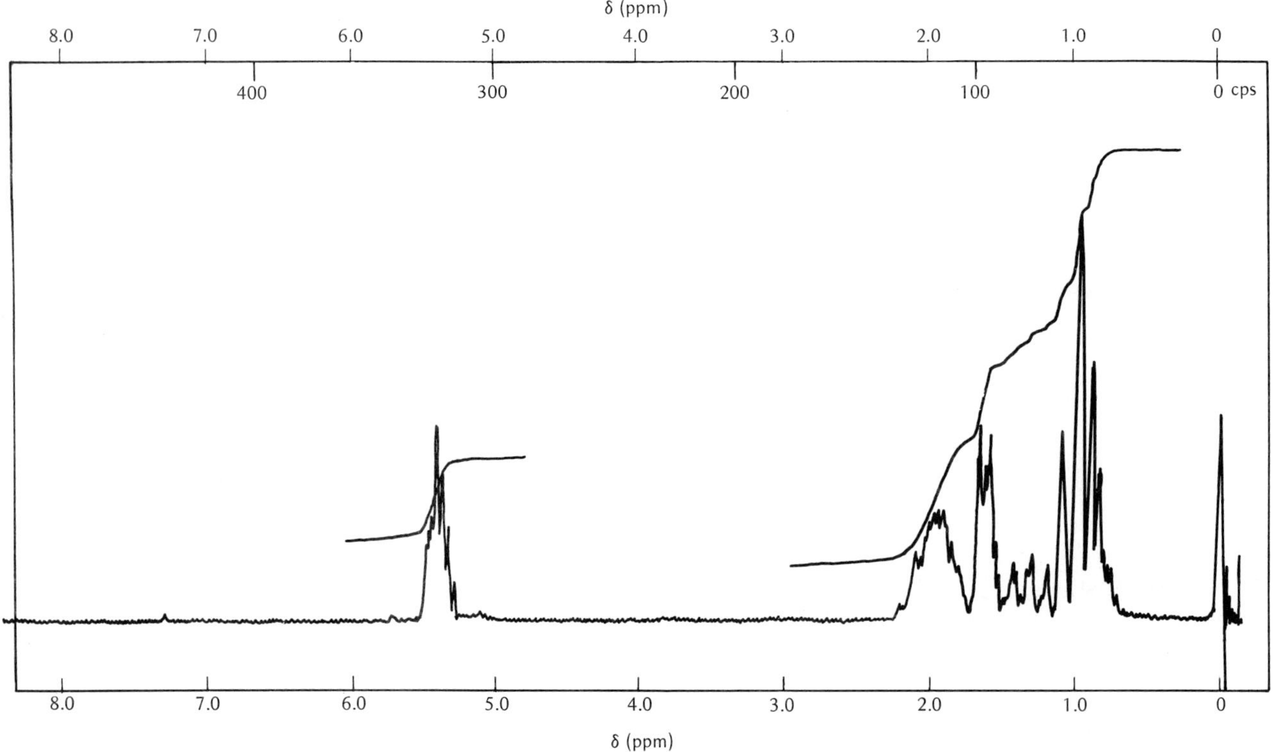

Fig. 2. NMR Spectrum of the Mixture of *trans*-2-Hexene and *trans*-3-Hexene Obtained as Described in Question 6.

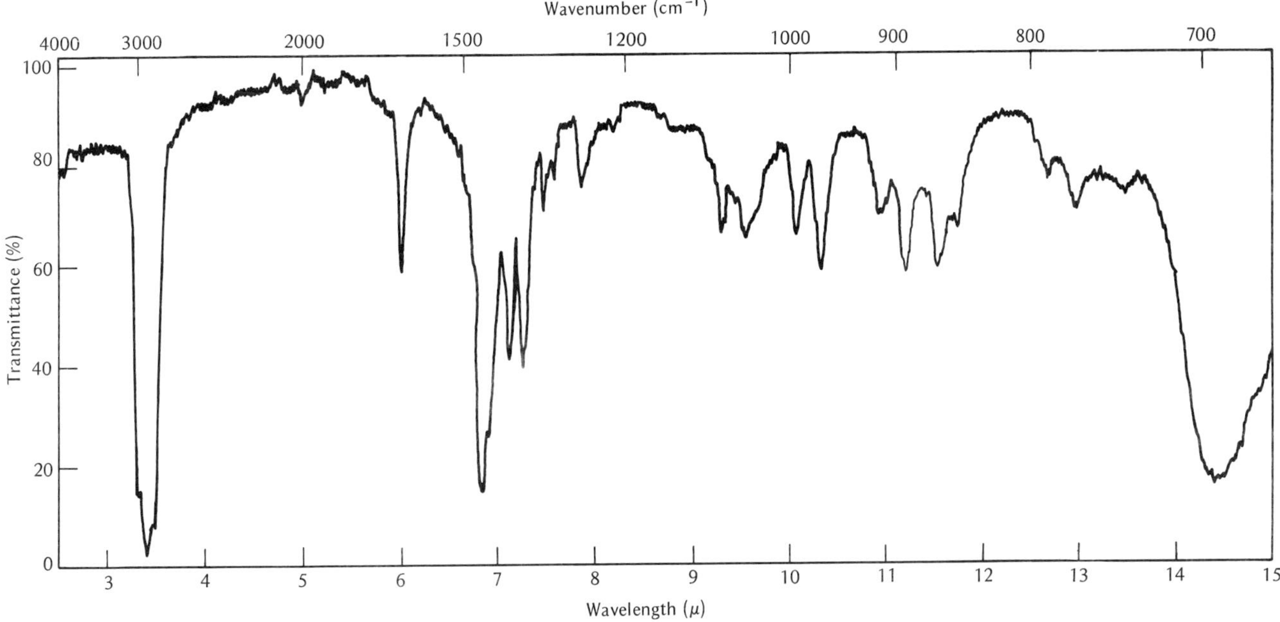

Fig. 3. Infrared Spectrum of *cis*-2-Hexene.

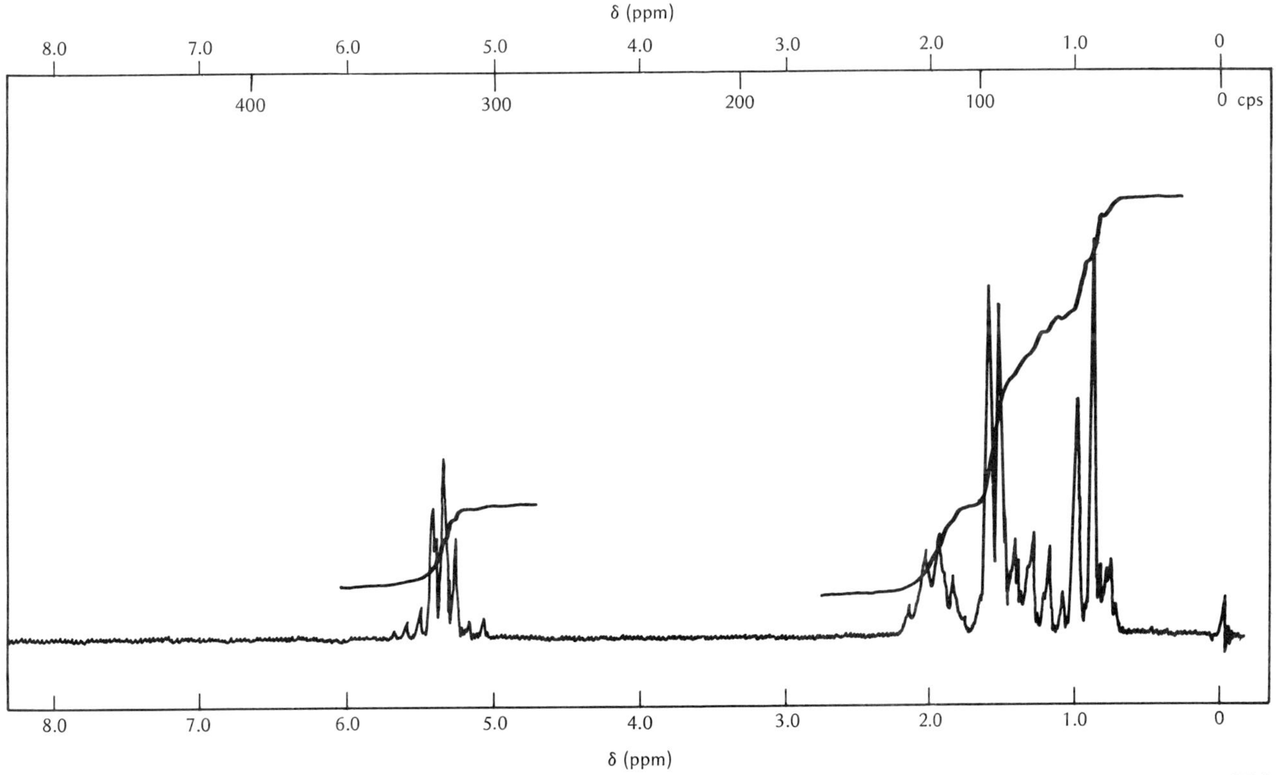

Fig. 4. NMR Spectrum of *cis*-2-Hexene.

Wavenumber (cm⁻¹)
4000
3000
2000
1500
1200
1000
900
800
700
100
80
60
40
20
0
Transmittance (%)
3
4
5
6
7
8
9
10
11
12
13
14
15
Wavelength (μ)

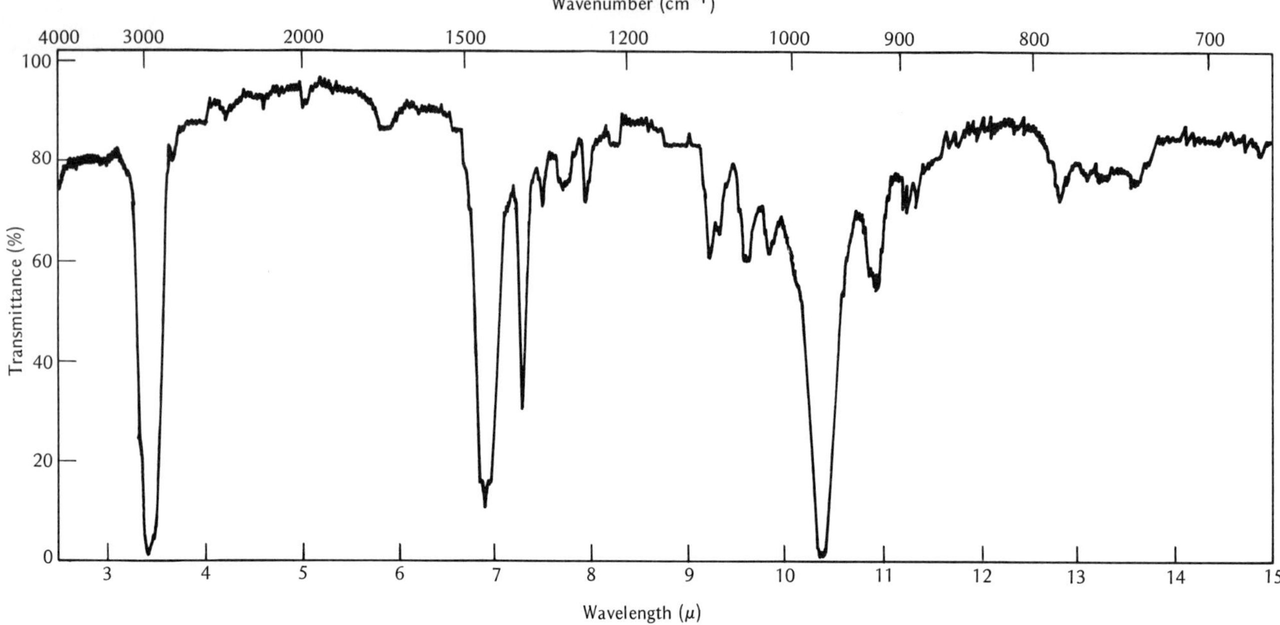

Fig. 5. Infrared Spectrum of *trans*-2-Hexene.

δ (ppm)
8.0
7.0
6.0
5.0
4.0
3.0
2.0
1.0
0
400
300
200
100
0 cps
8.0
7.0
6.0
5.0
4.0
3.0
2.0
1.0
0
δ (ppm)

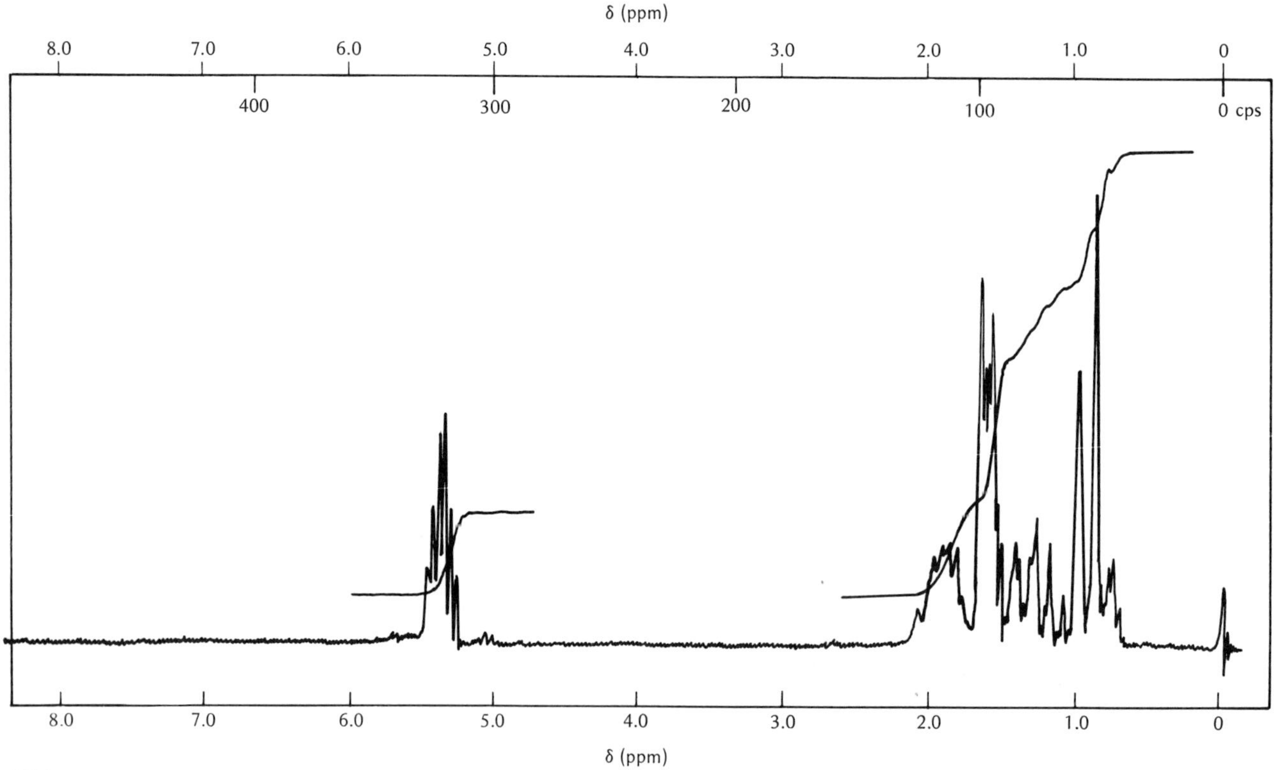

Fig. 6. NMR Spectrum of *trans*-2-Hexene.

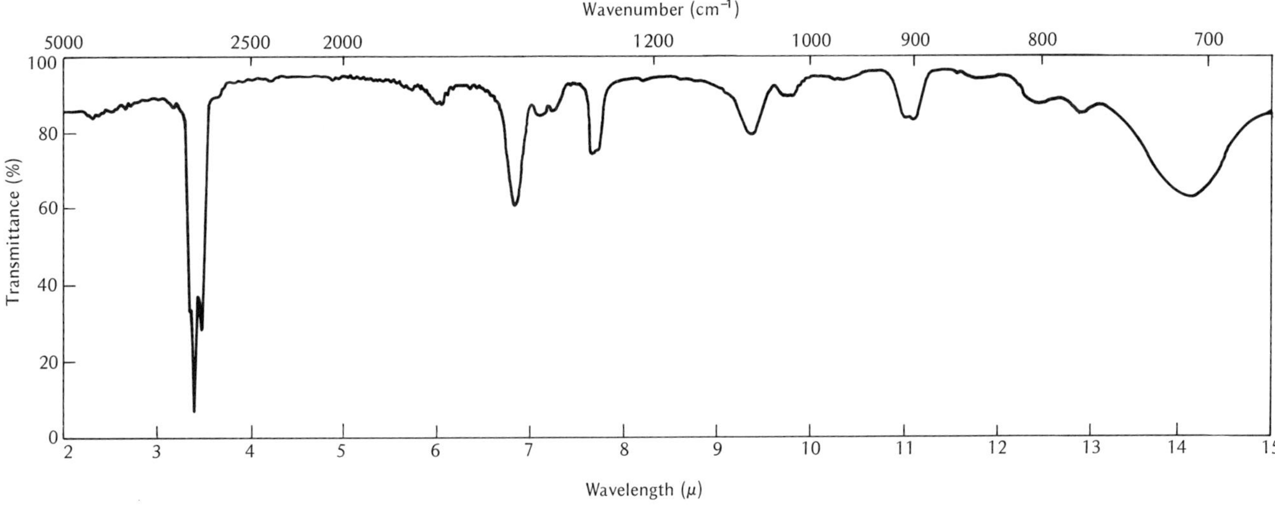

Fig. 7. Infrared Spectrum of *cis*-3-Hexene. (© Sadtler Research Laboratories, Inc., 1969.)

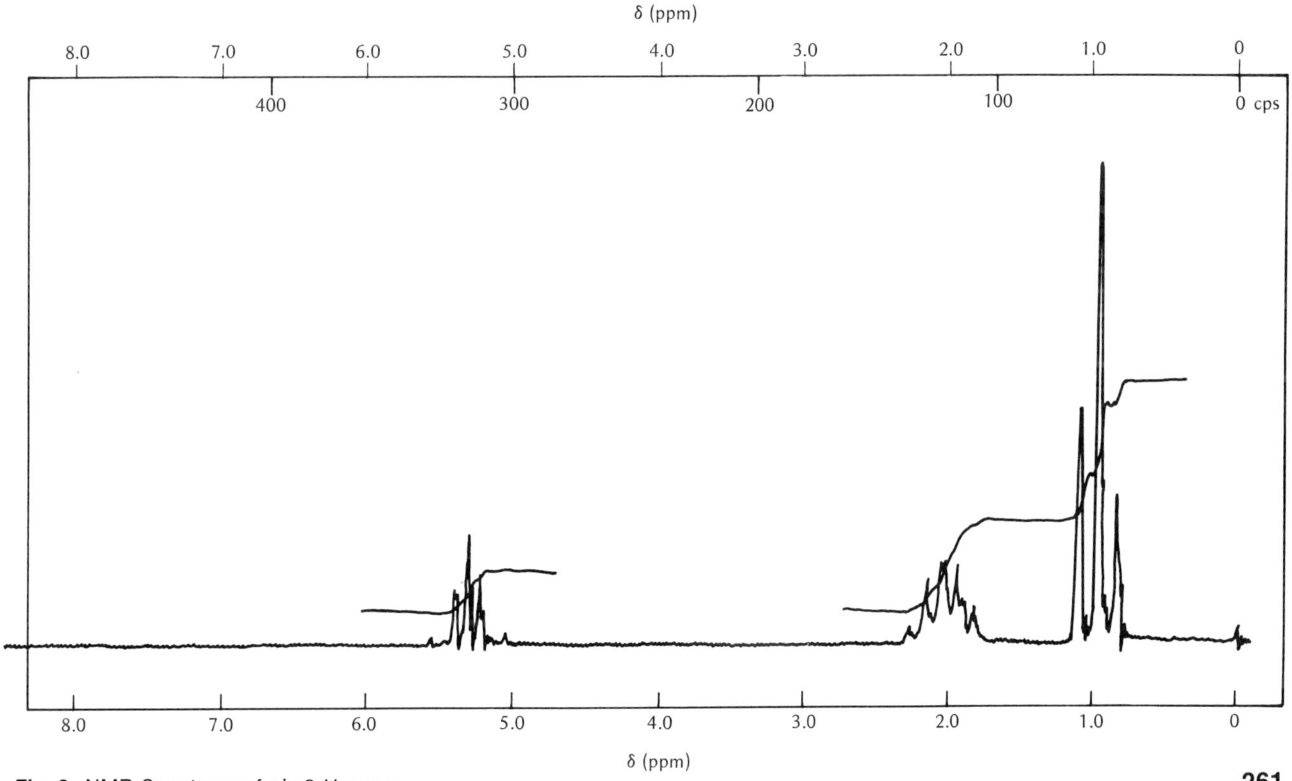

Fig. 8. NMR Spectrum of *cis*-3-Hexene.

Wavenumber (cm⁻¹)
5000
2500
2000
1200
1000
900
800
700
Transmittance (%)
100
80
60
40
20
0
2
3
4
5
6
7
8
9
10
11
12
13
14
15
Wavelength (μ)

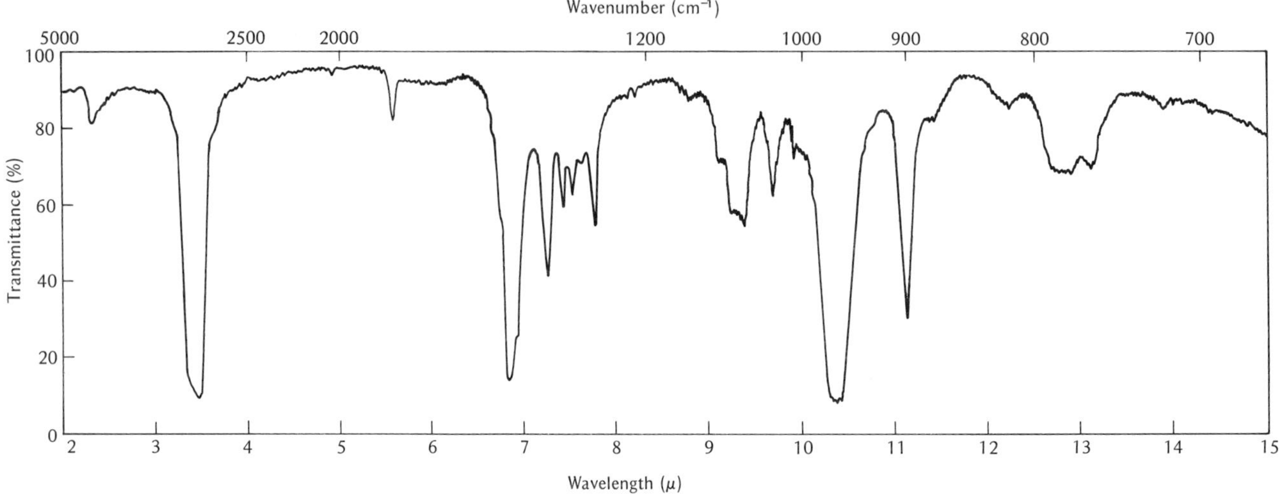

Fig. 9. Infrared Spectrum of *trans*-3-Hexene. (© Sadtler Research Laboratories, Inc., 1959.)

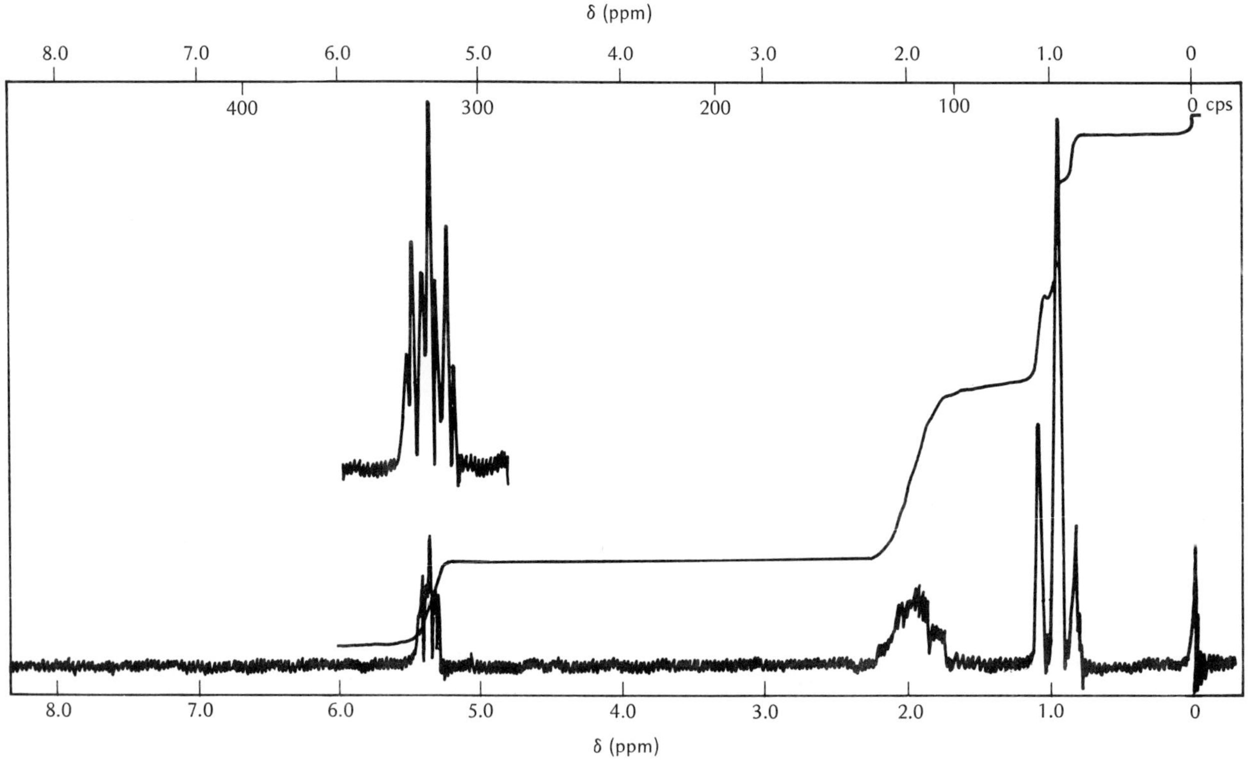

δ (ppm)
8.0
7.0
6.0
5.0
4.0
3.0
2.0
1.0
0
400
300
200
100
0 cps
8.0
7.0
6.0
5.0
4.0
3.0
2.0
1.0
0
δ (ppm)

Fig. 10. NMR Spectrum of *trans*-3-Hexene. (© Sadtler Research Laboratories, Inc., 1968.)

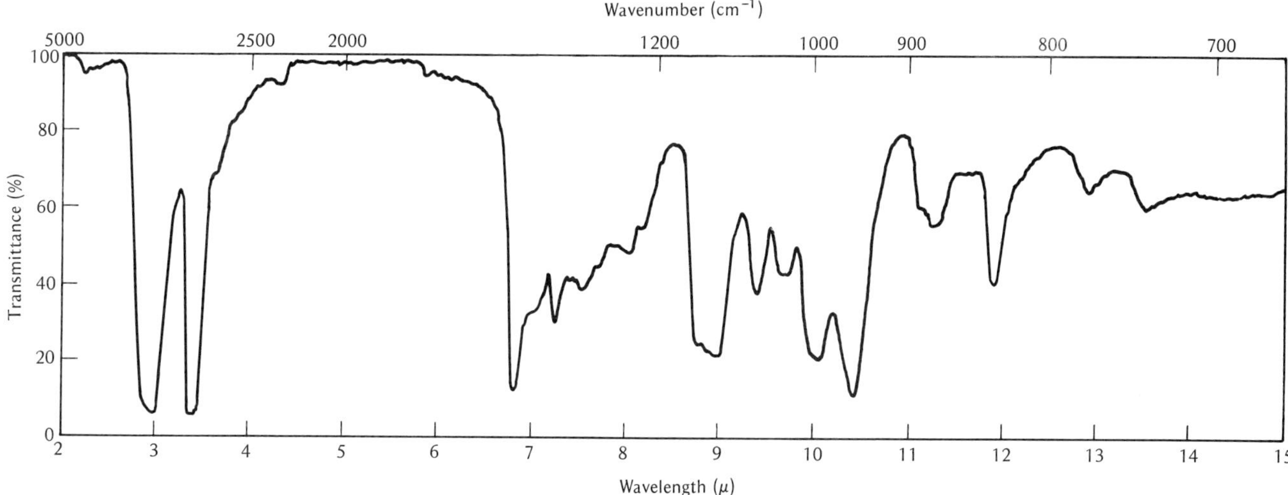

Fig. 11. Infrared Spectrum of 3-Hexanol. (© Sadtler Research Laboratories, Inc., 1954.)

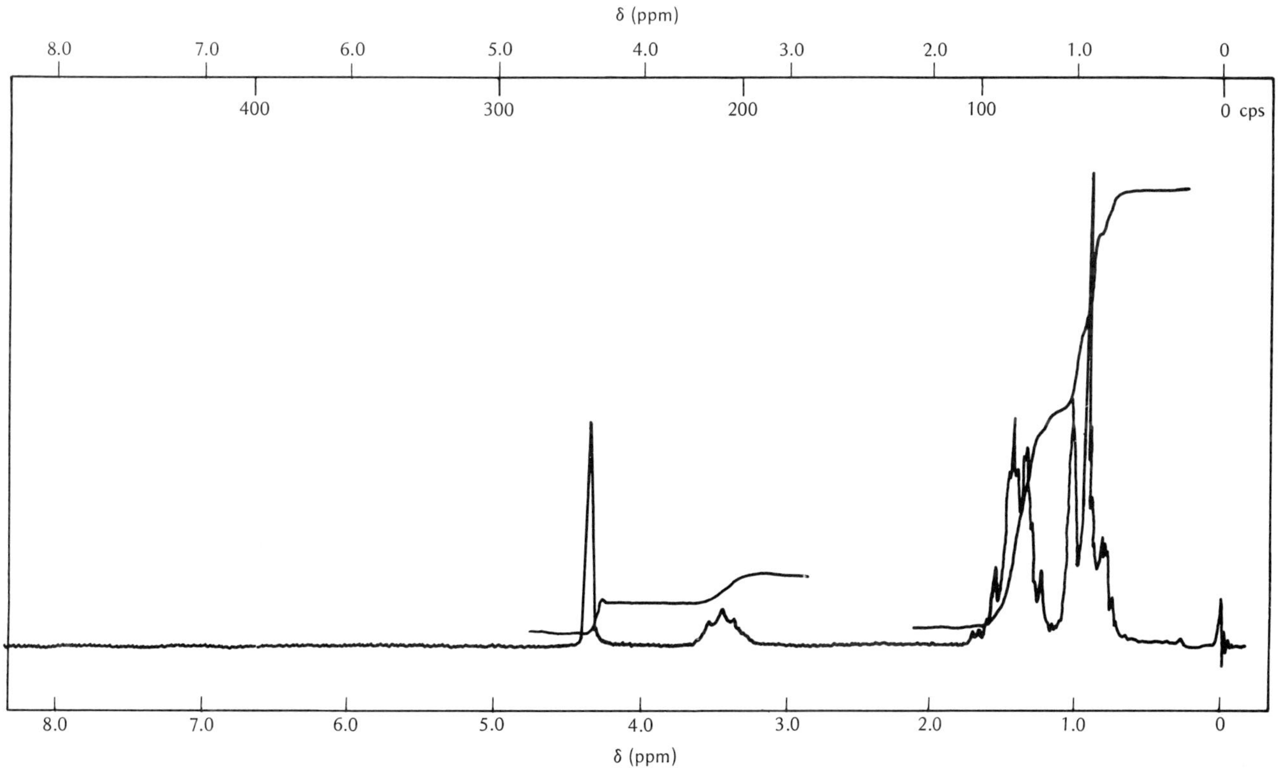

Fig. 12. NMR Spectrum of 3-Hexanol.

Wavenumber (cm⁻¹)
4000
3600
3200
2800
2400
2000
1800
1600
1400
1200
1000
800
700
Transmittance (%)
100
90
80
70
60
50
40
30
20
10
0
2.5
3
4
5
6
7
8
9
10
11
12
13
14
15
Wavelength (μ)

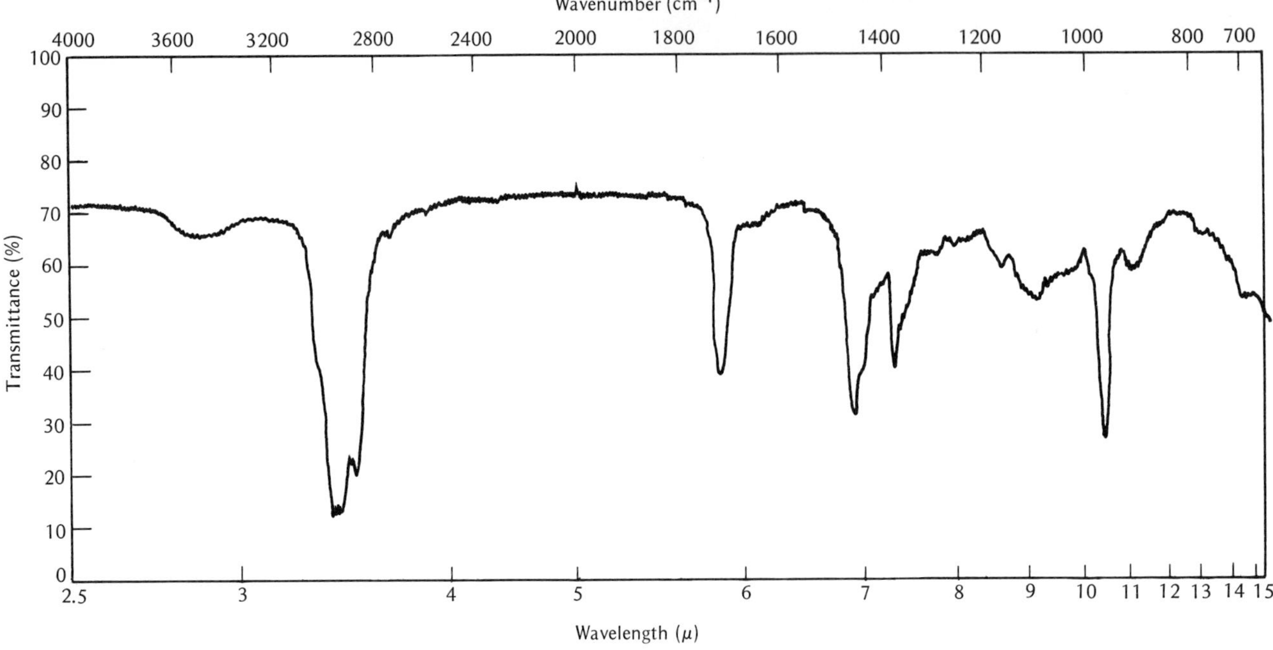

Fig. 13. Infrared Spectrum of the Product of Dehydration of 1-Hexanol (Question 7). This spectrum was taken on an instrument that is linear in wavenumbers instead of wavelengths. The wavenumber scale changes at 2000 cm^{-1} so that the space per unit is smaller at wavenumbers greater than 2000.

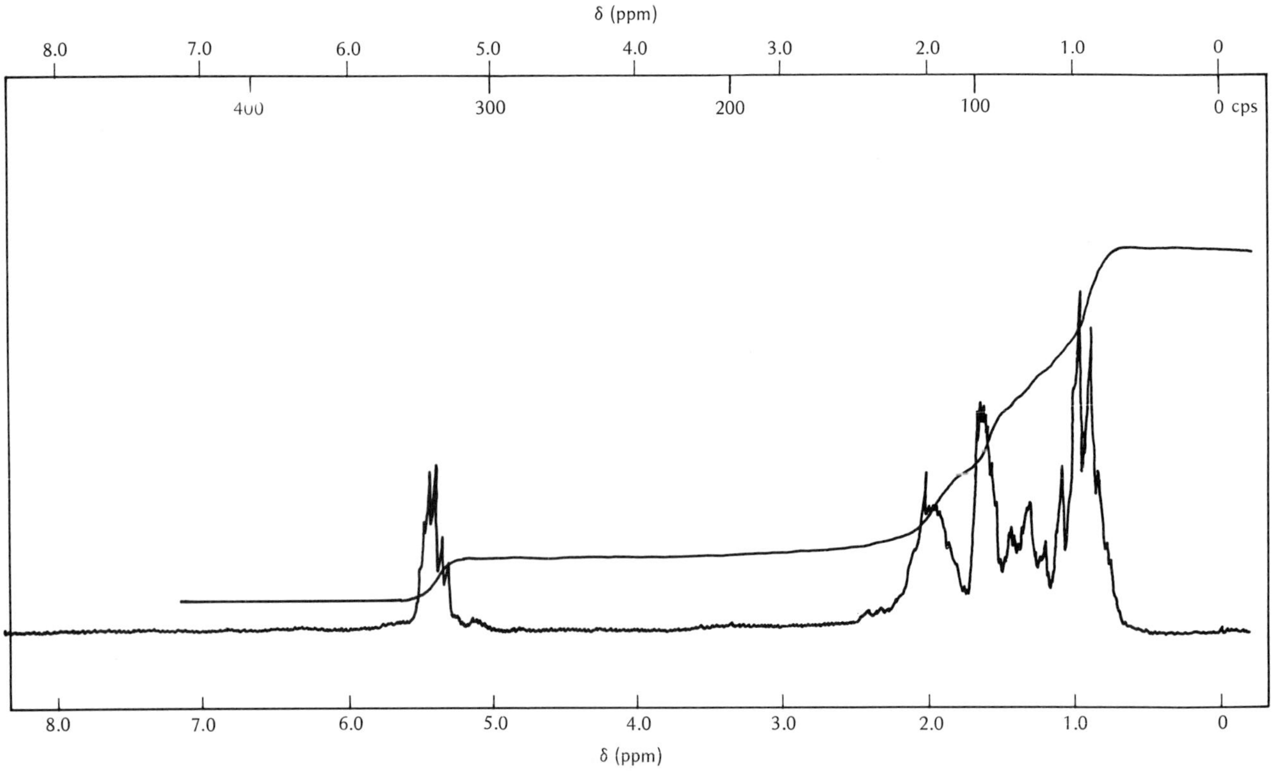

δ (ppm)
8.0
7.0
6.0
5.0
4.0
3.0
2.0
1.0
0
400
300
200
100
0 cps
8.0
7.0
6.0
5.0
4.0
3.0
2.0
1.0
0
δ (ppm)

Fig. 14. NMR Spectrum of the Product of Dehydration of 1-Hexanol (Question 7).

Wavenumber (cm^{-1})
5000 2500 2000 1200 1000 900 800 700
Transmittance (%)
100 80 60 40 20 0
2 3 4 5 6 7 8 9 10 11 12 13 14 15
Wavelength (μ)

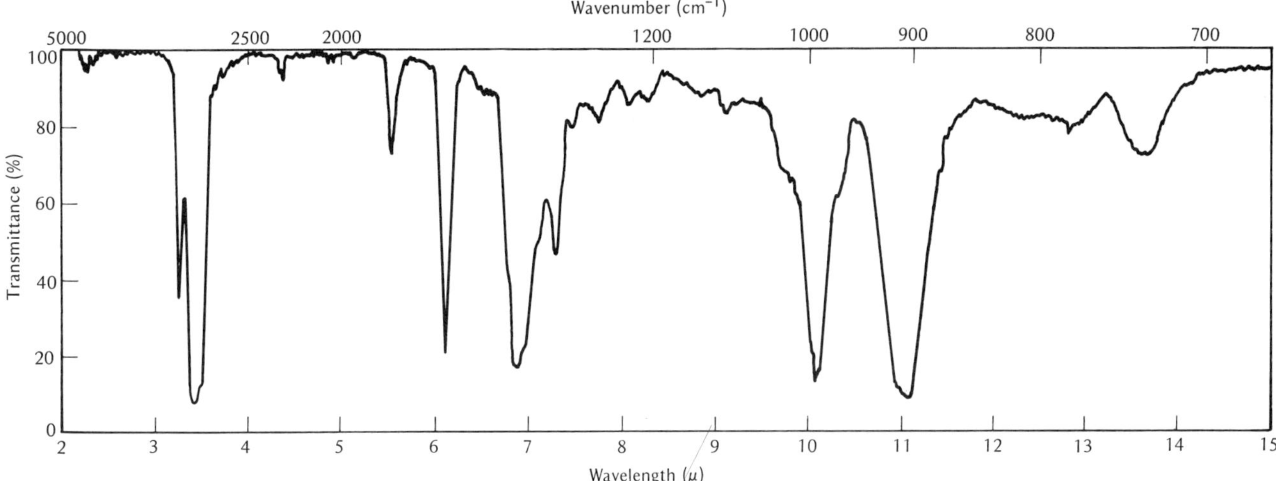

Fig. 15. Infrared Spectrum of 1-Hexene. (© Sadtler Research Laboratories, Inc., 1962.)

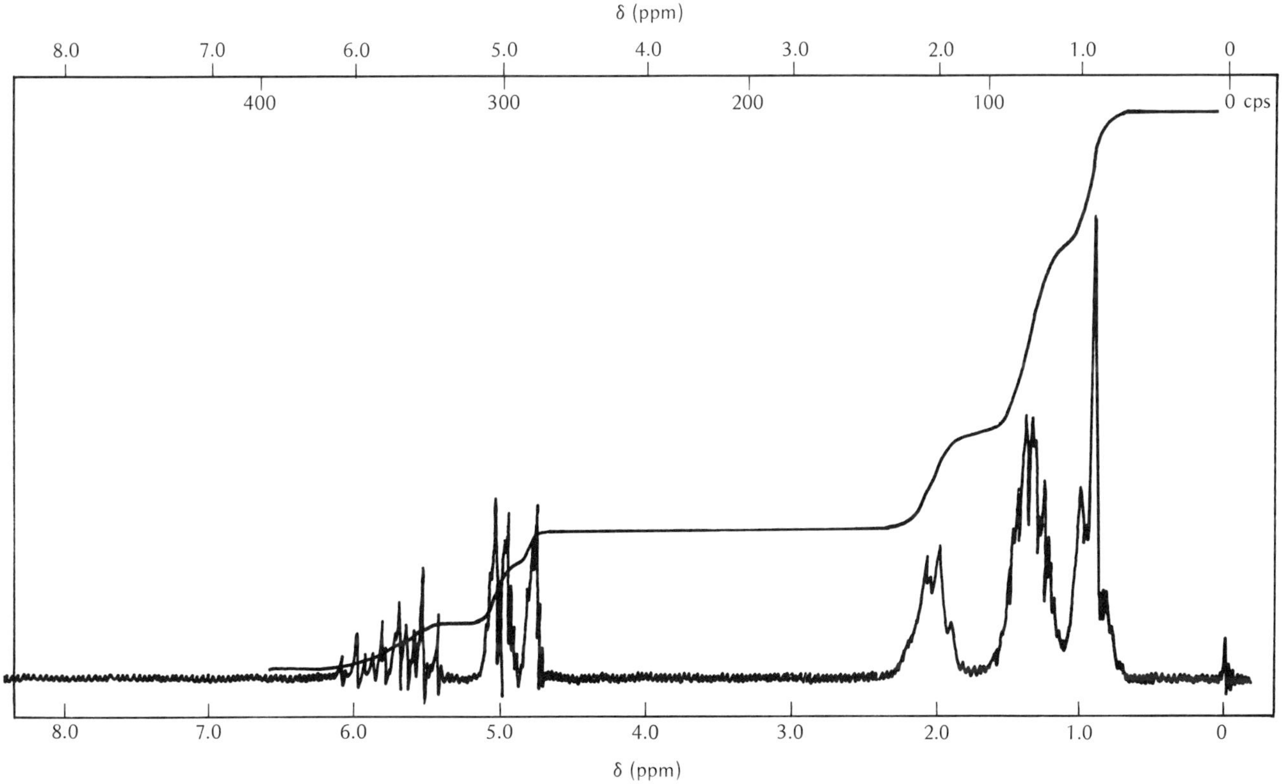

δ (ppm)
8.0 7.0 6.0 5.0 4.0 3.0 2.0 1.0 0
400 300 200 100 0 cps
8.0 7.0 6.0 5.0 4.0 3.0 2.0 1.0 0
δ (ppm)

Fig. 16. NMR Spectrum of 1-Hexene. (© Sadtler Research Laboratories, Inc., 1967.)

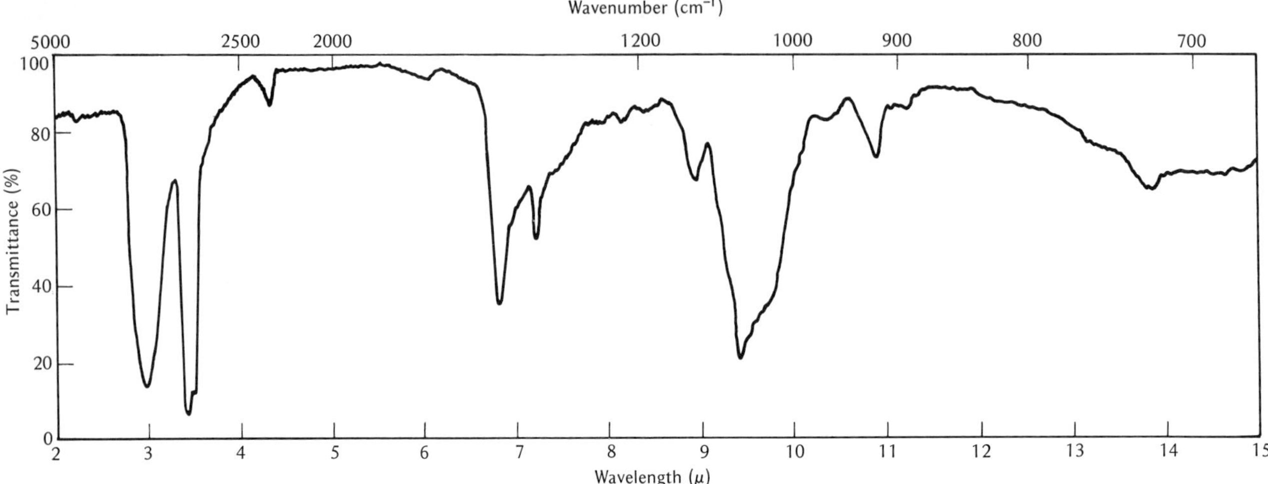

Fig. 17. Infrared Spectrum of 1-Hexanol. (© Sadtler Research Laboratories, Inc., 1967.)

δ (ppm)
8.0 7.0 6.0 5.0 4.0 3.0 2.0 1.0 0
400 300 200 100 0 cps
Acid added
8.0 7.0 6.0 5.0 4.0 3.0 2.0 1.0 0
δ (ppm)

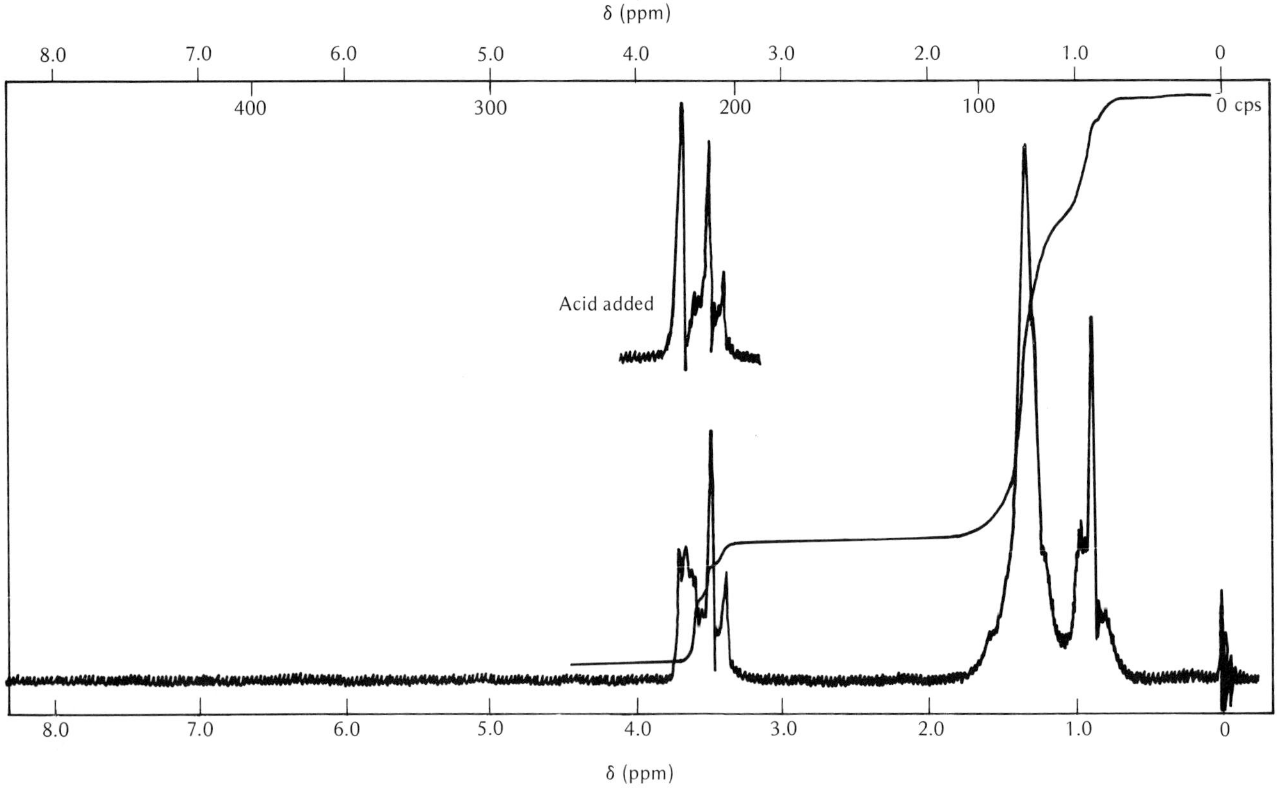

Fig. 18. NMR Spectrum of 1-Hexanol. (© Sadtler Research Laboratories, Inc., 1967.)

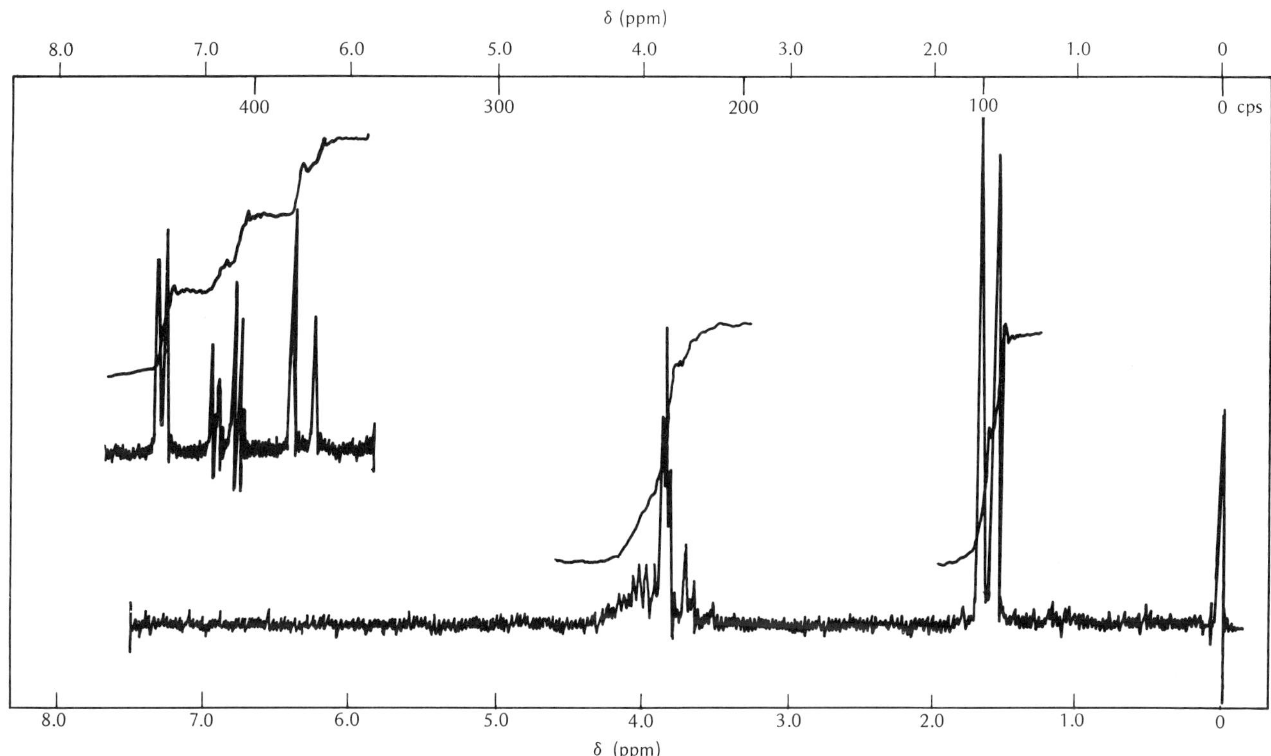

Fig. 19. NMR Spectrum of Adduct A from Addition of 2,4-Dinitrobenzenesulfenyl Chloride to Propylene. Solvent, 1:1 $(CD_3)_2CO$ and $CDCl_3$, sweep width 500. The downfield portion of the spectrum has been offset by 100 cps in order to show the aromatic protons on this scale.

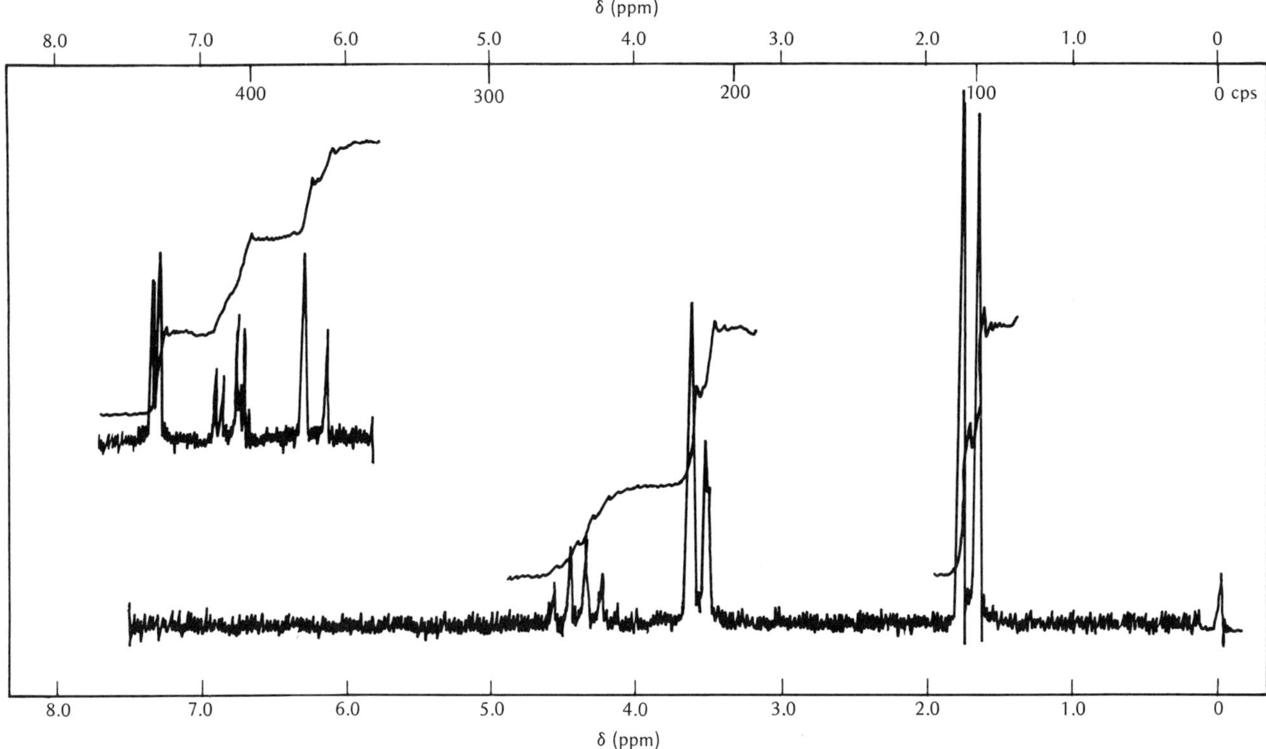

Fig. 20. NMR Spectrum of Adduct B from Addition of 2,4-Dinitrobenzenesulfenyl Chloride to Propylene. Solvent, 1:1 $(CD_3)_2CO$ and $CDCl_3$, sweep width 500. The downfield portion of the spectrum has been offset by 100 cps in order to show the aromatic protons on this scale.

Experiment 14

Organometallic Compounds, Grignard Reactions, and Hydroboration

226. Organomagnesium Compounds. Metallic magnesium reacts with most alkyl or aryl halides (fluorides excepted) in the presence of a weakly basic solvent, such as ether, to form a Grignard reagent. The precise nature of the reagent is complex, but it is usually written RMgX for convenience. The equilibrium below represents the facts more precisely.

$$2RMgX \rightleftharpoons R_2Mg{\cdot}MgX_2 \rightleftharpoons R_2Mg + MgX_2$$

There has been some controversy about the rate of attainment of this equilibrium and the relative amounts of the different compounds vary with structure, solvent, etc. X-ray crystallography shows that the crystals isolated from the Grignard reagent prepared from bromobenzene and magnesium in diethyl ether are monomeric with phenyl, bromine, and two ether molecules bonded tetrahedrally to the magnesium.

$$\begin{matrix} C_6H_5 \searrow & & \swarrow O(C_2H_5)_2 \\ & Mg & \\ Br \nearrow & & \nwarrow O(C_2H_5)_2 \end{matrix}$$

The situation in solution is far less certain and aryl Grignard reagents are often dimeric, or more highly associated, in ether.

The nature of the solvent is important for the successful preparation of Grignard reagents. Until recently, diethyl ether was almost always used for synthetic applications, although sometimes this solvent was removed after the reagent had been prepared, and replaced by one that boiled higher in order to increase the rate of certain slow reactions. Such unreactive halides as aryl chlorides or vinyl halides usually failed to react with magnesium in ether. Tetrahydrofuran, a slightly more basic ether than diethyl, permits the preparation of Grignard reagents from these unreactive halides as well as from most others. This solvent is less volatile than diethyl ether (bp 66° instead of 35°) and is safer to use. It has the disadvantage that it is completely soluble in water, which sometimes makes products of reactions of the Grignard reagent harder to isolate. If the reaction mixture is poured into water, tetrahydrofuran may act as a mutual solvent to keep products in solution. Tetrahydrofuran is about the same price as anhydrous ether. In most instances it can be used without further drying. It is specified in the present experiment for reasons of safety. The commercial material has usually been stabilized with 0.1% of hydroquinone to prevent peroxide formation. This antioxidant need not be removed.

Most Grignard reagents react smoothly with many carbonyl compounds by addition, although reduction and enolization are side reactions which are of increasing importance for more highly branched carbonyl compounds and reagents. An ester normally reacts with two molecular equivalents of the organomagnesium compound to yield, after hydrolysis, a tertiary alcohol in which two of the groups attached to the carbinol carbon are alike. The reactions are the following:

$$R'C\begin{matrix}\nearrow O\\ \searrow OC_2H_5\end{matrix} + RMgX \longrightarrow R'-\overset{\displaystyle OMgX}{\underset{\displaystyle OC_2H_5}{\overset{|}{\underset{|}{C}}}}-R \longrightarrow R'\overset{\displaystyle O}{\overset{\|}{C}}R + MgXOC_2H_5$$

$$R'\overset{\displaystyle O}{\overset{\|}{C}}R + RMgX \longrightarrow R'\overset{\displaystyle OMgX}{\overset{|}{C}}R_2 \xrightarrow[H_2O]{HX} R'R_2COH + MgX_2$$

It is seldom possible to isolate a reasonable yield of ketone in this reaction, even when the ester is present in excess, because most ketones are more reactive than the corresponding esters.

An ester is chosen to illustrate the reaction of carbonyl compounds with Grignard reagents because the product is more readily isolated than are products from most available ketones.

227. Preparation of a Grignard Reagent. Grignard reagents are usually prepared with stirring in an apparatus such as shown in Fig. 4 or Fig. 5, §26; most halides react more rapidly and give higher yields under these conditions. For experiments on a larger scale than the one below, stirring becomes essential. The reagent also reacts rapidly with moisture and with the oxygen of the air, so it is necessary to protect the solution by a dry, inert atmosphere if the reagent is to be stored or if the preparation and further reaction of the reagent takes place below reflux temperature. One usually depends upon solvent vapor from the refluxing reaction mixture to prevent air from attacking the reagent. The apparatus shown in Fig. 1

is satisfactory for the present experiment. The dropping funnel is mounted in the top of the condenser with a slotted cork that permits necessary relief of pressure from an otherwise closed system.

Bromobenzene is specified because with it the reaction usually starts at once. Chlorobenzene may be tried if the apparatus provided with stirrer is available, but greater attention must then be paid to the freshness of the magnesium and the dryness of all reagents and equipment. Ethyl bromide (1 or 2 ml) should be used to initiate reaction if the less reactive halide is employed.

Assemble the apparatus (200-ml flask) and be sure that all parts of it are dry (dry corks are important too). Place 3.9 g (0.16 g at.) of freshly cut, dry magnesium turnings in the flask and connect a calcium chloride tube at the top of the condenser to keep moisture out of the reaction. Provide an ice bath to cool the flask if the reaction becomes too vigorous. Do not interrupt the experiment from this point until the Grignard complex has been decomposed by dilute sulfuric acid as described below.

Prepare a solution of 26.7 g (0.17 mole) of dry bromobenzene in 50 ml of tetrahydrofuran. Remove the cork that carries the calcium chloride tube and add 25 ml of the solution to the magnesium; then replace the drying tube. The reaction should start promptly, as shown by the generation of heat. If the reaction does not begin within a minute, disconnect the flask and crush one or two pieces of the magnesium below the surface of the solution with a clean, dry glass rod. The trouble may be the result of old magnesium that has a corroded surface; the crushing operation serves to expost fresh surface. (Part of the beneficial results of stirring Grignard preparations probably arises from contact between the magnesium and the stirrer.) A crystal of iodine may also initiate reaction, or the reaction mixture may be refluxed with steam bath or hot plate. A little dry mercuric chloride may also be added to amalgamate the magnesium and hence activate it. Should all these devices fail to induce reation, it is probable that reagents or apparatus were not dry enough. In that event start over with more carefully dried equipment and chemicals. Do not add the remainder of the bromobenzene solution until the reaction is proceeding, since the reaction may then start suddenly, get out of hand as a result of the large amount of halide present, and expel foam out of the top of the condenser.

After the reaction starts, replace the drying tube by the dropping funnel with slotted cork and add the remainder of the bromobenzene solution dropwise. The addition can usually be completed safely in 20 min, but the objective should be to add the bromobenzene rapidly enough to maintain reflux without the application of external heat and yet not to overtax the condenser. Swirl the flask occasionally if a stirrer is not in use. The yields of Grignard reagents are usually better if the halide is added as rapidly as can be done without flooding the condenser. After addition is complete, replace the dropping funnel by the drying tube and reflux

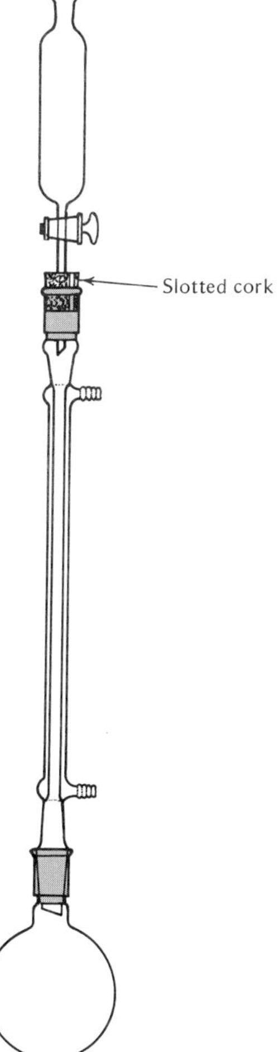

Fig. 1

the reaction mixture until all the magnesium has reacted (usually 15–20 min; a little finely divided magnesium left in the solution does no harm). The reagent should be used at once to prepare either triphenylcarbinol or benzoic acid as described in the next sections.

228. Triphenylcarbinol. A solution of 12 g (0.08 mole) of dry ethyl benzoate in 30 ml of tetrahydrofuran is added dropwise during about 10 min to the Grignard solution. The solution is not heated during the addition, but the heat of reaction is sufficient to maintain refluxing. Addition should be started while the Grignard solution is still warm. After addition is complete, the reaction mixture is allowed to stand for 15–20 min. Then carry out one of the following procedures as directed by your instructor.

PROCEDURE 1. Periods of reflux or standing in the above procedure should be utilized to assemble a steam-distillation apparatus (§79); a 1-liter flask should be employed. Place 100 g of ice, 150 ml of water, and 10–12 g of concentrated sulfuric acid in the 1-liter flask, and add the Grignard ★ reaction mixture with stirring.

Pass steam through the hydrolysis mixture, slowly at first while the tetrahydrofuran distills, then more rapidly. Unreacted ethyl benzoate and bromobenzene will steam-distill fairly rapidly, but biphenyl (the coupling product from the Grignard reagent and bromobenzene) comes over more slowly. Most of this will have distilled after 400–500 ml of distillate has been collected, although the distillate may still be milky. The triphenylcarbinol remains behind in the steam distillation flask, and when the passage of steam is stopped, special care must be taken so that the flask contents are not sucked back into the steam line (open the trap or disconnect the steam tube from the flask before the steam is completely turned off). Sometimes the triphenylcarbinol separates into oil globules that solidify and trap volatile impurities so they are not removed by the steam. If this occurs, interrupt the steam distillation (precaution against suck-back!) and break up the lumps with a flattened glass rod. Then continue the distillation.

Cool the contents of the steam distillation flask, filter with suction, wash the crude product with 10% sulfuric acid (break up lumps carefully so inorganic salts can be washed out) and several times with water. Press the precipitate as dry as possible, transfer to a 250-ml conical flask, and recrystallize from alcohol or ligroin (bp 66–75°) with decolorizing. The yield should be 10–15 g. Pure triphenylcarbinol melts at 162°, and material obtained here should melt not lower than 160°.

PROCEDURE 2. If the reaction has gone nearly to completion, the steam distillation may be omitted and a satisfactory product obtained by crystallization. Pour the Grignard reaction mixture with stirring into a 500-★ ml conical flask containing 150 ml of 10% sulfuric acid and 100 g of ice.

Dilute the mixture further with 100 ml of water. If an oil separates, cool the mixture and scratch with a glass rod until a solid is obtained. If too much starting material (bromobenzene and ethyl benzoate) remains, it may be impossible to induce crystallization. The mixture should then be steam-distilled as described in procedure 1 until these starting materials have been removed. One may avoid the loss of time involved in assembly of the special apparatus required for steam distillation by simple distillation of the aqueous hydrolysis mixture in an ordinary distillation apparatus. Some loss of product may result if organic matter clings to the inner walls of the flask and is decomposed by too strong heating. Bromobenzene and ethyl benzoate steam-distill fairly rapidly; no attempt is made to continue the distillation until biphenyl is removed. Triphenylcarbinol contaminated with biphenyl separates as a solid in the distilling flask when this is cooled.

Isolate the solid by suction filtration on a Büchner funnel, wash it with water, dry it briefly, and recrystallize from ligroin, bp 66–75°. Alcohol is not a satisfactory solvent for this recrystallization because it does not remove biphenyl. About 50 ml of ligroin should be required. The crude material may melt as low as 110–115°. If it contains inorganic salts, these will not dissolve and should be filtered out of the hot solution (not with suction). If the solution is colored, it should be decolorized (§§130–132). Evaporate the filtrate slowly on a steam bath until crystals of triphenylcarbinol begin to separate. Allow crystallization to proceed first at room temperature, then at 0°. Isolate the crystals as usual. Concentration of the mother liquor may yield a second crop, which should be combined with the first crop only if the melting point is as good.

229. Benzoic Acid. Prepare phenylmagnesium bromide as described in §227. Now place about 75 g of solid carbon dioxide (dry ice), coarsely crushed and as free as possible from particles of frost (H_2O), in a 600- or 800-ml beaker. Add slowly, with stirring, the ethereal solution of the Grignard reagent. The mixture becomes very thick. Finally, break up the almost solid mass, allow carbon dioxide gas to escape, and dissolve the material in a mixture of 20 ml of concentrated hydrochloric acid, 50–75 g of ice, and 100 ml of cold water. The cold acid may be added to the Grignard reaction mixture, but sometimes the reagent does not react completely with carbon dioxide so that, at first, addition must be cautious. Mix the material thoroughly and transfer it to a separatory funnel. Rinse the beaker with ordinary diethyl ether and add the rinse ether to the material in the separatory funnel. Add a total of about 50 ml of ether, shake cautiously, and separate layers. Extract the aqueous acid solution twice more with ether (25-ml portions) and combine the ether solution in the separatory funnel.

Extract the benzoic acid from the ethereal solution with aqueous alkali (carbonate or hydroxide). Clarify the alkaline extract with decolorizing carbon and add hydrochloric acid until precipitation is complete. Recrystallize the precipitate from water. Mp is 122°. Yield is 5–10 g.

230. Other Halides. Bromobenzene is not the only halide that can be used, but relatively few of the others have been tried in tetrahydrofuran. Many Grignard reagents give side reactions to a greater extent than the one from bromobenzene. Grignard reactions have been surveyed up to 1954 in M. S. Kharasch and O. Reinmuth, *Grignard Reactions of Nonmetallic Substances,* Prentice-Hall, Englewood Cliffs, N.J., 1954. The preparation of a number of arylmagnesium chlorides is described by Ramsden et al., *J. Org. Chem.,* **22,** 1202 (1957). H. Normant has reviewed alkenylmagnesium halides in Vol. II of *Advances in Organic Chemistry: Methods and Results,* R. A. Raphael, ed, Wiley–Interscience, New York, 1960. See also E. C. Ashby, "Grignard Reagents: Composition and Mechanism of Reaction," *Quarterly Reviews,* **21,** 259 (1967).

231. Procedure with Diethyl Ether. Phenylmagnesium bromide may be prepared in anhydrous diethyl ether by essentially the same procedure as for tetrahydrofuran, but much greater care must be taken to condense ether vapors and avoid all flames in the laboratory. There may be greater difficulty in getting the reaction to start. It is preferable to place 15 ml of ether and 4 ml of bromobenzene on the magnesium. After the reaction starts, an additional 70 ml of ether should be added all at once and the pure bromobenzene added dropwise. Addition of the bromobenzene must be slower in order to keep ether from going out of the top of the condenser. Usually it takes longer for the reaction to reach completion because the temperature is lower. After addition is complete, 30 min or more of refluxing may be necessary.

Triphenylcarbinol is prepared by dropwise addition of 0.08 mole of ethyl benzoate in 35 ml of anhydrous ether as before (slower addition necessary). The reaction mixture should be allowed to stand for 30 min before hydrolysis.

Benzoic acid is prepared as directed above.

232. 2-Methyl-2-hexanol. Synthesis of this compound involves the preparation of a *n*-butyl Grignard reagent, which is accomplished somewhat more readily than the formation of the phenyl derivative. The phenyl bromide and ethyl benzoate of §§227 and 228 are replaced by *n*-butyl bromide and acetone, the latter especially well dried with freshly prepared anhydrous magnesium sulfate.

Assemble a reflux apparatus using your 200-ml round-bottom flask; leave all rubber hoses off for now. Using a cool, but *not* luminous Bunsen flame, begin at the bottom of the flask and heat gently. Proceed upward slowly and follow the condensation until all the water vapor is chased out the top of the condenser. Fit the top with a cork holding a drying tube and allow to cool slowly. [The drying tube is prepared by placing a plug of glass wool into the bulb end, filling partially with drying agent, and placing a second plug of glass wool in the other end to hold the drying agent (Drierite)].

At this point all flames must be extinguished *and for the remainder of the period no flames will be allowed in the laboratory*. No ether will be dispensed until *all* flames are out. After your apparatus has cooled, connect the condenser tubing and begin the condenser flow. Place 2.4 g of clean, dry magnesium turnings in a clean, dry mortar and grind them a few strokes to provide a clear surface, free of oxidation. Place this in the round-bottom flask, add about 25 ml of anhydrous ether (preferably freshly dried over sodium), and add a *small* crystal of iodine to catalyze the reaction with butyl bromide. Suspend a small separatory funnel in the top of the condenser by means of a cork, notched so that the system is not closed, and place in it 13.5 ml (17 g) of pure *n*-butyl bromide. Allow about 1 ml of the *n*-butyl bromide to drop into the flask and wait for the reaction to begin, as evidenced by some bubbling and boiling of the ether. If there is a delay of more than a few minutes, warm the flask gently with your hand and wait for a few more minutes. If the reaction has still not started, do not add more *n*-butyl bromide, but consult your instructor instead.

When the reaction is started, add 75 ml of dry ether to the separatory funnel and swirl to mix the ether with the butyl bromide. Add this solution at a rate sufficient to keep the reaction going but not so rapidly that the boiling gets out of control. If the boiling begins to get out of control, cooling may be necessary, but do not cool so much that the boiling stops entirely. After the reaction has subsided and the ether no longer boils, cool the flask to ice-bath temperature.

Add in small portions, through the condenser as before, a solution of 5.8 g of acetone in 50 ml of dry ether, vigorously shaking the flask after each addition. After the addition is complete, reflux the ether for 15 min by use of a steam cone, shaking the flask vigorously from time to time. (Stop here unless you can get to the drying stage.) ★

Cool the reaction mixture in an ice bath and add, very slowly at first, a saturated solution of ammonium chloride. A solid will form during the addition; when the addition of more ammonium chloride solution no longer causes precipitation of a solid, decant the ether layer into a separatory funnel. Wash the pasty residue with 10 ml of ether, and add this by decantation to the other ether solution in the separatory funnel. Wash the combined ether extracts with 10 ml of water, separate the ether layer, dry over anhydrous potassium carbonate or magnesium sulfate, and distill fractionally from an ordinary distilling flask to remove the ether. Yield is about 50%. Boiling point of 2-methyl-2-hexanol is 140°.

233. Hydroboration-Oxidation. Having initially illustrated the powerful synthetic utility of organometallics with one that was developed about the turn of the century by Victor Grignard, let us further demonstrate their value with another type, alkylboranes, whose versatility in organic syntheses was largely developed by Professor H. C. Brown and his students during the past fifteen years.

Alkylboranes are readily accessible by the process of hydroboration; in the presence of ether solvent, the B—H bond in diborane adds readily across a carbon-carbon double bond in a concerted *cis* fashion to yield trialkylborane.

$$B_2H_6 + 6RCH{=}CH_2 \xrightarrow{\text{ether}} 2(RCH_2CH_2)_3B$$

The reaction proceeds in such a way as to place the boron on the least-substituted carbon atom. However, if the olefin is highly substituted and sterically hindered, the reaction may proceed only to the dialkylborane, R_2BH, or the monoalkylborane, RBH_2. One of the most useful reactions involving the resulting organoborane is the alkaline hydrogen peroxide oxidation to form alcohols.

$$(RCH_2CH_2)_3B + NaOH + 3H_2O_2 \rightarrow 3RCH_2CH_2OH + NaB(OH)_4$$

Thus, the two steps, hydroboration and oxidation, provide a convenient and useful method for the anti-Markovnikov hydration of olefins.

Although diborane is now commercially available, it is usually generated directly in the laboratory. The most common method involves the reaction of sodium borohydride with boron trifluoride etherate, usually in diglyme or tetrahydrofuran (THF).

$$3NaBH_4 + 4BF_3{-}OEt_2 \rightarrow 3NaBF_4 + 4EtO_2 + 2B_2H_6$$

The diborane can be used directly or can be stored for long periods in cold THF. For the infrequent user of diborane, it is generally most convenient to generate the diborane as needed. If the generation is performed in THF, the resulting $NaBF_4$ is insoluble and precipitates out and can be removed by filtration if desired.

234. 1-Hexanol (OR, **13,** 1). **Warning:** *The boron hydrides and boranes are dangerously flammable!*

The apparatus used here is the same as that for the preparation of Grignard reagent described in §227, except that a 500-ml flask, containing a magnetic stirring bar, is employed. The assembled apparatus should be flame-dried with a stream of nitrogen passing through. Place 1.7 g of sodium borohydride into the flask followed by 17.9 ml of 1-hexene and 90 ml of tetrahydrofuran. Flush the system briefly with nitrogen, after which the boron trifluoride etherate (8.5 g; **warning:** *avoid direct contact*) in 20 ml of tetrahydrofuran is added dropwise from the funnel into the well-stirred mixture over a period of 15–30 min, the temperature being maintained at 25°. Keep the flask at 25° for an additional hour before destroying the excess hydride with water.

The organoborane is oxidized at 40–50° (water bath) by the addition of 16.6 ml of 3 *N* sodium hydroxide followed by the dropwise addition of 16.6 ml 30% hydrogen peroxide (**warning:** *strong oxidant*). Allow the reaction to stir at 50° for an additional 10 min after the peroxide addition to insure complete oxidation. Saturate the reaction mixture with sodium chloride, and then separate the tetrahydrofuran layer and wash with saturated aqueous sodium chloride. Dry the extract over anhydrous magnesium sulfate and then distill the 1-hexanol, bp 158°. (2-Hexanol, bp 140°, is formed in this reaction also, but only to the extent of about 6%.)

Questions

1. What reagents would be required to synthesize each of the following products by Grignard method:
(*a*) isopropyl alcohol.
(*b*) *tert*-amyl alcohol.
(*c*) α-naphthoic acid.
(*d*) β-phenylethyl alcohol.

2. Why is such care taken to exclude from the Grignard reaction (*a*) water, (*b*) ethyl alcohol, both of which occur in common commercial ether?

3. Why is an ether required in this experiment?

4. Suppose that the tetrahydrofuran (sp. g. 0.88) used in the triphenylcarbinol experiment, §228, had been contaminated with 0.5% of water but that in spite of this the preparation and reaction of the Grignard reagent was successful. How much benzene would be produced in the reaction and what is the maximum yield of triphenylcarbinol that could be obtained (i.e., what is the theoretical yield of triphenylcarbinol assuming that all of the water reacted with phenylmagnesium bromide before it had reacted with ethyl benzoate)?

Experiment 15
Stereochemistry

Stereoisomers are isomeric compounds of identical gross structure that differ only in the arrangement of their atoms in space. Two types of stereoisomerism are encountered.

235. Geometrical Isomerism. In geometrical isomerism the stereoisomers differ because rotation about one or more bonds and suitable difference in attached substituents prevent conversion of one isomer to the other under the usual conditions at which the compounds exist. This type of isomerism is illustrated here by the preparation of maleic and fumaric acids. The structural feature responsible for the stereoisomerism is the double bond. The experiment involves a chemical reaction in which this bond is reversibly converted to a single bond which permits interconversion of the isomers. Geometrical isomerism is also found in suitably substituted cyclic compounds.

236. Maleic and Fumaric Acids. Maleic anhydride, an important industrial reagent used in manufacture of plastics, is readily hydrolyzed to maleic acid, whose molecule, like that of the anhydride, has the *cis* configuration. The product is converted by hydrogen chloride into fumaric acid, its *trans* geometric isomer. It is presumed that a molecule of hydrogen chloride undergoes 1, 4 addition, to the conjugated system, yielding the transitory intermediate represented in brackets below. Since the central structure of the molecule is no longer rigidly held in the *cis* position, the two halves rotate to give an equilibrium mixture of the two isomers. The more stable of these has the carboxyl groups *trans*, and the product is largely this isomer, fumaric acid.

```
                                           Cl                 O
                                           |                  ||
HC==CH           HC—CH               [    HC—CH    ]         HOC
|    |      →    |    |         →    [    |   ||   ]    →      |
OC   CO         HOC   COH            [  HOC   COH  ]          HC=CH
  \O/            ||   ||             [   ||    |   ]              |
                 O    O              [   O     O   ]              COH
                                     [         H   ]              ||
                                                                  O
```

Warning: *Maleic anhydride is toxic and a powerful irritant. It causes burns. Avoid contact with skin, eyes and clothing; do not breathe the concentrated vapor. In case of contact immediately flush skin or eyes with water for at least 15 min.*

In a 125-ml conical flask containing 30 ml of water, heated just to boiling, place 25 g of maleic anhydride. The anhydride first melts but soon combines with water and dissolves. Cool the solution under the tap to slightly below room temperature and allow it to stand for about 10 min. Collect the solid product on a Büchner filter and dry without attempting to wash the crystals. Maleic acid is soluble in water at 25° to the extent of 78.8 g in 100 ml. The melting point of maleic acid is 137–138°.

To the mother liquor, which contains a substantial residue of maleic acid, add 25 ml of CP concentrated hydrochloric acid, and reflux very gently for about 10 min. Crystals of fumaric acid soon separate from the hot solution. Cool the reaction mixture, collect the crystals and wash them with a little cold water. Recrystallize from hot water (about 12 ml of water required per gram of fumaric acid), with slow cooling. Combined yield of maleic and fumaric acids is about 25 g. Fumaric acid sublimes above 200°; its solubility in water at 25° is 0.70 g/100 ml.

Slow cooling of the maleic acid solution yields larger crystals (monoclinic prisms).

237. Optical Isomerism. A second kind of stereoisomerism, called optical isomerism, is the result of molecular dissymmetry. A dissymmetric object cannot be superimposed on its mirror image. When a compound and its mirror image are not superimposable, they are said to be enantiomers and have the property of rotating the plane of plane-polarized light. One enantiomer rotates the plane a characteristic number of degrees in one direction, and the other rotates it an equal amount in the other direction. Plane-polarized light constitutes a dissymmetric physical environment. Enantiomers have exactly the same physical and chemical properties except when they encounter dissymmetric physical or chemical environments. Thus it would be expected that the chemical reactivities of the two members of a pair of enantiomers would be different for reaction with another dissymmetric (chiral) molecule.

Enantiomers are optical isomers, but not all optical isomers are enantiomers. Dissymmetry in organic molecules usually results from the presence of a carbon attached to four nonequivalent groups; such a carbon is called an asymmetric carbon. Dissymmetry may also result from a dissymmetric group of atoms or from an asymmetric atom other than carbon. If there is more than one such center of asymmetry in a molecule, an increased number of optical isomers results and not all of these are enantiomers. Any two compounds that are optical isomers of one another, but are not enantiomers, are called diastereoisomers.

[§237] A mixture of equal numbers of enantiomeric molecules is optically inactive and is called a racemic modification. A process that separates the enantiomers is called a resolution. The properties of the various kinds of optical isomers are illustrated in the present experiment by resolution of racemic α-methylbenzylamine. Reaction of this racemic modification with (+)-tartaric acid yields the two salts I and II; these salts are diastereoisomers. The electrostatic bond that unites the hydrogen tartrate anion with the ammonium cation is as real for determining symmetry as the covalent bonds. Although the ammonium part of I is the mirror image of the ammonium part of II, the cationic parts are not mirror images. Thus one salt molecule is not the mirror image of the other.

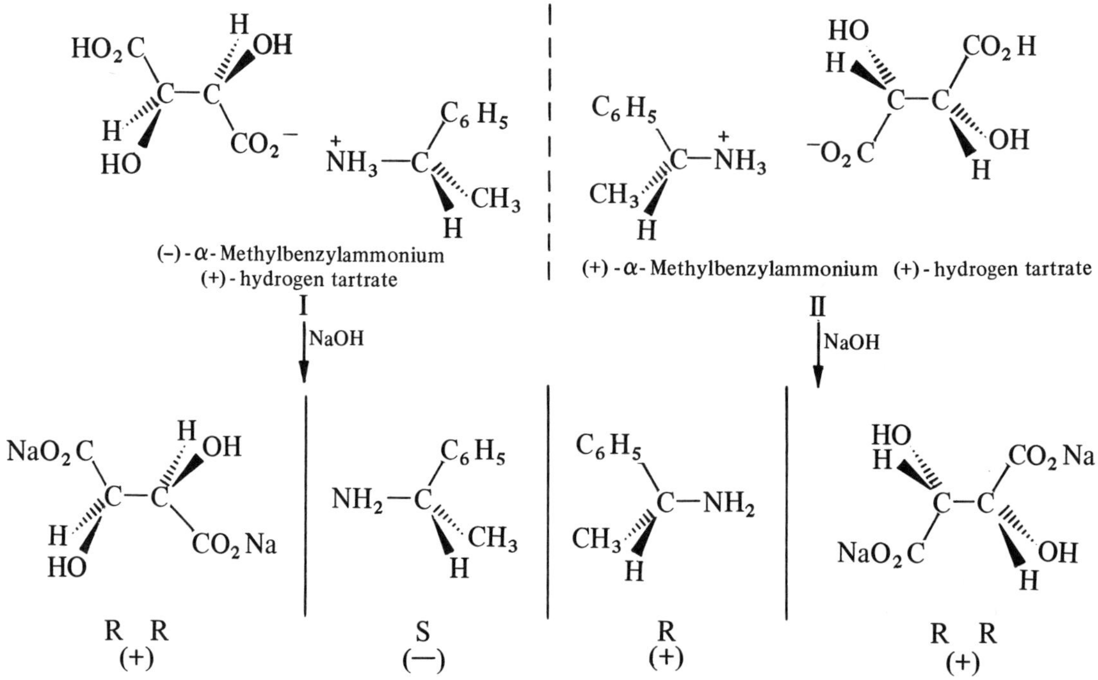

(−)-α-Methylbenzylammonium (+)-hydrogen tartrate
I
↓ NaOH

(+)-α-Methylbenzylammonium (+)-hydrogen tartrate
II
↓ NaOH

R R (+)	S (−)	R (+)	R R (+)

The configurations shown are the correct absolute configurations and the configuration of each asymmetric carbon is designated R or S.

These disastereoisomers do not have identical properties and in particular differ in solubility. They can therefore be separated by crystallization. Each disastereoisomer is then treated separately with sodium hydroxide to form water-soluble sodium tartrate and allow recovery of the pure (+) and (−) forms of α-methylbenzylamine.[1] This optically active amine is useful for resolution of racemic acids.

[1]Also called α-phenylethylamine.

238. Resolution of Racemic α-Methylbenzylamine,[1] $C_6H_5CH(CH_3)NH_2$. Add 38.0 g (0.253 mole) of (+)-tartaric acid to 500 ml of methanol in a 1-liter Erlenmeyer flask and heat the mixture almost to boiling on a steam bath. Add 30.5 g (29.0 ml, 0.252 mole) of (±)-α-methylbenzylamine cautiously to the hot solution. Too rapid addition will cause the mixture to boil over. Swirl the hot mixture until solution is complete and allow the solution to stand undisturbed for 24 hr or longer at room temperature. Prismatic crystals of the (−)-amine (+)-hydrogen tartrate should separate. Sometimes very fine needles are obtained and these give α-methylbenzylamine of lower optical purity ($[\alpha]_D^{25}$ −19° to −21°); simply redissolve the needles by warming the solution, and allow crystallization to occur again, preferably with seeding with a few of the prismatic crystals.

Collect the crystals by suction filtration and wash them with a little cold methanol. The yield should be about 22 g (~65%) after short air drying.

Concentrate the filtrate to 250 ml and allow crystallization to proceed again for 24 hr at room temperature.[2] Collect, wash, and air-dry the crystals as before (yield about 4 g); add this second crop to the first. Be sure to record all yields and calculate percentage yields. Place the filtrate in an evaporating dish (marked with your name and desk number) in the hood, and let it evaporate to dryness by standing until the next laboratory period.

Partially dissolve the product (25–26 g) in 85 ml of water and carefully add, with swirling, 15.0 ml of 50% sodium hydroxide solution. This operation may be carried out in a separatory funnel, but care must be taken to add the sodium hydroxide solution slowly and to cool the basic solution before carrying out the extraction. Add about 15 ml of ether to the cool solution and separate the ether solution of the amine. Wash the aqueous layer with two or three small portions of ether and combine the ether extracts and amine solution. Dry the combined ether solution in a 125-ml Erlenmeyer flask over anhydrous magnesium sulfate for 10 min with occasional swirling and filter it through filter paper in an ordinary funnel directly into the round-bottom flask from which the amine will be distilled (preferably a 50-ml round-bottom flask if the volume of solution is small enough, otherwise a 100-ml flask). Add a boiling chip and remove the ether on a steam bath in the hood. Distill the residue to obtain (−)-α-methylbenzylamine, bp 184–186°. There is usually quite a bit of foaming at the beginning of this distillation; for this reason it is better to use the Claisen distilling head rather than the simple three-way head, and to heat rather cautiously during the first part of the operation. The yield should be about 7.3 g (55%).

[2]The experiment may be shortened if no second crop is obtained and the recovery of the crude (+)-amine omitted. The amount of 50% sodium hydroxide solution should then be reduced to correspond to the weight of the salt actually used.

[§238] Determine the rotation of the product and calculate the specific rotation. The rotation is usually measured on the undiluted liquid (neat); but if there is not enough product to fill the available polarimeter tube, a solution may be used. The specific rotation reported by W. Theilaker and H. G. Winkler, *Ber.*, **87,** 691 (1954), for (−)-α-methylbenzylamine is $[\alpha]_{D}^{22} = -40.3°$. The product obtained in the present experiment is usually of slightly lower optical purity $[\alpha] \cong -38°$ because the salt was not recrystallized. Specific rotations at 20−22° in solution [calculated from molecular rotations reported by V. M. Potapov and A. P. Terent'ev, *Zhur. Obsh. Khim.*, **31,** 1003, 1720 (1961)] are $[\alpha]_{D}^{20\text{-}22}$ −26.3° (2%, methanol), −39.5° (2.41%, benzene), −35.4° (5%, isooctane), but the specific rotation for the neat amine used in these determinations was −39.1°.

The hydrogen tartrate of the α-methylbenzylamine from evaporation of the original methanolic filtrate to dryness (see above) is rich in the dextrorotatory enantiomer. Recover the remainder of the amine by sodium hydroxide treatment, extraction, and distillation as for the (−) isomer. Determine the specific rotation of the recovered amine. This amine is less optically pure than the (−) enantiomer. If time allows, it may be obtained in a relatively optically pure state as described by A. Ault, *J. Chem. Ed.*, **42,** 269 (1965).

Experiment 16

Carboxylic Acids

Carboxylic acids are among the most common and important organic compounds. Many general synthetic methods are known and several have already been illustrated in this laboratory manual. Procedures for hydrolysis of nitriles are found in Experiment 11 and for carbonation of a Grignard reagent in Experiment 14. The malonic ester synthesis is described in Experiment 12. The most general synthetic methods involve oxidation. Primary alcohols and aldehydes are readily oxidized to carboxylic acids, as discussed in Experiment 8, but procedures for these reactions are not included here. Methyl ketones or secondary alcohols, $RCHOHCH_3$, yield carboxylic acids in the haloform reaction, Experiment 24, with loss of the terminal carbon. Procedures are given below for oxidation of the side chain on a benzene ring and for cleavage of a carbocyclic ring with formation of a dicarboxylic acid.

239. Oxidation of an Aliphatic Side Chain. The direct oxidation of an alkyl group to the carboxyl stage would be questionable in the ordinary aliphatic domain, with hazard of general decomposition of the molecule involved. The process is feasible, however, when the alkyl group is attached to the relatively stable benzene ring. Regardless of their length or structure most aliphatic side chains (except for those with a tertiary carbon atom attached directly to the aromatic ring, or those having two or more aromatic rings attached to the same carbon) can be converted to a carboxyl group. Suitable oxidizing agents for laboratory use include alkaline permanganate or acidic dichromate, although other oxidants, such as hot dilute nitric acid, have been employed.

An important point to note here is that the aromatic nucleus survives conditions under which an aliphatic side chain will be attacked and destroyed. Side-chain oxidation, thus, demonstrates the remarkable stability of the aromatic system.

In the procedure given below toluene is oxidized to benzoic acid by potassium permanganate. Experiment 17 on acid derivatives describes the use of the product in a sequence of reactions. If you are to prepare enough benzoic acid to use in this sequence, the amounts of starting material and reagents must be increased by a factor of 10.

240. Benzoic Acid. In a 200-ml round-bottom flask place 75 ml of

$$CH_3 \quad \xrightarrow{KMnO_4} \quad CO_2H$$

water, 8.0 g of $KMnO_4$, and 2 ml of 10% sodium hydroxide solution. After warming to effect dissolution and while the mixture is warm, but not hot to touch, add 2.5 ml (2.25 g) of toluene and two boiling chips, and connect the flask to a reflux condenser. Moderately reflux the mixture for 45 min. As the oxidation progresses, MnO_2 will be precipitated and may cause irregular boiling, or "bumping." After the reflux time is over, cool the flask to room temperature, and carefully acidify with 10% H_2SO_4. *Move the flask to the hood,* and destroy any excess permanganate and manganese dioxide by slowly adding just enough sodium bisulfite. At this point the solution may have a suspension of white crystals of benzoic acid. (Check the pH of the solution; if it is not already strongly acidic, acidify to pH 1.) Recrystallize the product directly from this solution by heating the solution until these crystals are dissolved (adding more water if necessary), filtering or decanting the hot solution to remove the boiling chips and insoluble impurities, and cooling the mixture to room temperature or below. Filter off the recrystallized benzoic acid and spread it out to dry. When the product is thoroughly dry, weigh it to determine the yield, and determine the melting point. Pure benzoic acid melts at 122°.

Questions

1. What is the chemical reaction involved in the use of sodium bisulfite with the crude product in the oxidation of toluene by potassium permanganate?

2. Outline two other general methods for preparing aromatic carboxylic acids.

3. Write a balanced equation for the conversion of toluene and permanganate to benzoic acid and manganese dioxide in the presence of hydroxide ion.

241. Adipic Acid. This experiment illustrates the opening of a ring by oxidation. The carbinol group in cyclohexanol marks a vulnerable position in an attack by a strong oxidizing agent; in this case, nitric acid. If the organic reagent had been an open-chain secondary alcohol, presumably two molecules of carboxylic acid would have been formed.

Since the secondary alcohol in this case is cyclic, a single molecule of a dibasic acid is being produced instead. See §15 for further discussion.

Unfortunately, boiling nitric acid attacks cork seriously. This makes it highly desirable, though not necessary, to use standard-taper glassware for the main oxidation reaction.

In a 200-ml three-neck flask fitted with a mechanical stirrer, a thermometer and a 125-ml dropping funnel, place 100 g of 50% nitric acid (10.5 *N*). The apparatus of Fig. 5, §24, is suitable if the reflux condenser is replaced by a thermometer held in a slotted cork. The stirrer should have glass paddles and not the wire shown. Care must be taken to support the thermometer so it does not interfere with the stirrer. **Carry out the reaction in the hood.** Heat the acid nearly to boiling, start the stirrer, and begin the introduction of 25 g of cyclohexanol, contained in the dropping funnel. Add about 5–10 drops at first and no more until the reaction has started, as indicated by the evolution of oxides of nitrogen. With a pan of crushed ice at hand maintain the temperature of the mixture at about 85°, and add the remainder of the cyclohexanol as rapidly as possible, with temperature not exceeding 90°. After the addition is complete, stir for 30 min longer and pour the mixture carefully into a 400-ml beaker that is cooled in crushed ice. The crystalline product is collected on a suction filter, washed with ice water (10 to 20 ml), and dried in air. The acid may be recrystallized from concentrated nitric acid or water. Recrystallize a small sample from about twice its weight of water, dry, and determine the melting point.

Experiment 17

Carboxylic Acid Derivatives

242. General Discussion. Carboxylic acid derivatives are compounds in which the hydroxyl of the acid has been replaced by another group and which yields the acid by hydrolysis. The most common derivatives are acid chlorides, anhydrides, esters, amides, and salts. Nitriles are also considered to be carboxylic acid derivatives because they are readily hydrolyzed to acids. The relative ease of hydrolysis of these derivatives falls in the order acid chloride > anhydride > ester > amide.

Carboxylic acids can be converted to any of these derivatives but not always directly. Acid chlorides are prepared from carboxylic acids by use of acid chlorides of strong inorganic acids, for example, thionyl chloride, phosphorus trichloride, or phosphorus pentachloride. Anhydrides can be prepared from acid chlorides by reaction with carboxylic acid salts, but are more often obtained by anhydride–acid interchange, §245. Acid chlorides and anhydrides have few end uses, but are frequently employed to make other acid derivatives or for synthesis of many other classes of compounds. They are readily converted to esters by reaction with alcohols and to amides by reaction with ammonia or amines. The synthesis and hydrolysis of nitriles were exemplified in Experiment 11.

The present experiment describes a sequence of reactions in which the benzoic acid from Experiment 16 is converted into either benzoyl chloride or benzoic anhydride and these are converted to benzamide by reaction with concentrated ammonium hydroxide. Finally, the amide is converted to an amine, aniline, by the Hofmann rearrangement. Other procedures given include syntheses of succinic anhydride (for use in a Friedel and Crafts acylation, Experiment 30), of butyl acetate from acetic acid and butyl alcohol, and of various acetates from acetyl chloride or acetic anhydride; hydrolysis of an ester with recovery of both the alcohol and carboxylic acid components; and determination of the saponification equivalent of an unknown ester. Examination of fats and detergents is given in Experiment 18. Experiment 19 describes acetylation of phenols and aromatic amines.

243. Reactive Acid Derivatives. Acyl chlorides are more reactive acid derivatives than carboxylic anhydrides; in fact, some of their reactions are quite vigorous and even violent (e.g., CH_3COCl + cold H_2O react almost explosively). Therefore, the anhydride is often used in preference to the acyl chloride when a moderate reaction is desired. The disadvantage involved with the anhydride is the loss of essentially half of the starting anhydride during the reaction as the acid.

$$(R\overset{O}{\overset{\|}{C}})_2O + HY \longrightarrow R\overset{O}{\overset{\|}{C}}{-}Y + R\overset{O}{\overset{\|}{C}}OH$$

Procedures are given below for the preparation of both benzoyl chloride and benzoic anhydride, each of which can be converted to benzamide in the subsequent reaction. Your instructor will decide which acid derivative you will prepare.

244. Benzoyl Chloride. Thionyl chloride is usually the reagent of choice for conversion of carboxylic acids to acid chlorides because the other products in the reaction are gases, hence easily removed.

$$C_6H_5CO_2H + SOCl_2 \longrightarrow C_6H_5COCl + SO_2 + HCl$$

Phosphorus trichloride is often used to prepare acetyl chloride because the latter is volatile and can be purified by direct distillation from the reaction mixture leaving H_3PO_3 behind as a nonvolatile residue. Benzoyl chloride is difficult to separate from H_3PO_3. These two reagents are comparable in price if one takes into account molecular weights and the fact that 1 mole of PCl_3 converts 3 moles of a carboxylic acid to acid chloride, $SOCl_2$ only 1.

It is advisable to conduct this reaction in a hood; otherwise a gas trap must be used. Place 24.4 g (0.2 mole) of dry benzoic acid in a dry 100 ml round-bottom flask equipped with a water-cooled condenser connected to the gas trap, consisting of a regular adapter through the top of which, via a two-hole stopper, both a stream of water (can be exit water from condenser) and the byproduct gases (SO_2 and HCl) enter, with the bottom of the adapter being drained into the sink (see accompanying figure). Pour an excess of thionyl chloride (50 ml, 0.70 mole) into the reaction flask and reconnect the reflux condenser quickly. Heat the mixture to reflux for about 30 min (by then vigorous evolution of HCl and SO_2 should have ceased). Remove the reflux condenser and attach the flask to an apparatus for simple distillation. The receiving flask should be a suction flask (or see Fig. 2, §62) with a drying tube attached to the side arm and, in the absence of a hood, leading to the gas trap. After distilling excess

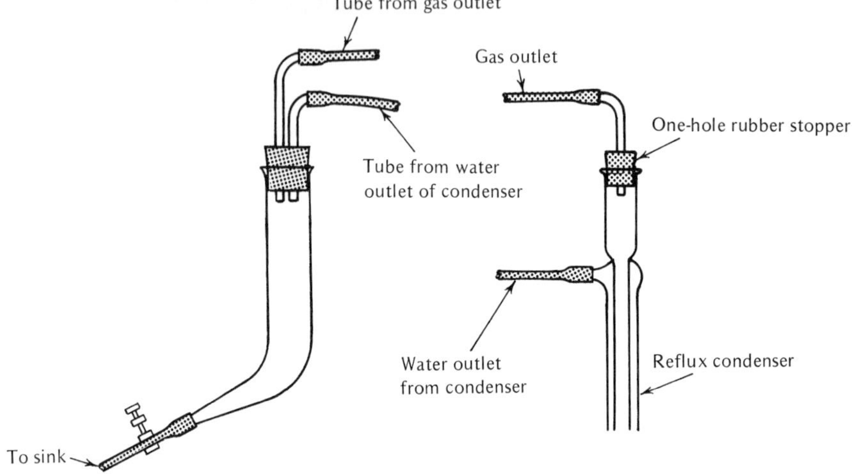

thionyl chloride (bp 79°), discontinue the water flow in the condenser and collect a second fraction boiling up to 191°. Discard this fraction in a sink in the hood. The third fraction, bp 192–198°, will be the product. Store in a glass-stoppered bottle and determine the yield of product. **Warning:** *Avoid exposure to benzoyl chloride vapors and keep from contacting the skin since it produces burns.*

245. Benzoic Anhydride. Acetic anhydride is commonly used as a dehydrating agent in the preparation of higher molecular weight anhydrides. An equilibrium is established between the anhydrides and acids:

$$2\,C_6H_5CO_2H + (CH_3CO)_2O \rightleftharpoons (C_6H_5C(=O)\!-\!)_2O + 2CH_3CO_2H$$

Acetic acid and excess acetic anhydride can be conveniently distilled from the reaction mixture leaving the higher molecular weight anhydride, which can be distilled, if volatile, or recrystallized.

Place 24.4 g (0.2 mole) of dry benzoic acid in a dry 200-ml round-bottom flask equipped with a water-cooled condenser and a calcium chloride drying tube. Add an excess (80 g) of acetic anhydride, replace the condenser, and reflux for two hours. After cooling and adding fresh boiling chips, attach the reaction flask to an apparatus for vacuum distillation. (Atmospheric distillation may be used, but vacuum is desirable to avoid very high temperatures; bp of benzoic anhydride is 360°.) Distill at 13 torr pressure, collecting the initial fractions with the condenser water on (up to 125°); turn off the water for the collection of the later fractions. Pure benzoic anhydride is collected at 205–210° and crystallizes when the flask is put in an ice bath (mp 42°).

246. Benzamide. Add 10 ml (12 g) of benzoyl chloride (or 19.2 g of

$$C_6H_5COCl \text{ or } (C_6H_5C(=O))_2O \xrightarrow{NH_4OH} C_6H_5CONH_2$$

benzoic anhydride) dropwise to 50 ml of cold concentrated ammonium hydroxide in a 250-ml Erlenmeyer flask, shaking vigorously after each addition. Pour the resulting mixture into 100 ml of water contained in a second Erlenmeyer flask. Collect the precipitated benzamide, recrystallize from water, and determine the yield and melting point. Pure benzamide melts at 132.5–133.5°.

247. Aniline—Hofmann Rearrangement of Benzamide. (OR, 3, 267).

$$C_6H_5CONH_2 \xrightarrow[Br_2]{KOH} C_6H_5NH_2$$

Prepare a solution of 10.2 g of potassium hydroxide and 10.8 g of bromine (use *care* in handling elemental bromine; see §218) in 100 ml of water. Pour this solution on 7.3 g (0.06 mole) of benzamide contained in a 250-ml Erlenmeyer flask. This mixture is finally poured into a solution of 14.4 g of potassium hydroxide in 25 ml of water contained in a 500-ml round-bottom flask. Maintain the temperature at 70–75° for about 45 min and then steam-distill the resulting aniline. (See §79 for setting up steam-distillation apparatus.) Pass steam at moderate speed through the reaction mixture. Note that at first the aniline comes over at so rapid a rate that the accompanying water is unable to dissolve all of it in the condenser. Accordingly, only globules of the free amine appear. Continue the distillation for a short time after the distilling liquid ceases to be turbid.

Saturate the distillate—aniline and water together—with clean, fine granulated salt (sodium chloride), and extract the aniline with two successive 15 ml portions of methylene chloride. Dry the combined methylene chloride solutions over anhydrous magnesium sulfate and filter into a 50-ml distilling flask. Add two small boiling chips and distill off the methylene chloride. Replace the water-cooled condenser with an air-cooled condenser and distill the aniline. Boiling point of aniline is 184.4°. Collect the fraction boiling at 180–185°.

248. Succinic Anhydride. The reaction of an acid chloride with a carboxylic acid forms an anhydride. If the acid chloride is from a different acid, the product is a mixed anhydride; with heat or in the presence of acids or bases, such a mixture tends to disproportionate to a mixture of symmetrical anhydrides. Cyclic anhydrides form readily from dibasic acids if formation of five- or six-membered rings is involved. The procedure given below for succinic anhydride is convenient because the product is obtained in high yield and purity directly from the reaction mixture. This anhydride can be made more economically by other methods. This procedure is taken from OS–CV **2** where other methods are also given. Since a considerable excess of acetyl chloride is used, the products are mainly succinic anhydride and acetic anhydride, but some acetic acid is also formed.

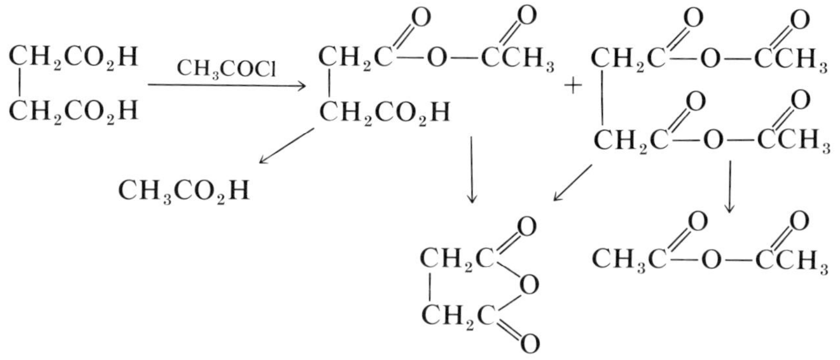

$$\begin{matrix} CH_2CO_2H \\ | \\ CH_2CO_2H \end{matrix} \xrightarrow{CH_3COCl} \begin{matrix} CH_2C(=O)\text{—}O\text{—}C(=O)CH_3 \\ | \\ CH_2CO_2H \end{matrix} + \begin{matrix} CH_2C(=O)\text{—}O\text{—}C(=O)CH_3 \\ | \\ CH_2C(=O)\text{—}O\text{—}C(=O)CH_3 \end{matrix}$$

$$CH_3CO_2H \qquad \begin{matrix} CH_2C(=O) \\ | \qquad\quad \rangle O \\ CH_2C(=O) \end{matrix} \qquad CH_3C(=O)\text{—}O\text{—}C(=O)CH_3$$

In a 200-ml round-bottom flask, with accessories as shown in the figure of §196, place a mixture of 11.8 g (0.1 mole) of finely divided succinic acid and 12 ml of acetyl chloride. Reflux on the water bath until active evolution of gaseous hydrogen chloride ceases; then remove the water bath, heat the reaction mixture over wire gauze to expel additional gas, and allow to cool. Without removing the reflux condenser, add 70 ml of chloroform and heat on the water bath until the product has dissolved. Use a few more milliliters of chloroform if complete solution has not been attained. Allow the solution to cool slowly to about 40°, and finally use an ice bath. Filter and dry the product, mp 119–120°.

249. Esterification. Alcohols react with carboxylic acids in the presence of strong acids such as sulfuric to give equilibrium mixtures with esters and water.

$$RCO_2H + R'OH \rightleftharpoons RCO_2R' + H_2O$$

This equilibrium is established very slowly in the absence of a strong acid, which acts as a catalyst through formation of the conjugate acid of the carboxylic acid. The conjugate acid quite rapidly undergoes nucleophilic attack by the alcohol. The equilibria involved are shown below:

$$RCO_2H + H_2SO_4 \rightleftharpoons \left[RC\begin{matrix} \diagup OH \\ \diagdown OH \end{matrix} \right]^+ + HSO_4^-$$

$$\begin{matrix} RCO_2H_2^+ \\ R'OH \end{matrix} \rightleftharpoons \left[RC\begin{matrix} OH \\ -OH \\ OR' \\ H \end{matrix} \right]^+ \rightleftharpoons \left[RC\begin{matrix} OH_2 \\ -OH \\ OR' \end{matrix} \right]^+ \rightleftharpoons \left[RC\begin{matrix} \diagup\!\!\diagup OH \\ \diagdown OR' \end{matrix} \right]^+ + H_2O$$

$$\left[RC\begin{matrix} \diagup\!\!\diagup OH \\ \diagdown OR' \end{matrix} \right]^+ + HSO_4^- \rightleftharpoons RC\begin{matrix} \diagup\!\!\diagup O \\ \diagdown OR' \end{matrix} + H_2SO_4$$

Acid-catalyzed hydrolysis of an ester involves the same equilibria.

The extent of esterification can be expressed as an equilibrium constant, K. The various intermediates that do not appear as starting materials or products are unnecessary in the calculation. From the simple equilibrium given at the outset

$$\frac{[\text{Ester}] \times [H_2O]}{[\text{Acid}] \times [\text{Alcohol}]} = K \quad \text{(equilibrium constant)}$$

The following are examples of K values for several esters of acetic acid, experimentally determined in systems at equilibrium. These numerical values may be regarded roughly as indications of the efficiency of the forward or esterification reaction in laboratory practice. For example, *n*-butyl acetate (4.24) is prepared with far greater efficiency in a simple laboratory experiment than *tert*-butyl acetate (0.0049).

Methyl	5.24	*n*-Butyl	4.24	3-Pentanol	2.01
Ethyl	3.96	Allyl	2.18	*tert*-Butyl	0.0049
n-Propyl	4.07	Isopropyl	2.35	Phenol	0.0089

Incidentally, the foregoing example illustrates the rule that primary alcohols are easily esterified, secondary with slight difficulty, and tertiary scarcely at all. Even the acid oxonium ion scarcely reacts with the tertiary alcohol.

The primary butyl alcohol used in this experiment is still only about two-thirds converted into the ester at equilibrium. This laboratory assignment therefore offers optional procedures: (*a*) simple reaction to equilibrium and purification of the ester (§250) and (*b*) esterification followed by special technique in displacing equilibrium, as in §251.

250. *n*-Butyl Acetate. Shorter Experiment. In a 200-ml round-bottom flask fitted with a water-cooled reflux condenser place 0.3 mole of *n*-butyl alcohol, 0.35 mole of glacial acetic acid, and 3 ml of concentrated sulfuric acid. Reflux the mixture over wire gauze for 30 min or more. ★

Transfer the reaction mixture (now at or near equilibrium) to a small distilling flask, add a boiling stone, and distill until nearly all of the liquid has gone over into the flask used as receiver.

Wash the crude ester product with several small portions of cold, saturated sodium carbonate solution until the ester layer no longer reddens blue litmus paper; then wash with 25 ml of cold water. Separate and discard the aqueous layer. Dry in a conical flask over about 6–8 g of anhydrous magnesium sulfate. At the next laboratory hour remove the drying agent, and distill the ester.

251. *n*-Butyl Acetate. Longer Experiment. Assemble apparatus as pictured in Fig. 11, §75, with use of a 200-ml round-bottom flask and a 50- or 100-ml graduated cylinder. Place 0.5 mole of *n*-butyl alcohol in the flask. Add 3 ml of concentrated sulfuric acid and a quantity of glacial acetic acid, 20% in excess of that required by theory.

Boil the mixture gently for 5 min or more, during which time the entire vapor product should reflux from the beads back into the boiler. By this time equilibrium will have practically been reached. Now heat more strongly so that a distillate will slowly accumulate. This distillate is a crude two-phase azeotropic mixture, in which water is a predominating component. When the distillate is cool (use ice if necessary) and has reached a volume of about 30 ml, stop the distillation and record in the notebook the volume of the aqueous or denser layer. This dense layer gives an index of the completeness of the esterification reaction. Separate and discard the aqueous layer. Return the lighter layer, which contains much unused alcohol, through a simple funnel into the top of the fractionating column.

Resume distillation, and once again carry out the above procedure of collecting distillate, cooling, recording, separating, and pouring back. This cyclic procedure may then be repeated until no further aqueous layer is obtainable. At this point the residue in the flask is largely ester plus excess acetic acid. ★

Wash the ester with several small portions of cold saturated sodium carbonate solution until the ester layer no longer reddens blue litmus paper, and finally wash with 25 ml of cold water. Dry over 12 g of anhydrous magnesium sulfate and distill, collecting the fraction boiling over the range 119–125°. Yield is over 70%.

A recyclization apparatus that simplifies this experiment has been described [M. B. Naff and A. S. Naff, *J. Chem. Ed.,* **44,** 680 (1967)].

252. Esters from Acid Derivatives. Acetyl chloride and acetic anhydride are readily converted into their esters by the mere addition, perhaps with gentle heating, of any ordinary primary or secondary alcohol. Such reactions are sometimes too vigorous to permit safely the complete admixture of the reagents all at once. Accordingly, one reagent is placed in a flask under reflux condenser and the second is slowly added.

After selecting a suitable alcohol, work out the details of the proposed experiment and submit them to the instructor for approval. The following problems should be considered:

1. Apparatus; examine diagrams of §§62, 196, 227, and 243, and consider possible combinations that will not permit the escape of offensive vapors into the laboratory.
2. Which of the two reagents should be placed first in the flask?
3. Should the quantity of alcohol equal, exceed, or fall below the value chemically equivalent to the amount of acid derivative used? See Chapter 3.
4. Which is the safer procedure, to keep the reagent in the flask cool or hot while the addition of the other is taking place?

253. Hydrolysis of Esters. In this experiment §§254 and 255 deal with hydrolysis as the primary means of identification of the two main sections of an ester molecule, with only incidental use of simple acidimetric titration. The second part (§256) describes the quantitative estimation of an "unknown" ester, with use of techniques variously known under the headings saponification number, ester number, or more directly, equivalent weight of an ester.

Some time in advance of this experiment, hand in a test tube with label bearing your name and the indication "Unknown Ester." A sample of ester will then be issued to you, and this should be handled as explained in §§254–255 or 256.

254. Alcohol Constituent. Place 5 ml of the ester (or 5 g if it is a solid) and 25 ml of 6 *N* sodium hydroxide in a 200-ml round-bottom flask equipped with a reflux condenser. Add a boiling stone and reflux the solution gently for 30 min, or until any possible oily layer disappears. Cool the solution, remove the condenser, and replace it with a well-fitted cork bearing a delivery tube of 7- or 8-mm glass tubing leading nearly to the bottom of a 6-in. test tube immersed in a beaker of ice. Collect about 5 ml of distillate in the test tube, and preserve the residue in the flask for use in §255. ★

To the distillate add solid potassium carbonate until the aqueous layer is saturated. Carefully remove the supernatant alcohol layer with a pipet fitted with a rubber bulb and add it to a little dry potassium carbonate in a small, dry test tube. Allow the alcohol to remain in this test tube until you are ready to prepare a derivative, namely the 3,5-dinitrobenzoate, as ★ described in §392.

255. Acid Constituent. Place a 1-ml sample of the alkaline residue from the procedure of §254 in a very small test tube and acidify with dilute hydrochloric acid. Since the esters chosen for this experiment are restricted to those that yield crystalline acids of low solubility, there should be an immediate precipitation of the acid.

Filter the alkaline reaction mixture, if necessary, to remove solid impurities. Acidify the filtrate with hydrochloric acid, and continue to add acid until a sample of the filtered solution yields no more crystalline product upon further acidification. Cool the mixture, collect the product on a suction filter, and recrystallize from boiling water.

After the crystals are dry, determine melting point and find the equivalent weight (or neutralization equivalent) by titrating two or more samples of about 1 g weight, weighed to the accuracy of about 0.005 g, with standard alkali solution of about 0.5 N concentration.

256. Saponification Equivalent. The equivalent weight of an ester is that weight in grams of the substance from which one equivalent of acid is obtainable by hydrolysis; or that quantity which reacts with one equivalent of alkali. In practice, such an equivalent weight is estimated by treating a known weight of the ester with a known quantity of caustic alkali used in excess. The amount of alkali remaining is then readily determined by titration of the reaction mixture with a standard acid solution. The amount of alkali used by the ester is thus revealed and with it the chemical equivalent of the ester.

Do not interrupt this experiment after the reaction of ester with alkali has been started. If the mixture is allowed to stand until another day, carbon dioxide from the air is likely to enter and react with the residual alkali, thus ruining the determination.

Unless notice is given to the contrary by the laboratory instructor, only a single determination is to be made. If check or duplicate runs are required, special notice should be given of the slight increase needed in quantity of alcoholic alkali solution.

After reading the directions for this experiment, prepare in advance in the notebook a blank form ready for entry of the appropriate numerical values as they should appear during the procedure. Follow this with calculation of the theoretical value after the instructor has reported the identity of the compound to you.

To 150 ml of 95% ethyl alcohol add approximately 6 ml (measured in a small graduate) of 50%, carbonate-free potassium hydroxide solution. If potassium hydroxide of suitable purity is not available, substitute 4 ml of 18 N (approximately saturated) sodium hydroxide solution. Sodium carbonate happens to be almost insoluble in very concentrated sodium hydroxide solutions, and the difficulties encountered with carbonates in alcohol are thus obviously avoided. Place the resulting alcoholic alkali solution in a bottle or flask, stopper securely, and shake well.

Pass a quantity of the alcoholic hydroxide solution through a clean buret and back into the stock bottle; and shake the solution again to ensure mixing with any water that may have been carried down from a newly washed buret. None of the solution is wasted. Now fill the buret to the mark with the alkali solution.

In a very small flask, or a weighing bottle, place a quantity of the unknown ester sufficient for three trials of the experiment. Three grams of material is a suitable amount for an ordinary single substance, 8 g if it is a true fat (olive oil, tallow, etc.). If the ester is a liquid, place a medicinal eyedropper along with the sample in the flask. If the ester is a low-melting semisolid material like coconut oil or butter fat, melt it and treat as an ordinary liquid, but go ahead with the work promptly before resolidification takes place. The eyedropper permits a neat transfer of the analytical sample without loss by spilling between weighings. If you have not had a course in quantitative analysis and do not thoroughly understand the significance of the word "quantitatively" in the following paragraphs, make the proper inquiries!

Weigh flask, ester, and dropper to 5 mg, using a balance of adequate sensitivity. If the ester is volatile (bp 100° or below), reverse the order of the next two operations (*a*), (*b*), to avoid loss by evaporation.

(*a*) Transfer quantitatively a sample of suitable size (0.8–1.2 g, or 30–50 drops, of a pure single ester; 2–3 g, or 60–100 drops, of a true fat) to a 200- to 500-ml round-bottom flask, and weigh the flask with dropper again. Do not use samples exceeding the values recommended, or the experiment may be a total failure.

(*b*) Transfer quantitatively about 49–50 ml of the alcoholic alkali solution into the round-bottom flask mentioned in (*a*).

Boil the mixture of ester and alcoholic alkali under a reflux condenser for 30 min or more. Even with high-boiling esters this minimum time allowance will carry the reaction so near to completion that any possible shortage is well within the experimental error of this analytical method.

While the reflux operation is taking place, bring a clean dry flask to the storeroom to obtain the standard acid, whose normal concentration will be reported to you by the storekeeper. After cleaning the second buret, rinse it with about 5 ml of the standard acid, allowing the rinsings to drain away as thoroughly as possible. Now fill the buret with a fresh supply of the standard acid. Use this setup to determine the concentration of the alcoholic alkali by titration, with phenolphthalein (5–10 drops) as an indicator. Addition of a little distilled water (20–30 ml) to the alcoholic mixture undergoing titration will prevent precipitation of potassium (or sodium) sulfate and will facilitate the process. Record in the notebook all titration data to the accuracy of 0.05 ml.

At the end of the half-hour of refluxing, the reaction mixture of ester and alkali may contain a mass of crystalline salt, nearly insoluble in alcohol. Sodium hydroxide is perhaps more likely to offend in this manner than the potassium reagent. It is barely possible, moreover, that the ester of a dibasic acid with a short molecular carbon chain (for example, diethyl oxalate, $C_2H_5O_2CCO_2C_2H_5$) might be only half-hydrolyzed, precipitating a mixed half-ester–half-salt, with a formula such as $C_2H_5O_2CCO_2^-$ Na^+. Should such a compound be left in solid form without further treatment, the analysis would be ruined. To avoid such a situation, proceed as follows: as soon as no further separation of the salt seems to be taking place, add through the reflux condenser just enough distilled water so that the crystals will dissolve at boiling temperature; but do not add more than 50 ml. Now reflux for 15 min, and then allow the reaction mixture to cool.

Pour 50 ml of distilled water through the reflux condenser (or 25 ml if water has already been added on account of crystallization). Now remove the condenser. Add 5–10 drops of phenolphthalein, and titrate the solution with the standard sulfuric acid without removal from the reaction flask.

After the titration is complete, add 2–3 ml of concentrated sulfuric acid to the reaction mixture. A volatile substance of characteristic odor may be apparent, or some liquid or solid may separate from the solution. Determine as closely as possible what this material is. If the sample is a fat instead of a single substance of definite composition, calculate its *saponification number;* that is, the number of milligrams of potassium hydroxide required to react with 1 g of the fat. The resulting value may be compared with known values given in the Appendix (§580). In view of the international custom in reporting milligrams of potassium hydroxide, report values in such terms even though sodium hydroxide may actually have been used.

Warning: *Do not leave any solution of caustic alkali standing in a buret beyond the period of immediate use at the laboratory hour, lest the surface of the glass be etched and the buret be seriously damaged.*

CALCULATIONS. Make the necessary calculations to show what quantity of alkali was removed by reaction with the ester. From this value calculate the equivalent weight of the ester. If the value so reported does not agree with the character of the sample analyzed, repeat the determination. There should be enough alcoholic standard solution left for one repetition, if care has been taken.

By use of diethylene glycol as a solvent instead of alcohol, the determination may be expedited [C. E. Redemann and H. J. Lucas, *Ind. Eng. Chem. Anal. Ed.*, **9**, 521 (1937)].

Questions

1. Give the order of reactivity toward nucleophilic substitution of the four carboxylic acid derivatives (acyl halide, anhydride, ester, amide) and explain why this order is observed.

2. Why is acetyl chloride significantly more reactive than benzoyl chloride?

3. Outline the preparation of *tert*-butylamine from isobutylene via a Hofmann rearrangement and discuss the particular usefulness of this rearrangement for preparation of primary amines in which the alkyl group is tertiary.

4. What relationship is there between the Hofmann rearrangement of amides and carbonium ion rearrangements?

5. Phosphorus pentachloride is often used to prepare *solid* acid chlorides. Why might this reagent be more suitable than phosphorus trichloride for this purpose? (Consider the physical properties of possible products).

6. Calculate the weight of toluene needed as a starting material to prepare 100 g of aniline by the synthetic sequence of §§240–247 assuming yields of 70% (§240), 85% (§244), or 70% (§245), 80% (§246, from either the acid chloride or anhydride), and 70% (§247).

7. Calculate the cost of preparing one mole of butyryl chloride from butyric acid (*a*) with thionyl chloride, assuming use of 15% excess and a yield of 85%, and (*b*) with phosphorus trichloride, assuming use of 0.4 mole of reagent for 1 mole of butyric acid and a yield of 85%.

8. One mole each of acetic acid and pure ethyl alcohol was refluxed until equilibrium was established. Titration of a sample of the equilibrium mixture indicated that 20 g of the acetic acid had remained unesterified. What is the equilibrium constant?

9. Suppose the equilibrium mixture of acid RCO_2H, alcohol $R'OH$, ester RCO_2R', and water contained equimolal quantities of all four reacting substances. What would be the equilibrium constant for this reaction?

10. If you were required to manufacture *n*-butyl acetate by direct esterification of acid and alcohol, on an industrial scale, what changes would you make in the plan of our "longer experiment," (§251) to eliminate the delay and inconvenience of pouring part of a distillate back into the top of the fractionating column?

11. What is the greatest possible quantity of phenyl acetate that could be prepared in a simple mixture of 1 mole of phenol and 2 of acetic acid, with the necessary catalyst?

12. Explain, with reference to the appropriate equation, why the operation of pouring back and redistilling should improve the yield of ester in the experiment of §251.

13. Suppose benzyl caproate were being prepared instead of butyl acetate. Suggest modifications of technique that you think would be nec-

essary, noting boiling points of all the compounds in the proposed experiment.

14. Why is sodium carbonate preferable to sodium hydroxide for removal of acid residues from ester preparations?

15. Outline a laboratory procedure for solving each of the following problems:

(*a*) Is the time of 30 min suggested for esterification in §250 adequate or excessive?

(*b*) Determine the equilibrium constant for esterification of *n*-butyl alcohol and acetic acid.

16. In the procedure of §256 (*a*) what is the chemical composition of the mixture just after the reflux operation but before titration, (*b*) just after titration, (*c*) after the excess of sulfuric acid has been added at the end of the experiment?

17. Calculate the equivalent weight of (*a*) methyl formate, (*b*) ethylene glycol oxalate, (*c*) ethyl acid phthalate, (*d*) ethyl succinate.

18. Why is a specially large sample weighed out when the saponification equivalent of a fat is being determined?

19. Ethylene glycol, $C_2H_6O_2$, is said to contain two alcoholic hydroxyl groups per molecule. How might the experiment described in §256 be used as part of a method of proving this fact?

20. Which substance would have the higher saponification number, glyceryl tributyrate or glyceryl tristearate? Why?

21. Infrared spectra for benzoic acid, benzoyl chloride, benzoic anhydride, ethyl benzoate, and benzamide are given in Figs. 1–5. Identify the characteristic infrared bands for each compound. How does the position of the carbonyl band vary in these different carboxylic acid derivatives?

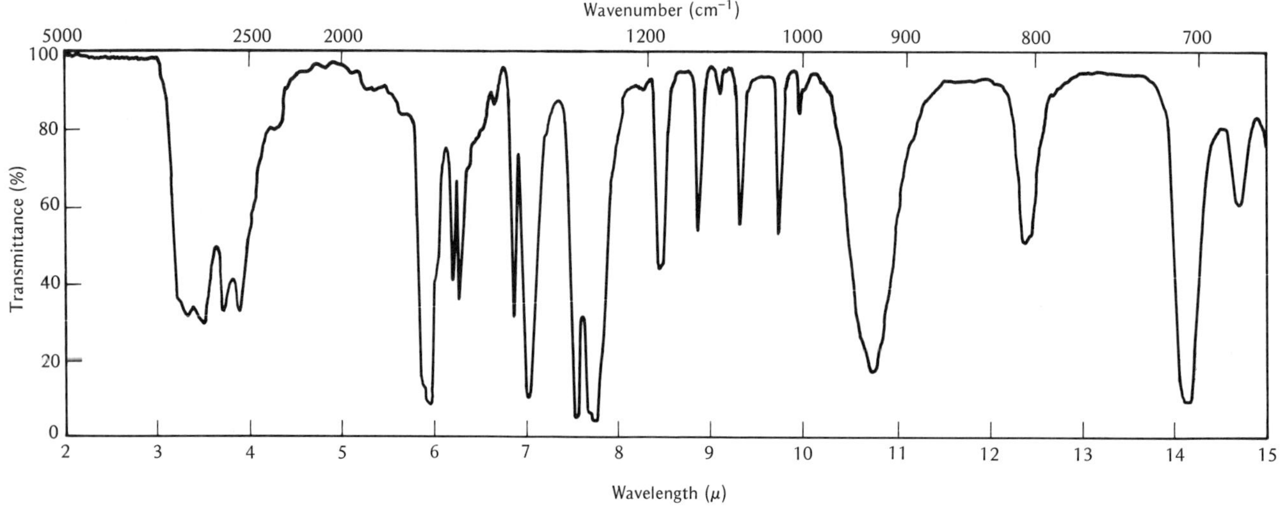

Fig. 1. Infrared Spectrum of Benzoic Acid. (© Sadtler Research Laboratories, Inc., 1967.)

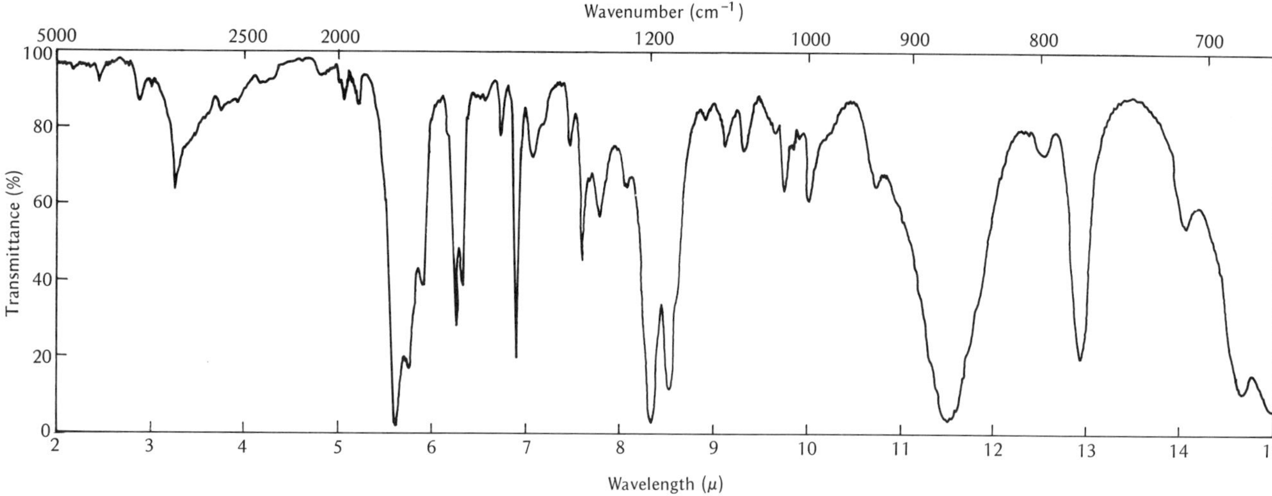

Fig. 2. Infrared Spectrum of Benzoyl Chloride. (© Sadtler Research Laboratories, Inc., 1959.)

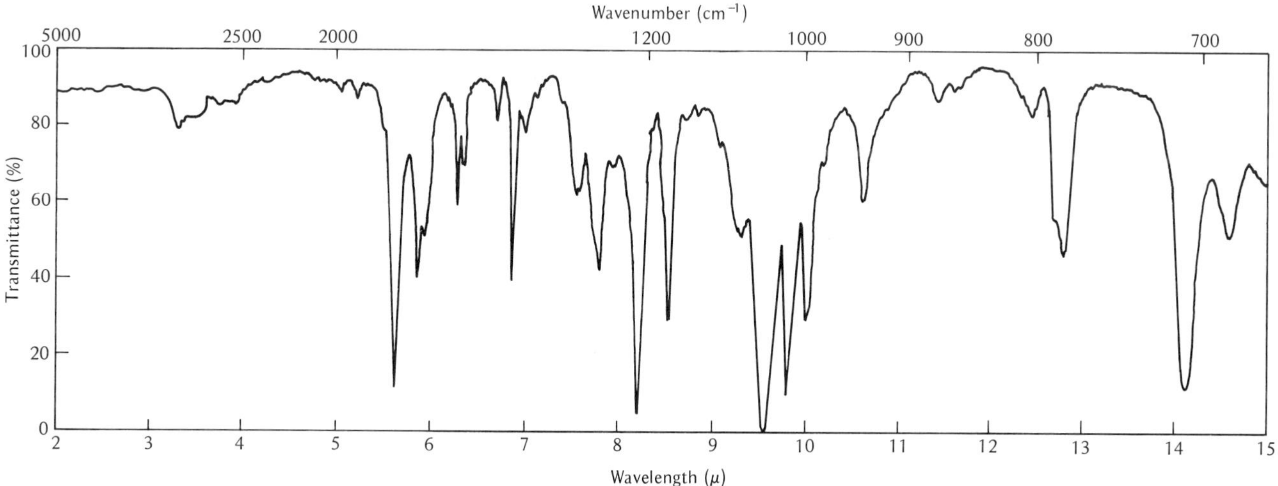

Fig. 3. Infrared Spectrum of Benzoic Anhydride. (© Sadtler Research Laboratories, Inc., 1962.)

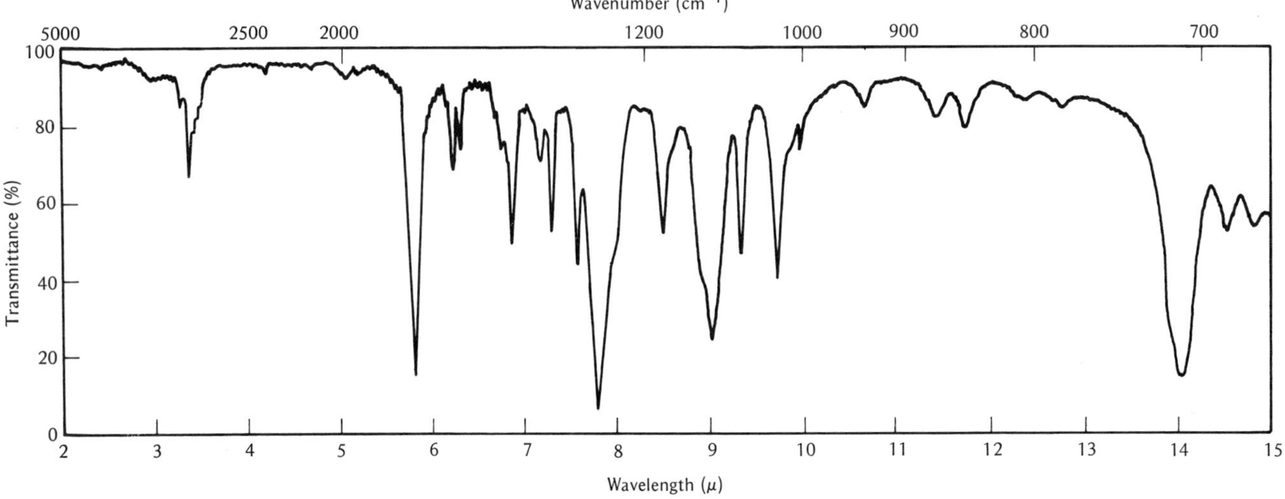

Fig. 4. Infrared Spectrum of Ethyl Benzoate. (© Sadtler Research Laboratories, Inc., 1962.)

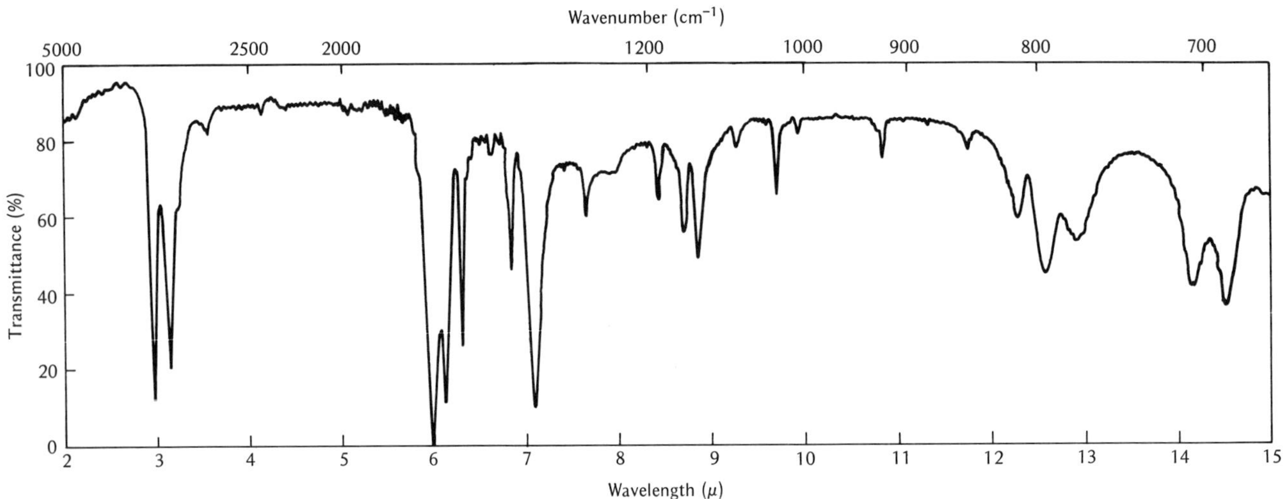

Fig. 5. Infrared Spectrum of Benzamide. (© Sadtler Research Laboratories, Inc., 1956.)

Experiment 18

Fats and Detergents

257. General Discussion. Fats, including vegetable oils as well as solid or semisolid materials like tallow and butterfat, are esters of glycerol and aliphatic straight-chain carboxylic acids (mainly C_{12} to C_{22}, saturated or unsaturated). Sodium or potassium salts of these acids, known commonly as soaps, were formerly the most important detergents or cleaning aids. In recent years synthetic detergents have been pushing soaps out of the market and now outsell soaps in the United States about seven to one.

Surfactants are surface-active agents that affect (usually reduce) surface tension in water solutions. Soaps are surfactants, but there are many surfactants that have cleansing action like soaps but are not salts of fatty acids. All of these surfactants (including soaps) are structurally unsymmetrical molecules containing both hydrophilic (water-soluble) and hydrophobic (oil-soluble) groups. Synthetic detergents are suitably formulated mixtures containing major amounts of nonsoap surfactants. Since 1965 the most important surfactant in synthetic detergents has been the sodium salt of an alkylbenzenesulfonate having a C_{11} to C_{13} (mainly C_{12}) alkyl group and called LAS (linear alkylated sulfonate). The nonbranched side chain of LAS makes the product biodegradable ("soft") and therefore less prone to foaming in sewage disposal plants. "Hard" detergents contain alkylbenzenesulfonates (ABS) with branched chains that make them resistant to biodegradation. These are no longer used in the United States but are still used in other countries. In 1969 the free world consumption of "soft" synthetic detergents was about 1.3 billion pounds and of "hard" about 600 million pounds. Sodium alkyl sulfates, $ROSO_3Na$ ($R =$ linear C_{12} to C_{18}) are also used in "soft" detergents. A variety of other surfactants find commercial use.

Synthetic detergents also contain a variety of additives that enhance the cleaning efficiency. Inorganic phosphates are cheap and effective as "builders" for this purpose and have been major components (up to 50% or more) in most synthetic detergents. A high phosphate content produces an environmental problem. Discharge of large amounts of phosphates into rivers and lakes causes excessive growth of algae and weeds. The oxygen content of water is thus depleted and fish no longer thrive. One fairly expensive solution to the problem is removal of phosphates during sewage treatment. Since phosphate nutrients also arise from human waste and agricultural fertilizer, this is the most effective solution. However, a major effort is being made to replace phosphates in synthetic detergents by other "builders." Sodium nitrilotriacetate, $N(CH_2CO_2Na)_3$, has been claimed to be an effective substitute.

The calcium, magnesium, and iron salts of the carboxylic acids of soaps are water insoluble and have no cleansing action in water solution. As a result soaps are ineffective in many natural waters because of the presence of these ions (hard water). The organic sulfonates and sulfates used in synthetic detergents give water-soluble calcium, magnesium, and iron salts, which are still effective surfactants. Thus synthetic detergents have an advantage over soaps for many uses.

258. Preparation of a Soap. In a 200-ml round-bottom flask fitted with a water-cooled reflux condenser, heat a mixture of 5 g of solid potassium hydroxide, 75 ml of alcohol, and 5 g of an animal or vegetable fat (or oil). Different students may well select different types of fat:

(*a*) Suet, tallow, or lard.
(*b*) Butterfat (butter that has been melted and freed from water and salt).
(*c*) Vegetable oil: cotton, corn, olive, rape-seed, sesame.
(*d*) Coconut oil.
(*e*) Drying oil: linseed, fish, poppy seed.
(*f*) Hard-hydrogenated vegetable oil.

When the reaction is complete (in not more than 30 min), no globules of oil are obtainable when a test sample of a few drops is mixed with water.

Distill the reaction mixture, recovering the alcohol, and dissolve the residue in about 50 ml of distilled water, obtaining a solution of "soft ★ soap" or potassium soap.

259. Reactions of a Soap. (*a*) To a 10 ml sample of the potassium soap solution add 10 ml of clean, saturated salt solution. The precipitate is "hard" or sodium soap, similar to common white household soaps of commerce.

(*b*) To another 10 ml sample add a few drops of calcium chloride solution and shake. Loss of detergent value is shown by the failure of this mixture to yield the customary soap foam or lather. Now add, gradually, trisodium phosphate until the detergency is restored.

(*c*) Acidify a second portion with dilute sulfuric acid and note the characteristics of the resulting precipitate both at room temperature and at 80° or above. The choice of a particular type of fat is particularly significant in this test. Notice odor, if present.

(*d*) Acidify the rest of the solution with dilute sulfuric acid and separate the insoluble organic product. Dissolve this in a few milliliters of carbon tetrachloride and add a few drops of a solution of bromine in carbon tetrachloride. Compare with results from the saponification of some other fat.

260. Glycerol. The filtrate from (*a*) above contains glycerol. Although the procedure is not profitable here, one could neutralize with sodium carbonate, evaporate to dryness, extract with absolute alcohol, and evaporate the alcoholic extract, leaving the relatively involatile glycerol as liquid residue.

261. Synthetic Detergents. Prepare 10 ml of an approximately 1% solution of a commercial synthetic detergent, and test this in the same manner as the soap sample in (*b*) above. For an experiment on analysis of phosphates in detergents, see G. S. King, Jr., and K. D. Kriz, *J. Chem. Ed.*, **48,** 551 (1971).

262. Trimyristin from Nutmegs. Although a fat usually contains glycerides of several fatty acids, a remarkable exception is seen in oil of nutmeg, which is principally the ester of one fatty acid, the C_{14} compound. The following example illustrates continuous extraction by the Soxhlet principle (§117) applied to nutmeg powder or meal, obtainable from wholesale grocers. (Simplification of method of G. D. Beal, OS, **6,** OS-CV, **1.**)

Provide 30 g (or 1 oz) of commercial nutmeg meal and a 500-ml flask containing 250 ml of diethyl ether. A Soxhlet extractor, with carefully bored cork, is fitted to this flask. Place either 15 g or the whole 30 g of nutmeg (as capacity of the extractor permits) in the Soxhlet porous thimble, and reflux the ether over a steam bath for 2 hr. If only 15 g of the meal has been handled, empty the thimble, refill with the second charge of nutmeg meal, and reflux for 2 hr more. ★

Cool the resulting ethereal solution of trimyristin, filter to remove possible suspended solids, and evaporate the ether by distillation through a condenser. **(Fire Precaution.)** See §63. Dissolve the fatty residue in 30 ml of boiling alcohol. Add decolorizing carbon and filter through a Büchner filter into a clean suction flask. If carbon comes through, refilter.

Upon cooling, trimyristin separates. Recrystallize from 45–50 ml of alcohol, with stirring of the hot solution as it cools (mp 55°).

1. Calculate the equivalent weight of trimyristin.
2. Write equations showing how you might convert trimyristin into myristic acid, and write practical directions which you think would be suitable for this project.
3. Calculate the saponification number of trimyristin.

For an experiment on a smaller scale without use of a Soxhlet extractor but including saponification of the trimyristin and use of tlc, see F. Frank, T. Roberts, J. Snell, C. Yates, and J. Collins, *J. Chem. Ed.*, **48,** 255 (1971).

Questions

1. If a few molecules of soap were allowed to be dispersed in a suspension of oil globules in water, where would you expect them to go, in agreement with rules of solubility?

2. Explain the primitive process of mixing tallow with water that has been passed through a bed of wood ashes, followed by boiling in an iron pot.

3. Soiled clothes are often contaminated with acidic substances. Why does this fact concern the laundry operator?

4. Why is trisodium phosphate more suitable than sodium nitrate as an aid in the use of soap under troublesome conditions?

5. How are LAS and ABS prepared commercially?

6. The surfactants described in §257 are classified as anionic. Describe cationic and nonionic surfactants briefly and give some commercial examples.

7. Briefly describe the environmental aspects of the detergent industry.

Experiment 19

Antipyretic Drugs and Acetylation

263. Acetylsalicylic Acid (Aspirin). The acetylation of the phenolic hydroxylic group in salicylic acid yields a substance of great value as a fever-depressant. The product is much less toxic than acetanilide (§319), formerly much used for the purpose.

$$C_6H_4(OH)(CO_2H) + (CH_3CO)_2O \longrightarrow C_6H_4(OCOCH_3)(CO_2H) + CH_3CO_2H$$

In a small flask heat a mixture of 3 g of salicylic acid, 6 g of acetic anhydride, and three drops of concentrated sulfuric acid to 50° for 10 or 15 min, using a water bath. Now cool the reaction mixture in an ice bath, with stirring, until it thickens to a crystalline, semisolid mass. Add 50 ml of water, stir until a thin sludge remains, and filter. The solid product is crude acetylsalicylic acid, or aspirin. Dry the product in the open air, and dissolve in the minimum volume of gently boiling ether, heated under reflux condenser over the steam bath. To the resulting solution add an equal volume of light ligroin (petroleum solvent) or petroleum ether and allow to stand on ice for several hours. Collect the resulting purified crystals on the Büchner filter and dry in the open air. The melting point of pure acetylsalicylic acid is not definite, on account of slight decomposition. Figures around 128–133° are likely to be obtained. Yield is 3 g.

Add a few drops of aqueous ferric chloride to a solution made by dissolving a few crystals of each of the following in 5 ml of water in a test tube.

(*a*) Salicylic acid.
(*b*) Pure aspirin.
(*c*) Your preparation of aspirin.

The appearance of a striking coloration indicates free phenolic hydroxyl groups.

For a more extensive experiment on the synthesis and reactions of aspirin, see D. A. Brown and L. B. Friedman, *J. Chem. Ed.*, **50,** 214 (1973).

264. *p*-Ethoxyacetanilide (Acetophenetidine, Phenacetin)

$$p\text{-}C_2H_5O\text{-}C_6H_4\text{-}NH_2 + (CH_3CO)_2O \longrightarrow p\text{-}C_2H_5O\text{-}C_6H_4\text{-}NHCOCH_3 + CH_3CO_2H$$

The acetyl derivatives of certain aromatic amines have had wide application as drugs that have more or less value as fever-depressant and analgesic (pain-reducing) remedies. Since the amine used in this experiment, *p*-phenetidine, discolors rapidly on standing, the procedure should not be interrupted at any point before recrystallization. Have all reagents ready for a continuous operation, as follows: sodium acetate, anhydrous or hydrated (0.025 mole); Norite or equivalent decolorizing carbon (2–3 g); acetic anhydride (2 ml); concentrated hydrochloric acid, 12 *N* (2 ml); and clear, colorless, or light-colored *p*-phenetidine (2.6 ml). It will probably be necessary to redistill any *p*-phenetidine that has been stored even one day after preparation.

In a 300- or 500-ml conical flask place 100 ml of water, and add the hydrochloric acid and *p*-phenetidine provided as above. Add the Norite, decolorize by boiling for a minute or two, and filter through a fluted filter. The filtrate, dilute *p*-phenetidine hydrochloride solution, should be colorless. Cool at once to room temperature and dissolve the acetic anhydride in the solution as rapidly as possible. Immediately add the sodium acetate, and shake thoroughly. Cool if necessary to 20° or below, filter, wash the crude crystalline phenacetin with cold water, and recrystallize from water. If the hot solution prepared for the recrystallization is colored, decolorize with carbon. The compound may also be recrystallized from dilute ethyl alcohol. Yield is 1.5 to 2 g according to the quality of the phenetidine. Melting point of pure *p*-ethoxyacetanilide is 137°.

Experiment 20

Amino Acids

265. Replacement of Halogen by the Amino Group. When an aliphatic α-chloro or α-bromo acid is treated with ammonia, the halogen atom is slowly replaced by the primary amino group. This is an example of nucleophilic substitution at saturated carbon. In the example below a primary halide, chloroacetic acid or its anion is attacked by NH_3 in an S_N2 displacement that yields glycine, an inner salt, $\overset{+}{N}H_3CH_2CO_2^-$, more commonly written $NH_2CH_2CO_2H$.

$$NH_3 + ClCH_2CO_2^- \longrightarrow \overset{+}{N}H_3CH_2CO_2^- + Cl^-$$

Amino acids are amphoteric and form salts with either strong acids or strong bases. Substitution reactions of secondary or tertiary α-halo acids can occur by either S_N1 or S_N2 processes and are complicated by the neighboring group effect of the carboxyl or carboxylate.

The presence of considerable amounts of ammonium carbonate leads to better yields in the synthesis of glycine [N. D. Cheronis and K. H. Spitzmueller, *J. Org. Chem.* **6,** 349 (1941)]. It is interesting that commercial ammonium carbonate often contains appreciable amounts of carbamate, $(H_2N\overset{O}{\overset{\|}{C}}O)^-NH_4^+$, which is both an amide and an ammonium salt.

266. Glycine (Aminoacetic Acid). In a 500- or 700-ml flask dissolve 90 g of powdered ammonium carbonate in 120 ml of warm water (50°–60°) and add 100 ml of concentrated ammonium hydroxide (sp. g. 0.90). To this solution add slowly, with swirling of the flask, a solution of 0.25 mole of chloroacetic acid (95–100%) in 35 ml of water. (Keep chloroacetic acid away from the skin.) Stopper the flask tightly and allow to stand, preferably in a warm place, for at least 24 hr.

At the next laboratory period distill the reaction mixture either in the manner suggested in the accompanying figure, or in whatever other way avoids the nuisance of ammonia fumes in the laboratory, until the volume

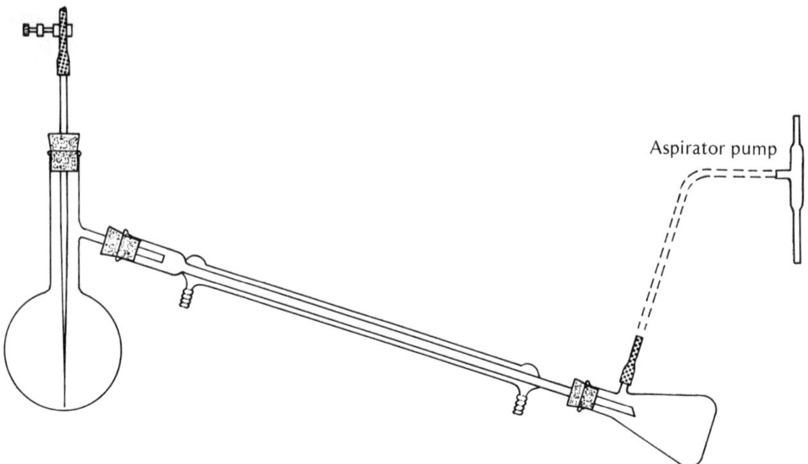

has been reduced to 75 ml. During the process, particularly toward the end, avoid heating at the margin of the liquid, which would cause charring. If the solution is not colorless, treat with about 1 g of decolorizing carbon. Add to the clarified filtrate 250 ml of methanol (95–100%), mix thoroughly, and set aside in a refrigerator for 6 hr or more for crystallization of glycine.

The resulting product of crude glycine contains ammonium chloride as a major impurity. This by-product may be largely removed by suspending the crystals in 50 ml of methanol for 5–10 min, and filtering. More thorough purification is accomplished by dissolving in water and reprecipitation with methanol. Yield is 13 g. The compound does not have a sharp melting point, but chars at about 235°.

The presence of any appreciable quantity of ammonium chloride, not acceptable to biochemists who have considerable use for glycine, is detected by tasting, which is quite safe, since neither the chloride nor glycine is toxic. Ammonium chloride has a sharp, salty taste, while pure glycine is sweet. Special quantitative determination of ammonia or chloride may also be carried out.

REACTIONS OF AN AMINO ACID

267. Amphoterism—Two Methods of Salt Formation. (*a*) Dissolve 1 g of glycine in 25 ml of hot water and drop in particles of cupric carbonate until the resulting reaction ceases. Heat to boiling, decant the clear blue liquid, and cool. Beautiful blue crystals of the sparingly soluble cupric salt separate, illustrating a reaction with the carboxy group.

(*b*) Dissolve 1 g of glycine in the smallest possible quantity (a few drops only) of concentrated CP hydrochloric acid that has been heated just to the boiling point with little or no loss of hydrogen chloride vapor. When the reaction mixture has cooled, crystals of amino acid hydrochloride separate, illustrating a reaction with the amine group.

268. Effect of Formaldehyde. Dissolve 0.5 g of glycine in 5 ml of water in a test tube. Add a drop of methyl red indicator solution. If the glycine is pure, a yellow color will be obtained, since glycine is nearly neutral in reaction. If a reddish or orange tint is observed, acidic impurities are present, causing a change in pH value to some point below 6. In such a case touch the solution cautiously with a stirring rod that is wet with very dilute ammonia or sodium hydroxide, thus obtaining a yellow color. Now treat a 3- to 5-ml sample of concentrated formaldehyde solution (15–35%) with methyl red, and neutralize it in the same way if necessary. Mix the two yellow solutions. The formaldehyde reacts with the amine group. Explain the change in color.

269. Reaction with Nitrous Acid. To a solution of about 0.5 g of sodium nitrite in 5 ml of cold water, add a few drops of acetic acid. A few minute bubbles appear, indicating slight decomposition of the nitrous acid formed. Now drop in a few crystals of glycine. What is the gas formed? In what way could this experiment be made quantitative? What difficulties might arise?

270. The Schotten–Baumann Reaction. When a primary amine is mixed with an acid chloride in the presence of aqueous alkali, the reaction to produce an amide is more rapid than that of ordinary alkaline hydrolysis.

$$RCOCl + NH_2CH_2CO_2H + NaOH \longrightarrow RCONHCH_2CO_2H + NaCl + H_2O$$

271. Hippuric Acid (Benzoylaminoacetic Acid). Dissolve 3 g of glycine in its equivalent of 1 *N* sodium hydroxide solution in a 200- or 300-ml flask. Place a piece of litmus paper in the flask and add in small portions a quantity of benzoyl chloride, about 15% in excess of that required by theory to react with the glycine. After adding each portion, stopper the flask and shake well until all the chloride has reacted. The mixture must be kept slightly alkaline, with the aid of additional sodium hydroxide if necessary. **Warning:** *Benzoyl chloride is very offensive; work in a hood or outdoors.* Cool the reaction mixture under the tap if it gets warm. Finally, eliminate paper or other solid material by filtration, and then place in the solution a piece of Congo Red indicator paper.

On a second piece of Congo Red paper place one drop of 6 *N* hydrochloric acid, and then allow a few drops of water to run upon the acid spot and spread downward on the paper. Notice the appearance of a deep purple shade, to be used for comparison.

Using an eyedropper, and stirring the reaction mixture effectively, add just enough dilute (6 *N*) hydrochloric acid to turn the Congo Red paper from deep red to purple, as described above. Cool the flask and allow the mixture to stand for 10 min or more.

Collect the resulting crystalline precipitate of hippuric acid (benzoylglycine) upon the Büchner filter and dry in a vacuum desiccator. When the crystals are thoroughly dry, place in about 25 ml of carbon tetrachloride and boil very gently in a beaker under a watch glass for about 5 min. Filter the mixture while hot, and repeat the treatment, which is virtually an extraction of benzoic acid from the preparation. Finally, wash the crystals with diethyl ether and dry again in the desiccator. Yield is 85% of the theoretical. Melting point of hippuric acid is 187°.

272. Synthesis in the Human Body. If benzoic acid or a suitable one of its salts is introduced into the human digestive system, the benzoyl radical of the substance soon combines with an equivalent quantity of glycine, yielding the amide, hippuric acid, which is then eliminated in the urine, from which it may be recovered by crystallization. Sodium benzoate of chemically pure or USP grade, commonly used as a food preservative, is selected as more palatable than the free acid.

Just before retiring at night drink a solution of 5 g of pure sodium benzoate in a half-glassful of water. Next morning preserve the urine and proceed with the laboratory experiment on that same day. For each 100 ml of urine add 25 g of clean solid ammonium sulfate and 1.5 ml of concentrated sulfuric acid, stirring until the salt is completely dissolved. Cool the mixture in a refrigerator for several hours, or until a maximum crystallization of hippuric acid has been attained.

Filter, and recrystallize the solid product from about 50 ml of hot water with the aid of decolorizing carbon, and boil the mixture 10 min or more. Cool the final filtered solution for an hour in an ice bath, and collect the purified hippuric acid upon a filter.

Question

The amounts of available ammonia from the ammonium carbonate and ammonium hydroxide specified above in §266 are far in excess of theoretical requirements. Why should such great excess be of particular importance in this experiment?

Experiment 21

Sugars—Monosaccharides and Disaccharides

273. Identification. Presence of an aldehyde group in certain sugars is revealed first by aid of reducing reactions with silver or copper reagents. A ketone group, usually inactive toward such reagents, also reduces silver and copper, provided hydroxyl is present in the adjoining α-position, as is normal in ketoses.

The particular sugar is more exactly identified by attachment of some relatively large molecular group, such as phenylhydrazine, yielding a readily crystallizable derivative, such as the osazone.

Warning: *Phenylhydrazine may cause dermatitis or irritation of the skin. If this reagent comes in contact with the skin, wash promptly with soap and water.*

274. Aldehyde Reactions. Prepare a solution of 1 g of glucose in 10 ml of water and provide for comparison about 10 ml of 5% formaldehyde solution.

SILVER MIRROR TEST. Prepare ammoniacal silver nitrate reagent as described in §381 and place a 10-ml portion of this reagent in each of two carefully cleaned test tubes. Add a few drops of the formaldehyde solution to one of the test tubes and 2 ml of glucose solution to the other. Note experimental details of §381, including the necessity of prompt disposal of any of the silver reagent that may be left over.

FEHLING'S SOLUTION. Prepare 10 ml of this reagent, just before use, by mixing 5 ml of "No. 1" (a 3% solution of cupric sulfate cryst.) and 5 ml of "No. 2" (a 15% solution of Rochelle salt in 5% aqueous sodium hydroxide). The resulting mixture should be dark blue and clear.

Heat 5 ml of the mixed solution to boiling, and gradually add glucose solution drop by drop. Cuprous oxide is precipitated. Repeat with formaldehyde. (If too much formaldehyde is used, the reduction may proceed even to the stage of metallic copper.)

SCHIFF'S REAGENT. Test glucose and formaldehyde solutions by addition of 1 ml of Schiff's reagent to each. This is a direct test for *free* aldehyde groups. See §194.

275. Osazones. Dissolve approximately 0.2 g of glucose in 5 ml of water in a test tube or provide the equivalent from a laboratory solution of glucose. Add 3 ml of phenylhydrazine reagent. Heat the mixture in a beaker of boiling water for 10–20 min. A finely crystalline yellow precipitate of D-glucosazone appears. Examine under a low-power microscope, or filter and determine melting point.

As a variation, fructose, lactose, maltose, or galactose may be converted into osazones. Both the time of formation of the osazone and the appearance of the crystals under the microscope vary decidedly from the glucose example previously cited.

Note. Phenylhydrazine reagent should be freshly prepared either from (liquid) phenylhydrazine base, which does not keep very well, or from the hydrochloride salt, as follows:

(*a*) Phenylhydrazine (2 ml), glacial acetic acid (2 ml), water (18 ml). Filter with aid of decolorizing carbon if the solution is not clear.
(*b*) Phenylhydrazine hydrochloride (2 g), crystalline sodium acetate (3 g), water (15 ml). Filter with the aid of decolorizing carbon.

276. β-D-Glucose Pentaacetate. This experiment illustrates a method of identifying sugars. A compound with several hydroxyl groups in the molecule, such as a sugar, is likely to have unsatisfactory melting behavior. Conversion into an ester brings the material into lower and more convenient melting-point range. The procedure described below should not be interrupted until the reaction mixture is poured over ice and the solid product crushed.

In a dry porcelain mortar grind together 5 g of dry glucose and 4 g of anhydrous sodium acetate. Place the powdered mass in a 200-ml round-bottom flask and add 25 ml of acetic anhydride. To avoid charring, it is desirable to allow the mixture to stand until the next laboratory period before heating is attempted. Now heat the mixture on the steam bath under the air-cooled reflux condenser until the solids are dissolved (about 30 min). Continue heating for 2 hr more and pour the reaction mixture into about 300 ml of ice water containing a few pieces of crushed ice. Break up the resulting solid product and allow the finely divided suspension to stand for at least 2 hr or until the next laboratory period.

Collect the solid product on the Büchner filter, wash with a little cold water, and press down until as nearly water-free as possible. Recrystallize from ethyl alcohol (about 20–25 ml). Yield is 6–7 g. Melting point of pure β-D-glucose pentaacetate is 132°.

Questions

1. What relation does the designation "β" in the name of this product have to the question as to which five of the six carbon atoms in glucose are acetylated?

2. How could you use your final product in a laboratory experiment to prove that just five acetyl groups had combined with the glucose nucleus?

277. Mucic Acid from Lactose. Vigorous oxidation of a sugar may produce carboxyl groups at both ends of the monosaccharide carbon chain. In this experiment a disaccharide first splits into two portions, each of which is oxidized to a six-carbon dibasic acid.

Place 15 ml of water in a 5-in. evaporating dish in order to be able to make an approximate estimate later to show when the reaction mixture shall have been evaporated down to that volume. Now add 40 ml more of water, 22 ml of concentrated nitric acid, and 7 g of lactose. Evaporate the resulting solution outdoors, in a hood, or with other means of disposal of acid fumes, until the volume is reduced to 15 ml. Cool the mixture, dilute with 25 ml of water, and collect the crystalline product on a suction filter.

The crude mucic acid may be purified by carefully dissolving it in the minimum quantity of dilute sodium hydroxide and reprecipitating with hydrochloric acid at temperatures not exceeding 25°. Yield is 2.5 g. Melting point (with decomposition) of mucic acid is 206°.

278. Disaccharide—Sucrose. Prepare a stock solution of 5 g of sucrose in 50 ml of water. It is best to use commercial cane or beet sugar, brought at the time from home or grocery, and not a specimen that has been exposed to acid fumes of a chemical laboratory.

To half of the stock solution in a test tube or small flask, add 1 ml of concentrated hydrochloric acid, and heat for 5 min or more in a hot water bath (70–90°). Cool the solution and neutralize the acid (litmus paper) with a few drops of sodium hydroxide. Preserve the untreated sugar solution free from acid.

Test small portions of treated and untreated sugar solution with each of the following reagents in the manner already described in §§273–275. Explain differences in reaction:

Ammoniacal silver reagent
Fehling's solution
Schiff's reagent
Phenylhydrazine reagent

[§278]

Questions

1. How do the results of experiments in §278 support the idea that the ketone and aldehyde groups of the two units in the sucrose molecule are tied together?

2. What do you think of the statement occasionally put out in diet literature that grape sugar is more wholesome than "artificial, refined" cane sugar?

3. In what way does the normal action of the stomach resemble any part of the above experiments?

Experiment 22

Polysaccharides

279. Large Molecules. Molecules of polysaccharides are so large that it is difficult from a purely mechanical standpoint for such materials to form a true solution in water, despite their high hydroxyl content. Some may form jellies or gumlike masses when treated with water; others are completely insoluble. In any case it is usually necessary to hydrolyze the polysaccharide before specific chemical testing.

280. Hydrolysis of Starch. Grind about 1 g of starch with 10 ml of cold water, pour the milky suspension into 200 ml of boiling water, and set aside to cool. Part of the material will pass into a pseudosolution that is colloidal in nature. Pour out about 5 ml of this solution and test with a very dilute solution of iodine in potassium iodide.

Now add 10 ml of concentrated hydrochloric acid to the remainder of the starch solution and heat the mixture to boiling. Every 5 min pour out 5 ml of the boiling solution, cool the sample, divide into two equal portions, and apply the iodine and Fehling's tests, respectively. Discontinue the periodic tests when no further change in results is noted.

281. Cellulose—Schweitzer's Reagent. To 5 ml of Schweitzer's reagent (2% solution of cupric hydroxide in concentrated ammonium hydroxide) add a piece of filter paper about the size of a 25-cent piece. Pour the resulting solution into dilute hydrochloric acid.

282. Cellulose Nitrate. In a 250-ml beaker place 20 ml of concentrated nitric acid (sp. g. 1.42) and add 20 ml of concentrated sulfuric acid. While this solution is still warm, add 0.5 g of absorbent cotton. Stir the mixture for just 3 min. Remove the cotton with a glass rod, freeing it as much as possible from acid, and immediately wash it with a large quantity of cold water. Squeeze the cellulose nitrate product as dry as possible and allow to dry in the open air. *Do not heat the material.*

Test the flammability of the dry cellulose nitrate by placing a small fragment on a brick or stone, and lighting with a burner, not a match; or use tongs.

Cellulose nitrate prepared in this way contains mostly lower nitrates rather than the more dangerous guncotton, or trinitrate. Test the solubility of the nitrate ester in a mixture of 5 ml of ether and 5 ml of ethyl alcohol. Pour a few drops of the resulting solution (collodion) into water. Allow the remainder to evaporate on a watch glass or evaporating dish. Finally, destroy all residues of cellulose nitrate before you leave the laboratory.

283. Cellulose Acetate. In a small conical flask place 0.5 g of absorbent cotton, 20 ml of glacial acetic acid, 6 ml of acetic anhydride, and 2 drops of concentrated sulfuric acid. Stir with a glass stirring rod until the cotton is well distributed through the mass of the solution and all air bubbles have been eliminated. Stopper the flask and let stand for 12 hr or more. With stirring, pour the reaction mixture in a thin stream into 500 ml of water. The cellulose ester should separate out in solid form; collect it on a Büchner filter, wash with water, and press the mass between filter papers or upon a porous plate until dry.

Dissolve half of the cellulose acetate in 20 ml of chloroform contained in a small beaker or test tube. It may require some time for the material to pass into solution. If the solution is turbid, probably water is present, in which case it should be dried with the aid of a small amount of calcium chloride in the usual way. Pour the dry solution upon a watch glass (out of doors) and let it evaporate slowly. When the chloroform has disappeared, add water, which helps to loosen the cellulose acetate film from the glass. Remove the product and test its flammability.

Questions

1. Explain why sucrose and starch will not react with Fehling's solution, although the products that they yield upon "digestion" readily react in such manner.

2. Why is cellulose nitrate more dangerous than the acetate?

3. What is the essential difference in molecular structure between starch and dextrin?

Experiment 23

Pinacol Condensation

284. Pinacol Hydrate. Fit a 500-ml round-bottom flask with an efficient reflux condenser and mount a dropping funnel in the top of the condenser by means of a slotted cork. A three-neck flask with one neck stoppered may also be used. Both this apparatus and the reagents must be dry. Now place 8 g of magnesium turnings and 100 ml of dry benzene in the flask and a solution of 9 g of mercuric chloride in 75 ml of dry acetone in the dropping funnel.

Add about one quarter of the mercuric chloride solution, and warm the reaction mixture on the steam bath until vigorous ebullition ensues. Provide also a cold-water bath to use in case the reaction tends to get out of control. Once started, the reaction will proceed vigorously without heat from the outside, and the balance of the mercuric chloride solution should be added fast enough to keep the reaction going briskly. After the acetone solution has all been run in, reflux the complete mixture, with the aid of the steam bath, for 1 hr. The flask should be shaken occasionally during this period. If the magnesium pinacolate forms a mass too stiff for shaking, cool the mixture and break it up with the aid of a stirring rod after about 30 min. ★

At the end of the 1-hr period add 20 ml of water through the condenser and boil the mixture for 30 min with frequent shaking. This converts the pinacolate into pinacol (soluble in benzene) and a precipitate of magnesium hydroxide. Filter the hot solution by suction, return the solid magnesium hydroxide to the flask, and reflux it with 50 ml of ordinary benzene for 5–10 min; then filter the solution as before. Pour the combined filtrates into a flask and distill the solution on a steam bath to one third of the original volume. Now add 30 ml of water, and cool in an ice bath with good stirring. The pinacol hydrate separates as an oil that soon solidifies; it should be collected only after thorough cooling and stirring to give maximum crystallization. Scrape the crude material onto a suction funnel, wash it with a little cold benzene, press it well with a spatula, and let it drain for about 5 min in order to remove as much benzene as possible. Transfer the material to a 200-ml round-bottom ★ flask and keep it stoppered securely until it is converted into pinacolone.

285. Pinacolone. Into the 200-ml round-bottom flask pour 80 ml of water and 20 ml of concentrated sulfuric acid. Attach a reflux condenser and boil for 15 min, observing carefully the changes that take place. Cool until boiling ceases, connect the flask with a bent tube to a condenser for ordinary distillation, and distill until no more pinacolone comes over. Separate the upper layer of pinacolone, dry it with calcium chloride, and record the yield.

286. Pinacolone Oxime. Into a test tube provided with a cold-finger condenser measure approximately 1 ml of crude pinacolone and 3 ml each of 5 *M* hydroxylamine hydrochloride solution and 5 *M* sodium acetate solution; then add 5 ml of alcohol in order to bring the oil completely into solution. Reflux the solution gently on the steam bath for 2 hr, by which time enough oxime should be formed for identification. The oxime usually separates as an oily layer, and on very thorough cooling and scratching, it can be caused to solidify. Collect the sample on a small filter, dry on a portion of filter paper, and determine melting point. Pure pinacolone oxime melts at 77–78°. It is soluble in cold dilute hydrochloric acid, and the odor is very apparent on boiling the solution. The oxime evaporates rather rapidly, even at room temperature, and it should not be left exposed to the air for more than a few hours.

Experiment 24

Carbanions: The Aldol and Related Reactions and the Haloform Reaction

287. Enolate Ions. Resonance-stabilized carbanions obtained by removal of α-hydrogens from carbonyl compounds are called enolate ions and were described in Experiment 12, which gave examples of their reaction in nucleophilic substitution at saturated carbon. Base-catalyzed carbonyl condensation reactions involve nucleophilic attack by enolate ions at unsaturated carbon. When attack is on the carbonyl group of an aldehyde or ketone, an addition product is obtained and the reaction is an aldol reaction. When attack is on an ester or other carboxylic acid derivative, a substitution reaction occurs. Both types are illustrated below.

The haloform reaction involves attack by enolate ions on halogens and is also illustrated.

288. Aldol Reaction. One of the most important reactions that many aldehydes and ketones undergo is illustrated by the base-induced self addition of acetaldehyde.

$$2CH_3CHO \xrightleftharpoons{\text{trace of base}} \underset{\text{aldol}}{CH_3CH(OH)CH_2CHO}$$

The product readily dehydrates, even on warming, to yield α,β-unsaturated carbonyl compounds. The synthetic utility of the aldol reaction is expanded by the fact that the crotonaldehyde thus formed may be reduced to *n*-butyl alcohol or butyraldehyde. With strong alkali, the products first formed may condense further to give resinous polymers. Both the addition reaction and the dehydration are reversible.

Mixed aldol condensations may also be effected. Formaldehyde and ArCHO, which have no α-H but do have an active carbonyl group, are very useful in such mixed condensations.

The aldol reaction can be applied to ketones, but in many cases the equilibrium is unfavorable. Reaction conditions must then be employed that permit continuous separation of the product from the starting material and catalyst. Condensation of acetone (bp 57°) to diacetone alcohol (bp 168°) is an example.

$$2(CH_3)_2CO \underset{}{\overset{Ba(OH)_2}{\rightleftharpoons}} (CH_3)_2\overset{\displaystyle OH}{\overset{|}{C}}CH_2COCH_3$$

Acetone is placed in a flask attached to a Soxhlet extractor (§117), and the barium hydroxide used as catalyst is placed in the thimble of the extractor. The acetone is refluxed and recondenses to fill the body of the Soxhlet where it is in contact with the catalyst. Conversion to the equilibrium mixture of acetone and diacetone alcohol is rapid. When the Soxhlet empties, this mixture is carried into the boiling flask. Continued refluxing replenishes the acetone in the Soxhlet, but the diacetone alcohol accumulates in the flask because it boils so much higher than acetone. After several days the flask contains mostly diacetone alcohol that can then be purified by distillation; a 71% yield has been obtained (OS–CV **1,** 199).

These are simple examples of many important reactions that are related to each other by the following mechanism:

$$H{-}\overset{|}{\underset{|}{C}}{-}\overset{|}{C}{=}O + :B^- \rightleftharpoons \left(:\overset{|}{\underset{|}{C}}{-}\overset{|}{C}{=}\ddot{O}: \longleftrightarrow {-}\overset{|}{C}{=}\overset{|}{C}{-}\ddot{\underset{..}{O}}: \right)^- + HB$$

$$H{-}\overset{|}{\underset{|}{C}}{-}\overset{|}{C}{=}O + \left(:\overset{|}{\underset{|}{C}}{-}\overset{|}{C}{=}\ddot{O}: \longleftrightarrow {-}\overset{|}{C}{=}\overset{|}{C}{-}\ddot{\underset{..}{O}}: \right)^- \rightleftharpoons$$

$$H{-}\underset{|}{\overset{|}{C}}{-}\underset{|}{\overset{:\ddot{O}:^-}{\overset{|}{C}}}{-}\underset{|}{\overset{|}{C}}{-}\overset{|}{C}{=}O \rightleftharpoons H{-}\underset{|}{\overset{|}{C}}{-}\underset{|}{\overset{OH}{\overset{|}{C}}}{-}\underset{|}{\overset{|}{C}}{-}\overset{|}{C}{=}O$$

289. Benzalacetone. In the present example, note that the product, benzalacetone, is a skin irritant. Wash immediately if the material should reach the skin.

$$C_6H_5CHO + CH_3COCH_3 \xrightarrow{NaOH} C_6H_5CHOHCH_2COCH_3$$

$$C_6H_5CHOHCH_2COCH_3 \longrightarrow C_6H_5CH{=}CHCOCH_3 + H_2O$$

In a small flask place 25 g of benzaldehyde, 50 ml of acetone, and 25 ml of water. With frequent shaking of the mixture, slowly add (over 15–20 min) from a dropping funnel 10 ml of 3 *N* sodium hydroxide solution, keeping the temperature below 35°. Allow to stand for a day or more. Next day add hydrochloric acid until the mixture is just acid to litmus. Isolate the oily layer, extract the aqueous layer with one 20-ml portion of chloroform or benzene, and combine the oil and extract. Dry the oil–extract mixture over calcium chloride, transfer to a Claisen flask, evaporate the solvent, and distill under reduced pressure. The fraction distilling at 150–160°/30 torr is preserved. It should soon crystallize, yielding a slightly yellow product good enough for use in other work. The substance may be redistilled or recrystallized from a very small volume of light petroleum solvent.

290. Perkin Synthesis. This reaction further illustrates the ability of an aromatic aldehyde to react with a carbon atom holding at least one reactive hydrogen atom, as in acetic anhydride. A C_3 side chain is thus established on the benzene ring leading (in the present example) to cinnamic acid.

This experiment requires an alkali acetate as an auxiliary reagent, and potassium acetate seems to be more efficient than the sodium salt formerly recommended. In the present directions the potassium acetate is synthesized in the procedure itself. Pyridine has been recommended as a catalyst, and it is suggested that some students test its value by adding a few drops to the original reaction mixture.

$$C_6H_5CHO + (CH_3CO)_2O \xrightarrow[CH_3CO_2K]{K_2CO_3 \text{ or}} C_6H_5CH{=}CHCO_2K$$

which upon acidification gives cinnamic acid, $C_6H_5CH{=}CHCO_2H$.

291. Cinnamic Acid. In a dry 200-ml round-bottom flask place 10 g of pure benzaldehyde, which must ordinarily be freshly distilled to eliminate impurity of benzoic acid. Add 15 g of freshly distilled acetic anhydride, with very small content of acetic acid, and 7 g of thoroughly dried potassium carbonate. Attach an air-cooled reflux condenser, on the top of which is mounted a small calcium chloride tube, and mix the reagents thoroughly by shaking.

Cautiously introduce the flask with attached condenser into an oil bath heated to 180°. Although the oil bath affords the best control of temperature, it is possible with watchful attention to conduct the reaction over wire gauze. Use of a glass-cloth heating mantle controlled by a variable laboratory autotransformer is particularly appropriate here. Foaming, caused by escaping carbon dioxide, will now ensue for 5 min or more. When the foaming has somewhat abated, heat the reaction mixture steadily for 30 min (see note at end of directions) at 180–190°. Now cool the flask to 100° or below, and wash the contents out into a beaker with a solution of 45 ml of 6 *N* sodium hydroxide in 150 ml of water. Heat the resulting alkaline mixture until all lumps are disintegrated, cool, and filter to remove possible traces of more or less solidified tarry matter.

Wash the filtrate with 10–15 ml of benzene and decolorize with carbon. To the clarified hot filtrate add an appropriate acid to free the cinnamic acid. If slow crystallization and larger crystals are desired, use 25 ml of glacial acetic acid and wrap the vessel with a towel. Otherwise, use 30 ml of concentrated hydrochloric acid. After crystallization seems to be nearly complete, cool the mixture in an ice bath, filter, and wash the product with ice water. The cinnamic acid may be recrystallized from boiling water. Yield 6–7 g, mp 133°.

Note. Experiments in which the time of the main reaction was cut below 30 min gave low yields. It is questionable, however, whether the long periods usually specified for this operation (3–8 hr) give sufficient additional yield to be worthwhile. Students are invited to test this point further and report to the instructor.

292. Claisen Condensation. The Claisen condensation, a close relative of the aldol reaction, is the base-catalyzed condensation of an ester with an active hydrogen compound, commonly another molecule of the same ester:

$$CH_3CO_2Et + CH_3CO_2Et \xrightleftharpoons[\text{(EtOH, small amt.)}]{\text{2Na}} (CH_3COCHCO_2Et)^- + Na^+ + EtONa + H_2$$

The reaction mixture must be neutralized to isolate the free keto ester. Better yields are generally obtained by using sodium ethoxide directly in place of sodium as the catalyst. Since this condensation is reversible, it is desirable to continuously remove the alcohol that is formed. This is usually accomplished by distillation.

The mechanism proposed for the Claisen condensation is closely related to that for the aldol reaction.

$$CH_3CO_2Et + EtO^- \rightleftharpoons :\bar{C}H_2CO_2Et + EtOH$$

$$:\bar{C}H_2CO_2Et + CH_3CO_2Et \rightleftharpoons CH_3\underset{\displaystyle OEt}{\overset{\displaystyle O^-}{\overset{|}{\underset{|}{C}}}}{-}CH_2CO_2Et \rightleftharpoons CH_3COCH_2CO_2Et + EtO^-$$

$$CH_3COCH_2CO_2Et + EtO^- \rightleftharpoons (CH_3COCHCO_2Et)^- + EtOH$$

The driving force for the overall reaction is furnished by the almost complete irreversibility of the last step, which in turn is a result of the much greater acidity of acetoacetic ester over ethyl alcohol. Esters that have only one α-hydrogen atom do not undergo the condensation in the presence of sodium alkoxide catalyst because the β-keto ester formed would not be sufficiently acidic to be transformed to its conjugate base to any significant extent. The equilibrium in such reactions may be forced to the right by a more basic catalyst such as triphenylmethyl sodium, mesitylmagnesium bromide, sodium hydride, and sodium amide.

A useful modification of the Claisen condensation was studied by Dieckmann and involves the internal condensation of esters of dicarboxylic acids:

$$EtO_2C(CH_2)_4CO_2Et \xrightarrow{NaH} \text{2-Carbethoxycyclopentanone (cyclic: } H_2C\text{–}C(=O)\text{–}CHCO_2Et\text{, } H_2C\text{–}CH_2)$$

Diethyl adipate

81%

2-Carbethoxycyclopentanone

293. Diethyl Adipate. Place 25 g of adipic acid, 65 ml of absolute alcohol, 25 ml of toluene, and 0.25 ml of concentrated sulfuric acid in a 250-ml distilling flask mounted in an oil bath and attached to a water-cooled condenser equipped with a curved adapter. Heat the bath to about 115°, whereupon a ternary azeotrope of alcohol, toluene, and water distills at about 75° (vapor temperature). Continue distillation until the vapor temperature reaches 78°.

To the resulting distillate add 30 g of anhydrous potassium carbonate; shake, filter, and return filtrate to the distilling flask. Distill again, and discontinue distillation when the temperature reaches 80°.

Distill the residual liquid (which contains the desired high-boiling ester) under reduced pressure, rejecting low-boiling alcohol and toluene. Any working pressure from 10 to 30 torr will be satisfactory. Boiling point at 20 torr, 130°; at other pressures, use information from §37.

294. 2-Carbethoxycyclopentanone. [References: OS–CV **2;** L. H. Klemm and H. Ziffer, *J. Org. Chem.*, **20,** 182 (1955)]. Carefully dry a 250-ml standard-taper three-neck flask equipped with reflux condenser, sealed stirrer, and dropping funnel. Displace the air in the system by dry nitrogen. Place 5.9 g of a 55% sodium hydride dispersion (see below for details and precautions) and 125 ml of dry, thiophene-free benzene in the flask. Add dropwise 25.3 g (0.125 mole) of diethyl adipate to the flask with stirring during one hour. The mixture may solidify and the stirrer may stop as a result of this; a little more benzene should then be added. Record all observations with care.

While the addition is in progress prepare a Bunsen valve. This consists of a short piece of clean, dry, new, rubber pressure tubing closed at one end and slit on one side so that gas can escape but air will not be drawn in (see sample on instructor's table). Install the valve on a short length of glass tubing inserted into a one-hole rubber stopper that fits the center neck of the reaction flask.

When addition of the ethyl adipate is finished, remove the condenser, stirrer, and dropping funnel. Close the two side necks with solid rubber stoppers and the center neck with the stopper carrying the Bunsen valve. Small amounts of hydrogen will continue to evolve from the reaction mixture and the valve permits this to escape without allowing air and moisture to enter. The reaction mixture should be allowed to stand until the next laboratory period (at least two days).

Add 1 ml of glacial acetic acid and work this into the precipitate with a flattened stirring rod. Cool the reaction mixture in an ice bath and add slowly 70 ml of 10% hydrochloric acid, which has been precooled to 0°. Add the acid very slowly at first so that the temperature remains between 0° and 10°. If the temperature rises, some of the product may be hydrolyzed and lost. Separate the benzene layer and wash the aqueous layer with several small portions of benzene. Wash the combined benzene solution once with water, once with 5% sodium bicarbonate solution, and once more with water; then dry it over anhydrous magnesium sulfate.

If equipment is available for vapor-phase chromatography or nmr spectrometry, the relative amounts of product (2-carbethoxycyclopentanone) and unreacted starting material (ethyl adipate) should be determined. The boiling points of these two compounds are too close together for easy separation by simple distillation (ethyl adipate: 138°/20 torr; 2-carbethoxycyclopentanone: 113–116°/20 torr). If your products contain less than 5% of starting material, distill off most of the benzene and distill the residue under reduced pressure, bp 90–95°/5–6 torr; 113–116°/20 torr. The yield reported on a large-scale run was 81%.

If an appreciable amount of starting material is present, it should be removed chemically. This can be accomplished readily because 2-carbethoxycyclopentanone, a β-keto ester, is relatively acidic and soluble in dilute potassium hydroxide; diethyl adipate is insoluble in aqueous base.

$$\text{(cyclopentanone)}\text{—}\overset{H}{C}\text{—}CO_2C_2H_5 + OH^- \rightleftharpoons \text{(cyclopentanone)}\text{—}\overset{-}{C}\text{—}CO_2C_2H_5 + H_2O$$

Cool the benzene solution in an ice bath to near 0° (benzene solidifies at about 5°) and add it from a dropping funnel slowly with stirring to 60 ml of 10% potassium hydroxide solution, which has been precooled to 0° in an ice–salt bath (maintain it at 0° throughout the addition). The potassium salt of 2-carbethoxycyclopentanone usually precipitates during the operation. As soon as any precipitate appears it must be redissolved by cautious addition of water (precooled to near 0°) with stirring and care to keep the temperature near 0°. The precipitate is hard to dissolve if any appreciable amount is allowed to accumulate; however, the amount of additional water added should be kept to a minimum. Separate the benzene layer, and wash it with 10 ml of a cold 5% solution of potassium hydroxide. Combine the basic aqueous washings with the original aqueous layer; the product is in this solution, which should be kept near 0° at all times to avoid hydrolysis of the ester group. Wash this aqueous solution with 50 ml of ether to remove any remaining diethyl adipate and benzene. Then run the cold aqueous solution slowly into 100 ml of ice-cold 10% acetic acid solution. Separate the new oil layer (which now contains the product) with the aid of a little ether and extract the aqueous layer once with ether. Wash the combined ether extracts once with 5% sodium carbonate solution, dry over sodium sulfate or magnesium sulfate, filter, remove the ether by distillation from a steam bath, and distill the residue under reduced pressure. The purification process usually reduces the yield by 15–25%.

Sodium hydride is supplied as a dispersion in purified mineral oil. This material usually contains 50–55% by weight of sodium hydride; each batch is labeled to show the sodium hydride content. If the material supplied is less than 55% by weight, calculate the correct weight of the dispersion to use. These dispersions have the appearance of grey powders slightly moistened with oil. They may be transferred like any mealy solid as long as general safety precautions are observed.

Warning: *Sodium hydride in oil will react vigorously in contact with water, lower alcohols, or aqueous solvents and evolve hydrogen. The hydrogen does not generally ignite, but suitable precautions should be taken to prevent ignition by other sources. Material from which the oil has been removed, either by adsorption or solvent washing, is to be considered extremely reactive, and may be pyrophoric. It cannot be exposed to air or moisture without danger of spontaneous ignition.*

Burning hydride can be extinguished by smothering with soda ash, or dry powdered dolomite limestone. Commercially available Ansul Du-gas extinguishers are recommended for laboratory and plant use.

Warning: *Sodium hydride in oil is very caustic on contact with exposed areas of the body, and suitable handling and protective equipment should be employed. Safety glasses are essential, and for large-scale use face shields and rubber or plastic-coated gloves are recommended. Any skin areas exposed to the material should be flooded with water to prevent caustic and thermal burns. Reasonable care, for which there is no substitute, will enable this chemical to be used with a minimum of hazard.*

295. Haloform Reaction. The haloform reaction—that is, the reaction of methyl ketones (or compounds that may be oxidized to them, such as $R{-}\overset{\overset{\large OH}{|}}{C}H{-}CH_3$) with a halogen in basic medium—is related to the aldol condensation in that the initial step is formation of an enolate ion. This is followed by rapid reaction with the halogen:

$$R{-}\overset{\overset{\large O}{\|}}{C}{-}CH_3 \xrightarrow{OH^-} \left[R{-}\overset{\overset{\large O}{\|}}{C}{-}\bar{\ddot{C}}H_2 \longleftrightarrow R{-}\overset{\overset{\large :\ddot{O}:^-}{|}}{C}{=}CH_2\right] \xrightarrow{X_2} R{-}\overset{\overset{\large O}{\|}}{C}{-}CH_2X + X^-$$

The resulting product, $R{-}\overset{\overset{\displaystyle O}{\|}}{C}{-}CH_2X$, is more acidic than the starting $R{-}\overset{\overset{\displaystyle O}{\|}}{C}{-}CH_3$ owing to the inductive effect of the halogen atom, and further reaction occurs spontaneously. Since with each substitution of halogen for hydrogen the remaining α-hydrogens are rendered more acidic, the same general reaction sequence is repeated twice more until all of the α-hydrogens on the methyl group are replaced:

$$R{-}\overset{\overset{\displaystyle O}{\|}}{C}{-}CH_2X \longrightarrow R{-}\overset{\overset{\displaystyle O}{\|}}{C}{-}CHX_2 \longrightarrow R{-}\overset{\overset{\displaystyle O}{\|}}{C}{-}CX_3$$

The combined inductive effect of three halogens makes the carbonyl particularly susceptible to nucleophilic attack and at the same time promotes cleavage of the bond between the carbonyl carbon and the trihalomethyl carbon by stabilizing the leaving group CX_3^-. Thus, at this point attack by OH^- occurs to give the salt of the carboxylic acid and the haloform.

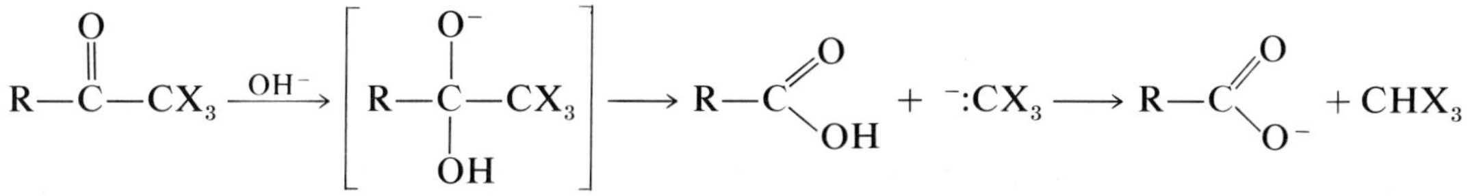

$$R{-}\overset{\overset{\displaystyle O}{\|}}{C}{-}CX_3 \xrightarrow{OH^-} \left[R{-}\overset{\overset{\displaystyle O^-}{|}}{\underset{\underset{\displaystyle OH}{|}}{C}}{-}CX_3 \right] \longrightarrow R{-}C\begin{smallmatrix}\nearrow O \\ \searrow OH\end{smallmatrix} + {}^-{:}CX_3 \longrightarrow R{-}C\begin{smallmatrix}\nearrow O \\ \searrow O^-\end{smallmatrix} + CHX_3$$

Note that all ketones with α-hydrogens undergo α-halogenation, but only those with a methyl group adjacent to the carbonyl groups will undergo carbon-carbon bond cleavage to give a haloform, since three halogens are required to cause cleavage. This fact is utilized in the iodoform test for methyl ketones, in which an unknown ketone is treated with iodine under basic conditions. Methyl ketones will yield iodoform, CHI_3, a crystalline yellow solid that precipitates from solution. The iodoform test can readily be demonstrated using a suitable methyl ketone such as acetone or the methyl *n*-amyl ketone derived from §215. The proper test procedure is outlined in the Qualitative Organic Analysis section, §381.

Occasionally the haloform reaction can be used to advantage in the synthesis of carboxylic acids, particularly when the corresponding methyl ketone is readily available. Usually the halogen used in this case is chlorine, since it is cheaper and is readily available as a 5% solution of sodium hypochlorite (Clorox, Purex, or similar commercial bleaching solution), which may be used directly. This method can be illustrated by the preparation of cinnamic acid from benzalacetone (§289), dimethylacrylic acid from mesityl oxide (OS-CV **3,** 302), or simply benzoic acid from acetophenone, as in the following section.

296. Benzoic Acid. To 80 ml of commercial 5% sodium hypochlorite solution in a 250-ml Erlenmeyer flask add 2 ml of acetophenone and swirl vigorously. When the initial reaction has subsided, evaporate the chloroform produced in the reaction by heating the flask on a steam cone. Destroy any remaining excess sodium hypochlorite by adding 1 ml of acetone, and heat an additional 10 min to evaporate the resulting chloroform and excess acetone. If necessary, the solution may be decolorized with decolorizing carbon (Darco, Norit, etc.) at this point. Acidify the hot solution to pH 1, cool to room temperature, and finally place in an ice bath to complete crystallization. Collect the crystals by suction filtration, wash them twice with 15-ml portions of cold water, dry them, and determine their melting point.

Questions:

1. What organic by-products would you expect in a reaction producing iodoform from (*a*) ethyl alcohol, (*b*) *sec*-butyl alcohol, (*c*) acetaldehyde, with iodine used as halogenating reagent.

2. How many moles of chloroform are theoretically obtainable from one mole of free chlorine?

3. Why is chloroform a good solvent for partially oxygenated substances?

4. The melting point of bromoform is almost as high in the presence of water as when dry, whereas in the case of phenol, the difference between presence and absence of water is much greater. Explain on the basis of melting-point theory.

5. The ir and nmr spectra for ethyl adipate and 2-carbethoxycyclopentanone are given in Figs. 1–4. Identify the characteristic ir bands for each compound and the signals for the various kinds of protons in the nmr spectra. Account for the spin-spin splitting patterns in the nmr spectra.

6. The nmr spectrum of a mixture obtained by following the procedure of §294 but omitting the purification through the potassium salt is given in Fig. 5. Determine the composition of the mixture.

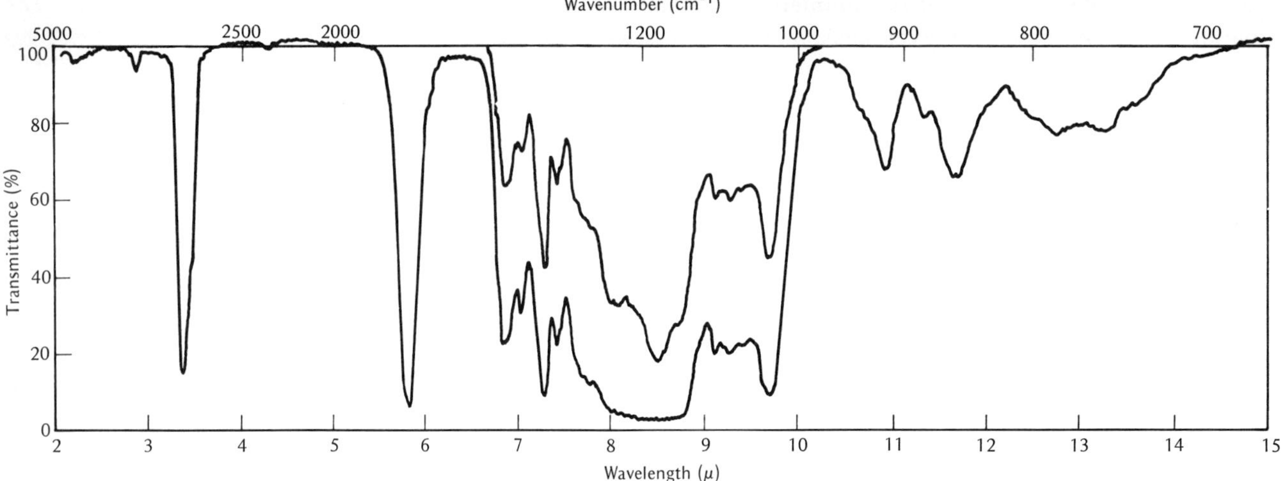

Fig. 1. Infrared Spectrum of Diethyl Adipate. (© Sadtler Research Laboratories, Inc., 1962.)

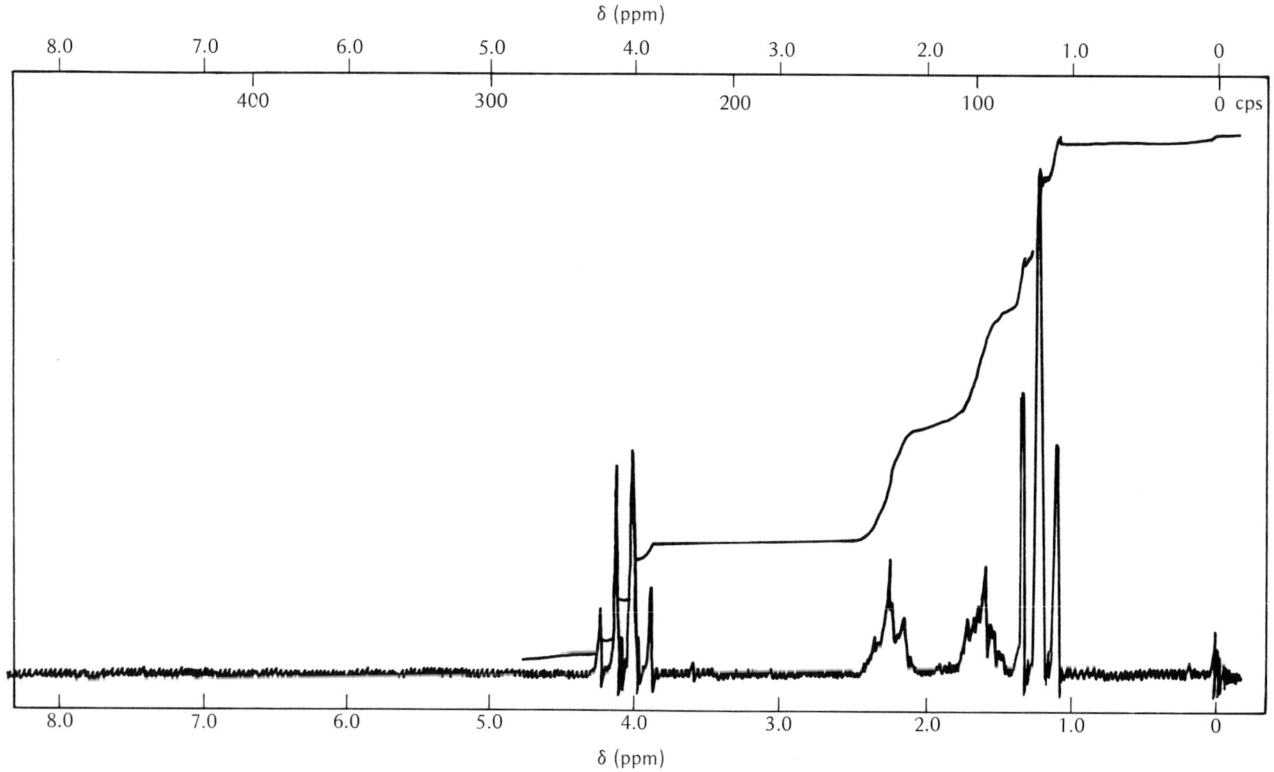

δ (ppm)

Fig. 2. NMR Spectrum of Diethyl Adipate. (© Sadtler Research Laboratories, Inc., 1969.)

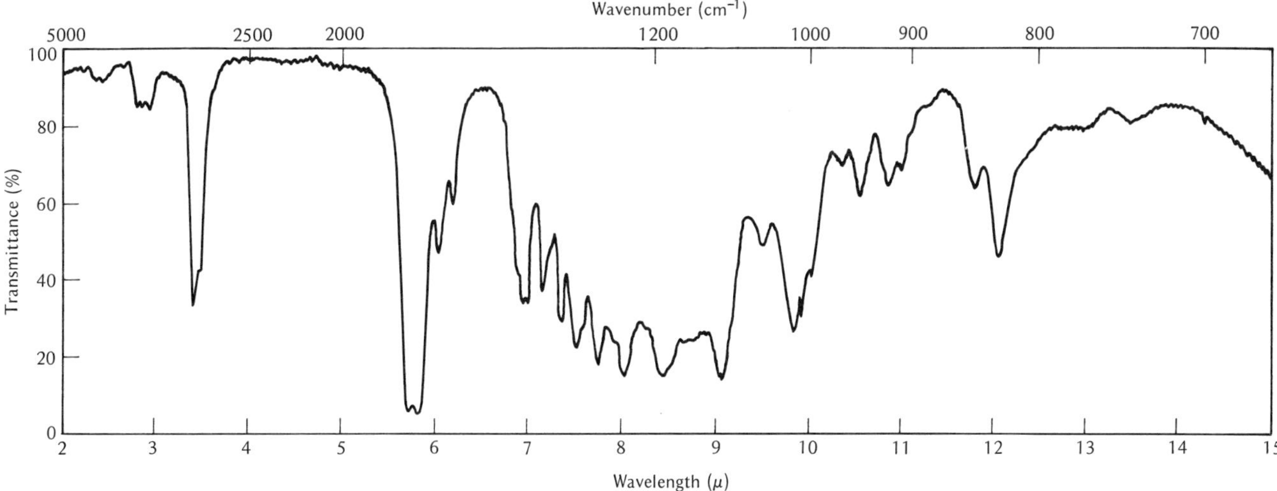

Fig. 3. Infrared Spectrum of 2-Carbethoxycyclopentanone. (© Sadtler Research Laboratories, Inc., 1958.)

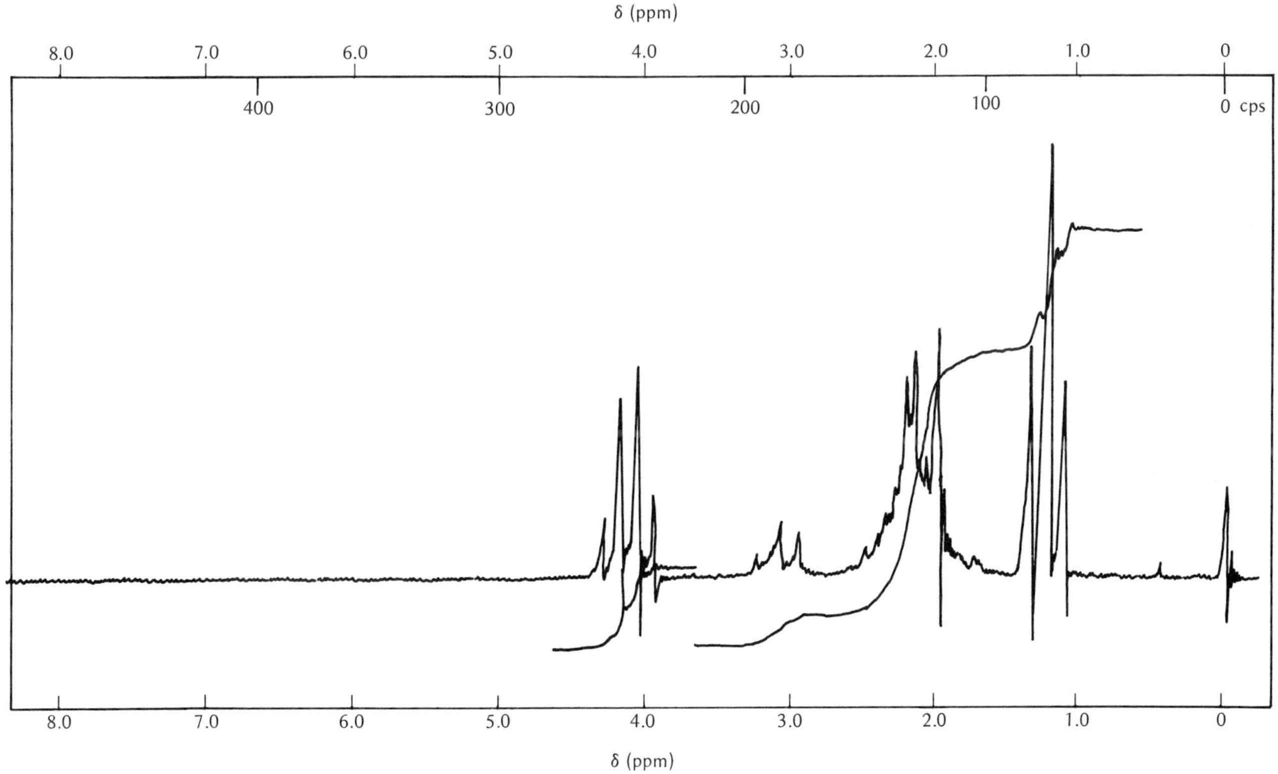

δ (ppm)
8.0 7.0 6.0 5.0 4.0 3.0 2.0 1.0 0
400 300 200 100 0 cps
8.0 7.0 6.0 5.0 4.0 3.0 2.0 1.0 0
δ (ppm)

Fig. 4. NMR Spectrum of 2-Carbethoxycyclopentanone.

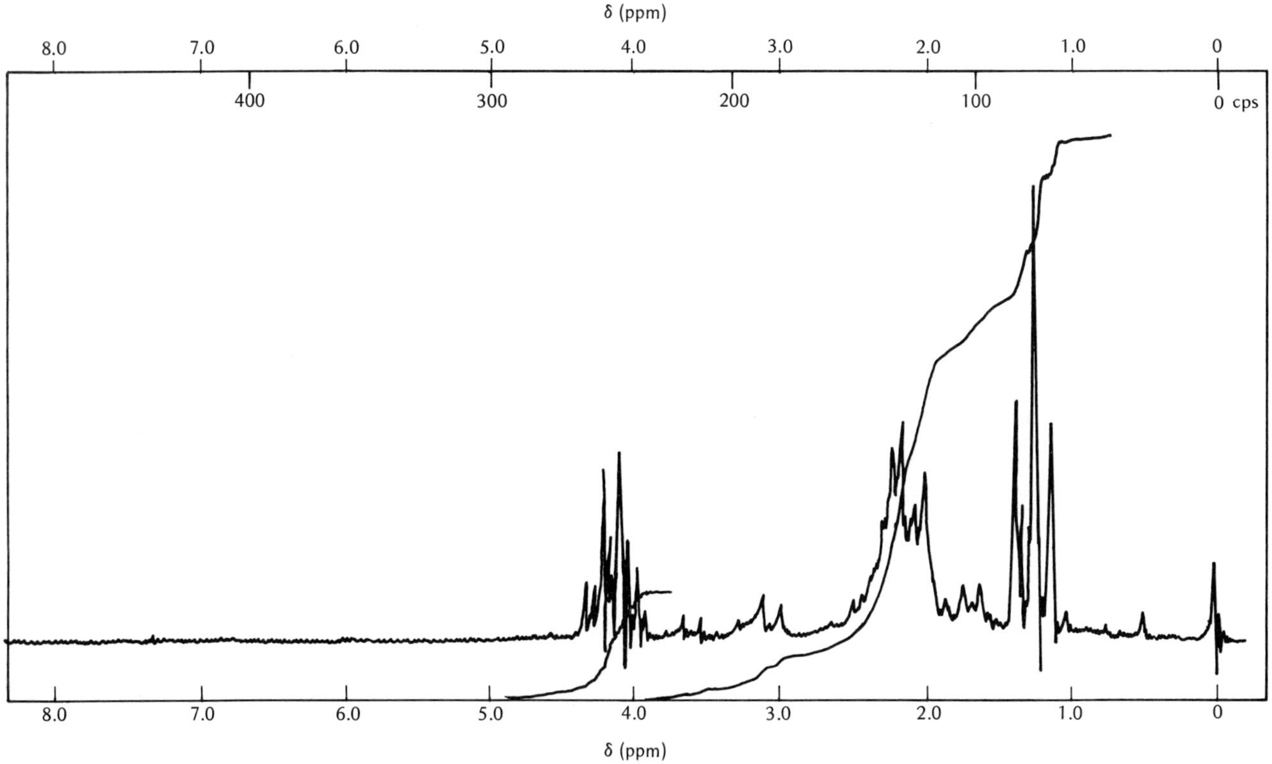

Fig. 5. NMR Spectrum of the Mixture of Starting Material and Product from §294.

Experiment 25

The Wittig and Related Reactions

297. General Comment and Mechanism. A general method for olefin synthesis involves conversion of a phosphonium salt to a carbanion in which the negative charge is stabilized by resonance involving the *d* orbitals of an adjacent positive phosphorus. Such a carbanion, called an ylide, is a nucleophile that can attack the carbonyl carbon of most aldehydes and ketones to form an olefin with elimination of a phosphine oxide. The mechanism is believed to involve a four-center transition state as shown. Triphenyl phosphine and a primary or secondary alkyl bromide are the most common starting materials.

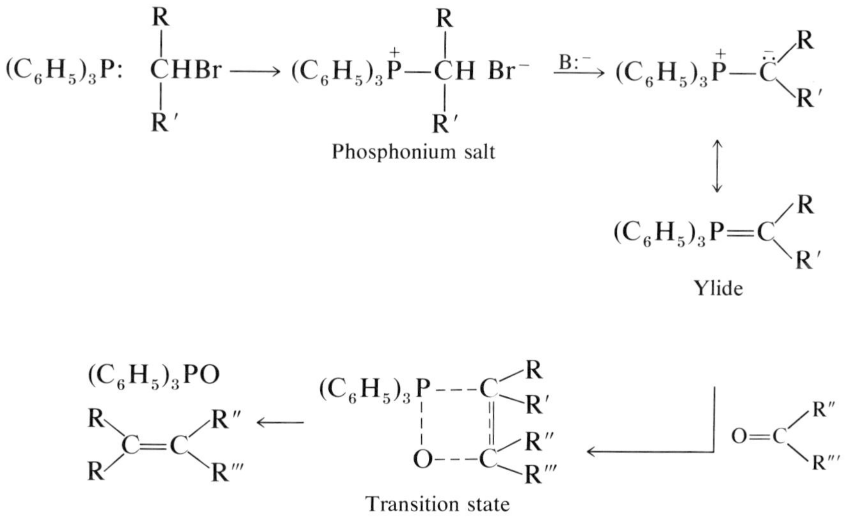

$$(C_6H_5)_3P\colon \; \mathrm{CHBr}(R)(R') \longrightarrow (C_6H_5)_3\overset{+}{P}-\mathrm{CH}(R)(R')\ \mathrm{Br}^- \xrightarrow{B:^-} (C_6H_5)_3\overset{+}{P}-\overset{-}{C}(R)(R') \longleftrightarrow (C_6H_5)_3P{=}C(R)(R')$$

Phosphonium salt

Ylide

$$(C_6H_5)_3P{=}C(R)(R') + O{=}C(R'')(R''') \longrightarrow \begin{array}{c}(C_6H_5)_3P---C(R)(R')\\ \vdots \qquad \vdots \\ O---C(R'')(R''')\end{array} \longrightarrow (C_6H_5)_3PO + (R)(R)C{=}C(R'')(R''')$$

Transition state

The method is excellent for the synthesis of olefins because yields are generally high, conditions mild, and isomeric (rearranged) olefins are not formed. Mixtures of *cis* and *trans* olefins are usually obtained.

The reaction is rather inconvenient as an experiment in an organic chemistry laboratory course because an inert atmosphere is usually required and the base most often employed is an organolithium, which is not too easily handled. A related alternative olefin synthesis, which appears to be less general but is easier to carry out, employs the less expensive triethyl phosphite instead of triphenyl phosphine. This or other phosphorous esters react with reactive organic halides, such as benzyl chloride, to give salts that are converted by heat to phosphonate esters (the Arbuzov reaction). Bases transform these esters to nucleophilic carbanions that attack carbonyl compounds giving olefins. Equations for application of these reactions to the preparation of 1,2-diphenylethylene and a procedure for the synthesis are given below. The product appears to be entirely the *trans* isomer.

$$C_6H_5CH_2Cl + P(OC_2H_5)_3 \longrightarrow [C_6H_5CH_2P(OC_2H_5)_3]^+Cl^- \xrightarrow{130-200^\circ}$$

$$C_6H_5CH_2\overset{\overset{\displaystyle O}{\|}}{P}(OC_2H_5)_2 + C_2H_5Cl \xrightarrow{CH_3ONa} [C_6H_5\bar{C}H\overset{\overset{\displaystyle O}{\|}}{P}(OC_2H_5)_2]Na^+ + CH_3OH$$

$$[C_6H_5\bar{C}H\overset{\overset{\displaystyle O}{\|}}{P}(OC_2H_5)_2]Na^+ + C_6H_5CHO \longrightarrow \begin{array}{l} C_6H_5CH-\overset{\overset{\displaystyle O}{\|}}{P}(OC_2H_5)_2 \\ \quad\;\; | \\ C_6H_5CH-O^-Na^+ \end{array} \longrightarrow$$

$$\begin{array}{ccc} C_6H_5 & & H \\ & C{=}C & \\ H & & C_6H_5 \end{array} + (C_2H_5O)_2PO_2Na$$

298. *trans*-Stilbene. Add 6 ml (6.6 g, 0.052 mole) of benzyl chloride to 9 ml (8.72 g, 0.053 mole) of triethyl phosphite in a 25- to 50-ml flask equipped with a water-cooled condenser, and reflux the solution in the hood for 1 hr. Both the benzyl chloride and the triethyl phosphite must be pure.

Triethyl phosphite boils at 50° at 13 torr, 63° at 24 torr, and 158° at 760 torr. It should be distilled under reduced pressure.

Warning: *Organophosphorus compounds are often toxic. Should any of these compounds come in contact with the skin, wash the affected area immediately and thoroughly with soap and water. Transfer of organophosphorus compounds should be done in a hood to avoid exposure to the vapors from these compounds. Volume measurements for this experiment may be made with a graduated pipet, but the starting materials should never be drawn into the pipet by mouth; either a suction pump or preferably a pipet filler should be used.*

Ethyl chloride is eliminated starting about 130° and the final temperature of the refluxing liquid approaches 200°. Cool the reaction mixture (phosphonate ester) to room temperature. Weigh 3.0 g (0.055 mole) of sodium methylate in a clean, dry, stoppered, 125-ml conical flask with care to protect this hygroscopic material from the moisture of the air as much as possible. Immediately add 25 ml of dry *N,N*-dimethylformamide and then the cool phosphonate ester. Rinse the phosphonate ester flask with an additional 25 ml of *N,N*-dimethylformamide and add this to the conical flask.

Cool the reaction mixture to below 20° by swirling the flask in a water–ice bath and add dropwise 5.2 ml (5.5 g, 0.052 mole) of benzaldehyde which is dry and free from benzoic acid. The rate of addition should be slow enough so that the temperature remains between 30° and 40°; the flask should be swirled during the addition. A graduated pipet can be used for the addition since the time required is short. After addition is complete remove the flask from the ice bath and allow it to stand at room temperature for 10 min. Add 20 ml of water with stirring, collect the crystals, and wash them on the Büchner funnel with cold aqueous methanol (a mixture of equal volumes of methanol and water, designated 1:1 v/v). A yield of 85% has been reported. Dry a small sample of the crystals and determine the melting point. Pure *trans*-stilbene melts at 126–127° cor. Recrystallize the product (the solvent pair ethanol–ethyl acetate has been used; see §100), dry the recrystallized material, and determine the melting point.

The procedure given above has been applied to several other aldehydes; see E. J. Seus and C. V. Wilson, *J. Org. Chem.*, **26,** 5243 (1961). Reaction with cinnamaldehyde, $C_6H_5CH{=}CHCHO$, gives the interesting conjugated diene 1,4-diphenyl-1,3-butadiene.

Experiment 26

Electrophilic Aromatic Substitution and Nitration

299. Electrophilic Aromatic Substitution. Electrophilic substitution is the most characteristic reaction of aromatic compounds. An aromatic nucleus has a system of electrons in π-orbitals that can be attacked by a powerful electron-seeking reagent (an electrophile). Reaction occurs by an addition-elimination mechanism in which the electrophile attacks the position or positions in the aromatic nucleus where the electron density of the π-system is greatest, forming a σ-complex. A proton is then lost to restore the π-system and complete the substitution.

Nitration is a typical and useful electrophilic aromatic substitution and is illustrated in the present experiment. The mechanism of nitration of benzene is formulated in the following equations. Nitronium ion, produced by reaction of sulfuric acid with nitric acid, is the attacking electrophile, and the proton lost from the benzene ring eventually ends up in a water molecule.

$$H_2SO_4 + HNO_3 \rightleftharpoons H_2NO_3^+ + HSO_4^-$$

$$H_2NO_3^+ \rightleftharpoons H_2O + NO_2^+$$

$$NO_2^+ + C_6H_6 \longrightarrow [C_6H_6NO_2]^+ \longrightarrow C_6H_5NO_2 + H^+$$

Other common electrophilic aromatic substitutions include sulfonation, chlorosulfonation, halogenation, the Friedel and Crafts reaction, and diazo coupling. These are illustrated in Experiments 27–30 and 34.

300. Nitrobenzene. Nitration occurs more cleanly if the nitrating agent is free from lower oxides of nitrogen, which are equivalent to nitrous acid. Such oxides may also be produced during the nitration by side reactions involving the oxidizing action of nitric acid. A small amount of urea is added to the reaction mixture to remove these oxides.

$$2HNO_2 + NH_2CONH_2 \longrightarrow 2N_2 + CO_2 + 3H_2O$$

Warning: *Nitro compounds are toxic and should not be allowed to escape into the atmosphere of the laboratory. See that the delivery tube of the condenser passes well into the mouth of the flask receiver. It is not permissible to allow distillates of liquids in the aromatic series to drip through the open air into beakers or bottles. This applies to aniline, toluidine, and other substances described in the following pages, as well as nitrobenzene.*

To 30 ml of concentrated nitric acid (sp. g. 1.42) in a small flask add slowly 30 ml of concentrated sulfuric acid, and cool the mixture in running water. Place 0.4 mole of benzene and 1 g of urea in a 500-ml flask and provide a thermometer (110° range). In view of the shaking required in the process, it is convenient first to mount the thermometer firmly in a cork fitted to the 500-ml flask and then to provide means for introducing the acid mixture. For this purpose a short air-cooled condenser mounted in a second hole in the cork is serviceable, or the cork may be channeled in a manner to permit addition of acid. Now add slowly in small portions the acid mixture prepared previously. From time to time cool the flask in running water so that the temperature of the reacting mixture ranges from 45° to 50°. *Frequent* and *vigorous* agitation is necessary to permit the acids and the benzene to come into contact with each other and react; otherwise they lie in two layers, one of which may become unduly warmer than the other. After all the acid has been added to the benzene, hold the reaction mixture, with occasionally shaking, at the temperature of 60° for 30–40 min. A hot-water bath and running water should be at hand for use as the situation demands.

Cool the reaction mixture; then separate and discard the acid layer. Wash the nitrobenzene layer first with 60 ml of water and then with a mixture of 5 ml of the laboratory solution of sodium hydroxide (or carbonate) and 25 ml of water. If the mixture after thorough shaking is acid to litmus, repeat the sodium carbonate treatment until a neutral or alkaline test is obtained. Finally, wash with 50 ml of water. ★

The washed product is likely to contain emulsified water. Remove this by heating gently with anhydrous calcium chloride (about 12 g) for a few moments and then cooling the mixture, with shaking, to room temperature. Allow the product to stand with the calcium chloride until it is dry, preferably for several hours or days.

Separate the dry nitrobenzene from the calcium chloride, transfer to a small distilling flask, and distill through an air-cooled condenser. Collect the distillate coming over between 200° and 208° (uncor.), and reject the high-boiling residue. Do not attempt to distill a higher-boiling fraction and do not overheat the small residue, lest there be a violent decomposition of higher nitro compounds present. If any large residue is left, it is likely that the reaction has been allowed to take place at too high a temperature, particularly at first when the acids were at maximum concentration. The product should not smell of nitric acid. Yield is 30 g.

Questions

1. Why is the acid added gradually to the benzene, instead of benzene to acid?

2. Why should one be more particular to avoid a high temperature at the start of the reaction rather than toward the close?

3. Is the correction for exposed stem of the thermometer of great significance here? Why?

4. Suppose (*a*) chlorobenzene and (*b*) benzoic acid were nitrated instead of benzene. What products would you expect?

301. Influence of Substituents on Electrophilic Aromatic Substitution. Electron-withdrawing substituents on a benzene ring retard electrophilic substitution and direct the entering substituent to the meta positions (with the exception of halogen substituents, which are ortho-para orienting); electron-donating substituents accelerate substitution and direct ortho-para. Because a methyl substituent is weakly electron-donating and a nitro group is strongly electron-withdrawing, substitution in *p*-nitrotoluene is slower than in benzene and requires both more concentrated acid and higher temperature. Both substituents direct the entering nitro group into the positions adjacent to the methyl.

302. 2,4-Dinitrotoluene—Semimicro Technique. In an 8-in. test tube place 4 ml of yellow fuming nitric acid (see §579) and gradually add (with mouth of test tube pointed away from the face) 4 ml of concentrated sulfuric acid. To this mixture add gradually, in four or five portions, 3 g of *p*-nitrotoluene. If the mixture should threaten to boil, dip the test tube in cold water.

> **Warning:** *Be very careful to keep fuming nitric acid off the skin, as it reacts vigorously and causes stains and serious burns.*

Heat the test tube and contents in a water bath at about 90° or 95° for 30 min, and then cool nearly to room temperature. Now pour the reaction mixture into about 100 ml of ice water containing a few pieces of ice. The solid that separates is crude 2,4-dinitrotoluene. Collect the product upon a Hirsch funnel (see §96) or upon a perforated, beveled-edge porcelain plate resting in an ordinary glass funnel. Wash well with cold water, and recrystallize from methyl alcohol. Yield is 3 g.

If this recrystallization were conducted on a larger scale, as with most other experiments in this manual, safety precautions would call for a reflux condenser during the heating process. With so small an operation as this experiment, careful work with an open flask heated on a steam or water bath, is satisfactory; but have a wet towel at hand if any flame is used in the vicinity, to extinguish a possible small fire. Pure 2,4-dinitrotoluene melts at 70.7°. Do not try to distill this substance, lest violent decomposition occur.

303. *m*-Dinitrobenzene. The foregoing directions (for dinitrotoluene) may be used, only with substitution of nitrobenzene for *p*-nitrotoluene as organic starting material (mp of *m*-dinitrobenzene, 89–90°).

Questions

1. Why is it better to use *p*-nitrotoluene in this experiment rather than the cheaper isomer, *o*-nitrotoluene?

2. Why not put in both nitro groups at one time, starting with toluene and using fuming nitric acid?

3. Why is the dinitrotoluene much more soluble in the highly oxidized nitrating acid than in the water into which it is later placed?

Experiment 27

Aromatic Sulfonic Acids

304. Sulfonation. Sulfonation is an electrophilic aromatic substitution reaction which is usually slower than nitration. The attacking electrophile is SO_3 or SO_3H^+. If electron-withdrawing substituents are present on the aromatic nucleus, fuming sulfuric acid is usually required. Unlike nitration, sulfonation is reversible and the sulfonic acid group on a benzene ring can usually be replaced by hydrogen on reaction with superheated steam.

If the aromatic compound that is to undergo the sulfonation reaction is an amine, there is first an immediate reaction to form the salt. Prolonged heating transforms such a salt into the amino-sulfonic acid, a compound whose structural formula may be represented directly as the name implies, or as the inner salt (zwitterion):

$$C_6H_5NH_2 \longrightarrow C_6H_5NH_3^+HSO_4^- \longrightarrow p\text{-}H_2NC_6H_4SO_3H \quad \text{or} \quad p\text{-}H_3N^+C_6H_4SO_3^-$$

305. Sulfanilic Acid. From aniline the rules of orientation would predict the formation of ortho and para sulfonic acids. At about 180°, however, the ortho derivative either does not form or is rearranged and transformed into the para compound. As a result, the single para compound, known as "sulfanilic acid," is obtained.

Place 0.2 mole of freshly distilled aniline in a 200-ml flask, using a funnel to avoid wetting of the neck of the flask with the aniline. Remove the funnel and cautiously add 30 ml of concentrated sulfuric acid. Heat the mixture in the open flask at 180–190°, thus allowing the water vapor to escape, until a sample of 2 or 3 drops yields no oily aniline when added cautiously to a few milliliters of dilute sodium hydroxide in a test tube. An oil bath is the most convenient method of maintaining the desired temperature, but a flask over wire gauze and burner may be used, provided the material is agitated from time to time and constantly watched. Three hours gives a fair yield; 4 hr should be ample. ★

A slightly simpler modification of this synthesis of sulfanilic acid has been described [N. Roegis, *J. Chem. Ed.*, **45,** 274 (1968)].

Cool the mixture to 50° or below and pour it with stirring into 200 g of cold water or of crushed ice. Collect the precipitated mass of crude sulfanilic acid upon the Büchner filter, wash with cold water, and dissolve in the least possible amount of boiling water. Add 2–3 g of decolorizing carbon to the boiling solution and filter through a fine-grained filter paper on the Büchner funnel. Upon cooling, the solution deposits crystals that contain two molecules of water of crystallization. The substance does not melt sharply, and no attempt should be made to estimate melting point. Yield is 15 g.

If the product is dark-colored (owing to inadequate control of the oil-bath temperature), it may often be purified to advantage in alkaline solution, as follows. Dissolve the impure solid in as small an amount as possible of warm 2 *N* sodium hydroxide solution, add decolorizing carbon, boil 3 min, and filter. To the filtrate then add an amount of concentrated hydrochloric acid slightly more than the quantity chemically equivalent to the sodium hydroxide used. Collect the resulting precipitate on a Büchner filter and recrystallize from the minimum volume of boiling water.

Care should be taken not to expose moist crystals of sulfanilic acid to air containing oxides of nitrogen during the drying process; otherwise the compound may be superficially diazotized and partially converted into colored material.

Questions

1. What is the general value of introducing the sulfonic acid radical into an aromatic compound?

2. What is the scientific basis for the test for completeness of reaction given in the sulfonation experiment which you conducted?

3. What effect does the entrance of the sulfonic acid group have upon the strength of aniline as a base?

306. 2-Methyl-5-nitrobenzenesulfonic Acid. In a 200-ml round-bottom flask suspended in a water bath at room temperature, place 5 g of *p*-nitrotoluene and 8 ml of (20%) fuming sulfuric acid, and mix well. After the mixture has stood for 10 min, slowly heat the bath to the boiling point, but not above 100°.

Now pour the mixture into 150 ml of water. Decolorize the resulting solution with carbon and neutralize the filtrate by the cautious addition of the requisite quantity of solid potassium carbonate. A change in color from light yellow to brownish marks the neutral point. Heat the resulting mixture until any precipitate dissolves. Upon cooling, the desired potassium salt crystallizes in yellowish brown needles along with a little potassium sulfate. Recrystallize from boiling water. The purified product is light yellow.

307. Derivatives. Interesting sequences of preparations may be arranged to follow this sulfonation experiment. The methyl group may be oxidized (see method of §240) to the carboxyl group, and the nitro group then reduced; or the oxidation step may be omitted. An amine thus prepared is then diazotized and converted into various azo dyes (Experiment 34) that are acidimetric indicators. If such sequence is contemplated, multiply the quantity of starting material by the factor of 4.

Experiment 28

Sulfanilamide

308. Chlorosulfonation

$$CH_3CONHC_6H_5 \longrightarrow CH_3CONHC_6H_4SO_2Cl \longrightarrow CH_3CONHC_6H_4SO_2NH_2 \longrightarrow H_2NC_6H_4SO_2NH_2$$

Provide an ice bath, and either work in a hood or use the inverted funnel arrangement described in §313. In a 200- or 250-ml conical flask place 25 ml of chlorosulfonic acid, which should be handled carefully because it may cause skin burns. Add gradually 10 g of dry, crystalline acetanilide (either the product from §319 or commercial material) using the ice bath to keep the reaction temperature between 10° and 20°. Heat the final reaction mixture on the steam bath for about 30 min, cool to room temperature, and pour with stirring into 150 to 200 ml of a thin slush of ice and water. Collect the granular product of *p*-acetaminobenzenesulfonyl chloride on a Büchner filter, wash with cold water, and use at once in the next procedure.

To the acid chloride just prepared, which has been returned to the reaction flask, add 35 ml of concentrated aqueous ammonia (sp. g. 0.90) and 35 ml of water (or use 70 ml of about 8 *N* ammonium hydroxide). Mix thoroughly, and heat gently to near the boiling point for 20–30 min. Cool the solution and add hydrochloric acid until the mixture is neutral to litmus. Filter, wash the product of *p*-acetaminobenzenesulfonamide with water, and carry to the next step, which is hydrolysis to sulfanilamide. ★

To the amide product just prepared, add 25 ml of 3 *N* hydrochloric acid, and boil cautiously, avoiding charring, for 15–20 min. If necessary, decolorize with carbon. Finally, add to the filtrate 10 g of sodium bicarbonate. Collect the product of sulfanilamide on a filter, wash with cold water, and dry; or recrystallize from hot water, using a stemless funnel in the process. Yield is 7 g; mp of pure sulfanilamide is 163°.

Warning: *Sulfanilamide is not safe for prescription by the layman, since serious physiological disturbances are caused in many cases. Do not use material prepared in this experiment for medicinal purposes.*

[§308] **Questions**

1. What is the objection to the presumably simple, direct procedure from an inexpensive starting material: sulfanilic acid $\longrightarrow$ sulfanilyl chloride $\longrightarrow$ sulfanilamide?

2. Why do we use sodium bicarbonate rather than sodium hydroxide to neutralize the solution in the last procedure?

Experiment 29

Bromobenzene and *p*-Dibromobenzene

309. Bromination. Elemental bromine is an electrophilic reagent capable of attacking electron-rich aromatic systems such as phenol or aniline. For benzene and less reactive aromatic compounds, a catalyst such as ferric bromide or anhydrous aluminum chloride is used. These are effective because they combine with bromine forming complexes in which one bromine atom has increased power to accept a pair of electrons from the aromatic system to form the σ-complex.

$$FeBr_3 + Br_2 \rightleftharpoons [FeBr_4]^-Br^+$$

$$[FeBr_4]^-Br^+ + C_6H_6 \longrightarrow [C_6H_6Br]^+ [FeBr_4]^- \longrightarrow C_6H_5Br + HBr + FeBr_3$$

In addition to the monobromo derivative, dibromobenzenes are produced; mostly the para isomer, a small amount of ortho, and a very small amount of meta. Although iron is specified here, some users of this manual have good success with the anhydrous aluminum chloride, of which 0.5–1 g is provided in a dry test tube.

Attach a dropping funnel and a water-cooled reflux condenser to a 500-ml three-neck flask and stopper the third neck. If a one-neck flask is used and the dropping funnel is mounted at the top of the reflux condenser by means of a slotted cork, special care must be taken to avoid bromine leakage and the experiment must be carried out in a hood. With the three-neck flask the experiment can be carried out in the open laboratory providing a suitable trap for hydrogen bromide is attached to the top of the reflux condenser (§198).

Place in the flask 0.35 mole of benzene and two or three shingle nails, or 2 to 3 g of small tacks.

Warning: *Do not substitute iron powder or excess of reagent ferric bromide, lest the high concentration of catalyst cause a runaway* *reaction and the escape of dangerous amounts of bromine vapor. Before handling bromine, see §218 dealing with the hazard involved in handling this very reactive liquid.*

After special care is taken that the stopcock of the dropping funnel is properly lubricated and securely fitted, preventing danger to fingers, pour 0.5 mole of chilled bromine, taken from the refrigerator, through a small funnel into the dropping funnel.

Gently heat the benzene nearly to the boiling point, and remove the burner. Start running in the bromine slowly, with constant caution not to allow the bromine to leak out at the stopcock. The addition may now be as rapid as possible without excessive ebullition of the reaction mixture.

On account of the great stability of the bromine derivatives of benzene, there need be no fear of interrupting this experiment at any point in the subsequent procedure. If the mixture is allowed to stand for a day or more, the residual bromine will be used up, and thus the next procedures will involve less vapor nuisance. If, however, the experiment is not subject to interruption, reflux the reaction mixture until red fumes of bromine in the flask have disappeared. It is interesting to have different students compare results with variation in time of standing, particularly as to relative quantities of mono and dibromo products obtained. Finally, ★ at the end of the period of bromination, boil the reaction mixture for 5 min.

Wash the crude reaction product with dilute (1–2 *N*) sodium hydroxide, followed by several portions of water. Either of two plans of purification may now be followed, according to time available. The first (§310), involving fractional steam distillation, gives a neater separation of the two desired compounds but is time consuming. Otherwise follow §311. In any case, note vapor-pressure data in §577 and warning in §300 regarding distillation.

310. Fractional Steam Distillation. The crude wet product is steam-distilled from a flask of at least 1500-ml capacity as described in §79. As soon as crystals appear in the condenser, change to a second receiver and continue distillation.

The first distillate is mainly bromobenzene with accompanying water. The second is principally *p*-dibromobenzene. Separate the first product, dry over anhydrous calcium chloride or magnesium sulfate, and distill from a small distilling flask, collecting the product coming over between 150° and 160°. The boiling point of pure bromobenzene is 156°. About 30 g of liquid, bp 150–160°, should be obtained. Recrystallize the solid dibromobenzene from ethyl alcohol containing 10–20% water (dissolve the solid in hot alcohol and add hot water). *p*-Dibromobenzene melts at 89°.

311. Ordinary Fractionation. After thorough washing with water to remove hydrobromic acid, separate the oil and dry over anhydrous calcium chloride, magnesium sulfate, or well-dried anhydrous sodium sulfate. Distill the product through a short distilling column insulated with glass wool or asbestos, and collect the liquid boiling at 150–160° as the bromobenzene fraction. It is fairly difficult to get a liquid boiling this high to distill smoothly unless the column is well insulated. The dibromobenzenes boil at about 220° and as the temperature at the top of the distilling column begins to rise above 160° there may not be enough vapors reaching the thermometer to maintain a steady temperature. Discontinue the distillation when crystals form in the sidearm or condenser. Combine the residue in the distilling flask with any crystals from the column or condenser, and recrystallize the crude dibromobenzene fraction (mainly *p*-dibromobenzene) as in §310.

Experiment 30

Friedel–Crafts Synthesis

312. Use of a "Lewis Acid" Catalyst. When an alkyl halide is treated with anhydrous aluminum chloride, the strongly electrophilic aluminum compound attracts a fourth electron-rich halogen atom to complete a group of four electron pairs, to form $AlCl_4^-$. Such possible reaction becomes feasible if the positive alkyl radical, about to be left behind, simultaneously has the opportunity to attack another electron-rich compound such as benzene.

In the first example the alkylating agent is *tert*-butyl chloride–$AlCl_3$ and the substrate is the very reactive arene, phenol. Because of the large size of the *tert*-butyl group and the strong ortho-para-orienting influence of the hydroxyl substituent, the product of this reaction is almost exclusively the para isomer.

$$C_6H_5OH + tert\text{-}BuCl(AlCl_3) \longrightarrow p\text{-}tert\text{-}BuC_6H_4OH + HCl$$

In the second example benzylbenzene, or diphenylmethane, is produced.

$$C_6H_6 + C_6H_5CH_2Cl(AlCl_3) \longrightarrow C_6H_5CH_2C_6H_5 + HCl$$

An important modification of the Friedel–Crafts reaction involves the use of acyl halides or carboxylic anhydrides in place of alkyl halides. Owing to complexation of aluminum chloride by oxygen-containing compounds, at least equivalent quantities of aluminum chloride are used in such acylations. In the final example a dicarboxylic anhydride, succinic anhydride, is used with benzene to form β-benzoylpropionic acid.

$$C_6H_6 + \underbrace{OCCH_2CH_2CO}_{O} \xrightarrow{AlCl_3} C_6H_5COCH_2CH_2CO_2H$$

313. *p-tert*-Butylphenol. To prevent the hydrogen chloride produced in this experiment from escaping to the air, either work in an efficient hood or make a miniature fume hood as follows. Support an inverted funnel at a height such that a 50-ml conical flask may be placed under the bell. By means of rubber tubing join the funnel stem to the sidearm of a filtering flask clamped into an upright position. Fit the flask with a stopper through which a piece of glass tubing extends about 2 in. above the stopper and to within about 1 cm of the bottom of the flask. Connect this glass tube to the aspirator by means of rubber tubing. This flask will serve as a trap should the water from the aspirator back up. The apparatus is shown in Fig. 1. The hydrogen chloride produced in the reaction will be drawn through the funnel and the trap and into the aspirator, where it will disperse and dissolve in the water discharge.

Place 4.7 g of phenol in a 50-ml conical flask, add to it 6.5 ml (5.5 g) of *tert*-butyl chloride, and swirl until the phenol is dissolved.

> **Warning:** *Phenol is poisonous and is rapidly absorbed through the skin! It will cause very painful skin burns and gives very little warning since it anesthetizes the skin as well. If you should get phenol on you, wash the affected area thoroughly with water, soap and water, and more water. Consult your instructor immediately. As a safety precaution everyone should wash his hands thoroughly just prior to leaving the laboratory.*

Place the flask under the funnel, start the aspirator, and add 0.5 g of anhydrous $AlCl_3$ in small increments. Protect the $AlCl_3$ from spattering of water and undue exposure to air since it reacts vigorously with moisture to give off HCl.

> **Warning:** *$AlCl_3$ can cause painful severe skin burns. Avoid all contact with this material—if contacted, wash with large amounts of water.*

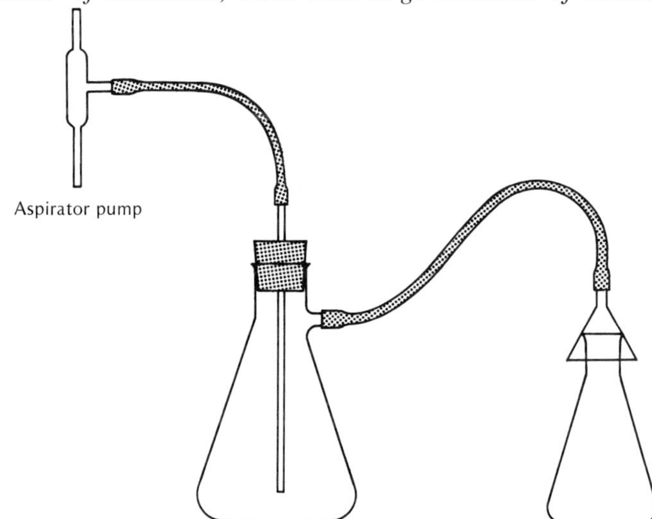

Fig. 1

If a new unopened bottle of commercial aluminum chloride is provided consult the instructor or storekeeper about danger from gas pressure in the sealed bottle.

Evolution of HCl should begin immediately. If the reaction temperature rises more than a few degrees above room temperature, as judged by placing your hand on the flask, cool the flask with ice or cold water. Swirl the mixture occasionally to keep the mixture in good contact with fresh catalyst. After about $\frac{1}{2}$ hr the mixture should solidify. When this occurs, add 25 ml of water and pulverize the solid. Suction-filter the solid and dry by drawing air through the funnel for about 5 min. Transfer the product to a 250-ml beaker and recrystallize from 60–80° petrolem ether, observing the normal precaution for recrystallizing from a flammable solvent.

Warning: *Observe the same precautions for handling the product as you did for handling the phenol. Avoid contact with the petroleum ether solution.*

(The use of decolorizing carbon is not advisable in this case, owing to the increased difficulty of manipulation.) Determine the melting point and yield of your product.

314. Diphenylmethane. Assemble apparatus as shown in the figure of §198 including a 500-ml round-bottom flask. Place in the flask 120 ml of dry, thiophene-free benzene that has been purified if necessary by agitation with concentrated sulfuric acid, washing, and drying. Add 34 ml of dry, redistilled benzyl chloride and provide 11 g of anhydrous aluminum chloride in a dry test tube.

Warning: *Observe the precautions given in the preceding section for handling aluminum chloride.*

Keep the reaction mixture continuously cool with the aid of an ice bath and add the aluminum chloride in several portions, waiting after each addition until the initial vigor of the reaction subsides. Reflux the final mixture on the water bath for 30 min, allow to cool, and add about 100 g of ice. Preserve the upper or benzene layer, and wash it with dilute hydrochloric acid and then with water. Dry rapidly with 3–4 g of calcium chloride, filter, and remove most of the benzene on the steam bath (**Fire hazard: no flames**).

The residue of crude diphenylmethane, containing traces of benzene, may now be distilled either (*a*) at ordinary pressure or (*b*) under reduced pressure at 20–40 torr, the latter option giving a product of slightly better appearance. Option (*a*) requires an ordinary air-cooled condenser; for option (*b*) see §65. See §300 for warning.

315. β-Benzoylpropionic Acid. Assemble in the hood an apparatus consisting of a 250-ml three-neck flask provided with sealed mechanical stirrer, reflux condenser, and dropping funnel (see §§24–28). The Nichrome wire stirrer described in §28 cannot be used here because the metal is attacked. If a hood is not available, use a vapor trap such as described in §228. Provide an oil bath to heat the flask. Place in the flask 6.8 g (0.068 mole) of succinic anhydride (see §248) and 35 g (0.45 mole) of dry, thiophene-free benzene. With stirrer in motion add all at once 20 g (0.15 mole) of powdered anhydrous aluminum chloride.

> **Warning:** *Observe the precautions given in §313 when handling aluminum chloride.*

Hydrogen chloride gas is evolved and the mixture becomes hot. Reflux with continued stirring for a half hour, then replace the oil bath by a cold-water bath.

Run through the condenser, slowly, 30 ml of water from a dropping funnel. Follow this with 10 ml of concentrated hydrochloric acid and remove the excess benzene by steam distillation. Transfer the residue to a 250-ml beaker, whereupon the crude β-benzoylpropionic acid separates as an oil, which soon solidifies. After cooling the product to 0°, filter and wash with 20 ml of cold 3 *N* hydrochloric acid.

Redissolve the crude β-benzoylpropionic acid in a solution of 7.5 g of sodium carbonate in 50 ml of water, boiling the mixture for about 15 min. Filter with aid of a mixture of decolorizing carbon and diatomaceous earth (Super-Cel). Acidify the clear, cool, colorless filtrate carefully, using about 13 ml of concentrated hydrochloric acid. Upon final cooling of the solution to 0° in an ice–salt bath, collect the crystalline product on the Büchner or Hirsch filter, wash well with water, and dry in a desiccator. Yield is 11 g (92% of theory), mp is 114–115°. See also §446.

Questions:

1. Compare and illustrate Friedel–Crafts alkylation and acylation reactions as regards relative (*a*) amounts of aluminum chloride, (*b*) tendencies to undergo rearrangement processes, (*c*) tendencies toward polysubstitution, and (*d*) orientations of substitution as on toluene.

2. Offer reaction sequencies for preparing α-tetralone from (*a*) benzene and succinic anhydride and (*b*) benzene and γ-butyrolactone.

Experiment 31

Reduction of Nitro Compounds

316. Effect of Acidity. Reduction of a nitro group by an active metallic reagent such as tin, iron, or zinc depends markedly on the hydrogen ion concentration of the aqueous medium in which the reaction occurs. In strongly acid solution the nitro group is converted into amine, in neutral or faintly acidic solution into the (aryl) hydroxylamine, and in alkaline solution into a hydrazo or azo derivative.

317. Aniline and Acetanilide. Nitrobenzene is reduced in hydrochloric acid solution to aniline. Since aniline is a much stronger base than nitrobenzene, the hydrochloride salt is immediately produced; but the addition of alkali after the main reaction is complete frees the desired amine, ready for separation, purification, or further reaction.

$$C_6H_5NO_2 \text{ (I)} \rightarrow C_6H_5NH_3^+Cl^- \text{ (II)} \rightarrow C_6H_5NH_2 \text{ (III)} \rightarrow C_6H_5N(H)COCH_3 \text{ (IV)}$$

In aqueous solution aniline buffered with alkali reacts almost instantaneously with acetic anhydride to form the substituted amide, acetanilide. In order that this reaction may proceed neatly in one clear solution, aniline is reconverted into the salt form [compound II above], where it remains while the acetic anhydride is being dissolved and thoroughly intermixed. Addition of mild alkali then rapidly produces in turn III and IV. There should be no delay while acetic anhydride is in contact with water (Method of Lumière and Barbier, 1905).

Warning: *The vapors of aniline are toxic and should not be allowed to escape into the atmosphere of the laboratory. See §300.*

318. Small-Scale Experiment. In a 200- or 300-ml flask fitted with a short air-cooled condenser, place 9 g of granulated (mossy) tin and 5 g of nitrobenzene. Add, through the condenser, 30 ml of concentrated hydrochloric acid in 5-ml portions. If the acid should boil actively, dip the body of the flask in a pan of cold water, moderating the temperature so that the reaction mixture is barely at the boiling point. Mix the reagents thoroughly from time to time by a swirling motion of the flask. As the reaction abates, place the flask in a boiling water bath for 10 min or more. Care should be taken that all the nitro compound is reduced. Hold the flask up to the light, with the surface of the solution above eye level. Look for possible oily drops of unchanged nitrobenzene at the surface. Any unreduced material will introduce a colored nitroamine addition compound into the amine preparation, while incompletely reduced material (such as azobenzene or azoxybenzene) will cause a yellow or orange coloration. The solution, containing reactive stannous chloride, should be distributed from time to time all over the inside walls and neck of the flask so as to reach all oxygen-containing material. The odor of nitrobenzene, noticeable during the first 30–45 min of the reaction period, finally disappears. If convenient, the mixture may be set aside until the next day, with still greater assurance of complete reduction. Additional delay of a week or two does no harm. In any case the newly formed aniline salt is likely to crystallize with tin chloride as a double salt as soon as the reaction mixture cools, but no complication is introduced by such an occurrence. ★

The reaction mixture is placed in the central flask of an apparatus for steam distillation; this flask should be of 1 or 2 liter capacity. Add 15 ml of water and then cautiously add 20 g of sodium hydroxide pellets. Care must be taken to avoid splashing that might throw the hot caustic alkali solution out of the flask. The alkali liberates aniline from its salt and precipitates stannous and stannic ions as hydroxides.

Pass steam at moderate speed through the reaction mixture. Note that at first the aniline comes over at so rapid a rate that the accompanying water is unable to dissolve all of it in the condenser. Accordingly, oily globules of the free amine appear.

319. Acetanilide. Soon the production of oily globules ceases, and subsequent distillation yields the balance of the product as a clear aqueous solution. When a total of 100 ml of distillate has been collected, it may be assumed that the organic product is completely distilled. To this distillate add a quantity of the dilute hydrochloric acid of the laboratory, just sufficient to react with all the amine, assuming 100% yield. The resulting solution shculd be clear and not above room temperature. Now dissolve 0.05 mole of acetic anhydride in the reaction mixture, and immediately add 0.05 mole of sodium acetate (either hydrate or anhydrous). Stir well, filter, wash the crystalline product with cold water, and dry. Yield is about 3 g; mp of pure acetanilide is 114°.

320. *p*-Toluidine. Substitution of *p*-nitrotoluene for nitrobenzene in §318 gives *p*-toluidine, which in turn is converted into aceto-*p*-toluidide, mp 152°. Since *p*-toluidine is a solid (mp 45°), it may crystallize in the condenser used in steam distillation, but it is easily melted by stopping the current of cooling water in the condenser jacket for a few moments. See §300 for warning about distillation.

321. Aniline with Ether Extraction. If supplies of material permit, the procedure of §318 may be run on a larger scale, using a 1-liter flask fitted with an air-cooled condenser. The following quantities of reagents are appropriate: 55 g of mossy tin, 30 g of nitrobenzene, 150 ml of concentrated hydrochloric acid, and 90 g of solid sodium hydroxide. The resulting steam distillate is treated as follows: saturate the distillate—aniline and water together—with clean, fine granulated or powdered salt, and extract the aniline with a total of 100 ml of ether, divided into such portions as you think advisable. See also §113. Dry the combined ethereal extracts with 6–8 g of coarsely broken pieces of potassium or sodium hydroxide (not powder). After the solution is dry, remove the drying agent, taking care to eliminate the few drops of highly concentrated caustic alkali solution that may have been produced.

It is now important to note the difference in technique when diethyl ether has been used, in contrast to that employing diisopropyl ether. Diethyl ether is dangerously flammable. Read §63 carefully before proceeding; then distill the ethereal aniline solution in the appropriate manner until the ether has been removed. See §115 for discussion of technique where a small volume of product is to be separated from a large volume of extract.

The ether recovered from the experiment should be placed in a stock bottle or in a container marked "Recovered Ether," or given to the next student requiring a supply.

After the ether has been expelled, replace the water-cooled condenser with an air-cooled condenser and distill the final product of aniline. If the distillate is of decidedly yellowish orange color, it may sometimes be improved by redistillation from the same distilling flask in the presence of a pinch of aluminum or zinc dust. Distillation under reduced pressure in the presence of the metallic dust is often effective. In any case, however, the product, even if colorless when freshly prepared, will become dark on standing. Yield is 20 g, bp of aniline is 184.4° (cor.).

If a larger amount of acetanilide is needed for the synthesis of sulfanilamide (§308), the steam distillate from the first paragraph of this section may be acetylated as in §319 after suitable modification of amounts.

Questions

1. Why is it necessary to add sodium hydroxide before steam distillation is carried out?

2. What can you judge of the vapor pressure of aniline, based on your experience in the steam distillation of the substance?

3. Why should nitrous acid vapors be excluded from an amine preparation?

4. Suppose you had used tin and hydrochloric acid in the preparation of an amine that was not volatile in steam, although it was soluble in water. Devise a method of isolating the product.

5. Tin has two valence changes, $Sn \longrightarrow Sn^{++} \longrightarrow Sn^{4+}$, during the aniline reaction. Why might this be considered favorable in contrast with zinc, which has only one valence change, $Zn \longrightarrow Zn^{++}$?

322. Synthetic Sequences Involving Reduction of Nitro Compounds. The reduction of aromatic nitro compounds is especially suitable for introductory laboratory work on synthetic sequences because the compounds are readily handled and delays between laboratory periods seldom cause trouble. One such series involves nitration of benzene (§300), reduction of nitrobenzene to aniline (§321), acetylation (§319), chlorosulfonation of acetanilide (§308), formation of *p*-acetaminobenzenesulfonamide (§308), and selective hydrolysis to remove the acetyl group without disturbing the sulfonamide function (§308).

Carry out calculations to determine the amounts of starting material needed to obtain 7 g of sulfanilamide as specified in §308, assuming that you will obtain the yields specified in the various procedures. Calculate the percentage yield for each reaction and the overall percentage yield for the sequence. When you actually carry out the reaction, start with somewhat more than the calculated amount in order to have a margin for error and enough material to save a small sample of each intermediate compound.

Another synthetic sequence involves benzene $\longrightarrow$ nitrobenzene $\longrightarrow$ phenylhydroxylamine $\longrightarrow$ nitrosobenzene. The reduction of nitrobenzene to nitrosobenzene and isolation of the nitroso compound is not practical because further reduction occurs or condensation takes place. Carry out the same calculations for this sequence as for the sulfanilamide series. Procedures for the last two steps are given in §§323 and 324.

Other interesting sequences involve reduction of nitrobenzene in alkaline solutions which leads to formation of N—N bonds. Reaction of nitrosobenzene with phenylhydroxylamine forms azoxybenzene, which can be isolated if a mild reducing agent such as arsenious oxide in sodium hydroxide solution is used (OS–CV **2**). Various combinations of metal and base yield azobenzene or hydrazobenzene. Reduction with magnesium to yield either of these products was described by A. I. Vogel, A. Watting, and J. Watting, *J. Chem. Ed.*, **35,** 40 (1958); procedures for these syntheses are given in §§325 and 326. Hydrazobenzene is readily oxidized by air. The azobenzene isolated has an *anti* configuration and can be isomerized to the *syn* compound by light [J. F. Janssen, *J. Chem. Ed.*, **46,** 117 (1969)]. If the isomerization is carried out in strongly acid solution, a further photochemical reaction occurs leading to benzo[*c*]cinnoline [R. F. Evans, *J. Chem. Ed.*, **48,** 768 (1971)]. Procedures for these reactions are given in Experiment 38.

323. Phenylhydroxylamine. In a 2-liter flask equipped with a mechanical stirrer and a thermometer, place 12.5 g of ammonium chloride, 400 ml of water, and 25 g of nitrobenzene. Stir the mixture vigorously, and during a 15-min period add 40 g of high-grade zinc dust. The addition of zinc dust should be rapid enough so that the temperature rises rapidly to about 55° or 60° and remains in this range until the zinc has all been added. When the temperature starts to fall, filter the reaction mixture, and saturate the filtrate with clean common salt. Wash the zinc oxide precipitate with 50 ml of warm water and treat the washings with salt also, thus getting a small additional yield. Phenylhydroxylamine crystallizes in feathery needles and is collected on filters after the crystallizing mixtures have stood for an hour in the ice bath. Proceed at once to convert the product into whatever secondary preparation is desired. See §324.

324. Nitrosobenzene. Dissolve 12 g of phenylhydroxylamine in 300 ml of ice-cold 6 *N* sulfuric acid contained in a 1.5- or 2-liter flask. Dilute the solution with 500 ml of ice water, and run in rapidly a solution of 12 g of sodium dichromate (dihydrate) in 200 ml of water, with intermittent shaking, and cooling of the flask in an ice bath. Nitrosobenzene is precipitated, collected on the Büchner funnel, washed with water, and distilled in steam. The resulting purified nitrosobenzene, while white or colorless in the solid state, is green when liquid. It may be recrystallized from a very small volume of alcohol; mp is 68°.

325. Azobenzene. Place 6.0 g (5.0 ml, 0.049 mole) of pure nitrobenzene, a small crystal of iodine, and 110 ml of anhydrous methanol in a 250-ml flask equipped with a water-cooled reflux condenser. Have a water ice bath at hand to cool the reaction mixture if the reduction becomes too vigorous. Add 3.0 g of magnesium turnings to the methanol solution and attach the condenser at once. If the quality of the magnesium turnings is satisfactory and the methanol anhydrous, the reaction should start almost at once. If no reaction is evident after 3 min, warm the flask gently on a steam bath until the reaction starts. Remove the flask from the steam bath as soon as reaction begins and allow it to continue as vigorously as possible without overtaxing the condenser. When the reaction has subsided (7 or 8 min) add an additional 2.5 g of magnesium turnings and when any further spontaneous reaction has subsided, heat the reaction mixture on a steam bath for 30 min. Cool, pour the reaction mixture into 150 ml of equal amounts of ice and water, add glacial acetic acid until acid to litmus, and cool in an ice bath with stirring to obtain maximum crystallization of azobenzene. Filter with suction, wash the product with cold water, and recrystallize from 90% ethanol. The yield is 2.5–3 g, mp 68°. Dry the product. It is suitable for use in Experiment 38.

326. Hydrazobenzene. Proceed exactly as for azobenzene, §325, but add 9.0 g of magnesium turnings in three equal portions instead of 5.5 g. Finally, heat the reaction mixture on a steam bath until it is colorless. More magnesium methoxide separates in this experiment than in the preparation of azobenzene and causes bumping. It may be necessary to place the flask in a pan of hot water at 75–85° to reduce bumping; the temperature of the bath can be maintained by placing it on a steam bath. Shake the flask frequently. The reduction is usually complete after 30 min. Disconnect the flask, stopper it at once, carry it to the hood, stir the reaction mixture to form a slurry, and pour it into a 1-liter beaker. Add 100 ml of water to the material in the beaker and use another 100 ml of water to rinse into the beaker any of the reaction mixture that remains in the flask. Stir the contents of the beaker and add glacial acetic acid slowly until the reaction mixture is acid to litmus; about 45 ml should be required. The magnesium methoxide dissolves and a pale yellow or white precipitate of hydrazobenzene remains. There may also be some unreacted magnesium at the bottom of the beaker and this should be left as much as possible in the beaker as the product is filtered with suction.

Wash the product thoroughly with water on the funnel to remove magnesium salts. Hydrazobenzene oxidizes rapidly in the air so avoid drawing air through the solid. This crude product may be used directly to prepare benzidine (§327) or it may be recrystallized from alcohol containing a little sulfur dioxide. Expose the product to air as little as possible during recrystallization and dry the crystals in a desiccator that was previously filled with nitrogen. The melting point of pure hydrazobenzene is 131°, but the product obtained here usually melts somewhat lower.

327. Benzidine. It was formerly common in a beginning laboratory course in organic chemistry to extend the synthetic sequence that started with nitrobenzene and led to hydrazobenzene by converting the latter to benzidine. The benzidine in turn might be converted to a bis diazonium salt, which yields a common indicator, Congo Red, when coupled with naphthionic acid [E. R. Kline, *J. Chem. Ed.,* **15,** 129 (1938)].

[§327] The Occupational Safety and Health Administration of the U.S. Department of Labor adopted on May 3, 1973, an emergency standard which prescribes work practices for employees exposed to any of the following carcinogens: 2-acetylaminofluorene, 4-aminodiphenyl, benzidine and its salts, bis(chloromethyl) ether, 3,3′-dichlorobenzidine and its salts, 4-dimethylaminoazobenzene, ethylenimine, methyl chloromethyl ether, 4,4′-methylenebis(2-chloroaniline), α-naphthylamine, β-naphthylamine, 4-nitrobiphenyl, *N*-nitrosodimethylamine, β-propiolactone. In view of these standards, which specify elaborate precautions that must be taken when any of these compounds are used or stored, benzidine should not be prepared or used in beginning organic chemistry laboratories.

Experiment 32

Diazotization and Preparation of Phenol

328. Diazotization. In each of Experiments 32–34, the first operation is the conversion of a primary aromatic amine into a diazonium salt, as illustrated in the simplest example involving aniline:

$$C_6H_5NH_2 + HCl \longrightarrow C_6H_5NH_3^+Cl^- \xrightarrow{HNO_2} C_6H_5N_2^+Cl^-$$

Conversion of the radical $—NH_3^+$ into the unstable $—N{\equiv}N^+$ is known as diazotization, which normally requires temperature control (0–5°). Do not interrupt any experiment involving diazotization from the time that the formation of the diazonium salt begins until that compound is completely transformed into the desired derivative, such as phenol, aryl bromide, dye, etc. Never attempt to keep a diazonium salt overnight, even in a refrigerator.

Since the nitrous acid required in diazotization is very unstable, it is prepared just at the instant it is needed, by cautious dropwise addition of sodium nitrite to strong acid. The aromatic amine already present reacts in a few seconds, forestalling the decomposition of the nitrous acid that would otherwise occur. Ordinarily the limits of temperature range 0–5° are observed. Temperatures above 5° may cause decomposition, while those below 0° involve too slow a reaction rate. If a beaker is used for diazotization, ice should be placed both inside and outside, since the reaction produces considerable heat and it is difficult to agitate a solution in a beaker sufficiently to effect thorough mixing and cooling. The process is more easily carried out in a flask if ice is not admissible in the reaction mixture.

Although the equation for diazotization requires, in theory, only two equivalents of strong acid (one to form the amine salt and a second to convert the nitrite into nitrous acid), at least one additional equivalent is normally required, in actual laboratory practice, to inhibit side reactions such as formation of an insoluble diazoamino compound.

Occasionally the amine is amphoteric, as in the case of sulfanilic acid, thus introducing the complication of an inner salt (zwitterion) and causing considerable change in the technique of diazotization.

In this first diazo experiment the diazonium group ($-N_2Cl$) is replaced by hydroxyl, yielding phenol. Simple heating accomplishes this result. Loss of N_2 from the diazonium salt produces the very unstable phenyl cation, which combines rapidly with water, the most available nucleophile. This forms the conjugate acid of phenol, which is in equilibrium with phenol:

$$(C_6H_5N_2)^+X^- \longrightarrow N_2 + (C_6H_5)^+X^- \xrightarrow{H_2O} (C_6H_5OH_2)^+X^- \rightleftharpoons C_6H_5OH + HX$$

It is not satisfactory to diazotize at room temperature on the presumption that the diazonium salt has to be heated anyway. Such technique causes excessive tar formation and gives low yield of phenol.

329. Phenol. In a 400-ml beaker place 150 g of crushed ice (200 g in warm weather) and 18 ml of concentrated sulfuric acid. Place the beaker in an ice bath and add with effective stirring 14 g (0.15 mole) of purified [i.e., light yellow or colorless (see §321)] aniline. After the aniline has been added, introduce in small portions with stirring, 12 g of sodium nitrite. The sodium nitrite should be finely powdered and introduced slowly enough so that the temperature of the reaction mixture remains between 0° and 5°. Insert a thermometer in a cork or rubber stopper and support it in the beaker slantwise so that the bulb is in the liquid, but you can still see the 0–5° range. Mechanical stirring is not essential but is preferable because it permits easier temperature control; if it is used, take care that the thermometer is not broken by contact with the blades of the stirrer. After the sodium nitrite is added allow the reaction mixture to stand in the ice bath (0–3°) for 10 min; then add 2 g of urea to remove excess nitrous acid.

Assemble a distillation apparatus consisting of a 1-liter distilling flask, efficient water-cooled condenser, and receiver of at least 500 ml capacity. Mount a dropping funnel on the distilling flask so that a solution in the funnel can be added dropwise or in a small stream directly into a solution in the flask. Place 300 ml of water in the flask and add 20 ml of concentrated sulfuric acid. Now heat the sulfuric acid solution to boiling, and pass in the diazonium salt solution through the funnel as fast as possible without flooding the mixture and thus interrupting the boiling or causing foam to rise up into the neck of the flask. Note carefully that the mere evolution of small bubbles of nitrogen is not boiling. In other words, keep the temperature up in the vicinity of 100° and see that steam is being expelled in quantity into the condenser.

Place no more than 25 ml of the diazonium salt solution in the dropping funnel at one time and keep the rest of the solution ice cold until it is needed to renew the supply in the funnel.

After all the diazonium solution has been run in, continue distillation until 350–400 ml of distillate has been collected. Saturate this distillate with common salt. Read discussion of extraction (§114) and use a 200-ml supply of ether in an effective manner to extract the phenol. Ethyl ether gives the best product, but isopropyl ether is safer and is acceptable. Now dry the combined extracts over anhydrous magnesium sulfate. ★

Warning: *Phenol is injurious to the skin and will cause painful burns. If the substance gets upon the skin, wash at once thoroughly with soap and water. If irritation develops from exposure to phenol, consult the instructor or a physician.*

In view of the fact that the extract contains far more solvent than phenol, read §115, describing special technique in separating the components of such a mixture. Now remove the ether by an appropriate mode of distillation. Replace the condenser with one of air-cooled type and distill the residue. The fraction boiling between 176° and 182° is collected as phenol. Yield is 8 g, bp of pure phenol is 182°, mp 41°. Purity of phenol (with water as impurity) may be estimated from the melting point [L. R. Pollack, *Anal. Chem.* **19,** 241 (1947)].

330. Reactions of Phenol. Dissolve approximately 1 g of phenol in 15 ml of water, and use this solution in tests (*a*) to (*e*).

(*a*) Test the reaction of phenol with neutral litmus paper.

(*b*) Add one drop of dilute sodium hydroxide and one drop of phenolphthalein indicator to 10 ml of water. Treat 3 ml of this solution with 1 or 2 ml of your phenol solution.

(*c*) Repeat (*b*), using methyl orange instead of phenolphthalein, and explain difference in behavior.

(*d*) Add bromine water slowly, with shaking, to 15 ml of the phenol solution until a faint yellow color persists in the solution. The bromine water should be prepared by addition of 10 g of bromine to 100 ml of water in which has been dissolved 15 g of potassium bromide. Add about 50 ml of water to the reaction mixture, shake vigorously to break up any lumps, filter the suspension, and wash the precipitate with a dilute solution of sodium bisulfite. Recrystallize from water–ethanol mixture. Write an equation for the reaction.

This procedure is suitable for preparation of solid derivatives of many phenols. Water-insoluble phenols may be dissolved in ethanol, acetone, or dioxane, and bromine water added as above.

(*e*) Add a few drops of 1% ferric chloride solution to 3 ml of the phenol solution.

Questions

1. Suppose benzylamine were substituted for aniline in this experiment. What difference in behavior of the reagents would you expect?

2. Why is phenol more soluble in sodium hydroxide than benzyl chloride would be? (Two reasons.)

3. Why does benzoic acid dissolve freely in sodium bicarbonate solution, whereas phenol does not?

4. What objection would there be to mixing the nitrite and sulfuric acid, and then adding this mixture to the aniline and ice?

5. How could you modify this experiment to prepare bromobenzene instead of phenol?

6. Why does the undissolved phenol in the original reaction distillate remain as a liquid, whereas the melting of phenol is 10° or 20° higher than the temperature of the distillate and receiver?

7. Which do you think could be more easily extracted from aqueous solution, phenol or aniline?

8. In the distillation of the original reaction mixture, why does not the volatile component, water (bp 100°), nearly all pass over first, and finally the less volatile component, phenol (bp 182°)?

9. Why is it desirable that the ethereal extract be slowly distilled, instead of rapidly, as the temperature approaches 170°?

Experiment 33

Replacement of the Diazonium Group

331. Sandmeyer Reaction. In this reaction the diazonium group is replaced by chlorine or bromine, by reaction with the corresponding cuprous halide and hydrohalogen acid.

$$ArN_2X \xrightarrow{CuX + HX} ArX + N_2$$

For replacement by bromine an aryldiazonium sulfate in excess sulfuric acid may be treated with cuprous bromide and sodium bromide; it is less common to prepare the diazonium bromide directly in hydrobromic acid. Cuprous iodide is not necessary for replacement by iodine; the diazonium sulfate or chloride is simply treated with potassium iodide solution. The reaction of a diazonium salt with cuprous cyanide and excess cyanide ion to yield an aryl nitrile is also a Sandmeyer reaction.

Precautions. This experiment should not be interrupted until the steam distillation has been completed. Prepare the parts of the steam distillation apparatus for a 1-liter flask during the previous laboratory period so that the apparatus can be assembled in a minute or two. Plan the experiment carefully.

332. *p*-Chlorotoluene. Dissolve 48 g of copper sulfate ($CuSO_4 \cdot 5H_2O$) and 12.5 g of sodium chloride in 150 ml of hot water in a 1-liter flask. Add slowly a solution of 10 g of sodium bisulfite and 6.8 g of sodium hydroxide in 75 ml of water to the hot solution, with swirling. Cuprous chloride precipitates as a white powder. Stir or shake to break up any lumps and leave the flask in ice until the separate operation of diazotization has been completed. The solid should be covered with the mother liquor because cuprous chloride darkens on exposure to air.

Dissolve 16 g of *p*-toluidine by suspending it in 25 ml of water in a 250-ml conical flask and adding 15 ml of concentrated hydrochloric acid (heat if necessary). Add 22.5 ml more hydrochloric acid, mix, and cool to 0° in an ice bath containing a little salt. Place a little finely divided ice in the suspension and add slowly from a dropping funnel a solution of 10.5 g of sodium nitrite in 30 ml of water. Stir or swirl the flask during the addition and regulate the rate of addition to maintain a temperature of 0–5° in the diazotization mixture. After all the sodium nitrite has been added, allow the solution to stand for several minutes and test for the presence of excess nitrous acid with starch-iodide paper. Because concentrated hydrochloric acid alone liberates iodine from potassium iodide after brief standing, it is best to place a stirring rod in the solution, withdraw only the amount that clings to the rod, dilute this with a little water, and carry out the test on the dilute solution. Add a little urea to the diazonium salt solution, to destroy excess nitrous acid, and allow the reaction mixture to stand in the ice bath while the cuprous chloride solution is prepared.

Decant the supernatant liquid from the cuprous chloride, wash the solid once with water by decantation, and dissolve in 70 ml of cold concentrated hydrochloric acid. Cool the solution thoroughly in ice and add the diazonium solution to it with care (the solution should not wet the neck of the flask). Stir or swirl the flask during the addition. A solid addition product separates from the mixture, which is now allowed to warm up to room temperature. Nitrogen is evolved. Allow the mixture to remain at room temperature for 15–30 min while the flask is arranged to permit steam distillation. A steam bath should be provided to heat the flask. When the evolution of nitrogen has almost stopped, heat the flask on the steam bath. After the reaction mixture is hot, start the steam distillation. Continue to steam-distill until very little water-insoluble material comes over.

Extract the distillate with a little ether and wash the ether layer with 10% sodium hydroxide solution to remove the *p*-cresol that is formed by hydrolysis of the diazonium salt. Finally wash with a sodium chloride solution, dry over calcium chloride, and distill. The fraction boiling at 160–162° should be collected. Pure *p*-chlorotoluene boils at 162° (cor.) and melts at 7°.

Experiment 34

Azo Dyes

333. Coupling Reaction. Diazonium salts undergo reactions with highly reactive aromatic compounds, such as phenols or amines, without loss of nitrogen. This coupling reaction is an electrophilic substitution at the reactive para or ortho positions of the phenol or amine. Para substitution usually occurs almost to the exclusion of ortho unless the para position is blocked. The product is a colored azo compound.

Control of pH is important for this reaction. The diazonium group is not a powerful electrophile and will not react with free phenol. Instead, conditions must be sufficiently basic to convert the phenol to phenolate ion, which reacts rapidly. In practice, coupling with phenols is usually carried out at pH 5–9.

$$C_6H_5-\overset{+}{N}\equiv NX^- + C_6H_5-O^- \longrightarrow C_6H_5-N=N-C_6H_4-OH + X^-$$

In acidic solution an amine is largely in the unreactive ammonium salt form. Since coupling will occur only with free amine, the solution must be made more basic than is necessary for diazotization. A pH range of 3.5–7 is usually chosen in practice. With primary and secondary amines the coupling reaction can also occur at the nitrogen atom to yield a diazoamino compound (I) as well as on the ring to form the aminoazo derivative (II). Usually I forms more rapidly, but this reaction is reversible; II is more stable. Diazoamino compounds cannot form with tertiary aromatic amines.

$$\underset{\text{I}}{C_6H_5-N=N-\overset{H}{N}-C_6H_5} \rightleftharpoons C_6H_5-N_2X + C_6H_5-NH_2 \rightarrow \underset{\text{II}}{C_6H_5-N=N-C_6H_4-NH_2}$$

Coupling reactions with amines are frequently carried out in buffered solutions.

Azo compounds are colored. Many commercial dyes are prepared by the coupling reaction. The experiment below illustrates the formation of a well-known indicator. Additional illustrations of the coupling reaction are given among the special experiments, §§473–475.

334. Amphoteric Compounds. The diazotization of an aromatic amino sulfonic acid leads to a diazonium inner salt instead of the usual chloride or sulfate. Occasionally such inner salt may even be isolated in crystalline form. In this example sulfanilic acid, which is conventionally represented as $HSO_3C_6H_4NH_2$, is designated by the inner-salt formula $^-SO_3C_6H_4NH_3^+$. Nitrous acid in the presence of sulfuric acid causes the three terminal hydrogen atoms to be replaced by an atom of nitrogen. Accordingly the substance is converted without change of ionic charge into the diazonium inner salt $^-SO_3C_6H_4N_2^+$. Addition of the coupling agent dimethylaniline, followed by sodium hydroxide, yields methyl orange:

$$^-SO_3C_6H_4N_2^+ + C_6H_5N(CH_3)_2 + NaOH \longrightarrow NaSO_3C_6H_4N{=}NC_6H_4N(CH_3)_2 + H_2O$$

Treatment of methyl orange with a strong acid instantly transforms it into a red compound often called "helianthin." It is supposed that the proton from the strong acid unites with one of the azo nitrogen atoms and that the necessary rearrangement of electrons—or bonds—in the molecule is responsible for the color change.

$$NaSO_3C_6H_4N{=}N{-}C_6H_4{-}N(CH_3)_2 + HCl \longrightarrow {^-SO_3C_6H_4}\underset{H}{N}{-}N{=}C_6H_4{=}N(CH_3)_2^+ + NaCl$$

335. Methyl Orange. In order to avoid any excess of a reagent that could decompose and produce tar to contaminate the dye, weigh the quantities of solid reagents carefully, to the accuracy of 0.05 g or better.

Dissolve 0.02 mole of sulfanilic acid (note whether the supply is monohydrate or anhydrous) in about 50 ml of a solution of sodium carbonate containing 0.02–0.025 mole of that reagent, warming the mixture slightly to expedite solution. Test the solution to see that it is alkaline; add 0.02 mole of sodium nitrite, stir, and cool to about 25°. Diazotization does not occur in this basic solution that now contains sodium sulfanilate and sodium nitrite.

To about 75 g of crushed ice in a 400-ml beaker add enough hydrochloric acid of 6 or 12 *N* concentration to furnish 0.06 mole of HCl. Now add the solution containing sodium sulfanilate and sodium nitrite (prepared previously) in a fine stream with effective stirring to the cold hydrochloric acid solution. In a few minutes the resulting diazonium salt will be partially precipitated. Diazotized sulfanilic acid is somewhat more stable and less soluble than many other diazonium salts, but it should be maintained at 0–10°. Allow the mixture to stand in an ice bath or refrigerator while you prepare a solution of dimethylaniline (0.02 mole) in glacial acetic acid (0.02 mole).

The critical portion of the experiment is the coupling reaction, which must be carried out with effective stirring. Add the dimethylanilinium acetate solution slowly with constant stirring to the cold suspension of the diazonium salt. A dull reddish purple mass should appear. Stir vigorously to form a suspension and add 60 ml of 1 *N* sodium hydroxide very slowly over a period of 10 min while stirring is continued (mechanical stirring may be used). If the sodium hydroxide is added too rapidly, free dimethylaniline will probably separate as an oily phase. Such misfortune would then leave a chemically equivalent amount of diazonium salt unreacted. This excess of diazonium salt, upon warming to room temperature or above, would tend to change in part to brown tar, which would contaminate the otherwise beautiful crystalline orange dye. Brown or golden-brown color is then observed.

When the coupling reaction is complete, heat the reaction mixture to boiling, set it aside to cool and crystallize, and filter off the methyl orange. To a test-tube sample of the dye suspension, add a slight excess of dilute hydrochloric acid, heat to boiling, cool, and collect the product of helianthin. Do not attempt to determine the melting point of either methyl orange or helianthin because both are salts and do not melt cleanly.

336. 1-Phenylazo-2-naphthylamine. Earlier editions of this manual gave directions for the synthesis of 1-phenylazo-2-naphthylamine by the coupling reaction of phenyldiazonium chloride and β-naphthylamine. The discovery that β-naphthylamine is carcinogenic and the regulation of work practices involving this compound by the Occupational Safety and Health Administration of the U.S. Department of Labor (see §327) dictate that this compound not be used in a beginning laboratory of organic chemistry. This experiment has therefore been abandoned.

Students and instructors are warned that care is also necessary in the choice of azo dyes for synthesis in laboratory courses. Butter Yellow, p-$(CH_3)_2NC_6H_4N{=}NC_6H_5$, is carcinogenic and on the list of regulated compounds (§327). Certain other azo dyes once approved for use in food, drugs, or cosmetics have been removed from the approved list because they may be carcinogenic. For example, 1-phenylazo-2-naphthylamine, CI 11380, was one of the certified food colors, but it is no longer approved. Results of testing for carcinogenicity have been rather contradictory and for a while it was permitted for external use (in cosmetics). See footnote 3, §190, for information about *The Colour Index* and about certification of dyes for use in food, drugs, and cosmetics.

Experiment 35

Quinones

337. Benzoquinone. This pungent, volatile, solid cyclic ketone, commonly known by the simple general name "quinone," is readily obtained from hydroquinone (quinol) by direct oxidation. If one is not concerned with industrial economy, the elegant method of H. N. McCoy (1937), described below, is more convenient than the more widely known procedures using chromium and manganese compounds as oxidizing agents.

$$C_6H_4(OH)_2 + BrO_3^- + H^+ \longrightarrow OC_6H_4O + Br^- + H_2O$$

In a 200- or 250-ml flask place 10 g of hydroquinone, 5.5 g of potassium bromate, 5 ml of 1 *N* sulfuric acid, and 100 ml of water. Slowly heat the mixture to 60°. The solids dissolve, and the reaction starts promptly. Soon the greenish black crystalline quinhydrone appears, indicating that part of the material has been oxidized. Without further supply of heat from outside, the temperature rises spontaneously to about 75°.

After about 15 min the oxidation is complete, as indicated by a change in color to the bright yellow of quinone. Now heat the mixture to 80° to dissolve the quinone completely, and finally cool it in an ice bath to 0°. Collect the crystalline quinone on a filter, wash with ice water to eliminate salts, and dry. Yield is 8 g; mp 115.7°. Note that quinone has an appreciable vapor pressure in the solid state.

338. Quinhydrone. Quinone and hydroquinone combine instantly, mole for mole, at ordinary temperatures to form the addition compound quinhydrone ($OC_6H_4O \cdots HOC_6H_4OH$). This compound is a π-molecular complex or charge transfer complex in which the π-electrons of electron rich hydroquinone are donated to a certain extent to quinone, which is a Lewis acid (electron acceptor). Hydrogen bonding between hydroxyl hydrogens and quinone oxygens also contributes to the bonding. Quinhydrone is soluble in water at 25° to the extent of about 4 g per liter, and the dissolved compound is about 93% dissociated into its components. The quinhydrone electrode is useful to measure pH.

Dissolve 0.5 g of hydroquinone in 30 ml of lukewarm water (30–40°) in a small beaker or flask. Into this solution pour with stirring a solution of 0.5 g of quinone (approximately an equimolar quantity) in 8–10 ml of ethyl alcohol. Allow to stand in a cool place without further agitation. Shiny greenish black crystals of quinhydrone separate. Collect on a filter, wash rapidly with a little cold water, and dry. Do not attempt to recrystallize, as partial decomposition, difficult to detect, may occur.

If one wishes to prepare quinhydrone in quantity economically, it is best to add to a solution of commercial hydroquinone just half of the amount of a suitable mild oxidizing agent (ferric ion) theoretically required to produce quinone. Quinhydrone then crystallizes directly. See M. Trénel and C. Bischoff, *Angew. Chem.*, **42,** 288 (1929).

339. Anthraquinone. The preparation of purified anthracene obtained in §182 is suitable for use in a reaction of direct oxidation. Chromic acid attacks the two most central hydrogen atoms, numbered as 9 and 10, replacing them with oxygen. The resulting derivative, anthraquinone, is thus structurally two benzene rings connected by two ketone groups, $C_6H_4(—CO—)_2C_6H_4$.

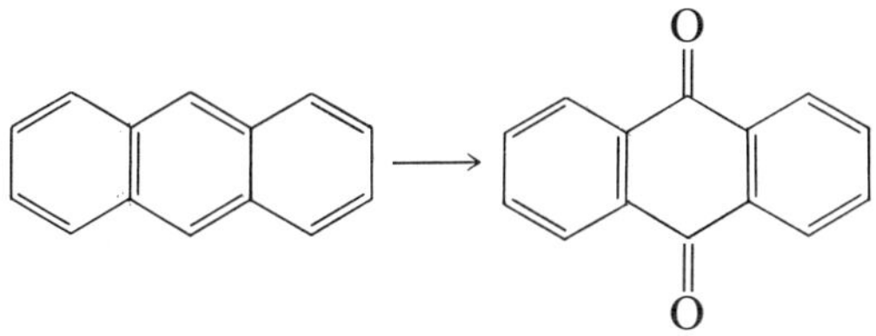

In a small flask equipped with a reflux condenser, place 2 g of purified anthracene and 20 ml of glacial acetic acid. Dissolve 3 g of chromic anhydride in a mixture of 10 ml of glacial acetic acid and 10 ml of water. See §340 for method of preparing the anhydride.

Heat the anthracene mixture to boiling, and slowly add the chromic solution through the top of the condenser. (If the preparation is attempted on a large scale, a two-neck flask and dropping funnel should be provided.)

Boil the reaction mixture for 10 min and filter by suction. The solid residue is crude anthraquinone. Wash it with water, dry on a steam bath, in an oven, or in the open air, and sublime in a manner similar to that employed with anthracene. It is preferable to use a smaller beaker in which to vaporize the anthraquinone. The product may also be recrystallized from glacial acetic acid.

340. Preparation of Chromic Anhydride. Dissolve 6 g of sodium dichromate (dihydrate) in 15 ml of water. To this solution add 25 ml of concentrated sulfuric acid in a fine stream, with stirring. Cool the mixture to room temperature to permit adequate crystallization. Mount a common

[§340] conical funnel in a suction flask, and place in the funnel a piece of glass wool just large enough to block the passage into the stem. Quickly pour out the chromic anhydride mixture upon this funnel, and suck down as much mother liquor as possible. No washing should be attempted.

Without further treatment the crystalline mass of chromic anhydride is used in the oxidation of anthracene. If the work is done carefully, the yield will be greater than 3 g, but the excess will do no harm, and the whole product may be used as described above in §339.

Questions

1. Why would you expect the central rather than some of the outer hydrogen atoms of anthracene to be removed?

2. Draw a vapor-pressure curve for anthracene, based on your observation of its behavior in the laboratory during the process of sublimation.

3. What becomes of the chromic anhydride?

4. Why was glacial acetic acid employed?

Experiment 36

Phthalein Dyes and Indicators

341. Formation of the Phthalein Structure. When a phenol is heated with phthalic anhydride, one might expect an ester such as phenyl phthalate to be produced. Long before any such conventional esterification could occur, however, the carbonyl group of the anhydride reacts with two molecules of the phenol at the highly reactive para position, yielding a **phthalein,** with by-product of water. This condensation is acid catalyzed and occurs by electrophilic attack of the conjugate acid of the anhydride on the para position of the phenol. Zinc chloride is the acid employed in the fluorescein synthesis and the $ZnCl_2$–anhydride complex is the active electrophile.

Phthaleins that still possess phenolic hydroxyl groups have different colors in acidic and basic solutions. They are valuable as indicators. The phthalein from resorcinol has a strong fluorescence and was appropriately named fluorescein. It is readily brominated to yield a tetrabromide in which the four hydrogens ortho to hydroxyl groups have been replaced by bromines. The product is eosin, a brilliant red dye used in biological laboratories and in formulation of red ink.

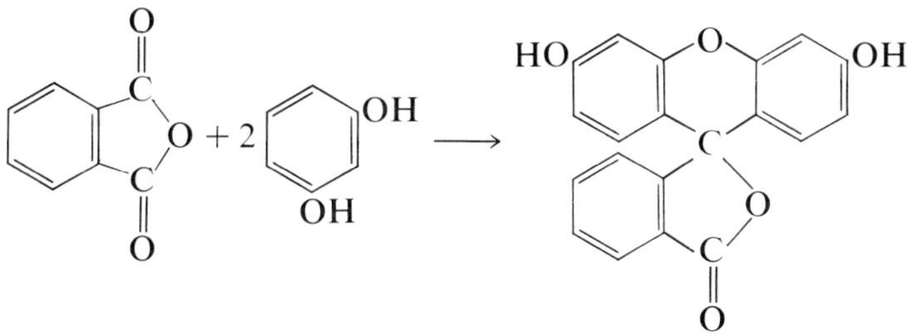

O + 2 OH OH ⟶ HO O OH C O C O

Resorcinol Fluorescein

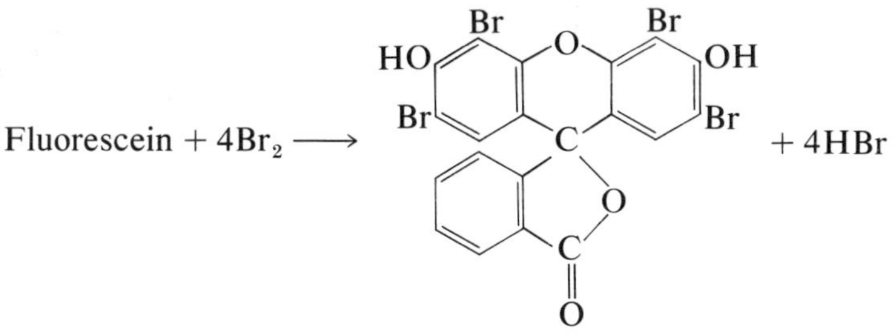

Fluorescein + $4Br_2 \longrightarrow$ Eosin $+ 4HBr$

Eosin

If one of the carboxylic groups in the original phthalic anhydride is replaced with the sulfonic group, a closely analogous derivative may be made. Such a product is known as a sulfonphthalein. Although the mixed carboxylic–sulfonic anhydride (supposedly needed for this synthesis) is not commercially available, it happens that a closely allied substance, benzoicsulfonic imide (saccharin), serves the purpose.

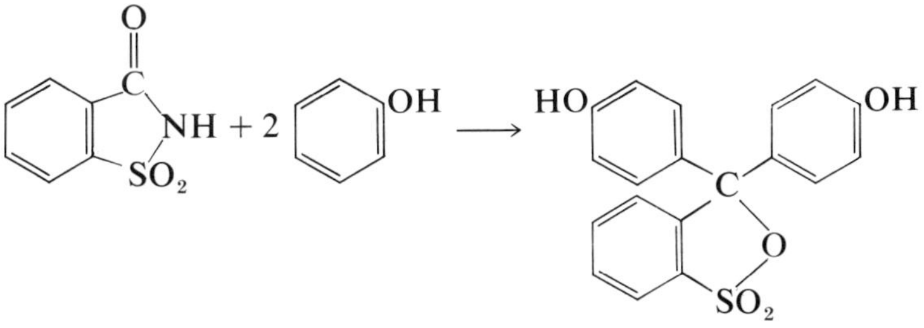

Saccharin + 2 phenol $\longrightarrow$ Phenol Red

Saccharin Phenol Red

342. Fluorescein. In a conical flask of 300- to 500-ml capacity, place 0.05 mole of phthalic anhydride and 0.1 mole of resorcinol. Suspend the flask in an oil bath and heat the oil to a temperature of about 180°. While the oil is being heated, weigh out rapidly 5–7 g of anhydrous zinc chloride and immediately grind it into a coarse powder in a small mortar. (*Note.* The zinc chloride must not be exposed to the air longer than is absolutely necessary. If the laboratory supply is obviously wet or hydrated, dry a 10-g portion by fusing in a porcelain dish.)

As soon as the zinc chloride is ready, stir it into the phthalic anhydride–resorcinol mixture, which by this time has probably melted. Keep the oil at the temperature of 180°, and stir the mixture every few minutes. Excessive heating may permit phthalic anhydride to sublime out of the flask, with resulting loss of yield. The reaction is complete when the solution becomes so viscous that further stirring is not practicable. The dark red mass consists principally of a mixture of fluorescein with zinc chloride, together with basic zinc salts resulting from partial hydrolysis of the original zinc salt at high temperatures.

Allow the oil to cool to 100°, and add to the reaction mixture 100 ml of water and 5 ml of concentrated hydrochloric acid. Now raise the temperature of the oil until active boiling of water ensues. Stir the mixture from time to time after the temperature of the oil passes 110°. Careful attention is required at this time to prevent the dilute acid from boiling over. Continue the boiling until the reaction mixture has been disintegrated and all zinc salts dissolved. The insoluble residue, which should be crushed into small pieces, is fluorescein. It is collected on a Büchner filter, washed well with water, and dried in the open air or on a steam bath. Yield is 13 g.

343. Eosin. Take a flask, containing 10 g of fluorescein and 40 ml of alcohol, to a hood, and add 8 ml of bromine, drop by drop from a dropping funnel.

> **Warning:** *See that the stopcock of the funnel is well lubricated in advance, and be careful not to pull out the plug lest bromine reach the fingers. Should bromine get upon the skin, at once wash with water and have the burn treated by a physician.*

During the 15 min normally required for addition of the bromine, note that the soluble dibromofluorescein first forms. Later the tetrabromide (eosin) is formed. Being only slightly soluble in alcohol, this substance crystallizes out. After the reaction mixture has stood 2 hr or more, the product is collected on a filter, washed with alcohol, and dried in the open air. Yield is 15 g.

344. Sodium Eosin. Grind together in a mortar 10 g of eosin and 2 g of anhydrous sodium carbonate, and place the mixture in a 200-ml flask or beaker. Add 10 ml of alcohol and 10 ml of water, and warm on the water bath until evolution of carbon dioxide ceases. Add 40 ml of alcohol, heat to boiling, and filter through a fluted filter into a beaker. (A stemless funnel is desirable.) Sodium eosin slowly crystallizes from the filtrate. It may be necessary to aid the process of crystallization by scratching the walls of the beaker with a glass rod.

Other salts of eosin may be prepared in crystalline form. See various manuals of organic chemistry for the preparation of potassium and ammonium eosin; or devise methods of preparing other salts.

345. Red Ink. If the faint green fluorescence of eosin is not considered objectionable, an acceptable red writing ink may be made according to the following recipe:

Gum arabic (gum acacia)	2 g
Water	50 ml
Thymol	0.2 g
Sodium eosin	2.5 g
Aqueous NaOH (0.1 *N*)	50 ml

Soak the gum arabic in the water overnight; or pulverize it, stir rapidly into the water, and mix thoroughly. As soon as the gum is completely dispersed in the water, add the mixture to a second solution (which is prepared in advance by dissolving the thymol) and next, the sodium eosin in the sodium hydroxide solution. Finally, filter the completed mixture through a fluted filter.

The gum aids in permitting the ink to flow in a desirable manner from the pen, while the thymol serves as a preservative and disinfectant. The amount of sodium eosin may, of course, be varied to suit personal preference.

Questions

1. In what way does fluorescein differ from phenolphthalein?

2. Draw the structures for phenolphthalein in acid solution and in basic solution.

346. Sulfonphthaleins. When 1,2-benzisothiazolin-3-one 1,1-dioxide (insoluble saccharin)

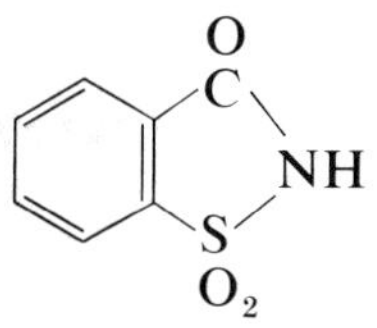

O
C
NH
S
O_2

is substituted for simple phthalic anhydride, a series of phthaleins of striking character is developed. Several of the well-known Clark–Lubs indicators are of this class, including the simplest of the series, which is phenolsulfonphthalein, or Phenol Red.

347. Phenol Red. In a small flask mix 10 g of insoluble saccharin (the free imide containing no sodium), 25 g of phenol, and 12 ml of concentrated sulfuric acid. Heat the mixture in an electric oven at 120° for 48 hr. Dissolve the resulting red, viscous reaction product in 200 ml of water to which is added enough sodium carbonate to neutralize the acids present. Steam-distill the solution until excess phenol is expelled, filter, and acidify with hydrochloric acid.

The precipitated Phenol Red may be purified by redissolving in dilute sodium carbonate and reprecipitating with acid. For further details, see R. Freas and E. A. Provine, *J. Amer. Chem. Soc.*, **50,** 2014 (1928). See also §476.

Experiment **37**

The Diels–Alder Reaction

348. Pericyclic Reactions. The Diels–Alder reaction is an example of a concerted cycloaddition reaction. Such reactions involve a concerted, cyclic transition state in which two or more bonds are made or broken at the same time; the reaction is highly stereospecific and is controlled by the necessity for preservation of orbital symmetry during the transformation. Concerted cycloadditions are one type of pericyclic reaction, the broad reaction class that includes chemical transformations with the above characteristics. Students should study pericyclic reactions in a standard textbook in connection with the present experiment.

Cycloaddition of an olefin to the terminal positions of a conjugated diene was discovered and extensively developed by Otto Diels and Kurt Alder in many papers, the first of which was published in 1928. For this reasearch they received the Nobel prize in chemistry in 1950. The reaction is a general and useful synthesis of six-membered rings. The olefin is often activated by electron-withdrawing groups, but this is not always necessary. The experiment described below involves such an olefin, which adds to the central ring of anthracene. This ring can be regarded as possessing a conjugated diene system. The stereochemistry of the reaction is shown in the equation.

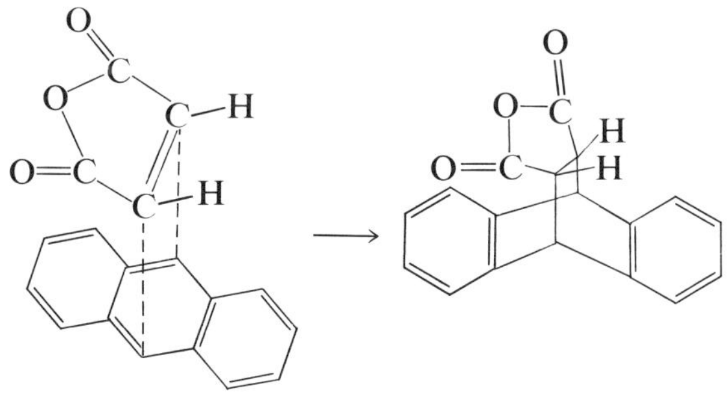

349. Anthracene-9,10-endo-α,β-succinic Anhydride. In a small round-bottom flask fitted with a reflux condenser place 2 g of light yellow, re-

[§349] sublimed anthracene, as available from the experiment described in §182, 5.6 g of maleic anhydride, and 40 ml of benzene.

> **Warning:** *Maleic anhydride causes burns. Avoid contact with skin, eyes, and clothing. Avoid exposure to concentrated vapor. In case of contact, immediately flush skin or eyes with plenty of water for at least 15 min.*

Shake the mixture with gentle heating until the solids are completely dissolved, and reflux over a water bath for 2–3 hr. Allow to cool slowly to room temperature; then warm slightly to dispel any turbidity due to slight crystallization of excess maleic anhydride.

Filter, and wash the pale yellow crystalline product with 5 ml of cold benzene. Dissolve in 75–85 ml of boiling ethyl acetate, evaporate to about 25 ml, and allow the residual solution to cool. Colorless granular crystals of the product are obtained. Yield is 2 g; mp, 262–263° (decomp.). The dried product should be bottled promptly. Exposure to moist air tends to cause hydrolysis of the anhydride portion of the molecule, a condition indicated by an opaque white coating on the crystals.

Questions

1. Represent structurally a Diels–Alder reaction of acrolein and isoprene.

2. Show how anthraquinone could be synthesized by use of the Diels–Alder reaction.

3. How may the diene synthesis be used to prove the structure of an unsaturated aliphatic compound?

4. Show how the synthesis could be used to prove the structure of 2-methyl-3-propylnaphthalene, a compound not described in the literature.

5. How could the Diels–Alder reaction be of use in the analytical laboratory of a synthetic-rubber manufacturing plant?

Experiment 38

Photochemistry and Thin-Layer Chromatography

350. Photochemistry. Absorption of light by organic compounds leads to activated molecules that lose their excess energy in a variety of ways. Utilization of this energy to bring about chemical transformations is photochemistry, a field of growing importance in organic chemistry. The photochemical transformations described in this experiment are *trans-cis* isomerization of azobenzene and photoaddition of maleic anhydride to benzene. The procedures are taken from J. F. Janssen, *J. Chem. Ed.*, **46,** 117 (1969), and R. E. Bozak and V. E. Alvarez, *ibid.*, **47,** 589 (1970). A chapter on photochemistry in a textbook of organic chemistry should be studied in connection with this experiment.

Photochemical transformations often lead to mixtures of compounds of comparable molecular weight and are frequently studied on a small scale. Thin-layer chromatography (tlc) is usually suitable to follow such reactions, as illustrated here. Read Chapter 9, and especially §125, before starting this experiment.

351. Plates for Thin-Layer Chromatography. Wash six microscope slides with soap and water, rinse them with tap water, then with distilled water, and finally with methanol. Allow them to dry while standing on end. Coat them by dipping two at a time in a suspension of Silica Gel G[1] (35 g in 100 ml of pure chloroform) available in a wide-mouth, glass-stoppered bottle on the reagent shelf. Hold the slides back-to-back with forceps, stir the suspension with the slides, and withdraw them with a smooth, steady motion. The speed with which the slides are withdrawn controls the thickness of the silica gel layer; rapid withdrawal leads to thin layers. Separate the slides and place them, coated side up, on a

[1]Adsorbasil, available from Applied Science Laboratories, P. O. Box 440, State College, Pennsylvania, 16801, was recommended, but other commercially available silica gels for tlc, which contain plaster of Paris but no fluorescent indicator, are suitable.

paper towel to dry. After all six slides have been coated, hold each one in the steam from a steam bath for 5 sec to set the plaster of Paris binder. Prior to use activate the adsorbent by heating the slide for 10 min over a hot plate (medium heat). A square piece of wire gauze with the corners bend downward can serve as a tray to support the slides over the hot plate. Activated slides can be stored conveniently in a desiccator, preferably in a microscope slide box.

352. Photoisomerization of Azobenzene. *trans*-Azobenzene absorbs ultraviolet light to give activated molecules that return either to the *cis* or *trans* ground states on loss of excess energy. The composition of the resulting mixture depends on the wavelength of light employed. Irradiation of benzene solutions of azobenzene with light of 365 nm wavelength produces mixtures containing over 90% of the thermodynamically less stable *cis* isomer, an example of optical pumping. The mixture produced in sunlight contains a smaller excess of *cis* isomer.

$$C_6H_5{-}N{=}N{-}C_6H_5 \ \underset{}{\overset{h\nu}{\rightleftharpoons}} \ C_6H_5{-}N{=}N{-}C_6H_5$$

trans or *anti* — *cis* or *syn*

Place a small spot of a solution containing 1 g of *trans*-azobenzene (§325 or commercial material) in 50 ml of benzene on each of two activated tlc slides [or strips cut from commercial precoated sheets (§125)]. A clean melting-point capillary may be used for the spotting. Touch the end of the tube to the azobenzene solution and allow the solution to fill the bottom 5 mm by capillary action. Touch the tube to the tlc slide or strip 1 cm from the bottom (centered right to left). Place one slide in the dark in your desk and expose the other to sunlight for an hour or two. Less time is required if the slide is exposed to an ultraviolet lamp, but care must be taken to mount such a lamp so that your eyes are never exposed directly to the light source.

Develop the chromatograms on both slides (irradiated and nonirradiated) in a chamber containing 3:1 cyclohexane–benzene (v/v) to a depth of 0.5 cm. A beaker or wide-mouth bottle capped with a watch glass can serve as the developing chamber. A strip of paper towel or filter paper lining the inner wall of the chamber serves to maintain an atmosphere saturated with vapors of the solvent. Allow the solvent front to migrate to within 1 cm of the top of the silica layer, mark the position of the front with a scratch, and remove the plates from the chamber. Determine and record the R^f values (see §125). The more polar *cis* isomer moves more slowly. If pure *trans*-azobenzene was available as starting material, the *cis* isomer should not appear on the slide kept in the dark. Note the relative areas of the spots.

The original reference can be consulted for a discussion of the mechanism of the photoisomerization [Janssen (see §350)]. A modified experiment, which includes isolation of the isomers and determination of spectra, is also given.

353. Photocyclization of *cis*-Azobenzene. It was originally planned to include as a second part of the experiment on photochemistry a photocyclization of *cis*-azobenzene described by R. F. Evans [*J. Chem. Ed.*, **46,** 117 (1969)]. One of the products in this cyclization is benzidine. The carcinogenic nature of this compound and the regulation of work practices involving it have already been discussed (§327). It was with regret that we abandoned our plans to include this beautiful experiment in the laboratory manual. In our opinion it must now be considered as unsuitable for a beginning organic chemistry laboratory.

354. Photoaddition of Maleic Anhydride to Benzene. In this photoinduced Diels–Alder reaction, maleic anhydride is activated photochemically so that it will undergo cycloaddition to benzene. A photosensitizer is necessary, and benzophenone is used. The product of addition of one maleic anhydride molecule leaves a reactive 1,3-diene system that undergoes a normal Diels–Alder reaction with a molecule of maleic anhydride that need not be activated. The following equations represent the synthesis.

$$(C_6H_5)_2CO \xrightarrow{h\nu} (C_6H_5)_2CO^*$$

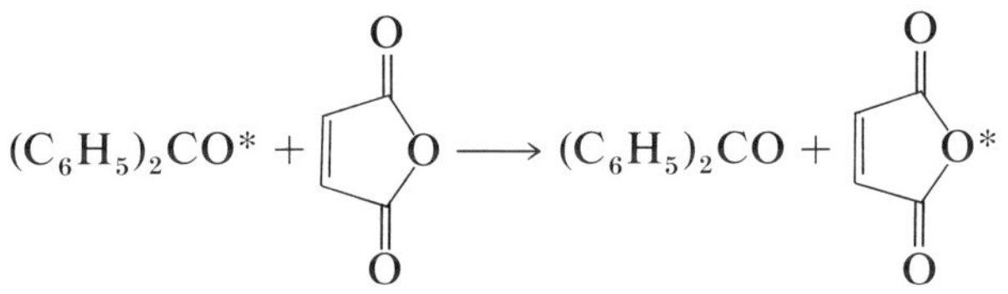

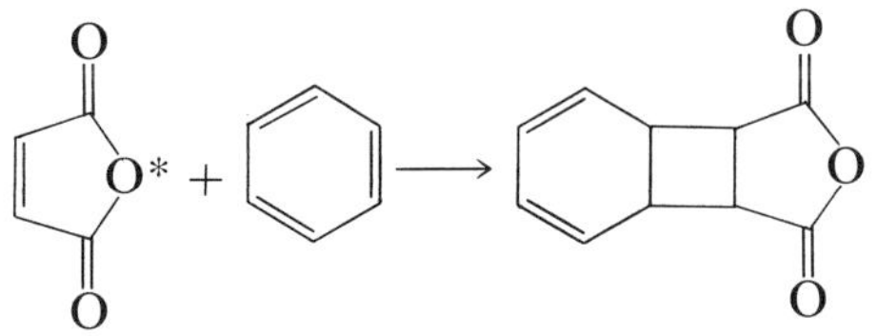

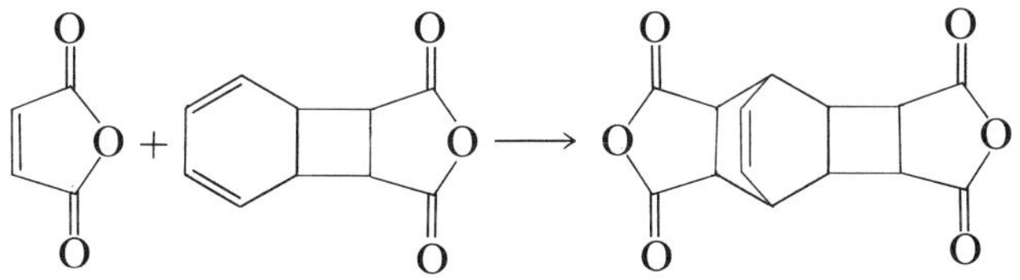

$(C_6H_5)_2CO^*$ + … ⟶ $(C_6H_5)_2CO$ + …

I

The product (I) bears the systematic name tricyclo[4.2.2.$0^{2,5}$]dec-7-ene-3, 4,9,10-tetracarboxylic 3,4:9,10-dianhydride. It has been patented [D. Bryce-Smith, G. Andrew, and V. Brian, *U. S. Patent* 3,480,532, Nov. 25, 1969; *Chem. Abstr.*, **72,** 31319 (1970)] and is said to be useful in forming polyimides and as a crosslinking agent in polymerizations. The procedure described below was devised by R. E. Bozak and V. E. Alvarez [*J. Chem. Ed.*, **47,** 589 (1970)].

Prepare a solution of 5.0 g of maleic anhydride and 2.0 g of benzophenone in 100 ml of benzene by refluxing the mixture on a steam bath for 15 min.

Warning: *Maleic anhydride causes burns. Avoid contact with skin, eyes, and clothing. Avoid exposure to concentrated vapor. In case of contact immediately flush skin or eyes with plenty of water for at least 15 min.*

Usually the maleic anhydride contains some maleic acid, which is insoluble in benzene and should be removed by filtration (use a stemless funnel and fluted filter paper; see §101). Expose this solution to direct sunlight for about 7 days. This can be done conveniently if the solution is placed in a 90- by 40-mm glass Petri dish or crystallizing dish covered by a watch glass (both of borosilicate glass) that fits so that the container can be sealed with masking tape. The watch glass should be placed convex side up, and the depth of the solution in the Petri dish should not exceed about 3 cm. Wrap aluminum foil around the bottom and sides of the dish to a height of 2 cm. The solution can also be exposed in a conical flask or Florence flask.

Crystals gradually separate from the solution during the reaction. Filter the reaction mixture with suction (No. 2 filter paper), wash the crystals with benzene, and dry them by continued suction for about 20 min. The yield should be about 1 g. The mp is 355–358° in an inert atmosphere, and the compound begins to turn brown about 300°. Do not attempt to determine the mp unless a suitable block (§108) is available.

Experiment 39

A Synthetic Sequence

355. Diphenylacetic Acid from Benzaldehyde. Preparation of diphenylacetic acid starting from benzaldehyde involves a sequence of steps, each of which occurs in excellent yield and is interesting as an illustration of a general reaction. The benzoin condensation is a reaction confined to aromatic aldehydes and cyanide ion is a specific catalyst. The product can be oxidized to an α-diketone, benzil, by mild oxidizing agents, which oxidize aldehydes to carboxylic acids (Tollen's reagent, Fehling's solution, etc.), but nitric acid, a more general oxidizing agent, is cheap and effective. Rearrangement of benzil yields benzilic acid, which is reduced to diphenylacetic acid with hydrogen iodide.

Save about 1 g of each product in the sequence. Recrystallize this sample, dry it carefully, and turn it in to your instructor in a neatly labeled sample vial. Use the rest of each product for the next step of the synthesis. Finally, purify the entire sample of diphenylacetic acid and turn it in to your instructor in a suitably labeled bottle or vial. Calculate the percentage yield for each step in the sequence and the overall yield of diphenylacetic acid based on benzaldehyde.

356. Benzoin. In a 500-ml round-bottom flask equipped with a water-cooled reflux condenser place a solution of 3 g of sodium cyanide (**warning:** *poison!*) in 30 ml of water. Add 30 ml of acid-free or freshly distilled benzaldehyde and 50 ml of 95% alcohol. Reflux the mixture from 30 to 45 min on the water bath. Upon cooling, crude benzoin crystallizes and is isolated on the Büchner filter. Wash the product first with a few milliliters of alcohol and then with several portions of water. Benzoin may be recrystallized from alcohol, but such purification is not necessary in the preparation of benzil.

357. Benzil. In the same apparatus used for the preparation of benzoin, place the crude benzoin with three times its weight of concentrated nitric acid (sp. g. 1.42). With occasional shaking of the flask, heat the reaction mixture for 1–2 hr, or until evolution of oxides of nitrogen ceases. This operation should be conducted in a hood, or outdoors, or with protection of a funnel and beaker of alkali as pictured in §198. (Note carefully the exact position of the funnel; not too deeply immersed.) Now pour the mixture into four or five times its volume of ice water and filter. Wash the yellow precipitate of crude benzil on the Büchner filter. Set aside about 1 mg (estimated) of this preparation for use as seed crystals and use the remainder for the subsequent experiment combining recrystallization with test of supersaturation.

358. Supersaturation Phenomena. In a large conical flask (500–700 ml) place the crude benzil and an amount of 95% alcohol at the rate of 6.5 ml for each gram of benzil used. Dissolve the benzil in the alcohol with the aid of the water bath, being very careful to leave no residual solid in any part of the vessel. Stopper the flask loosely, and set it aside to cool in a place where it will not be disturbed. When the solution has reached room temperature, seed it by dropping in a minute particle of solid benzil, taking care not to move or to disturb the flask. Beautiful yellow needles of purified benzil appear, and are eventually isolated and dried (mp of benzil, 95°). The velocity of crystallization, form and size of crystals, etc., may be varied neatly by slight changes in the 6.5-ml value specified.

359. Benzilic Acid—A Molecular Rearrangement. When benzil is heated in a solution of potassium hydroxide in dilute alcohol, a phenyl group shifts from one ketone group to the other. A molecule of water is added, and the substance benzilic acid (diphenylhydroxyacetic acid) is produced in the form of its potassium salt.

$$C_6H_5COCOC_6H_5 \longrightarrow (C_6H_5)_2COHCO_2K \longrightarrow (C_6H_5)_2COHCO_2H$$

In a 500-ml round-bottom flask fitted with a reflux condenser place 20 g of benzil, 55 ml of alcohol, 20 g of solid potassium hydroxide, and 40 ml of water. Reflux on the water bath for 15 min; then pour the mixture into an evaporating dish and allow to stand overnight. On a subsequent day, filter the mixture and wash the product of potassium benzilate with a little alcohol.

Dissolve the potassium benzilate in the minimum quantity of warm water and filter, using decolorizing carbon if the solution is turbid and discolored. Acidify the clear filtrate with concentrated hydrochloric acid, isolate the resulting benzilic acid, and recrystallize all or part from water if purification seems desirable. Yield is 12 g; mp of pure benzilic acid is 150°.

360. Diphenylacetic Acid. Place 30 ml of glacial acetic acid, 1.8 g of red phosphorus, and 0.6 g of iodine in a round-bottom flask. (**Warning:** *Handle the phosphorus carefully.*) Swirl the reaction mixture and let it stand for 10 min to allow the iodine to react with the phosphorus. Then add 0.5 ml of water and 10 g of benzilic acid. Attach a reflux condenser and reflux the mixture for 2 hr without interruption. Filter the hot solution to remove excess phosphorus. This should be done in a hood if possible because the acid vapors are acrid. Pour the warm filtrate slowly into a cold solution of 2.5 g of sodium bisulfite in 100 ml of water. The sodium bisulfite solution should be acid to litmus to prevent solution of some of the diphenylacetic acid as its sodium salt. Pass a little sulfur dioxide through the solution (in a hood) if it is not acidic. Collect the precipitate of diphenylacetic acid with suction, wash it with water, and recrystallize it from 50% aqueous alcohol (6–7 ml/g of product) with decolorizing. Yield, 7–8 g, mp 144–145°.

Experiment 40

Heterocyclic Series and Skraup Synthesis

361. Quinoline. Acrolein condenses with aniline at both ends of the aliphatic chain, causing ring closure and the formation of a heterocyclic compound. A small amount of oxidation is necessary to complete the process. Glycerol is an inexpensive source of the acrolein reagent which presumably is the active material in the ring closure.

$$C_6H_4(H)(NH_2) + CHO{-}CH{=}CH_2 \longrightarrow \text{quinoline}$$

In a 2-liter round-bottom flask equipped with a water-cooled reflux condenser, place successively, in the order named, 10 g of ferrous sulfate (heptahydrate), 108 g of anhydrous glycerol, 27 g of redistilled aniline, and 21 g of redistilled nitrobenzene. (Anhydrous glycerol may be prepared by heating commercial glycerol to 180° in an open evaporating dish.) Mix the reagents thoroughly and add slowly, with shaking, 50 ml of concentrated sulfuric acid. The resulting aniline sulfate should nearly all remain in solution. Provide an ice bath in convenient position so that the reaction flask may be immediately immersed and promptly chilled if there should be any sign of violent ebullition.

With the reflux condenser in place, cautiously heat the flask over wire gauze until the mixture boils. For some time the heat of reaction is sufficient to maintain the boiling temperature; later, a burner is employed. Boil the mixture for 3–4 hr; then cool the mixture below 100°, add 50 ml of water, and pass a current of steam through the material until the excess nitrobenzene has been expelled.

To the residue in the steam flask add a cool solution of 85 g of sodium hydroxide in 250 ml of water. Again pass steam through the material, this time collecting the distillate of water, quinoline, and aniline in a second 2-liter flask used as a receiver. This distillation is continued as long as oily products come over. Add 20 ml of concentrated sulfuric acid to the mixed distillate. Cool this solution to 0° and add a concentrated solution of sodium nitrite until a drop of the reaction mixture colors a potassium iodide–starch mixture a deep blue or purple instantly. As this diazotization process approaches completion, the reaction becomes slow. It is not considered to be complete until the aniline mixture, after a 5-min interval with stirring, gives, instantly, a positive test for excess nitrous acid. From 6 to 8 g of sodium nitrite will be required.

With the aid of a water bath, heat the diazotized solution for an hour and then add a solution of 40 g of sodium hydroxide in 150 ml of water, cooling the flask under the tap. Steam-distill the mixture, and saturate the distillate with common salt. Extract with ether, dry the extract over potassium hydroxide, filter, and remove the ether by distillation. (See §§63 and 115 for precautions and technique.) Distill the ether-free quinoline through an air-cooled condenser. Yield is 30 g. Boiling range is about 235–238°. Although colorless when pure, quinoline from this process has a pale yellow tint. Reduced-pressure distillation of the final product is desirable but not necessary.

Questions

1. How might one determine whether the aldehyde end of such a molecule as acrolein combined with the ortho position, as shown structurally above, or in reverse position with the amine group?

2. Why was so unusual an oxidizing agent as nitrobenzene employed?

3. Why was the mixture diazotized?

4. What is the purpose of the 40 g of sodium hydroxide used after diazotization?

Experiment 41

Synthetic Polymers

362. Monomers and Polymers. When molecules (monomers) unite with each other either through repeated addition or through repeated condensation (involving the splitting off of a simple molecule such as water or hydrogen chloride per new bond formed), the resulting high-molecular-weight product is referred to as a polymer. The polyaddition mode of polymerization will be illustrated by the formation of poly(methyl methacrylate), a colorless, transparent resin of the Lucite type.

$$x\mathrm{CH_2{=}C(CH_3){-}CO_2CH_3} \xrightarrow[\text{peroxide}]{\text{light and/or}} \left(\!\!-\mathrm{CH_2{-}C(CH_3)(CO_2CH_3)}-\!\!\right)_x$$

Methyl methacrylate

Polycondensation will be illustrated with the formation of a polyamide, Nylon 66, a well-known textile fiber, which, when introduced over thirty years ago, almost put the silk industry out of business.

$$x\mathrm{Cl{-}\overset{O}{\overset{\|}{C}}{-}(CH_2)_4{-}\overset{O}{\overset{\|}{C}}{-}Cl} + x\mathrm{H_2N(CH_2)_6NH_2} \xrightarrow[-\mathrm{HCl}]{\mathrm{OH^-}} \left[-\mathrm{\overset{O}{\overset{\|}{C}}{-}(CH_2)_4\overset{O}{\overset{\|}{C}}NH(CH_2)_6NH}-\right]_x$$

Adipoyl chloride — Hexamethylenediamine — Nylon 66

363. Radical Polymerization. Olefins in which the double bond is terminal ($\mathrm{CH_2{=}\overset{|}{C}{-}}$) and activated by such groups as carbalkoxyl or phenyl polymerize readily by a radical mechanism. The radicals required for initiation can be produced by irradiation of the olefin; or a compound such as benzoyl peroxide, which decomposes at fairly low temperatures to yield radicals, can be used as catalyst. The polymerization is a radical chain reaction that is terminated by interaction of any two radicals in the reaction mixture to give combination or disproportionation. The process is formulated below for polymerization of methyl methacrylate initiated by radicals from benzoyl peroxide. Poly(methyl methacrylate) terminates entirely by disproportionation at polymerization temperatures above 60° and partly by each mechanism at lower temperatures.

Initiation:

$$C_6H_5\overset{O}{\overset{\|}{C}}OO\overset{O}{\overset{\|}{C}}C_6H_5 \longrightarrow 2C_6H_5\overset{O}{\overset{\|}{C}}O\cdot \longrightarrow 2C_6H_5\cdot + 2CO_2$$

$$R\cdot + CH_2{=}\overset{CH_3}{\overset{|}{C}}CO_2CH_3 \longrightarrow RCH_2\overset{CH_3}{\overset{|}{\underset{\bullet}{C}}}CO_2CH_3 \; (R\cdot = C_6H_5\overset{O}{\overset{\|}{C}}O\cdot \text{ or } C_6H_5\cdot)$$

Propagation:

$$RCH_2\overset{CH_3}{\overset{|}{\underset{\bullet}{C}}}CO_2CH_3 + CH_2{=}\overset{CH_3}{\overset{|}{C}}CO_2CH_3 \longrightarrow RCH_2\overset{CH_3}{\overset{|}{\underset{\underset{CO_2CH_3}{|}}{C}}}{-}CH_2\overset{CH_3}{\overset{|}{\underset{\bullet}{C}}}CO_2CH_3 \cdots$$

Termination:

Disproportionation:

$$2 \sim\!\sim CH_2\overset{CH_3}{\overset{|}{\underset{\bullet}{C}}}CO_2CH_3 \longrightarrow \sim\!\sim CH{=}\overset{CH_3}{\overset{|}{C}}CO_2CH_3 + \sim\!\sim CH_2\overset{CH_3}{\overset{|}{C}}HCO_2CH_3$$

Combination:

$$\longrightarrow \sim\!\sim CH_2\overset{CH_3}{\overset{|}{\underset{\underset{CO_2CH_3}{|}}{C}}}{-}\overset{CH_3}{\overset{|}{\underset{\underset{CO_2CH_3}{|}}{C}}}CH_2 \sim\!\sim$$

One of the initiating radicals can be involved in termination in place of one of the growing chains.

364. Poly(methyl Methacrylate). Poly(methyl methacrylate) is an important commercial polymer because it is inexpensive and is outstanding in stability and optical properties, particularly in clarity and light transmission. It seldom causes an allergic reaction and thus can be used in contact lenses, artificial eyes, dentures, etc. Poly(methyl methacrylate) is prepared commercially by bulk or suspension polymerization; the former is illustrated in the present experiment. Closely related polymers such as polyacrylates, polyacrylonitrile, and poly(acrylic acid) are also important; commercial products are often copolymers obtained from mixtures of monomers.

365. Methyl Methacrylate. Although methyl methacrylate is readily available, it may be prepared by dehydration of methyl α-hydroxyisobutyrate as described below. The commercial starting materials are acetone and HCN, which yield acetone cyanohydrin (OS–CV 2). Methanol and sulfuric acid convert the CN to carbomethoxyl and dehydrate at the same time. A suitable experiment for a longer laboratory course involves this sequence. The easiest way to obtain the monomer is to depolymerize Lucite or some other commercial form of poly(methyl methacrylate) (§362).

366. Dehydration of Methyl α-Hydroxyisobutyrate. Provide a dry 500- or 1000-ml round-bottom flask with a water-cooled reflux condenser ready-fitted for immediate use. Counterpoise this flask on a balance, and weigh into it rapidly 47 g of phosphorus pentoxide. If there is any question as to quality of the pentoxide, with possibility of water already added to that reagent, include a slight excess of the material. Now place the flask immediately under the reflux condenser.

With clamps loosened slightly to permit agitation of the flask by hand, add through the condenser in a few small portions, with shaking, 59 g of methyl α-hydroxyisobutyrate. When the reaction subsides, use a glass rod to mix the reagents more thoroughly.

Provide accessories for steam distillation to fit the reaction flask. When the reaction is complete, steam-distill the mixture as fast as the slight foaming will permit until 50–60 ml of distillate has been collected. Now separate the methacrylate layer, discarding the aqueous layer, wash with 50 ml of cold water, and dry over anhydrous magnesium sulfate or sodium sulfate. Yield at this point is 25 ml or more.

To the crude distillate add 0.25 g of hydroquinone as an inhibitor, and distill through a small Hempel column or equivalent. Boiling point of the pure ester is 100°; material distilled over the range of 98–102° should be satisfactory.

367. Polymerization. In a small test tube bent as shown in the accompanying figure, place enough dry methyl methacrylate monomer so that the level of the liquid in the tube will be approximately as shown by the dotted line. Now pass a very slow stream of nitrogen or other inert gas through the monomer liquid for several minutes to remove dissolved oxygen. Stopper tightly and place the tube, firmly clamped in proper horizontal position, in direct sunlight. In five to seven days, if bright sunlight is available, the material polymerizes to form a transparent hemicylinder that one might use as a magnifying lens for a slide-rule scale. If direct sunshine is not available, dissolve a few milligrams (avoid excess!) of benzoyl peroxide in the liquid before exposure to the light.

In another test tube place 6–8 ml of the monomer, and add about 30–40 mg of benzoyl peroxide. Insert a thermometer into the tube and heat in a water bath at about 85° until rapid polymerization occurs. Note temperature changes, and do not leave the thermometer too long in the tube. Repeat the experiment with about 5 ml of the monomer, about 25 mg of benzoyl peroxide, and 0.5 g of hydroquinone.

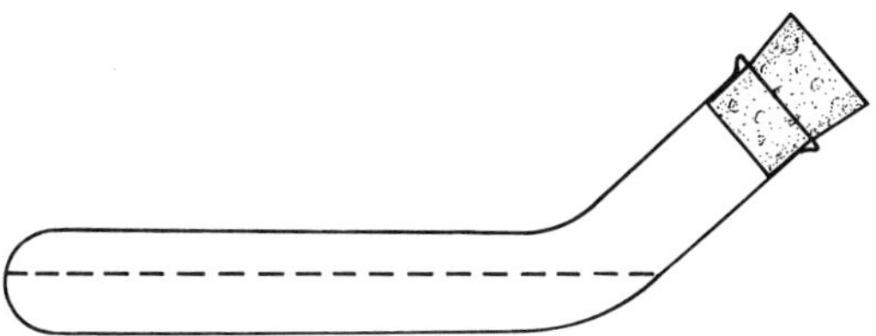

In the main experiment, omission of the process of flushing with inert gas retards polymerization.

Warning: *The storage and dispensing of benzoyl peroxide require special care* [N. V. Steere, *J. Chem. Ed.*, **41,** A445 (1964)].

368. Depolymerization of Poly(methyl Methacrylate). In a 100- to 200-ml distilling flask equipped with a water-cooled condenser place 25 g of Lucite or Plexiglas fragments. With use of a "soft" blue gas flame, cautiously heat the bulb of the flask, keeping the flame moving during the heating. Above 300° the polymer breaks down into monomer, which distills in the region of 100–110°. Redistill; the temperature will be about 102°. If there is to be any delay in the use of the product, add 0.1 g of hydroquinone as inhibitor. If a carbonaceous residue is left in the flask so that cleaning is difficult, note §6.

369. Nylon. Into a 150-ml beaker add 10 ml of an aqueous solution that is 0.50 *M* in hexamethylenediamine and 0.50 *M* in NaOH. Very carefully pour 10 ml of 0.25-*M* solution of adipoyl chloride in cyclohexane down the sides of the beaker so as not to mix the two layers. An irregular film will form immediately at the interface between the organic and water layers. When the film is disturbed, new, smooth film reforms at the interface. Using a pair of tweezers or wire hooked at the end, gently free the walls of the beaker from polymer strings, engage the polymer mass at the center, and slowly raise the tweezers or wire so that the polyamide forms continuously into a rope. The rope may be washed, dried, and examined. A less sophisticated experiment than this nylon rope trick involves vigorously stirring the two-phase system to form chunks of nylon. A small amount of this washed and dried polymer may be melted in a test tube by gentle heating. If one touches the melt with a glass rod and slowly withdraws it, fiber will result.

For other experiments on interfacial polymerization see §528.

Questions

1. Represent two alternative syntheses of methyl methacrylate, starting in each case with a simple monohydroxy alcohol.

2. Write equations showing the function of peroxide in polymerization of the monomer.

3. How does the hydroquinone function?

4. What role does light play in the photochemical method of polymerization as described above?

5. Why does omission of the process of flushing with inert gas retard the polymerization of methyl methacrylate?

6. Which monomers would produce the following polymer?

$$\left[-OCH_2CH_2O\overset{\overset{\displaystyle O}{\|}}{C}-C_6H_4-\overset{\overset{\displaystyle O}{\|}}{C}- \right]_x$$

An Introduction to Qualitative Organic Analysis

370. Background Information. Qualitative organic analysis represents one of the best ways to consolidate one's knowledge of organic chemistry. Classical qualitative organic analysis helps in learning many of the reactions of organic compounds and many of the techniques of organic chemistry using comparatively simple and readily available equipment. Also, rather than involving a mere following of directions, qualitative analysis requires the student to exercise his own judgment in the identification of the organic "unknown." Of prime importance are originality in planning one's work and the ability to make correct deductions from careful observations (note that even such common physical properties as color and odor, which may be distinctive, can provide valuable clues). In a sense, qualitative organic analysis borders on research in that one is asked to identify an "unknown."

In this elementary treatment of qualitative organic analysis, occupying only a small fraction of the time that would be allotted to a regular course in the subject, only reasonable unknowns should be employed, so as to avoid those that may be rare, dangerous, or ambiguous in their behavior toward characterization tests or derivatization procedures. The substances issued are arbitrarily restricted, for the sake of economy of reagents as well as simplicity of procedures, to the following: alcohols, phenols, aldehydes, ketones, acids, esters, amides, nitriles (and possibly simple organic salts and aromatic hydrocarbons, halides, and nitro compounds). In each case the unknown should contain not more than one salt- or ester-forming group and not more than one aromatic ring. Thus a nitrotoluidine or a glycol might be issued, but not sodium salicylate or an aminobenzoic acid. An attempt has been made to cull out the more complex compounds from the table in §402 (upon which the instructor may wish to expand, with suitable references). To simplify this treatment further, an effort has been made to [bracket] those compounds or sections that would appear to be less appropriate for a beginning level course. Also, it may be pedagogically advisable to start with simple unknowns, drawn from the more common major classes of compounds—for example, an alcohol unknown, followed by a carboxylic acid, followed by an aldehyde or ketone, followed by an amine—before proceeding to general unknowns. Prelaboratory lecture-discussions are advisable to acquaint the students more clearly with the upcoming procedures and relevant chemistry. As a rule, the following six steps will be followed in the identification procedure:

(*a*) Preliminary examination: physical [and burning] characteristics of the unknown.
(*b*) Purification and determination of physical constants: melting point and/or boiling point.
(*c*) Qualitative tests for the elements: nitrogen, chlorine, [bromine, iodine, and sulfur].
(*d*) Solubility tests: Solubility in water, 5% aq. NaOH, 5% aq. $NaHCO_3$, 5% aq. HCl, and H_2SO_4.
(*e*) Classification tests: functional group reactivity.
(*f*) Preparation of derivatives: proof of identity.

GENERAL PROCEDURE

371. Preliminary Examination. Simply observing whether the organic unknown is a solid or a liquid reduces the number of compounds under consideration.

Warning: *Impure "solids" may appear as liquids.*

[Ignition of a small amount of the unknown can also be informative. This procedure involves a gentle heating of a drop of liquid or about 0.1 g of solid on a crucible cover or a small spatula with a burner and then a stronger heating with ignition of the sample. Observations should be made concerning the melting of a solid and the nature of the flame. Aromatic or highly unsaturated aliphatic compounds burn with a sooty, yellow flame; aliphatic hydrocarbons burn with a yellow, but less sooty flame. A blue flame or less colored flame often suggests the presence of oxygen in the compound, whereas nonflammability may indicate a very high oxygen or halogen content. Any residue remaining after ignition should be tested by adding a drop of distilled water and then checking with pH paper. An alkaline residue indicates the unknown to be a metal salt such as that of sodium or potassium.]

Odor is another characteristic of many organic compounds, which may be useful in their identification. *Caution* must be exercised in smelling organic compounds since some have very disagreeable odors and others have very irritating effects on the mucous membranes. Still other compounds may have a very pleasant odor (e.g., esters), but most organic vapors are toxic if inhaled in large quantities. Description of odors is difficult; therefore, the student should familiarize himself with the odors of several classes of compounds in order to make use of this method of identification.

[The purity of an organic unknown can many times be judged by its color. Since most pure organic compounds are colorless or white, a brown or other dark color may indicate that impurities are present. On the other hand, color may implicate the presence of a highly conjugated unsaturated structure.]

372. Purification and Determination of Physical Constants. The student should make certain that his unknown is pure before he proceeds with the identification. As a criterion of purity he will choose one or more physical properties, usually the melting point if the substance is a solid and the boiling point if it is a liquid. The determination of the boiling point and/or melting point is, in any event, a necessary part of the identification. If an unknown is found to be contaminated (as indicated by a broad melting-point or boiling-point range), it should be purified before these properties are recorded.

The simplest and most satisfactory method of purifying most liquids is by distillation (see Chapter 6). In organic qualitative analysis the amount of sample is usually not large; therefore, a small distilling flask (10 or 25 ml) is used. It is equipped with a thermometer and boiling chip and connected to a condenser and receiver or, more simply, connected through an adapter to a test tube immersed in an ice bath. Over a low flame, normal distillation is carried out until a constant distilling temperature is reached (as much as 5–10% of the sample may be collected as forerun). Then without interrupting the distillation, the receiver is changed (a new test tube attached) and the essentially constant-boiling material is collected, but short of the point of dryness. The boiling point recorded for the liquid is the boiling point of the constant-boiling fraction.

Recrystallization from a suitable solvent is the most common method of purification of solid substances, as discussed earlier in Chapter 8 and Experiment 2. It is *always* sound practice to attempt the recrystallization on a small scale (a few crystals in 2–3 ml of solvent) before risking the entire sample.

In searching the literature or the lists of compounds from which the unknown may be drawn, it is well to consider all compounds melting and/or boiling within 5° of your value(s) as possibilities.

Two additional physical constants, the density and the index of refraction, may be employed for identification of a compound. The density of a liquid is determined with the aid of a small vessel referred to as a pycnometer. After weighing equal volumes of the sample and of water (i.e., filling the pycnometer first with one and then the other) at the same temperature, t, the absolute density, D_4^t, of the sample is found from the equation

$$D_4^t = \frac{\text{weight of sample}}{\text{weight of water}} \times 0.999973 \text{ g/ml (the density of water at } 4^\circ)$$

The index of refraction is a function of the velocity of light in air relative to the velocity of light in the substance in question. Because of the difference in the speed of light in different media, a beam of light bends on passing obliquely from air into the sample. The degree of bending can be precisely determined, most commonly by the Abbe refractometer, which is scaled to read the refractive index directly. An example is the refractive index of water, $n_D^{20} = 1.3333$, where 20 is the temperature at which the refractive index, n, was measured, and D indicates the nature of the monochromatic light employed in the measurement (the sodium D line).

Both the refractive index and the density values are useful to check on the purity of a compound as well as for identifying an unknown. Impurities have significant effects on these physical constants.

373. Qualitative Tests for the Elements. Organic compounds may often be distinguished conveniently from one another according to the elements they contain. The first step in the identification of a pure organic substance is a qualitative elemental analysis. The importance of this analysis cannot be overemphasized, since the subsequent tests performed in an attempt to identify a substance may be misleading and of no value if the elemental tests have not been performed correctly. *Distilled* water is to be used throughout in these tests. All apparatus prior to use should be rinsed with distilled water after being cleaned.

Fusion with Sodium

$$\text{Unknown} \xrightarrow[\Delta]{\text{Na}} \text{NaCN, NaX [Na}_2\text{S, NaCNS, etc.]}$$

Support a small dry Pyrex test tube (10 by 75 mm) on a ring stand by inserting it through a small hole in an asbestos square. Place a piece (no larger than a small pea) of freshly cut sodium in the tube and heat with a burner until the sodium vapor rises about 4–5 cm in the test tube. Drop a small amount of the substance (about 0.05 g) directly onto the molten sodium.

> **Warning:** *There may be a slight explosion;* **remember** *that goggles are to be worn at all times in the laboratory.*

Heat the tube to redness for about 1 min. (If the unknown is very volatile, mixing it with an equal amount of powdered sucrose before adding it to sodium is advisable.) Allow the test tube to cool slightly and add about 1 ml of alcohol to decompose any unreacted sodium. Evaporate the excess alcohol and heat the tube until red hot. Then plunge the red hot test tube into about 20 ml of cold, *distilled* water in a small beaker, causing the tube to crack up and allowing the salts to dissolve in the water. Break up any lumps with a stirring rod and heat to boiling to complete the solution process. Filtration of the resultant mixture should yield a clear filtrate on which the following tests are performed.

TEST FOR NITROGEN

$$18CN^- + 3Fe^{++} + 4Fe^{3+} \longrightarrow \underset{\text{Blue}}{Fe_4[Fe(CN)_6]_3} \downarrow$$

Adjust the pH of 1 ml of fusion filtrate to 13 (test with an indicator paper such as Hydrion E), and add 2 drops each of saturated ferrous ammonium sulfate solution and 30% potassium fluoride solution. Then boil the solution gently for about 30 sec and acidify while hot by careful dropwise addition of 30% sulfuric acid until the precipitate of iron hydroxide just dissolves. Avoid an excess of acid. The appearance of a brilliant blue precipitate of Prussian Blue indicates the presence of nitrogen in the original unknown. If the amount of cyanide ions produced in the filtrate was insufficient to form a precipitate (suggesting an incomplete sodium decomposition), a blue or greenish blue solution may be observed. Detection of the blue precipitate may best be accomplished by collecting and washing with water on white filter paper.

TEST FOR HALOGEN

$$AgNO_3 + NaX \longrightarrow NaNO_3 + AgX \downarrow$$

Acidify 2 ml of the fusion filtrate with a few drops of dilute nitric acid, boil gently for several minutes to expel hydrogen cyanide [or hydrogen sulfide] if nitrogen [or sulfur] was present in the unknown, and then add a few drops of aqueous silver nitrate solution. A heavy precipitate indicates the presence of chlorine [bromine and/or iodine], but a faint turbidity is probably the result of halide impurities in the unknown, the reagents, or the glass of the sodium-fusion tube. [Silver chloride is white; bromide, pale yellow; and iodide, yellow. However, all these darken in the light, and a positive differentiation of the halides in this way is not possible.]

[To distinguish among the halogens, acidify 3 ml of the fusion filtrate with dilute sulfuric acid and boil for a few minutes. After cooling to room temperature, add 1 ml of carbon tetrachloride and then 1 drop of freshly prepared chlorine water (a commercial laundry-type bleach may be used, but the solution must be kept acidic to litmus paper.) If the carbon tetrachloride solution turns purple, iodine is indicated.

$$2H^+ + ClO^- + 2I^- \longrightarrow \underset{\text{Purple}}{I_2(CCl_4)} + Cl^- + H_2O$$

To this same solution add more chlorine water drop by drop, shaking after each addition. The purple color will fade

$$I_2(CCl_4) + 5ClO^- + H_2O \longrightarrow \underset{\text{Colorless}}{2IO_3^-} + 5Cl^- + 2H^+$$

and the production of a reddish brown color in the carbon tetrachloride indicates the presence of bromine.

$$2Br^- + ClO^- + 2H^+ \longrightarrow \underset{\text{Reddish brown}}{Br_2(CCl_4)} + Cl^- + H_2O$$

If neither iodine nor bromine was indicated by these tests and there was a distinct white precipitate on treating the original fusion filtrate with silver nitrate, then the unknown contained chlorine.]

A simpler but not as conclusive test for halogen is the Beilstein test. A loop of copper wire is first freed of any possible halide salts by holding in a blue flame. The absence of green color in the flame shows that the wire is ready for the test. The loop is cooled, dipped into the liquid or powdered solid under investigation, and returned to the flame. A flash of green color in the flame indicates that the unknown substance contains chlorine, bromine, or iodine. If the substance is volatile, it may be difficult to obtain more than a fleeting trace of the green color, which indicates only a momentary reaction of copper with halogen.

[TEST FOR SULFUR

$$Pb(OAc)_2 + Na_2S \longrightarrow \underset{\text{Black}}{PbS\downarrow} + 2NaOAc$$

Acidify a 2-ml portion of the fusion filtrate with dilute nitric acid and then add 2 or 3 drops of lead acetate solution. A black or brown precipitate (of lead sulfide) indicates the presence of sulfur. Sometimes this test gives better results if carried out as follows: Acidify a 2-ml portion of the fusion filtrate and warm while a piece of filter paper moistened with lead acetate solution is held above the liquid in the tube. A brown or black coloration on the paper indicates sulfide.]

374. Solubility Tests. Another routine procedure is the examination of certain solubility properties of the unknown. Since different types of organic compounds differ in their solubilities in water, dilute acid, dilute base, and sulfuric acid, solubility of the unknown in one or more of these reagents may offer clues as to (1) the nature of the functional group(s) present, e.g., acidic or basic character, and (2) the molecular weight of the compound. For instance, benzoic acid, which is insoluble in cold water, will dissolve in 5% aq. NaOH since the resulting sodium benzoate is soluble in water. Likewise *p*-toluidine is insoluble in water but will readily dissolve in 5% aq. HCl since the *p*-toluidinium chloride is water soluble. As an only approximate generalization, in many mono-functional homologous series those compounds having less than five carbon atoms are water soluble, while those having more than five carbon atoms are water insoluble. Students should study Chapter 12 in connection with these solubility tests.

The solubility tests are summarized in Fig. 1. Compounds are divided at the outset into two groups according to their solubility in water. Further subdivisions are developed by the use of the other solvents. In considering solubility, a substance is *arbitrarily* said to be "soluble" if it dissolves to the extent of 3 g/100 ml (i.e., 0.1 g/3 ml) of solvent. In *no* case should the solvent systems be *heated* to induce solubility.

Testing for the solubility class of the unknown is normally done in an ordinary test tube (16 by 150 mm) and with vigorous shaking of the mixture. Quantities can best be estimated as follows. The level of 3 ml of solvent will be nearly 2 cm high in this test tube. One tenth of a gram of liquid solute is approximately equivalent to 2 drops from an ordinary eye dropper. One tenth of a gram of solid solute may be weighed out individually on a suitable balance or may be estimated by weighing out 1 g and then mechanically dividing it into ten approximately equal piles.

If the compound dissolves in water, one does not test for its solubility in 5% HCl, 5% NaOH, or H_2SO_4. Water-soluble compounds include monofunctional compounds, such as simple alcohols, esters, aldehydes, ketones, acids, and amines, that have approximately five or fewer carbons. To determine whether the unknown is an acid or an amine, the solution should be tested for acidity or basicity with litmus

Figure 1. Classification of Organic Compounds According to Solubility Behavior

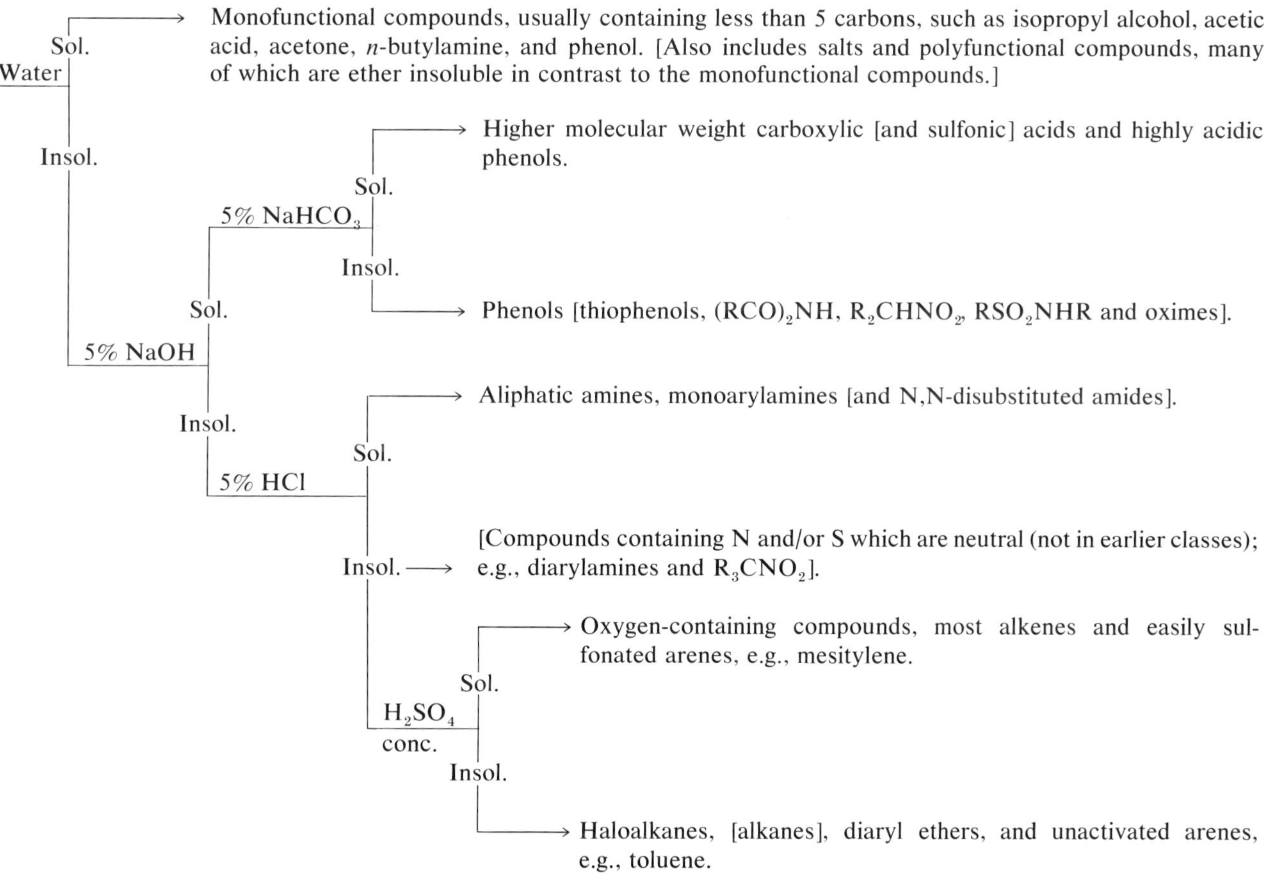

paper. [If your list of unknowns has been expanded to include salts and polyfunctional compounds, such as sugars and amino acids, these too will be water soluble. In this event a distinction can generally be made on the basis of ether solubility. The monofunctional compounds are ether soluble, whereas salts and many polyfunctional compounds are ether insoluble.]

Water-insoluble acidic compounds are detected by their solubility in 5% aq. NaOH. Further differentiation is made with the use of the weakly basic 5% $NaHCO_3$. In general, compounds soluble in both media consist of carboxylic [and sulfonic] acids and those substituted phenols which are more acidic than carbonic acid. The use of the ferric chloride test will indicate a phenol [a positive elemental analysis for sulfur may implicate the sulfonic acids]. There is no really good classification test for the carboxylic acid group (although an indirect test is described in §377; if this group is thought to be present, it is well to go directly to a neutralization-equivalent determination and to derivatization. [Indefinite behavior is observed with long-chain acids, which dissolve slowly in these reagents with the formation of a soap and a milky instead of a clear solution.] Solubility in aq. NaOH but insolubility in $NaHCO_3$ solution is limited primarily to simple phenols, which, also can be detected by use of ferric chloride.

Compounds failing to dissolve in water and aq. NaOH but soluble in dilute HCl generally contain nitrogen, since the most common type of organic base is the amine group. If the elemental analysis shows no nitrogen and the compound is soluble in 5% HCl, both the elemental analysis and the solubility tests should be checked. If an amine contains more than one aryl group, such as diphenylamine, the compound is no longer soluble in 5% HCl. The Hinsberg test (§378) is one of the most reliable tests for distinguishing the classes of amines.

[If a compound is soluble in 5% NaOH, testing it with 5% HCl is still advisable. This procedure will detect amphoteric compounds such as the amino acids.]

Compounds that are neither acidic nor basic but contain [sulfur or] nitrogen would be expected to be soluble in concentrated sulfuric acid; therefore this solubility test would serve no purpose. The most common members of this miscellaneous neutral group of compounds are di- or triarylamines [oximes, nitriles, nitro compounds, and amides.]

Most compounds containing an oxygen functional group or an easily sulfonated aromatic ring (such as mesitylene) and most olefins are soluble in 100% H_2SO_4.

The completely insoluble compounds are difficult to derivatize, since they do not contain a suitable functional group. In general, this class consists of unactivated arenes, halides, diaryl ethers [and alkanes].

Solubility behavior can contribute valuable information to the identification of an unknown. It should be noted, however, that these general rules of solubility have exceptions (e.g., phenol is soluble in water even though it has six carbon atoms). The student should therefore use solubility characteristics as tentative guidelines only.

Preliminary Report. To avoid loss of time through mistaken observation, it is recommended that at this point the student submit to the instructor his data on the physical constants and properties, elemental composition, solubility behavior, etc., of the unknown before attempting classification tests.

375. Classification Tests. The data in the preliminary report may suggest the type of functional group that is present. (A table of compounds from which your unknowns may be drawn, including information useful for their identification, is in §402.) More specific information concerning the presence or absence of common functional groups may be gained by employing a number of classification reagents. When each of the following tests is applied to an unknown compound, a control test should be carried out on a *known* compound to provide a basis for recognizing and distinguishing between positive and negative tests.

376. Weakly Acidic Compounds. A compound that dissolves in dilute aq. NaOH but not in dilute aq. $NaHCO_3$ is likely to be a phenol. Whether such a compound or a water-soluble compound is a phenol or whether it is an alcohol can be verified by the ceric nitrate test and the ferric chloride test which follow.

CERIC NITRATE TEST

$$\underset{\text{Yellow}}{(NH_4)_2Ce(NO_3)_6} + ROH \longrightarrow \underset{\text{Red, Brown, Green}}{(NH_4)_2Ce(OR)(NO_3)_5} + HNO_3$$

Water-soluble compounds are tested by adding, with shaking, 4 or 5 drops of the liquid unknown (4 or 5 drops of an aqueous solution of a solid) to 0.5 ml of the ceric nitrate reagent diluted with 3 ml of distilled water. Note the color. If the unknown is water insoluble, the ceric nitrate reagent (0.5 ml) should be dissolved in 3 ml of dioxane and the solution clarified with 3 or 4 drops of water if a precipitate forms. The unknown is then added as above.

A positive test for phenols is a brown to greenish brown precipitate in aqueous solution or a red to brown solution in dioxane. Alcohols give a red color. Larger compounds, particularly those with more than ten carbon atoms, produce too little color for the test to be useful.

The ceric nitrate reagent is prepared by dissolving, with warming, 200 g of $(NH_4)_2Ce(NO_3)_6$ in 500 ml of 2 *N* HNO_3.

FERRIC CHLORIDE TEST. Prepare a dilute aqueous solution of the unknown (0.03 g of a solid or 1 drop of a liquid in 1 ml of water) and add up to 3 drops of 1% aq. $FeCl_3$. (If the unknown is water insoluble, methanol may be used as a solvent.) Observe the color change at the instant of addition and contrast with the color produced by adding the same amount of ferric chloride solution to pure solvent. Most phenols [and enols] give a positive test (usually purple color); however, a number of them do not give colors.

[The observation has been made that some phenols that do not give color changes in aqueous or alcoholic solutions will produce color in chloroform, especially after a drop of pyridine is added. The procedure is the same as just described except that chloroform is used as the solvent for the unknown and for the 1% ferric chloride solution. The color change should be observed when 1 drop of pyridine is added to the test solution.]

377. Strongly Acidic Compounds. Compounds that dissolve in 5% aq. $NaHCO_3$ but give negative results to the ferric chloride and ceric nitrate tests (thereby eliminating highly acidic phenols) are most likely carboxylic acids [or sulfonic acids]. [The presence of a sulfonic acid would require a positive test for sulfur in the elemental analysis.] The presence of a carboxylic acid functional group may be ascertained as follows.

Place 0.1 g of the unknown in a 15-cm. test tube and add 6 drops of thionyl chloride. Immerse the test tube in boiling water and boil for 2 min. Add 1 ml of *n*-butyl alcohol and replace the tube in boiling water for 1 min. If a precipitate forms, add more *n*-butyl alcohol dropwise until it dissolves. Cool the tube and add 1 ml of water to hydrolyze any excess thionyl chloride. Add 1 ml of 5% hydroxylamine hydrochloride ($NH_2OH{\cdot}HCl$) in 95% ethanol and enough 5 *N* KOH in 80% ethanol to make the solution alkaline to litmus. Heat the mixture to boiling, then cool the tube. Acidify with dilute HCl and add 10% aq. $FeCl_3$ dropwise. A bluish red (magenta) color is indicative of a carboxylic acid.

$$RCO_2H \xrightarrow{SOCl_2} RCOCl \xrightarrow{n\text{-BuOH}} RCO_2Bu\text{-}n \xrightarrow{NH_2OH}$$

$$3RCONHOH \xrightarrow{FeCl_3} \left[RC\begin{matrix} \nearrow O \searrow \\ \searrow NH{-}O \nearrow \end{matrix} Fe \right]_3 + 3\ HCl$$

DETERMINATION OF NEUTRALIZATION EQUIVALENT. A useful determination for acidic compounds is the neutralization equivalent of the unknown substance, which is carried out as follows.

Either the student or his instructor should weigh out a sample of the unknown acid (0.2–0.5 g) on an analytical balance to the nearest milligram and dissolve it in 50 ml of freshly boiled water (or in an alcohol–water mixture if it is insoluble in water) contained in a 250-ml Erlenmeyer flask. A drop of phenolphthalein is added and the solution titrated with standard aq. NaOH to a faint pink color. The sample of acid taken should be large enough to require at least 20 ml of $0.1N$ NaOH. If an alcohol–water mixture is used as solvent, the solvent mixture should be titrated to a faint pink before dissolving the acid. The following equation is used to determine the equivalent weight:

$$\begin{aligned}\text{Normality of aq. NaOH} \times \text{ml of aq. NaOH} &= \text{milliequivalents of NaOH}\\ &= \text{milliequivalents of unknown acid}\\ &= \frac{\text{grams of unknown acid}}{\text{equivalent weight of acid}} \times 1000\end{aligned}$$

378. Basic Compounds. Basic compounds are those water-soluble compounds which turn litmus blue and those compounds which dissolve in 5% aq. HCl. The solubility classification depends upon (*a*) the molecular weight of the compound and (*b*) the availability of the lone pair(s) of electrons. For example, diphenylamine (ϕ_2NH) is a very weak base owing to the importance of resonance forms as shown below. In this manner, the electron pair is delocalized, making it less available for donation to an acid and causing the compound to be insoluble in dilute HCl.

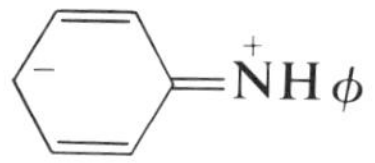

$$\text{(}^{-}\text{C}_6\text{H}_5\text{ ring)}{=}\overset{+}{N}H\phi$$

HINSBERG TEST. Only amines will be used as basic unknowns. The student, therefore, should determine whether the amine is primary, secondary, or tertiary via the Hinsberg test, which is outlined below.

Primary amine

$$RNH_2 + \phi SO_2Cl \xrightarrow{KOH} \phi{-}\overset{O}{\underset{O}{\overset{\|}{\underset{\|}{S}}}}{-}\overset{-}{N}R\ K^+ \xrightarrow{HCl} \phi{-}\overset{O}{\underset{O}{\overset{\|}{\underset{\|}{S}}}}{-}NHR$$

Soluble in base — Insoluble in acid

Secondary amine

$$R_2NH + \phi SO_2Cl \xrightarrow{KOH} \phi{-}\overset{O}{\underset{O}{\overset{\|}{\underset{\|}{S}}}}{-}NR_2 \xrightarrow{HCl} \phi{-}\overset{O}{\underset{O}{\overset{\|}{\underset{\|}{S}}}}{-}NR_2$$

Insoluble in both basic and acidic media

Tertiary amine

$$R_3N + \phi SO_2Cl \xrightarrow{KOH} \text{N.R.}^1 \xrightarrow{HCl} R_3\overset{+}{N}H\overset{-}{Cl}$$

Soluble in acid

[1]Tertiary amines are not completely inert to sulfonyl chlorides, but with reasonable control of conditions, the amine should be recovered unchanged [see C. R. Gambill, T. D. Roberts, and H. Schechter, *J. Chem. Ed.*, **49,** 287 (1972)].

To 0.5 ml (10 drops) of the unknown in a test tube, add 10 ml of 10% aq. KOH and 15 drops of benzenesulfonyl chloride. *Shake thoroughly* and note any reaction. Warm gently (do not boil) until the odor of benzenesulfonyl chloride cannot be detected. (The solution should still be basic at this point.) Cool the tube and note any solid precipitate or organic phase, the solubility of which in 5% aq. HCl should be tested for. Since certain high-molecular-weight or cyclic primary amines may form benzenesulfonamides that have only low solubility in aqueous alkali, the separated alkaline solution should be acidified with hydrochloric acid; a precipitate at this point indicates the sulfonamide of such a primary amine.

379. Neutral Compounds. Neutral compounds include alcohols, aldehydes, ketones, several carboxylic acid derivatives, hydrocarbons, alkyl and aryl halides, nitro compounds, etc. These several classes are treated individually in §§380–389.

380. Alcohols. The ceric nitrate test discussed under weakly acidic compounds (§376) may be used for the detection of alcohols. If the unknown is an alcohol, as indicated by this test and solubility tests, the student is urged to use the Lucas test to differentiate among primary, secondary, and tertiary alcohols.

Lucas Test

$$ROH + HCl(ZnCl_2) \longrightarrow RCl + H_2O$$

To 0.5 ml of the alcohol, add quickly 3 ml of the Lucas reagent (prepared by dissolving 16 g of anhydrous $ZnCl_2$ in 10 ml of conc. HCl with cooling). Close the tube with a cork and shake vigorously; then allow the mixture to stand. Observe the test tube for the first 5–10 min, then occasionally for 1 hr.
Note the time required for a second layer of the insoluble alkyl chloride to form (will show up as a milky liquid upon shaking).

Warning: *The Lucas reagent is highly corrosive and the stoppered test tube may build up pressure on shaking.*

Tertiary alcohols [also allyl alcohols, although these may appear elusive because of appreciable solubility of the chloride] react quite rapidly (usually in less than 1 min), secondary alcohols in 5–10 min, and primary alcohols react quite slowly; in some cases no reaction can be detected after 1 hr. If the alcohol has a high molecular weight, it may not be soluble in the reagent and the student may confuse the result for that of a tertiary alcohol. Consequently, it is important to check the list of possible unknowns carefully (this test is applicable only to alcohols bearing approximately six or fewer carbon atoms).

THE CHROMIC ANHYDRIDE TEST FOR PRIMARY AND SECONDARY ALCOHOLS. To 1 ml of acetone in a small test tube, add 1 drop of the liquid or 10 mg of the solid unknown. Then add 1 drop of chromic acid–sulfuric acid reagent (10 g of CrO_3 in 10 ml of conc. H_2SO_4 and carefully diluted with 30 ml of H_2O) and note any color change within 2 sec; run a control on acetone alone. A positive test consists in the production of an opaque suspension with a green to blue color. Tertiary alcohols give no visible reaction in 2 sec, the solution remaining orange in color. Disregard any changes after 2 sec.

381. Aldehydes and Ketones: 2,4-DINITROPHENYLHYDRAZONE. The presence of a carbonyl group (i.e., an aldehyde or ketone) may be detected by the use of 2,4-dinitrophenylhydrazine as indicated by the equation below, where R and R′ may be alkyl, aryl, or hydrogen.

$$R{-}\overset{\overset{\displaystyle O}{\|}}{C}{-}R' + NH_2NH{-}C_6H_3(NO_2)_2 \longrightarrow \underset{\textbf{I}}{\begin{matrix} R \\ R' \end{matrix}\!\!>C{=}N{-}NH{-}C_6H_3(NO_2)_2}$$

The hydrazone (I) may be yellow to red, depending upon the structure of the original carbonyl compound, and may be recrystallized and used as a derivative. (It should be cautioned that the reagent may react with other types of compounds, e.g., reactive esters, to give a spurious positive test).

Add a solution of 1 or 2 drops of the compound to be tested in 2 ml of 95% ethanol to 3 ml of 2,4-dinitrophenylhydrazine reagent. Shake vigorously, and, if no precipitate forms immediately, allow the solution to stand for 15 min.

In difficult cases it may be desirable to try the preparation of a dinitrophenylhydrazone in diethylene glycol dimethyl ether as described by Shriner, Fuson, and Curtin (ch. 10, p. 254). See §403.

The reagent is prepared by dissolving 3 g of 2,4-dinitrophenylhydrazine in 15 ml of concentrated sulfuric acid. This solution is then added, with stirring, to 20 ml of water and 70 ml of 95% ethanol. The solution is mixed thoroughly and filtered.

TOLLENS' TEST. The Tollens' test may be used to differentiate between aldehydes and ketones. This test should be run routinely on an unknown if a positive test was observed with 2,4-dinitrophenylhydrazine. Aldehydes will be oxidized by ammoniacal silver nitrate in accord with the equation given below, to give a silver mirror on the sides of the vessel, while ketones will not.

$$\overset{\overset{\text{O}}{\|}}{\text{RCH}} + 2\text{Ag(NH}_3)_2\text{OH} \longrightarrow \text{RCO}_2{}^-\text{NH}_4{}^+ + 2\text{Ag}\downarrow + 3\text{NH}_3 + \text{H}_2\text{O}$$

The Tollens' reagent is prepared by placing 2 ml of 5% aq. $AgNO_3$ solution in a *thoroughly clean* test tube and adding a drop of 10% NaOH solution. The brown precipitate of silver oxide formed will dissolve when a 2% ammonia solution is added drop by drop, with vigorous shaking. Avoid an excess of ammonia. Add a few drops of unknown compound and heat gently for 1 or 2 min. It is imperative that the student run a control on this test, as certain ketones and other classes of compounds will deposit a silver mirror particularly if heated long enough.

IODOFORM TEST

$$\overset{\overset{\text{O}}{\|}}{\text{RC}}\text{CH}_3 + \text{I}_2 + \text{NaOH} \longrightarrow \text{RCO}_2\text{Na} + \underset{\text{Yellow}}{\text{CHI}_3\downarrow}$$

A positive test is indicative of the structures given below and involves the appearance of yellow crystals of iodoform.

$$\overset{\overset{\text{O}}{\|}}{\text{RC}}\text{CH}_3 \qquad \text{R—}\overset{\overset{\text{OH}}{|}}{\text{C}}\text{HCH}_3 \qquad (\text{R} = \text{alkyl, aryl, or H})$$

In a small Erlenmeyer flask dissolve 0.5 ml of the unknown in 5 ml of water (if the unknown is insoluble in water, dissolve it in dioxane) and add 5 ml of 10% NaOH. Add dropwise, with shaking, a 10% solution of I_2 in aq. KI until a definite brown color persists. If a precipitate of iodoform does not form after 5 min, warm the solution to 60° in a beaker of water. The reagent is prepared by dissolving 20 g of KI and 10 g of I_2 in 100 ml of H_2O.

If the brown color is discharged, add more I_2/KI solution until the brown color persists for 2 min. Add a few drops of NaOH solution to remove the excess I_2, dilute with water (by filling the test tube) and allow to stand for 10 min.

Iodoform is a yellow crystalline solid, mp 119°.

382. Esters: HYDROXAMIC ACID TEST. Esters react with hydroxylamine to form hydroxamic acids, which form colored ferric chelates with ferric chloride (see equations, §377). It is particularly important that a preliminary test be run for confirmation in this classification test since many compounds give colored complexes with ferric chloride. The preliminary test should be run as follows. Dissolve 1 drop of a liquid or 0.05

g of a solid unknown in a minimum of ethanol (~1 ml) in a small test tube. Add to this 1 ml of 1 *N* HCl and then 1 drop of 5% aq. $FeCl_3$. If a red, orange, green, blue, or violet color develops and persists, the hydroxamic acid test should not be used for the detection of the acyl group.

Dissolve 1 drop of liquid or 0.05 g of solid in 1 ml of 5% hydroxylamine hydrochloride solution in 95% ethanol, and add to this 0.2 ml of 6 *N* NaOH. Heat this mixture to boiling, then cool slightly, and add 2 ml of 1 *N* HCl. Any cloudiness at this point should be removed by adding about 2 ml of 95% ethanol. Add 1 drop of 5% aq. $FeCl_3$ and observe the color. The color produced (if any) often disappears, so more ferric chloride solution should be added, if necessary, until the color persists. A positive test is noted if the color produced is a deep red to purple (shades of magenta) as compared to the yellow color observed in the preliminary test above.

Some amides also give a positive test with this procedure, but the color is usually a light magenta. For a modification of this procedure applicable to amides, see §385.

Saponification

$$RCO_2R' + {}^-OH \longrightarrow RCO_2^- + R'OH$$

This classification test should be used only after aldehydes have been excluded by previously listed tests since condensation reactions can occur with those aldehydes which have α-hydrogens and the Cannizzaro reaction with those with no α-hydrogens. A detailed procedure for the hydrolysis of esters (§396) and derivatization of both components (acid and alcohol portion) is given under preparation of derivatives.

383. Unsaturated Compounds. Unsaturation in a compound can be detected by the use of two common reagents, bromine in carbon tetrachloride and potassium permanganate solution. Both reagents are decolorized by simple olefins and simple acetylenes.

Bromine in Carbon Tetrachloride

$$>C=C< + Br_2 \longrightarrow >C(Br)-C(Br)<$$

$$-C\equiv C- + 2Br_2 \longrightarrow -CBr_2-CBr_2-$$

Add, dropwise with shaking, a 5% bromine-in-carbon-tetrachloride solution to 0.1 g of unknown compound dissolved in 2 ml of carbon tetrachloride, until the bromine color persists. The decolorization of the bromine solution *without hydrogen bromide evolution* constitutes a positive test for unsaturation. Detection of hydrogen bromide can be accomplished by blowing gently across the mouth of the test tube, a visible fog indicating the presence of hydrogen bromide, or by holding moist litmus paper at the mouth of the test tube.

Some unsaturated compounds substituted by electronegative groups react very slowly or not at all, so the following test should be run in conjunction to verify the absence or presence of unsaturation.

POTASSIUM PERMANGANATE SOLUTION (BAEYER TEST)

$$3\ \rangle C{=}C\langle\ + 4H_2O + 2KMnO_4 \longrightarrow 3\ \rangle\underset{OH}{\underset{|}{C}}{-}\underset{OH}{\underset{|}{C}}\langle\ + 2KOH + 2MnO_2 \downarrow$$

$$R{-}C{\equiv}C{-}R' + 2KMnO_4 \longrightarrow RCO_2K + R'CO_2K + 2MnO_2 \downarrow$$

Add, dropwise with shaking, 2% aq. $KMnO_4$ to 0.1 g of unknown dissolved in 2 ml of ethanol until the purple color of permanganate persists. A brown precipitate of MnO_2 will appear as well as the dissipation of the purple permanganate color, if oxidation has occurred. Care must be taken in the interpretation of results since all easily oxidizable compounds decolorize this reagent.

384. Neutral Compounds Containing Nitrogen [and/or Sulfur]. These compounds include amides, nitriles, and nitro compounds, as the most common members, and can be distinguished from the preceding neutral compounds by a positive test for nitrogen (§373).

385. Amides and Nitriles: HYDROXAMIC ACID TEST. A positive test is observed with amides and nitriles when the hydroxamic test procedure for esters is modified.

$$RCN + H_2NOH \longrightarrow R\overset{NH}{\overset{\|}{C}}{-}NHOH \xrightarrow{FeCl_3} \left[R{-}C\overset{\diagup\!\!\diagup N(H)\searrow}{\underset{H\diagup N{-}O\diagup}{}}\right]_3 Fe + 3\ HCl$$

$$R\overset{O}{\overset{\|}{C}}NH_2 + H_2NOH \longrightarrow R\overset{O}{\overset{\|}{C}}{-}NHOH \xrightarrow{FeCl_3} \left[R{-}C\overset{\diagup\!\!\diagup O\searrow}{\underset{H\diagup N{-}O\diagup}{}}\right]_3 Fe + 3\ HCl$$

A preliminary test also should be run in this instance; propylene glycol should be used as the solvent and the addition of hydrochloric acid omitted. Again, if a persistent red to violet color is obtained, the hydroxamic acid test should not be used.

Prepare a mixture of 2 ml of 1 M hydroxylamine hydrochloride in propylene glycol, 1 drop of liquid unknown (0.05 g of solid), dissolved in a minimum of propylene glycol, and 1 ml of 1 N KOH, heat to a gentle boil for 2 min and allow to cool to room temperature. Then add 5% $FeCl_3$ in 95% ethanol dropwise but not more than 1 ml. A positive test is indicated by a red to violet color, yellow is negative, and brown colors and precipitates are inconclusive.

386. Nitro Compounds: FERROUS HYDROXIDE TEST

$$\underset{\text{Blue-green}}{RNO_2 + 6Fe(OH)_2} + 4H_2O \longrightarrow RNH_2 + \underset{\text{Brown (rust colored)}}{6Fe(OH)_3}$$

Both aliphatic and aromatic nitro compounds are reduced by ferrous hydroxide to amines, normally within a minute. A change in the color of the precipitate from blue-green to reddish brown is considered a positive test, but a slight darkening of the ferrous hydroxide or a greenish precipitate is negative.

To 1 drop of liquid (or about 0.02 g of solid) unknown in a test tube, add 1.5 ml of ferrous sulfate reagent and 1 ml of alcoholic potassium hydroxide solution. Natural gas is bubbled through this solution for about 30 sec by placing a glass tube to the bottom of the test tube. (This removes as much air as possible from the solution.) Stopper the tube with a cork, shake, and then observe the color after 1 min.

The two reagents that are needed for this test are prepared as follows. Ferrous sulfate reagent: ferrous ammonium sulfate (15 g) is dissolved in 300 ml of distilled water, which has been boiled recently to drive out dissolved oxygen; concentrated sulfuric acid (2 ml) and an iron nail are added to retard oxidation by air. Potassium hydroxide solution: potassium hydroxide (17 g) is dissolved in 18 ml of freshly boiled, distilled water; this solution is then added to 135 ml of 95% ethanol.

387. Other Neutral Compounds. Other important classes of neutral compounds include the alkyl and aryl halides (indicated by a positive test for halogen, §373) and the unreactive hydrocarbons.

388. Alkyl Halides. Two reagents are useful in determining the substitution on the carbon containing the halogen atom in an alkyl halide: alcoholic silver nitrate solution and sodium iodide in acetone.

ALCOHOLIC SILVER NITRATE SOLUTION

$$RX + AgNO_3 \xrightarrow{EtOH} RONO_2 + AgX\downarrow$$

This reaction is considered to proceed via an S_N1 mechanism. Two general orders of reactivity observed are RI > RBr > RCl and benzyl, allyl > tertiary > secondary > primary. Simple aromatic and vinyl halides fail to react, while ionic halides (e.g., amine salts) react instantly.

Add 1 drop of unknown halogen compound to 2 ml of a solution of 2% $AgNO_3$ in ethanol. If no reaction is observed after 5 min, heat the solution to boiling. If a precipitate has formed during the heating, add 2 drops of 5% aq. HNO_3 to the solution. Silver salts of organic acids are dissolved by nitric acid, but silver halides remain insoluble.

The following halogen compounds give a precipitate within 5 min without heating: allyl, benzyl, and tertiary halides; alkyl iodides and monobromoalkanes; 2,4,6-trinitroaryl halides. Primary and secondary chlorides, 2,4-dinitrophenyl halides, *gem*-dibromides, and *gem*-tribromides give a precipitate with a silver nitrate solution only after boiling.

SODIUM IODIDE IN ACETONE

$$RCl + NaI \xrightarrow{\text{acetone}} RI + NaCl\downarrow$$

$$RBr + NaI \xrightarrow{\text{acetone}} RI + NaBr\downarrow$$

The utility of this test lies in the fact that sodium chloride and bromide are insoluble in acetone, while sodium iodide is soluble. The nucleophilicity of iodide ion promotes an S_N2 displacement process, causing the observed relative reactivity, primary > secondary > tertiary > aryl and vinyl.

Add 2 drops of the unknown chlorine or bromine compound (or 0.05 g of solid dissolved in a minimum amount of acetone) to 1 ml of sodium iodide–acetone solution (8 g of sodium iodide dissolved in 50 ml of acetone). Shake the resulting mixture and then allow to stand for 4 min. If no precipitate has formed in this time, heat the solution in a water bath for 8 min, then cool to room temperature, noting any precipitate that forms.

Primary bromides give a precipitate at room temperature within 4 min. Primary and secondary chlorides, and secondary and tertiary bromides react within 8 min at 50°. Tertiary chlorides do not react within a reasonable time, while simple aryl and vinyl halides do not react at all.

389. Hydrocarbons and Aryl Halides. Compounds that have been found to be inert toward all the test reagents include the alkanes and less reactive aromatic compounds. These two types of compounds can be distinguished by taking advantage of the ability of the aromatic ring to undergo electrophilic substitution.

FUMING SULFURIC ACID ($H_2SO_4 \cdot SO_3$)

$$ArH + SO_3 \xrightarrow{H_2SO_4} ArSO_3H$$

Aromatic compounds will be sulfonated by this reagent and dissolve, while the aliphatic hydrocarbons will float on top. Carefully add a small amount (0.5 ml of a liquid, 0.4 g of a solid) of the unknown compound to a clean, dry test tube containing 2 ml of fuming sulfuric acid. Shake the mixture vigorously, and then allow it to stand for a few minutes. **Warning:** *Avoid any direct contact with this highly corrosive reagent.*

ALUMINUM CHLORIDE WITH CHLOROFORM

$$3ArH + CHCl_3 \xrightarrow{AlCl_3} Ar_3CH \text{ (converted to colored complexes)} + 3HCl$$

This test offers a means of distinguishing among the different aromatic systems on the basis of color formations. Alkanes do not react and give either no color or only a pale yellow color. Characteristic colors produced are the following: benzene and its homologs, orange to red; aryl halides, orange to red; biphenyl, purple; naphthalene, blue; anthracene, green; and phenanthrene, purple.

Dissolve 3 drops of a liquid unknown (0.1 g of a solid) in 2 ml of dry chloroform in a clean dry test tube. After shaking vigorously, incline the test tube to wet the walls with the mixture. Add a small portion (about 0.5 g) of *anhydrous* aluminum chloride so that some of the solid falls on the moistened sides of the test tube. Note the color in both the solution and on the wall.

390. Preparation of Derivatives. When the preliminary tests have been completed, the identity of the unknown will have been narrowed to but a few possibilities. The final decision will generally depend on the preparation of a suitable derivative and a comparison of its melting point with the known melting points of the corresponding derivatives of the other compounds under consideration.

Suitable derivatives for the purpose of identification generally are stable solids that melt in the range 50–250° and are readily prepared in good yields. For the purposes of this course, the directions given here will probably prove adequate. The texts by Shriner, Fuson, and Curtin and by Vogel listed in §403 are recommended if additional information is required. The original chemical literature should be consulted for further details when necessary.

Generally, there will be several derivatives recorded for a single substance. One makes that derivative whose properties are distinctly different from those of the corresponding derivatives of the other possibilities for the given unknown. For example, in attempting to decide between 1-propanol (bp 97°) and 2-butanol (bp 99°) one finds that their 3,5-dinitrobenzoates melt at 74° and 75°, respectively, too close for a clear-cut distinction; on the other hand, the α-naphthylurethans melt at 80° and 97°, respectively. In this case the α-naphthylurethan is clearly the derivative of choice. When more than two possibilities are under consideration, it may be necessary to prepare more than one derivative in order to have a clear-cut distinction among them.

The quantities of reagents employed should be measured or weighed (at least approximately). The weighing of small solid quantities may conveniently be carried out on a triple beam balance; the measuring of liquids may be accomplished by means of a 1-ml graduated medicine dropper. If refluxing is required, it may be done in a test tube or small Erlenmeyer flask attached to a reflux condenser.

The precipitated product should be filtered *by suction* and washed. Drying may be achieved by sucking air through the washed precipitate for 15 min or by allowing the precipitate to remain on a watch glass or porous plate.

For recrystallization, small-scale tests should be made to find a suitable solvent, that is, a solvent in which the substance is not too soluble and in which it has a marked temperature coefficient of solubility (see §§98–106). Remember that solvents suggested in the general procedure sometimes are not suitable for specific compounds.

The supposed derivative should not be taken for granted. It may deserve confirmation, as by mixed melting points, to show that the original unknown substance or derivatizing agent has not simply been recovered unchanged.

Procedures will be described for the preparation of solid derivatives of the following classes of compounds:

1. Acids (carboxylic)
2. Alcohols
3. Aldehydes
4. Amides
5. Amines
6. Esters
7. Halides (alkyl)
8. Hydrocarbons (aromatic)
9. Ketones (see aldehydes)
10. Nitriles (see amides)
11. Nitro compounds
12. Phenols

391. Acids (Carboxylic): *p*-BROMOPHENACYL ESTERS. *p*-Bromophenacyl bromide is a useful reagent for forming characteristic derivatives of many carboxylic acids. The phenacyl bromide is a comparatively active alkylating agent, so that esters are readily formed from the sodium salts of acids.

$$RCOONa + Br\text{—}C_6H_4\text{—}\overset{\overset{\large O}{\|}}{C}\text{—}CH_2Br \longrightarrow R\text{—}\overset{\overset{\large O}{\|}}{C}OCH_2\text{—}\overset{\overset{\large O}{\|}}{C}\text{—}C_6H_4\text{—}Br + NaBr$$

Carefully neutralize 1 g of the unknown acid, dissolved or suspended in 5 ml of water, with 10% sodium hydroxide solution, using 1 drop of phenolphthalein as an indicator. Make the solution slightly acidic to litmus by adding a small amount of the unknown acid. (Hydrolysis of the phenacyl halide to the phenacyl alcohol occurs if the solution is alkaline.) Add to this solution 10 ml of ethanol and 1 g of *p*-bromophenacyl bromide. **Warning:** *Phenacyl halides are lachrymatory.* Avoid the use of an excess of the phenacyl halide, since it cannot easily be removed from the ester. Reflux the solution at least 1 hr. Reflux for 2 hr if dibasic (two CO_2H units per molecule) and 3 hr if tribasic with an appropriately larger quantity of the phenacyl halide. If a solid precipitates before refluxing is completed, add just enough ethanol to redissolve the solid. Allow the solution to cool to room temperature at which time the ester separates. The separation of the ester from solution may be slow, so sufficient time should be allowed for the oils that often separate to change to the crystalline state. The ester is then recrystallized from aqueous ethanol or acetone.

ANILIDES. Most carboxylic acids yield water-insoluble anilides, which serve to identify them.

$$RCO_2H + SOCl_2 \longrightarrow RCOCl \xrightarrow{\phi NH_2} R\overset{\overset{\displaystyle O}{\|}}{C}{-}NH\phi$$

Mix 1 g of the dry acid or its sodium salt with 4 ml of thionyl chloride and 1 drop of pyridine in a test tube under a reflux condenser equipped with a calcium chloride drying tube. Heat the mixture in a boiling water bath for 30 min *in the hood* to complete formation of the acid chloride. Replace the condenser with a bent tube and distill off the excess thionyl chloride from a water bath, condensing it in a test tube cooled in an ice bath. After cooling the undistilled mixture, add 2 g of aniline dissolved in 30 ml of benzene and warm that mixture on the steam bath for 2 min with shaking. Decant the benzene solution into a separatory funnel and wash with excess 5% HCl to remove any remaining aniline, followed by 3–5 ml of water. Evaporate the benzene and recrystallize the anilide from water, ethanol, or benzene–petroleum ether (bp 60–80°).

The *p*-toluidide derivative is prepared by substituting *p*-toluidine for aniline.

392. Alcohols. Urethans are the most readily prepared derivatives of water-insoluble alcohols and phenols, while the benzoate esters are generally more satisfactory as derivatives when the alcohols are water soluble or contain trace amounts of water. Tertiary alcohols tend to be dehydrated by isocyanates and may fail to yield urethans. Preparation of ester derivatives from tertiary alcohols is also difficult, since these alcohols are converted to the corresponding chlorides by acyl chlorides. However, esters can be prepared when the reaction is run in the presence of a base or a metal that removes the HCl as fast as it forms.

α-NAPHTHYLURETHANS

$$\text{ROH} + \alpha\text{-C}_{10}\text{H}_7\text{N=C=O} \longrightarrow \alpha\text{-C}_{10}\text{H}_7\text{NH}-\overset{\text{O}}{\overset{\|}{\text{C}}}-\text{OR}$$

In a flame-dried test tube fitted with a drying tube containing calcium chloride pellets, place a mixture of 1 g of the *anhydrous* alcohol or phenol and 0.5 ml of α-naphthyl isocyanate. (To insure that the alcohol is anhydrous, it should be dried over a suitable drying agent before use, e.g., anhydrous sodium or magnesium sulfates.) Anhydrous pyridine (2 or 3 drops) should be added to catalyze the reaction if the unknown is a phenol. Shake the reaction mixture; if an immediate reaction does not take place, warm on a steam bath for 5 min. An ice bath should be used to cool the reaction mixture; if necessary, scratch the walls of the test tube with a stirring rod to induce crystallization. Collect this solid, dissolve in 5–10 ml of hot petroleum ether (bp 100–120°), and filter hot to remove any di-α-naphthylurea. (The di-α-naphthylurea is a by-product arising from any water in the system, mp 297°.) Cool the filtrate in an ice bath and filter the crystals. If no crystals form, evaporate half the solution and re-cool in ice.

The phenylurethans may be prepared by substituting phenyl isocyanate in the procedure above. Generally, the naphthyl derivative is preferred because (*a*) naphthyl isocyanate is less lachrymatory and is less easily decomposed by water and (*b*) the melting points are generally higher than those of the corresponding phenylurethans.

3,5-DINITROBENZOATES

$$3,5\text{-}(O_2N)_2C_6H_3\overset{\text{O}}{\overset{\|}{\text{C}}}\text{Cl} + \text{ROH} \longrightarrow 3,5\text{-}(O_2N)_2C_6H_3\overset{\text{O}}{\overset{\|}{\text{C}}}\text{OR} + \text{HCl}$$

The pyridine method of preparation has been observed to give the desired derivative readily. Add about 0.5 g of 3,5-dinitrobenzoyl chloride to a mixture of 1 ml of alcohol or phenol and 3 ml of anhydrous pyridine in a test tube. Allow the initial reaction to subside and then gently heat the mixture for a minute using a low flame. Without cooling, pour this reaction mixture into 10 ml of water and stir vigorously. Collect the precipitate, then return to the test tube and wash with 5 ml of 5% aq.

Na_2CO_3. (The solid should be broken up with a stirring rod to give maximum contact with the wash solution.) Filter the precipitate and recrystallize from alcohol.

Benzoates and *p*-nitrobenzoates can also be prepared by the above procedure.

393. Aldehydes and Ketones: 2,4-DINITROPHENYLHYDRAZONES

$$R-\overset{O}{\overset{\|}{C}}-R' + H_2N-NH-C_6H_3(NO_2)_2 \longrightarrow \begin{matrix}R\\R'\end{matrix}\!\!>C=N-NH-C_6H_3(NO_2)_2 + H_2O$$

(R′ may be hydrogen)

These yellow to red solid derivatives are conveniently prepared by adding 15 ml of the 2,4-dinitrophenylhydrazine reagent (preparation given in §381) to 0.5 g of the unknown carbonyl compound dissolved in 20 ml of 95% ethanol, heating to reflux and then cooling to room temperature. If no crystallization occurs after 10 to 15 min, allow the solution to stand overnight or dilute the solution with 10% sulfuric acid. Filter the precipitate and recrystallize from alcohol or an alcohol–water mixture.

In some cases preparation of the phenylhydrazone may be preferred to the dinitrophenylhydrazone for identification purposes. Add 0.5 g of the carbonyl compound to a solution of 1 ml of phenylhydrazine in 0.5 ml of glacial acetic acid and 0.5 ml of water. Heat the mixture to boiling and, after cooling, shake thoroughly to hasten the crystallization of the product. After it has completely precipitated, filter the product and recrystallize from aqueous alcohol.

OXIMES

$$\begin{matrix}R\\R'\end{matrix}\!\!>C=O + H_2NOH \longrightarrow \begin{matrix}R\\R'\end{matrix}\!\!>C=N-OH + H_2O$$

(R′ may be hydrogen)

Add 2 ml of 10% aq NaOH to a solution of 0.5 g of hydroxylamine hydrochloride in 3 ml of water. (The hydroxylamine base, a substance that does not keep well, is freed by the sodium hydroxide.) Immediately add the unknown aldehyde or ketone; if it is not completely soluble in this basic solution, a small quantity of ethanol should be added to just give a clear solution. After 10 min of heating on a water bath with shaking, place the solution in an ice bath and scratch the walls of the test tube with a stirring rod to induce crystallization. (A few milliliters of water may be added to the cold solution to promote crystallization.) Collect the oxime product by filtration and recrystallize from water or an ethanol–water mixture.

394. Amides and Nitriles. Amides and nitriles are hydrolyzed under either acidic or basic conditions to the acid and amine components, from which suitable solid derivatives are prepared to fully characterize the compound.

BASIC HYDROLYSIS

$$\text{RCN or } \text{R}\overset{\overset{\displaystyle O}{\|}}{\text{C}}\text{NR}'_2 \xrightarrow[H_2O]{OH^-} \text{RCO}_2^- + \text{NH}_3 \text{ or } \text{HNR}'_2$$
$$(\text{R}' = \text{alkyl, aryl, or H})$$

Boil a mixture of 1 g of the unknown amide or nitrile and 10 ml of 10% aq. NaOH in a distilling flask, and collect ammonia or any volatile amine in dilute aq. HCl. Neutralize this acidic solution with dilute aq. NaOH and prepare a suitable solid amine derivative (§395). If the amine is not volatile, extract the basic hydrolysis solution with ether, dry the ether extracts over potassium hydroxide pellets, filter, and evaporate the solvent. This residue is then used to prepare a solid amine derivative.

Make the basic hydrolysis solution, which remains after the amine has been removed, slightly acidic with hydrochloric acid. Collect the acid by filtration if a precipitate has formed, or extract with ether or dichloromethane if no solid has precipitated. Removal of the solvent by evaporation leaves the acid component, which is characterized by preparing a suitable solid derivative (§391).

ACIDIC HYDROLYSIS

$$\text{RCN or } \text{R}\overset{\overset{\displaystyle O}{\|}}{\text{C}}\text{NR}'_2 \xrightarrow[H_2O]{H^+} \text{RCO}_2\text{H} + \text{NH}_4^+ \text{ or } \text{H}_2\overset{+}{\text{N}}\text{R}'_2$$

Heat a mixture of 1 g of the unknown nitrile or amide and 10 ml of concentrated sulfuric or hydrochloric acid in a water bath at 50° for about 30 min and then carefully pour into a flask containing approximately 30 ml of water. Reflux the resulting aqueous solution gently for 30 min to 2 hours; then allow to cool. Isolate the acid by filtration if it is a solid, or extract from the acidic hydrolysis solution with ether or dichloromethane if it is a liquid. A suitable solid derivative is then prepared for the acid (§391).

Make the acidic hydrolysis solution, which remains after the acid has been removed, alkaline with sodium hydroxide to liberate the amine. Distill the amine from the aqueous solution or extract with ether if it is not volatile enough to distill. The amine is then derivatized by the procedures described in the next section.

395. Amines. Suitable derivatives for primary and secondary amines are the benzamides and benzenesulfonamides. Tertiary amines cannot form these derivatives; therefore the methyl iodide adducts or picrates are prepared.

BENZAMIDES

$$\mathrm{RR'NH} + \mathrm{C_6H_5}\overset{\mathrm{O}}{\overset{\|}{\mathrm{C}}}\mathrm{Cl} \longrightarrow \mathrm{C_6H_5}\overset{\mathrm{O}}{\overset{\|}{\mathrm{C}}}\mathrm{-NRR'} + \mathrm{HCl}$$

(R′ may be hydrogen)

A number of procedures are available for the preparation of benzamides, but only two will be presented here.

Pyridine method. Dissolve the amine unknown (0.5 g) in 5 ml of anhydrous pyridine and 10 ml of dry benzene. Add 0.5 ml of benzoyl chloride to this solution dropwise. After heating this solution for 30 min in a water bath at 60–70°, pour into 100 ml of water. Separate the layers using a separatory funnel, extract the aqueous layer once with 10 ml of benzene; and then combine the two benzene layers. After washing with small amounts of water and then 5% aq. Na_2CO_3, dry the benzene extracts over a small amount of magnesium sulfate for a short time. Filter off the magnesium sulfate and reduce the volume of the benzene solution to 3 or 4 ml by evaporation on a steam bath. Add about 20 ml of hexane (petroleum ether) to this mixture to cause crystallization. Filter the solid, wash with a small portion of hexane on the filter, and then recrystallize from one of the following solvents: ethanol, aqueous ethanol, cyclohexane–hexane mixtures, or cyclohexane–ethyl acetate mixtures.

Schotten–Baumann method. Add 1 ml of benzoyl chloride, dropwise with shaking to a solution of 0.5 g of the unknown amine in 10 ml of 5% aq. NaOH. Filter the resulting precipitate, wash with water, and recrystallize from water or aqueous alcohol.

BENZENESULFONAMIDES

$$\mathrm{RR'NH} + \mathrm{C_6H_5SO_2Cl} \longrightarrow \mathrm{C_6H_5SO_2NRR'} + \mathrm{HCl}$$

(R′ may be hydrogen)

Schotten–Baumann method. Treat 1.0 g of the unknown amine with 20 ml of 10% aq. NaOH and add 3 g of benzenesulfonyl chloride in small portions with constant shaking. Warm gently until all of the excess sulfonyl choride is hydrolyzed (§378). Acidify the solution with dilute hydrochloric acid and filter off the sulfonamide. This solid is then recrystallized from alcohol or aqueous alcohol.

Weakly basic amines, such as, the nitroanilines, generally react so slowly with the acid chloride that most of the sulfonyl chloride is hydrolyzed before a reasonable amount of sulfonamide is formed; indeed, *o*-nitroaniline gives little or no sulfonamide under the conditions of the Hinsberg test (§378). Excellent results may be obtained in such cases by carrying out the following procedure.

Pyridine method. Reflux a mixture of 1 g (1 ml) of the amine, 2–3 g of benzenesulfonyl chloride, and 6 ml of pyridine in a test tube equipped with a reflux condenser for 30 min. Pour the reaction mixture into 10 ml of cold water and stir until the product crystallizes. Filter off the solid and recrystallize it from alcohol or aqueous alcohol.

METHIODIDES

$$RR'R''N + CH_3I \longrightarrow \left[R{-}\overset{R'}{\underset{R''}{\overset{|}{\underset{|}{N^+}}}}{-}CH_3 \right] I^-$$

Allow a mixture of 0.5 g of the amine and 0.5 ml of colorless methyl iodide to stand in a test tube for 5 min. If no reaction occurs, warm gently in a water bath for 5 min and cool in an ice bath. Scratch the walls of the test tube with a glass rod to induce crystallization. Recrystallize the solid product from absolute alcohol, ethyl acetate, acetone, glacial acetic acid, or alcohol–ether.

PICRATES

$$R{-}\overset{R'}{\overset{|}{N}}{-}R'' + 2,4,6\text{-}(O_2N)_3C_6H_2{-}OH \longrightarrow R{-}\overset{R'}{\underset{H}{\overset{|}{\underset{|}{N^+}}}}{-}R''\ {}^{-}O{-}C_6H_2(NO_2)_3\text{-}2,4,6$$

Picrates are prepared by adding a solution of the unknown amine (0.3–0.5 g) in 10 ml of 95% ethanol to 10 ml of a saturated solution of picric acid in 95% ethanol and heating to boiling. (If solution of the amine in the alcohol was incomplete after thorough mixing, the solid should be filtered off and the filtrate used in the picrate preparation.) Cool the reaction mixture slowly to room temperature, filter the solid, and recrystallize from ethanol.

If difficulty is experienced in the above procedure, an alternate preparation may be found in Vogel (1956), 422.

396. Esters

$$RCO_2R' + KOH \longrightarrow RCO_2K + R'OH$$

An ester is characterized by derivatization of both acid and alcohol portions. Only isolation of these components by basic hydrolysis is discussed here, since preparation of alcohol and acid derivatives has already been described (§§391 and 392). The following procedure can be used for all esters except those which have a very high-boiling alcohol component. (High-boiling alcohols cannot be distilled from the reaction mixture in a pure state since diethylene glycol boils at 244°.)

Introduce a mixture of diethylene glycol (3 ml), potassium hydroxide pellets (0.6 g), and water (0.5 ml) into a small round-bottom flask; heat over a low flame until the solution is homogeneous; and then allow to cool. Add 1 ml of the unknown ester and two boiling chips to this solution. Shake the resulting mixture thoroughly and then boil under a water-cooled condenser until the ester layer has dissolved or one liquid and one solid phase are present (usually 3–5 min). Allow the reaction flask to cool and then equip for a simple distillation, collecting the alcohol on strong heating. This distillate is then used to prepare the alcohol derivative, normally without further purification or drying (§392).

The acid portion (as the potassium salt) of the ester remains in the residue left after distillation. Add 10 ml of water to the cool residue, which may be a suspension, agitate the resulting mixture thoroughly, and acidify with 6 *N* H_2SO_4. If the acid precipitates, it should be collected on a filter and washed with a couple of 3-ml portions of cold water. If no precipitate has formed after a short period of standing, the aqueous solution should be extracted with ether or dichloromethane. These extracts should be dried over anhydrous magnesium sulfate; the drying agent filtered off, and the solvent removed by evaporation on a steam bath. A solid acid can be used as a derivative directly by determining its melting point, but a suitable solid derivative must be prepared for a liquid acid (§391).

397. Halides. These compounds are relatively inert, but can be made quite reactive when the Grignard reagent (RMgX) is prepared from them. The Grignard reagent is then treated with an isocyanate (usually phenyl or α-naphthyl) to form an amide after hydrolysis.

$$RX + Mg \longrightarrow RMgX \xrightarrow{ArN{=}C{=}O} R\overset{\overset{\displaystyle OMgX}{|}}{C}{=}N{-}Ar \xrightarrow{H_2O} R\overset{\overset{\displaystyle O}{\|}}{C}{-}NHAr$$

Prepare the Grignard reagent as previously described (§§226 and 227), using reduced quantities of magnesium (0.3 g) and the halide (1 ml). Add the ether solution (diethyl ether or tetrahydrofuran) of the Grignard reagent slowly, with shaking, to a solution of 0.5 ml of phenyl (or α-naphthyl) isocyanate in 10 ml of dry ether. Allow this mixture to stand for 15 min and then add 20 ml of water containing 1 ml of conc hydrochloric acid. After shaking thoroughly, separate the layers. Dry the ether layer over magnesium sulfate, filter, and evaporate the ether on a steam bath. The residue is recrystallized from ether, petroleum ether, or methanol.

398. Hydrocarbons (Aromatic). The preparation of derivatives of aromatic hydrocarbons takes advantage of the reactivity of the aromatic ring in one instance and the inertness to oxidation in another.

NITRATION

$$ArH + HNO_3 \xrightarrow{H_2SO_4} ArNO_2 + H_2O$$

Warning: *Many compounds react violently with the nitrating reagent or may produce dangerously explosive products.*

Mix 1 g of the unknown compound with 4 ml of conc. H_2SO_4. Add dropwise 4 ml of fuming HNO_3, shaking after each addition. Heat the resulting mixture for 5 min in a water bath held at about 80° and then pour onto 25 g of ice. Collect the precipitate and recrystallize from ethanol. Dinitro derivatives may be produced by this procedure; in fact, the more highly alkylated hydrocarbons (e.g., mesitylene and the xylenes), may yield trinitro derivatives. To minimize polynitration, this procedure may be modified, particularly for easily nitrated compounds, by the use of conc. instead of fuming HNO_3 and of lower temperature (45–50°).

OXIDATION (SIDE CHAIN). This procedure is only used for compounds having an alkyl side chain that can be oxidized to yield the corresponding benzoic acid.

The oxidizing solution is prepared by dissolving 7 g of sodium dichromate in 15 ml of water and slowly adding, with shaking, 10 ml of conc. H_2SO_4 (cooling if necessary). Add the unknown compound (2–3 g) to the oxidizing solution along with a couple of boiling chips and mix well. Attach a reflux condenser to the flask and heat gently until the reaction starts. Remove the flame and cool the reaction mixture if necessary. Return the flame to the flask after the initial reaction has subsided and reflux for 1–2 hr. After cooling for a few minutes, pour the reaction mixture into 25 ml of water and collect the precipitate. Break up the precipitate, place in 20 ml of 5% H_2SO_4, and heat on a steam bath with stirring. (This dissolves the chromium salts.) Cool the acid solution and filter. Wash the solid that is collected with two 10-ml portions of cold water and then dissolve in 20 ml of 5% aq. NaOH. Filter this solution and reprecipitate the acid by pouring into 25 ml of 10% H_2SO_4 with stirring. Collect the resulting solid, wash with cold water, and recrystallize from benzene or ethanol.

399. Nitro Compounds. Nitro compounds are reduced to primary amines by active metals in acid solution. The amine is then derivatized by the procedure described in §395.

$$RNO_2 \xrightarrow[HCl]{Sn} RNH_2$$

Add 1.5 g of the unknown nitro compound to a flask containing 4 g of granulated (mossy) tin and, if the nitro compound is insoluble in water, add 5 ml of 95% ethanol. Then equip the flask with a reflux condenser and add 45 ml of 10% aq. HCl in 4 ml portions through the condenser, shaking thoroughly after each addition. Warm this mixture in a water bath or steam bath for 15–20 min and then decant from any solid tin salts into a mixture of 15 g of ice and 3 g of sodium hydroxide. Stir the resulting mixture vigorously and add dropwise 40% aq. NaOH to dissolve any remaining tin salts. Extract this aqueous solution three to five times with 10-ml portions of ether. Combine the ether extracts, dry over potassium hydroxide pellets, and filter. Remove the solvent by distillation or evaporation, leaving the primary amine as the residue. A suitable solid derivative is then prepared (§395).

400. Phenols. α-Naphthyl isocyanate reacts smoothly with monohydric, but not polyhydric (more than one hydroxyl group), phenols to give α-naphthylurethans. The 3,5-dinitrobenzoates also form readily. These have been discussed previously under alcohols (§392). The sodium salts of phenols in most cases react readily with chloroacetic acid to form aryloxyacetic acids. Phenols are brominated under mild conditions to form solid derivatives.

ARYLOXYACETIC ACIDS

$$ArOH + NaOH \longrightarrow ArONa \xrightarrow{ClCH_2CO_2H} ArOCH_2CO_2H$$

Dissolve 1 g of the unknown phenol in 5 ml of 35% aq. NaOH and add to this 1.5 g of chloroacetic acid with vigorous shaking. If some of the sodium salt of the phenol should precipitate, add up to 5 ml of water to dissolve it. Heat this solution in a gently boiling water bath for 1 hr, then cool to room temperature, and dilute with water (10–15 ml). Add dilute hydrochloric acid to this solution until it is acid to Congo Red paper. Extract this acidic solution with 50 ml of ether and then wash the ether extract with 10 ml of cold water. Treat the ether layer with 25 ml of 5% aq. Na_2CO_3 and shake thoroughly to extract the derivative into the aqueous sodium carbonate layer. Separate the layers and acidify the sodium carbonate layer with dilute hydrochloric acid to precipitate the aryloxyacetic acid. Collect the precipitate by filtration and recrystallize from water.

BROMINATION. Phenols that contain no deactivating groups are easily brominated by an aqueous bromine solution. Generally the product is a polybromo compound, which is recrystallized from ethanol or ethanol–water mixture.

$$C_6H_5\text{–OH} + 3Br_2 \longrightarrow 2,4,6\text{-}Br_3C_6H_2\text{–OH} + 3HBr$$

To a solution of phenol (1 g) in ethanol, water, acetone, or dioxane (10 ml), add a brominating solution (8 g of KBr dissolved in 50 ml of water with 5 g of Br_2 then added) with swirling and shaking just until a yellow color appears in the reaction mixture. Then dilute the reaction mixture with 50 ml of water and collect the resulting precipitate on a filter. Break up the precipitate on the filter and wash with water and then a dilute aq. $NaHSO_3$. (This latter wash removes any unreacted bromine.) Recrystallize the product from an appropriate solvent.

401. Spectroscopy. The most powerful techniques for identifying organic compounds involve the use of infrared, nuclear magnetic resonance, ultraviolet, and mass spectroscopy. The extent to which these are used in beginning or intermediate organic laboratories depends on the availability of equipment and the interests of the particular instructor. Even if spectrometers are not available for use by undergraduates, the use of spectroscopy for identification of organic compounds can become part of the introductory course by supplying students with spectra for their unknowns. The procedure to be followed will be specified by the instructor.

Infrared and nuclear magnetic resonance spectroscopy are briefly discussed in Chapter 13 which should be studied carefully in connection with the laboratory work on identification of organic compounds. That chapter also includes references to all of the spectroscopic methods mentioned above.

INFRARED SPECTROSCOPY. Some of the practical details for determination of infrared spectra are given in §149. **Warning:** *Be especially careful to avoid contact of water with the sodium chloride cells.* Salt plates should be picked up only by the edges, preferably with gloves, so that moisture from the fingers does not cloud or etch the polished surfaces. Cells and plates should always be cleaned carefully after each use by washing with dry chloroform or with the dry solvent used to prepare the sample solution; they should then be dried in a stream of dry nitrogen and stored in a desiccator. Samples should be carefully dried before determination of infrared spectra. The instructor will discuss the particular instrument you will use for the measurement.

NMR SPECTROSCOPY. Good spectra can be obtained only if samples are of high purity and free from traces of lint, dust, or other foreign matter (especially impurities having magnetic properties). Nonviscous liquids may be used without solvent, but solids or viscous liquids must be dissolved in a suitable solvent to give a solution of low viscosity (10–25% by weight is a common concentration range). Tetramethylsilane (about 1%) is usually added as an internal standard unless it is insoluble; other standards may then be used (see the references in Chapter 13). Carbon tetrachloride is a common solvent; deuterated chloroform ($CDCl_3$) is useful when the sample is insoluble in CCl_4. The sample tube is made of tubing that has thin, uniform walls; variation of wall thickness leads to distortion of the magnetic field. For very careful work the tube should be degassed to remove dissolved oxygen (which leads to line broadening) and sealed under reduced pressure. The instructor will demonstrate the determination of the spectrum on the particular instrument to be used.

402. Tables of Unknowns and Their Derivatives. The tables in this section can be greatly expanded, if the instructor desires. Additional sources are given in §403.

Alcohols

	Boiling Point, °C	*Melting Point of Derivative, °C*		
		α-Naphthylurethan	*3,5-Dinitrobenzoate*	*Phenylurethan*
Methanol	65	124	107	47
Ethanol	78	79	93	52
2-Propanol	83	106	122	76
t-Butyl alcohol	83	101	142	136
Allyl alcohol	97	109	48	70
l-Propanol	97	80	74	57
2-Butanol	99	97	75	65
[*t*-Amyl alcohol	102	71	117	42]
2-Methyl-l-propanol	108	104	87	86
3-Methyl-2-butanol	113	109	76	68
3-Pentanol	116	95	99	49
l-Butanol	116	71	64	61
3,3-Dimethyl-2-butanol	120		107	77–8
Ethylene chlorohydrin	131	101	92	51
4-Methyl-2-pentanol	131	88	65	143
Isoamyl alcohol	132	67	62	57
[l-Hexanol	156	59	58	42]
4-Heptanol	156	80	64	—
Cyclohexanol	160	128	112	82
2-Methylcyclohexanol	165	155	114	91

Acids

		Melting Point of Derivative, °C			
Liquids	*Boiling Point, °C*	*p-Bromophenacyl Ester*	*Amide*	*Anilide*	*Toluidide*
Propionic	140	63	80–1	106	126, 123
Acrylic	140	—	84–5	104	141
Isobutyric	155	77	129	105	109
n-Butyric	163	63	115	95	75
Solids	*Melting Point, °C*				
o-Toluic	104	57	140	125	144
m-Toluic	110	108	94	125	118
p-Chlorophenylacetic	105	—	175	164	190
trans-Cinnamic	133	145	144	153, 109	168
trans-α-Chlorocinnamic	137	—	121	118	116
m-Nitrobenzoic	140	132	143	153	162
o-Chlorobenzoic	142	106	142	114	132
Adipic	153	154	220	240	241
Salicylic	157	140	142	136	156
m-Chlorobenzoic	158	116	134	122	—
p-Toluic	179	153	160	144	160
2,4-Dinitrobenzoic	183	158	203	—	—
3-Iodobenzoic	187	128	186	—	—
Succinic	188	211 (di)	260d (di) 157 (mono)	230 (di) 148 (mono)	254–5 (di) 178–80 (mono)

	Boiling Point, °C	Melting Point of Derivative, °C	
		Oxime	2,4-Dinitrophenylhydrazone
Propionaldehyde	50	40	154
Glyoxal, $HC(=O)-C(=O)H$	50	178	328
Acetone	56	59	126
n-Butyraldehyde	74	—	122
2,2-Dimethylpropanal	75	41	209
2-Butanone	80	152	117
2-Pentanone	102	58	144
3-Pentanone	102	69	156
Pentanal (valeraldehyde)	103	52	106
3,3-Dimethyl-2-butanone (pinacolone)	106	74	125
2,4-Dimethyl-3-pentanone	125	34	95
4-Methyl-3-penten-2-one (mesityl oxide)	130	49	203
Hexanal	131	51	104
Cyclopentanone	131	56	142
4-Heptanone	145	—	75
Cyclohexanone	155	90	162
n-Heptaldehyde	156	57	108

Amines

	Boiling Point, °C	*Melting Point of Derivative, °C*			
		Benzenesulfonamide	*Benzamide*	*Methiodide*	*Picrate*
2-Aminobutane	63	70	76		140
Isobutylamine	69	53	57		150
Piperidine	106	94	48		152
Di-*n*-propylamine	110	51			75
Ethylenediamine	116	168 (di)	244 (di)		233 (di)
Pyridine	116			117	167
n-Hexylamine	128–130	96	40		126
N-Ethylpiperidine	128				167.5
Morpholine	130	118	75		146
Cyclohexylamine	134	89	149		
N,*N*-Dimethylbenzylamine	181				93
Aniline	183	112	160		180
Benzylamine	184	88	105		194
N,*N*-Dimethyl-*o*-toluidine	185				122 (116)
N-Methylaniline	196	79	63		145
β-Phenylethylamine	198	69	116		174 (167)
o-Toluidine	199	124	145		213
N-Methyl-*N*-ethylaniline	201			125	134
m-Toluidine	203	95	125		200
o-Chloroaniline	207	129	99		134
o-Ethylaniline	211		147		195
N,*N*-Diethylaniline	217			102	142
N-Ethyl-*p*-toluidine	217	66	40		
N-Ethyl-*o*-toluidine	218	62	72		

Carboxylic Acid Derivatives

	Boiling Point, °C	Melting Point of Acid, °C	Boiling Point of Acid, °C	Melting Point of Derivative of Parent Acid, °C			
				Amide	*Anilide*	*p-Toluidide*	*p-Bromophenacyl Ester*
Methyl propionate	79	−21	140	80–1	106	126, 123	63.4
Propionyl chloride	80	−21	140	80–1	106	126, 123	63.4
Acetonitrile	81	16.6	118	82	114	153, 147	86.0
Propionitrile	97	−21	140	80–1	106	126, 123	63.4
Butanoyl chloride	101	−6	163	115	96, 97	75	63
Methyl *n*-butyrate	102	−6	163	115	96, 97	75	63
Isobutyronitrile	108	−46	155	129	105	109	77
3-Methylbutanoyl chloride	115	−30	176	136	109.5	106–7	68
Isobutyl acetate	117	16.6	118	82	114	153, 147	86
n-Butyronitrile	118	−6	163	115	96, 97	75	63
Pentanoyl chloride	127	−34.5	186	106	63	74	75
Methyl *n*-valerate	128	−34.5	186	106	63	74	75
Isobutyl propionate	138	−21	140	80–1	106	126, 123	63.4
Acetic anhydride	138	16.6	118	82	114	153, 147	86
n-Valeronitrile	141	−34.5	186	106	63	74	75
Isoamyl propionate	160	−21	140	80–1	106	126, 123	63.4
Capronitrile	164	−4	205	100, 101	94	74	72
p-Tolunitrile	66	179–80	275	160, 158	144–5	160, 165	153
Phenyl benzoate	70	122.4	249	128	160	158	119
Propionamide	80	−21	140	80–1	106	126, 123	63.4
Acetamide	82	16.6	118	82	114	153, 147	86
n-Butyramide	115	−6	163	115	96, 97	75	63
m-Nitrobenzonitrile	118	140	—	143	154	162	132
Succinic anhydride	120	188	235d	157 (mono) 260d (di)	148.5 (mono) 130 (di)	179–80 (mono) 254–5 (di)	211 (di)
Benzamide	128	122.4	249	128	160	158	119
Isobutyramide	128	−46	155	128	105	109	77
o-Toluamide	140	104–5	259	140	125	144	57
o-Chlorobenzamide	142	142, 140	—	142	114, 118	131	106
Cinnamamide	144	133	300	144	153, 109	168	145.6
m-Nitrobenzamide	144	140	—	143	154	162	132
Phenylacetamide	156	76.5	256.5	156	117–8	135–6	89
p-Toluamide	160	179–80	275	160	144–5	160, 165	153
3-Nitrobenzoic anhydride	161	140	—	143	154	162	132
o-Nitrobenzamide	176	146	—	176	155	—	107
p-Chlorobenzamide	179	243, 240	—	179	194	—	126
m-Iodobenzamide	186	187	—	186	—	—	128
p-Bromobenzamide	190	251–3	—	189–90	197	—	—

403. References. To supplement the tables in §402, there are several books that contain lists of organic compounds by functional class (e.g., hydrocarbons, alcohols, etc.). Within each class, arrangement is by melting point or boiling point. Such lists are very useful in many research and industrial problems encountered by the organic chemist, and form the basis for courses in qualitative organic analysis. A particularly useful compilation of such data is the following:

Handbook of Tables for Organic Compound Identification (3rd ed.), Compiled by Zvi Rappoport, The Chemical Rubber Co., Cleveland, Ohio, 1967.

Textbooks for courses in qualitative organic analysis usually include fairly extensive lists arranged similarly. They also contain more detailed information about solubility tests, classification reactions, and the preparation of derivatives than is included in the present treatment. Some of these are listed below:

R. L. Shriner, R. C. Fuson, and D. Y. Curtin, *The Systematic Identification of Organic Compounds* (5th ed.), John Wiley & Sons, New York, 1964.

N. D. Cheronis, J. B. Entrikin, and E. M. Hodnett, *Semimicro Qualitative Organic Analysis* (3rd ed.), Wiley-Interscience, New York, 1965.

A. I. Vogel, *A Textbook of Practical Organic Chemistry* (3rd ed.), John Wiley & Sons, New York, 1956.

T. C. Owen, *Characterization of Organic Compounds by Chemical Methods*, Marcel Dekker, New York, 1969.

Other useful compilations with different arrangements of compounds include:

Dictionary of Organic Compounds, Sir Ian Heilbron et al., Oxford University Press, 1965-present. (Alphabetical arrangement)

Melting Point Tables of Organic Compounds, W. Utermark and W. Schicke, Wiley-Interscience, New York, 1963.

Beilstein's *Handbuch der Organischen Chemie*. See §155.

Special Experiments

404. Preparations from the Literature. One aim of the laboratory course in organic chemistry is to develop in a prospective chemist a clear understanding of syntheses and reactions, which will enable him to select reasonable conditions for any transformation he may wish to carry out. The student should gain confidence in his ability to carry reactions through to successful conclusions on the first attempt. As experience is gained in the operations and techniques of organic chemistry, less detailed directions will be needed. After a number of experiments have been carried out according to the carefully tested directions presented in the laboratory manual, further experience can be obtained by looking up syntheses or procedures in major publications of organic chemistry. Finally the student may well undertake syntheses for which no good directions exist, or of new compounds not prepared previously. For the latter it is best, at first, to use general reactions for which related examples have been carefully worked out.

This section of the laboratory manual contains a list of experiments that appear to be suitable for the first step beyond the use of laboratory manual directions. Each student should choose a different synthesis in consultation with his laboratory instructor and should read more than the single reference given before attempting the experiment. *Chemical Abstracts* abbreviations are used except as noted below:

Organic Syntheses—cited as OS with volume number.

Organic Syntheses, Collective Volume—cited as OS–CV with volume number.

Organic Reactions—cited as OR with volume number.

Systematic Organic Chemistry (4th ed.), W. M. Cumming, I. V. Hopper, and T. S. Wheeler, Constable and Co., London, 1950—cited as Cumming, Hopper and Wheeler.

A Textbook of Practical Organic Chemistry (3rd ed.), A. I. Vogel, John Wiley & Sons, New York, 1956—cited as Vogel (1956).

The examples chosen are merely illustrative; no attempt has been made to survey the above publications, or others devoted mainly to organic synthesis, for suitable procedures. A few syntheses of historical interest have been included. Syntheses from *Organic Syntheses* are usually on a moderately large scale but can easily be scaled down.

In recent years an increasing number of excellent experiments suitable for use in the beginning organic chemistry laboratory have been published in the *Journal of Chemical Education*. Many of these have been included in the following sections because they supplement the procedures in the book and permit an instructor to find experiments that reflect the special emphases in the course he is teaching. A number of these experiments are more complicated or offer experience with special techniques. Experiments in various kinds of chromatography, especially in glc (not specifically illustrated in this book), are included. A number of physical organic chemistry experiments are also listed. A few nmr experiments are cited.

Special attention is paid to synthetic sequences and a number of these are presented. These supplement the experiments in the book (§§222, 242–247, 322, and 355–360).

ALIPHATIC COMPOUNDS—GENERAL

405. Wurtz Reaction. Synthesis of an alkane from an alkyl halide and metallic sodium; example: *n*-decane [*J. Chem. Ed.,* **7,** 2712 (1930)].

406. Alkyl Halides. Alkyl iodides. Ethyl or methyl iodide is prepared in quantity by a semicontinuous process utilizing both red and yellow phosphorus. Special apparatus assembly; a good experiment to set up for visitors' or "open house" day in a college laboratory [*J. Amer. Chem. Soc.,* **41,** 789 (1919); OS–CV **2**]. *n*-Hexyl fluoride. Replacement of the bromine of *n*-hexyl bromide by fluorine, by use of potassium fluoride (OS–CV **4**). For hexyl bromide, see Alkyl and Alkylene Bromides (OS–CV **1**). Ethyl chloride. Preparation of an extremely volatile material [*J. Chem. Ed.,* **17,** 461 (1940)].

407. Halogenation of Alkanes. Bromination of alkanes [J. Warkentin, *J. Chem. Ed.,* **43,** 331 (1966)]; chlorination of 2,3-dimethylbutane [J. H. Markgraf, *ibid.,* **46,** 610 (1969)]; chlorination of 1-chlorobutane [P. C. Reeves, *ibid.,* **48,** 636 (1971)]; chlorination of hydrocarbons [H. B. Cutter and H. C. Brown, *ibid.,* **21,** 443 (1944)]; chlorination of cyclohexane [F. E. Condon and M. Sokoloff, *ibid.,* **36,** 554 (1959)]. Some of these involve glc.

408. Neopentyl Alcohol. Formation and decomposition of a hydroperoxide from diisobutylene (OS **40**).

409. *tert*-Butyl Acetate. Special technique is required for the preparation of an ester from a tertiary alcohol (OS–CV **3**).

410. Ethyl Formate and Other Simple Esters. Interesting variations on simple esterification [*J. Chem. Ed.*, **27**, 245 (1950)].

411. Ketenes. Methods for preparing these interesting compounds and procedures are given in OR **3**. See also OS–CV **4** for dimethylketene; bromination of an acid through its acid halide is also illustrated here. For ketene itself, see OS–CV **1**; *J. Org. Chem.*, **5**, 122 (1940); OS **45**.

412. Acetylenes. A number of excellent procedures illustrating dehydrohalogenation and reactions of metallic derivatives of acetylenes are available. These include reactions in liquid ammonia. See OR **5** (where earlier references to OS are listed along with other specific procedures), stearolic acid (OS–CV **4**), diphenylacetylene (OS–CV **4**), 4-pentyn-1-ol (OS–CV **4**), *n*-butylacetylene (OS–CV **4**), and phenylacetylene (OS–CV **4**). Other recent OS procedures for acetylenes appear to be less suitable for students.

413. Triallylamine. Preparation of a tertiary amine [*J. Chem. Ed.*, **28**, 191 (1951)].

414. Methylation of Amines. Eschweiler–Clark modification of the Leuckart reaction utilizing formic acid–formaldehyde. Benzylamine is converted to dimethylbenzylamine [S. H. Pine, *J. Chem. Ed.*, **45**, 118 (1968)].

415. Preparation and Isomerization of Allyl Thiocyanate. D. W. Emerson, *J. Chem. Ed.*, **48**, 81 (1971).

416. Reformatsky Reaction. Use of zinc instead of magnesium in a reaction suggestive of the Grignard method; example, ethyl β-phenyl-β-hydroxypropionate (OR **1**).

417. Cyanohydrin Reaction. Mandelic acid from benzaldehyde via mandelonitrile (OS–CV **1**).

418. Mannich Reaction. An interesting condensation reaction. Several straightforward procedures are given in OR **1**.

419. Acetylacetone. A condensation (OS–CV 3).

420. Tetraacetylethane. Coupling of an active methylene compound (OS–CV **4**).

421. Aldol Condensation. All of OR **16** is given over to this condensation. Procedures supplement those given in Experiment 24.

422. Resolution of *dl*-Histidine. Uses an ion-exchange resin [A. J. Bosch, *J. Chem. Ed.,* **46,** 691 (1969)].

423. Diels–Alder Reaction. Many useful procedures in OR **4** and OR **5.** The preparation of triptycene through the intermediate benzyne is an interesting example (OS–CV **4**). **Warning:** See §576. A procedure for generation of 1,3-butadiene from 3-sulfolene and its use in Diels–Alder condensation has been described [T. E. Sample, Jr., and L. F. Hatch, *J. Chem. Ed.,* **45,** 55 (1968)].

424. Knoevenagle Condensation. Several procedures for this useful reaction are given in OR **15.**

425. Urea. The historic experiment of Wöhler and Liebig whereby an inorganic compound was converted into an "organic" compound without aid of a life process (H. J. Lucas and D. Pressman, *Principles and Practice in Organic Chemistry,* John Wiley & Sons, New York, 1949, p. 310). For an interesting discussion of Wöhler's discovery see T. O. Lipman, *J. Chem. Ed.,* **41,** 452 (1964).

426. 2-Chlorocyclohexanone. Chlorination of a ketone (OS–CV **3**).

427. Cyclobutane Chemistry. F. T. Williams Jr., and S. C. Baber, *J. Chem. Ed.,* **41,** 563 (1964).

428. Beckman Rearrangement. Several procedures are given in OR **11.**

429. Favorskii Rearrangement. Several procedures in OR **11.** Methyl cyclopentanecarboxylate (OS–CV **4**).

430. *Cis* and *Trans* 4-*tert*-Butylcyclohexanols. I. A. Kaye, *J. Chem. Ed.,* **43,** 535 (1966).

431. Stereospecific Synthesis of *Cis* and *Trans* 1,4-Disubstituted Cyclohexanes. R. S. Monson, *J. Chem. Ed.,* **48,** 197 (1971).

432. Stereochemical Correlations: Norbornane and Camphor Series. J. H. Markgraf, *J. Chem. Ed.,* **44,** 36 (1967); J. H. Markgraf and P. Leung, *ibid.,* **47,** 707 (1970).

433. Stereochemistry of Menthol. H. C. Dunathan, *J. Chem. Ed.,* **40,** 205 (1963).

434. Carbenes. The simplest carbene experiments for organic chemistry students involve generation of dichlorocarbene in a two-phase system (chloroform–50% aqueous sodium hydroxide) in the presence of an olefin and a phase-transfer catalyst such as tricaprylmethyammonium chloride. The carbene adds to the olefin forming 1,1-dichlorocyclopropanes [M. Makosza and M. Wawrzyniewicz, *Tetrahedron Lett.*, 4659 (1969); C. M. Starks, *J. Amer. Chem. Soc.*, **93,** 195 (1971); M. Newman and Zia ud Din, *Syn. Comm.*, **1,** 247 (1971)]. Other procedures for halocarbenes are given in OR **13.** Various reactions of diazoacetic esters, some of which involve generation of carbenes, are described in OR **18.** See W. Kirmse, *Carbene Chemistry* (2nd ed.), Academic Press, New York, 1971, for a comprehensive review with many references. An interesting experiment for students examines transannular carbene reactions [S. S. Hecht, *J. Chem. Ed.*, **48,** 340 (1971)].

AROMATIC COMPOUNDS—GENERAL

435. Iodination of Aromatic Compounds. Replacement of the diazonium group is the most common reaction for introducing iodine into an aromatic nucleus. See iodobenzene (OS–CV **2**), *p*-iodophenol (OS–CV **2**). Direct iodination cannot be accomplished unless the hydrogen iodide is removed as it is formed (iodobenzene, OS–CV **1;** *p*-iodoaniline, OS–CV **2;** tetraiodophthalic anhydride, OS–CV **3**). Iodine monochloride also effects iodination (2-hydroxy-3,5-diiodobenzoic acid, OS–CV **3;** 5-iodoanthranilic acid, OS–CV **1**). An activated chlorine or bromine on an aromatic ring can be replaced by iodine (2,4-dinitroiodobenzene, OS **40**).

436. Aromatic Fluorine Compounds. Synthesis by the diazo route with the aid of fluoroboric acid (fluorobenzene, *p*-fluorobenzoic acid, and 4,4′-difluorobiphenyl, OS–CV **2**).

437. Nitration of 4-Nitrophenol. R. E. Pearson, *J. Chem. Ed.*, **46,** 692 (1969).

438. 2-Nitroresorcinol. R. E. Schaffrath, *J. Chem. Ed.*, **47,** 224 (1970).

439. Separation of Ortho and Para Isomers. Ortho- and paranitrophenols, illustrating theory of chelation in the separation [Cumming, Hopper, and Wheeler; L. O. Binder, *J. Chem. Ed.*, **23,** 285 (1946)].

440. Picric Acid. Introduction of three nitro groups into the phenol molecule. Antiseptic; also a high explosive (Cumming, Hopper, and Wheeler).

441. 2,4,6-Tri-*tert*-butylphenol. Alkylation of a phenol with an alkene [B. G. Somers and C. D. Cook, *J. Chem. Ed.*, **32,** 312 (1955)].

442. β-Benzoylacrylic Acid. Friedel and Crafts synthesis (OS–CV **3**).

443. Derivatives of 1,2-Diphenylethane. L. F. Fieser, *J. Chem. Ed.*, **31,** 291 (1954).

444. Isomerization of Xylenes. K. J. Harbison, *J. Chem. Ed.*, **47,** 837 (1970).

445. Substituent Effects in Aromatic Electrophilic Substitution. P. R. Ferguson, *J. Chem. Ed.*, **48,** 405 (1971).

446. Clemmensen Reduction. The keto acid described in §315 is reduced by amalgamated zinc and hydrochloric acid, yielding γ-phenylbutyric acid (OS–CV **2** and OR **1**).

447. Fries Reaction. A molecular rearrangement (OR **1**).

448. 9-Anthraldehyde. Formylation with *N*-methylformanilide (OS–CV **3**).

449. *n*-Butylbenzene. Wurtz–Fittig reaction (OS–CV **3**).

450. Alkaline Fusion. Preparation of phenols by fusion of aryl sulfonates (arenesulfonates) with caustic alkali; example: β-naphthol (Cumming, Hopper, and Wheeler).

451. *p*-Nitrobenzaldehyde. Oxidation of a methyl group (OS–CV **2;** see also OS–CV **4,** 713).

452. 2-Naphthaldehyde. The Sommelet reaction and bromination with *N*-bromosuccinimide [H. M. Doukas, *J. Chem. Ed.*, **31,** 12 (1954)].

453. Phenylnitromethane. An aliphatic nitro compound (with one aryl substituent radical), which may actually be prepared in two distinct tautomeric forms (OS–CV **2**).

454. *p*-Anisonitrile. R. F. Smith and A. C. Bates, *J. Chem. Ed.*, **46,** 174 (1969).

455. The Smiles and Related Rearrangements. Several procedures in OR **18.**

456. *p*-Phenylacetophenone. Bromination of this substance produces *p*-phenylphenacyl bromide, a useful reagent for identification of organic acids [N. L. Drake and J. Bronitsky, *J. Amer. Chem. Soc.*, **52,** 3715 (1930).]

457. Phenanthrenequinone. Chromic acid oxidation of phenanthrene (OS–CV **4**).

458. 9-Fluorenecarboxylic Acid. A benzilic acid rearrangement (OS–CV **4**).

459. Selective Reduction of Dinitrobenzenes. J. P. Idoux and W. Plain, *J. Chem. Ed.,* **49,** 133 (1972).

HETEROCYCLIC COMPOUNDS—GENERAL

460. Benzimidazole. Reagent prepared from *o*-phenylenediamine and formic acid. Used for making crystalline derivatives of aliphatic acids in qualitative analysis [W. O. Pool, H. J. Harwood, and A. W. Ralston, *J. Amer. Chem. Soc.,* **59,** 178, (1937)].

461. 3-Carbethoxycoumarin. A Knoevenagel reaction (OS–CV **3**).

462. Benzofurazan Oxide. Hypochlorite oxidation of *o*-nitroaniline [OS–CV **4**; C. W. Schimelpfenig, *J. Chem. Ed.,* **36,** 570 (1959)].

463. 2-Furoic Acid. Catalytic oxidation of furfural by oxygen (OS–CV **4**).

464. 2-Acetothienone. Friedel and Crafts reaction on thiophene (OS–CV **3**).

465. 2-Iodothiophene. Iodination in the presence of mercuric oxide (OS–CV **2**).

466. 4-Chlorobutyl Benzoate. Opening of a heterocyclic ring (OS–CV **3**).

467. 2-Arylimidazo [2,1-*b*] benzothiazoles. H. Alper, *J. Chem. Ed.,* **47,** 223 (1970). See H. Alper, A. E. Alper, and A. Taurens, *ibid.,* **47,** 222 (1970) for a related experiment on ring-chain tautomerism.

468. 3-Phenylsydnone (OS **45**).

469. 2-Methyl-4*H*-3,1-benzoxazin-4-one and Related Compounds. D. R. Eckroth, *J. Chem. Ed.,* **49,** 66 (1972).

470. Hantzsch Pyridine Synthesis. B. E. Norcross, G. Clement, and M. Weinstein, *J. Chem. Ed.,* **46,** 694 (1969).

DYES, PIGMENTS, INDICATORS, PERFUME INGREDIENTS

471. Mauve. An attempt to repeat Perkin's synthesis [W. H. Cliffe, *J. Soc. Dyers and Colourists,* **72,** 563 (1956)].

472. Erythrosine. Officially approved red coloring material for food products; iodine derivative of fluorescein [M. Gomberg and D. L. Tabern, *Ind. Eng. Chem.,* **14,** 1115 (1922)].

473. Indigo. Classical example of a synthetic dye replacing a more costly, naturally occurring dye. **Malachite Green,** an important dye of the triphenylmethane series, from benzaldehyde. **Para Red,** a brilliant red pigment that may be used in powdered form in paints or may be synthesized in the porous fibers of a textile (ingrain process) to produce a color that is hard to remove [H. J. Lucas and D. Pressman, *Principles and Practice in Organic Chemistry*, John Wiley & Sons, New York, 1949; see also Vogel (1956)].

474. Methyl Red. A valuable acidimetric indicator, with sharp transition from lemon yellow to brilliant carmine red at a pH value near 7. This is an azo dye (OS–CV **1**).

475. Nitrazine Yellow. Substitute for litmus, but with different colors [H. Wenker, *Ind. Eng. Chem.,* **26,** 350 (1934)].

476. Sulfonphthalein Indicators. A group of acidimetric indicators with color changes over a wide range of pH values. Usually these indicators have two distinct colors (in higher and lower pH conditions), being thus much more suitable for pH determination than the simpler phthaleins like phenolphthalein (see §347). For additional examples see Vogel (1956), 989–990.

EXAMPLES. **Bromcresol Green, Chlorphenol Red, and Cresol Purple** (I. M. Kolthoff and C. Rosenblum, *Acid-Base Indicators,* 124, Macmillan Publishing Co., Inc., 1937). **Bromphenol Blue** [W. R. Orndorff and F. W. Sherwood, *J. Amer. Chem. Soc.,* **45,** 486 (1923)]. **Thymol Blue** [W. R. Orndorff, and R. T. K. Cornwell *J. Amer. Chem. Soc.,* **48,** 981 (1926)].

477. Thymolphthalein. Indicator similar to phenolphthalein, color change blue to colorless [R. Jakinowicz, *Ber.,* **28,** 1876 (1895)].

478. Dibromofluorescein. Coloring material used in certain types of lipstick (M. A. Phillips, *J. Chem. Soc.,* **1932,** 724).

479. Thermochromic Compounds. A. Ault, R. Kopet and A. Serianz, *J. Chem. Ed.,* **48,** 410 (1971).

480. *β*-Phenylethyl Alcohol. Used in synthesis of components of perfumes. A variety of syntheses have been reported [C. S. Leonard, *J. Amer. Chem. Soc.*, **47,** 1774 (1925); R. F. Nystrom and W. G. Brown, *ibid.*, **69,** 2548 (1947); R. C. Huston and A. H. Agett, *J. Org. Chem.*, **6,** 123 (1941)].

481. *β*-Naphthyl Methyl Ether (Nerolin). Example of formation of a methyl ether from dimethyl sulfate (Cumming, Hopper, and Wheeler).

COMPOUNDS OF INTEREST IN MEDICINE

Several of these are given in the experiments in the book, Experiments 19 and 28.

482. Chloramine-T. Antiseptic drug (sodium *p*-toluenesulfonchloramide) whose virtue lies in the "oxidizing" chlorine on nitrogen (Cumming, Hopper, and Wheeler).

483. Hexachlorophene—Manufacturing the Great Clean-all. [A. L. Moyé, *J. Chem. Ed.*, **49,** 770 (1972)].

484. Benzyl Benzoate. Reaction of metallic sodium with benzyl alcohol and benzaldehyde gives a high yield of the ester. Used as a local anaesthetic (OS–CV **1**).

485. DDT. The synthesis of this controversial but still extremely important insecticide has remained much the same for decades [E. L. Bailes, *J. Chem. Ed.*, **22,** 122 (1945); S. F. Darling, *ibid.*, 170; M. Fields, J. Gibbs, and D. E. Walz, *Science*, **112,** 591 (1950); Vogel (1956)].

486. Benzocaine. Ethyl *p*-aminobenzoate [Vogel (1956)]. For *p*-aminobenzoic acid see C. B. Kremer, *J. Chem. Ed.*, **33,** 71 (1956).

487. 8-Hydroxyquinoline. Antiseptic and fungicide, also an analytical reagent [F. E. King and J. A. Sherred, *J. Chem. Soc.*, **1942,** 415].

488. Thin-Layer Chromatography of Drugs. R. L. Neman, *J. Chem. Ed.*, **49,** 834 (1972).

489. *α*-Naphthaleneacetic Acid. A plant-growth hormone; also an intermediate in preparation of the highly active nasal decongestant Privine [Y. Ogata and J. Ishiguro, *J. Amer. Chem. Soc.*, **72,** 4302 (1950)].

490. Caffeine. Caffeine is not only a constituent of tea and coffee, but is present in certain soft drinks, proprietary medicines, etc. Isolation from tea or coffee may present difficulties, and several experiments involving the other sources have been described [A. L. Moyé, *J. Chem. Ed.,* **49,** 194 (1972); J. A. Laswick and P. H. Laswick, *ibid.,* **49,** 708 (1972); J. W. Pavlik, *ibid.,* **50,** 134 (1973)]. An experiment on analysis of APC tablets, which contain caffein, has also appeared [V. T. Lieu, *J. Chem. Ed.,* **48,** 478 (1971); E. S. Hanrahan, *ibid.,* **46,** 511 (1969)].

491. Atophan. 2-Phenylquinoline-4-carboxylic acid, also called Cinchophen. No longer a recommended drug [Vogel (1956)].

492. Antipyrin. 1-Phenyl-2,3-dimethyl-5-pyrazolone, an analgesic and antipyretic [Vogel (1956)].

Naturally Occurring Compounds and Derivatives

493. Orange Oil. F. H. Greenberg, *J. Chem. Ed.,* **45,** 537 (1968).

494. Essential Oils. O. Runquist, *J. Chem. Ed.,* **46,** 846 (1969).

495. Pigments from some Marine Specimens. C. W. J. Chang and J. C. Moore, *J. Chem. Ed.,* **48,** 408 (1971).

496. Incorporation of Glucose into Chelidonic Acid. B. A. Bohm, *J. Chem. Ed.,* **43,** 539 (1966).

497. Chloroplast Pigments. A detailed experiment on chromatography [H. H. Strain and J. Sherma, *J. Chem. Ed.,* **46,** 476 (1969)].

498. Arabinose from Mesquite Gum. Isolation of an aldopentose from a natural source (OS–CV **1**).

499. Furfural (furfuraldehyde or 2-furaldehyde). Reaction of ground corn cobs with dilute sulfuric acid gives the heterocyclic derivative (OS–CV **1**).

500. Cystine from Hair. Acid hydrolysis of the protein in hair yields the sulfur-bearing amino acid cystine (OS–CV **1**).

501. Tyrosine from Silk. Preparation of a naturally occurring amino acid by acid hydrolysis (Cumming, Hopper, and Wheeler).

502. Camphor. Students in subtropical regions may test the technique of steam sublimation by use of fresh young leaves of the ornamental camphor tree, common on streets of southern California. A 5-gal can is filled with leaves gathered in early springtime. Steam is passed through the mass, thence to a condenser. Yield is small, but a beautiful product crystallizing directly from condensing steam, mp about 100°.

503. Methyl Myristate and Methyl Palmitate. From bayberry wax (OS–CV **3**).

504. *d*-Glucosamine Hydrochloride. Hydrolysis of chitin from crab shells (OS–CV **3**).

505. Juglone: an Organic Chemistry–Ecology Interaction Experiment. R. G. Jesaitis and A. Krantz, *J. Chem. Ed.*, **49,** 436 (1972).

METALLOORGANIC COMPOUNDS

506. Ferrocene. OS–CV **4** and OR **17.**

507. Preparation and Resolution of *N,N*-Dimethyl-α-ferrocenylethylamine. G. W. Gokel and I. K. Ugi, *J. Chem. Ed.*, **49,** 294 (1972).

508. Organotin Compounds. Triphenyltin hydride: synthesis and nmr spectrum [C. W. Allen, *J. Chem. Ed.*, **47,** 479 (1970)]; tetraethyltin (OS–CV **4**).

509. Diethylzinc. A mixture of ethyl bromide and iodide is converted into the zinc derivative (OS–CV **2**).

510. Arsonation. Introduction of the arsonic acid group on an aromatic system by replacement of the diazonium. Procedures are given for **phenylarsonic acid** and other compounds (OR 2). The arsonic acid group can also be introduced into highly reactive aromatic systems such as aniline by direct arsonation (analogous to sulfonation) with arsenic acid. Aniline yields **arsanilic acid,** a compound made famous by Ehrlich and coworkers and used in the manufacture of medicinal arsenicals. (OR **2;** OS–CV **1**; OS–CV **2**; Cumming, Hopper, and Wheeler.)

Warning: *Most arsenic compounds are toxic.*

511. Organomercury Compounds. *o*-Chloromercuriphenol (OS–CV **1**). Oxymercuration-demercuration [R. Gibbs and W. P. Weber, *J. Chem. Ed.*, **48,** 477 (1971)].

Warning: *Many organomercury compounds are poisonous, and some people are very sensitive to almost all mercury compounds.*

512. Mesitylene Tricarbonyl Molybdenum. R. J. Angelici, *J. Chem. Ed.*, **45,** 119 (1968).

SPECIAL REAGENTS

513. Aluminum Isopropoxide. Preparation and use in reduction of an aldehyde to an alcohol by the Meerwein–Ponndorf–Verley procedure. Discussion and various procedures (e.g., crotonaldehyde to crotyl alcohol, OR **2** and Vogel. Chloral to trichloroethyl alcohol (OS–CV **2**) and heptaldehyde to 1-heptanol (OS–CV **1**). Benzyl alcohol from benzaldehyde [Q. P. Cole, *J. Chem. Ed.,* **28,** 142 (1951)].

514. *N*-Bromosuccinimide. Free radical bromination of *p*-toluic acid. [D. L. Tuleen and B. A. Hess, Jr., *J. Chem. Ed.,* **48,** 476 (1971)].

515. Aluminum *tert*-Butoxide. Use in Oppenauer oxidation of cholesterol to cholestenone [OS–CV **3**; Vogel (1956)].

516. Dimethylglyoxime. Widely used as a reagent for detection and estimation of nickel [R. Adams and O. Kamm, *J. Amer. Chem. Soc.,* **40,** 1281 (1918)].

517. Lead tetraacetate. This salt of "plumbic" lead has special usefulness in oxidative splitting of a glycol [Cumming, Hopper, and Wheeler; also W. S. McClenahan and R. C. Hockett, *J. Amer. Chem. Soc.,* **60,** 2062 (1938)].

518. 3,5-Dinitrobenzoic Acid. Valuable reagent for identification of alcohols, since most of its esters are solids with satisfactory melting points (OS–CV **3**). See §392 for use of this reagent.

519. Polyphosphoric Acid. Uses of this reagent and appropriate procedures are described in *Advances in Organic Chemistry: Methods and Results,* **1.**

SPECIAL TECHNIQUES

520. Reduction with Lithium Aluminum Hydride, Sodium Borohydride, or Other Metal Hydrides. These reducing agents are now so common that they should probably not be classified under special techniques.

> **Warning:** *Lithium aluminum hydride is a potentially dangerous chemical. It has been known to ignite spontaneously; chances of this are diminished by handling the reagent with a porcelain spatula. Never look directly into a container of lithium aluminum hydride.*

For a variety of procedures see OR **6** and Vogel (1956). An experiment in which a safer reducing agent, sodium bis (2-methoxyethoxy) aluminum hydride, replaces lithium aluminum hydride has been described [A. S. Kushner and T. Vaccariello, *J. Chem. Ed.,* **50,** 154 (1973)].

521. Catalytic Hydrogenation. This well-known procedure is easily carried out if equipment is available. Some care is necessary in filtering spent catalyst from reaction mixtures because it may catch fire and ignite the solvent if air is drawn through finely divided catalyst on a suction filter. Examples: benzalacetophenone to benzylacetophenone, maleic acid to succinic (OS–CV **1,** 61, 101); also many others (see the Type of Reaction index in collective volumes of OS). For example of a semimicro procedure for reduction of benzaldehyde, benzophenone, and salicylaldehyde at ordinary pressures see N. D. Cheronis and N. Levin, *J. Chem. Ed.,* **21,** 603 (1944).

522. Catalytic Dehydrogenation. A number of different dehydrogenation reactions are known. A convenient apparatus for dehydrogenation of primary alcohols to aldehydes or secondary alcohols to ketones in the vapor phase is described by E. Allison, R. Gorsich, and L. O. Binker, *J. Chem. Ed.,* **32,** 209 (1955). A liquid-phase dehydrogenation of dodecanol to dodecanal has also been described [A. Halasz, *J. Chem. Ed.,* **33,** 624 (1956)].

523. Dehydrogenations with Quinones. This special technique has been reviewed in *Advances in Organic Chemistry: Methods and Results,* **2.** Syntheses are given for the following high-potential quinones: tetrachloro-1,2-benzoquinone, 2,3-dichloro-5,6-dicyano-1,4-benzoquinone, and 3,5,3′,5′-tetrachloro-4,4′-diphenoquinone. A procedure for the dehydrogenation of 1,1-dimethyltetralin to 1,2-dimethylnaphthalene is also given. Chloranil may be used for such dehydrogenations; its synthesis is described in Cumming, Hopper, and Wheeler.

524. Photochemical Reactions. Experiment 38 describes the photochemistry of azobenzene and photoaddition of maleic anhydride to benzene. A photochemical synthesis of tetraphenylethyleneglycol (benzpinacol) from benzophenone is well known (OS–CV **2**). For a photochemical isomerization of methyl maleate (*cis*) to methyl fumarate (*trans*) see O. Grummitt, *J. Chem. Ed.,* **18,** 477 (1941). Other experiments include flash photolysis [D. M. Goodall, P. W. Harrison and J. H. M. Wedderburn, *J. Chem. Ed.,* **49,** 669 (1972)] and an experiment that combines photochemistry with organic qualitative analysis [M. R. van DeMark and P. L. Kumler, *ibid.,* **50,** 512 (1973)].

525. Electrolytic Preparations. Electrolytic reductions: *o*-aminophenol from *o*-nitrophenol [J. E. Weber and A. E. Meister, *J. Chem. Ed.,* **27,** 571 (1950)]; dichloroacetic acid from trichloroacetic acid [P. E. Iversen, *ibid.,* **48,** 136 (1971)]; *p*-phenylenediamine from *p*-nitroaniline (Cumming, Hopper, and Wheeler). Electrolytic oxidation to synthesize iodoform is described by L. M. Lariviere and J. E. Weber, *J. Chem. Ed.,* **45,** 54 (1968); K. Helle et al., *ibid.,* **46,** 518 (1969). The Kolbe electrolytic synthesis is another example of the use of electrolysis. See *Advances in Organic Chemistry: Methods and Results,* **1,** for a review and procedures; also preparation of dimethyl octadecanedioate and 2,7-dimethyl-2,7-dinitrooctane (OS **41**).

526. Sealed-Tube Techniques. For suitable experiments see A. Grummitt and J. Fink, *J. Chem. Ed.,* **27,** 538 (1950); also preparation of 1,1,3-trichlorononane by addition of chloroform to 1-octene (OS **45**).

527. Reactions in Liquid Ammonia. See earlier under acetylenes. Also α,β-diphenylpropionic acid (OS **40**).

528. Polymerization Experiments. Interfacial polymerizations: in addition to the one described in Experiment 41, careful directions for a nylon from sebacoyl chloride and hexamethylenediamine are given by P. W. Morgan and S. L. Kwolek, *J. Chem. Ed.,* **36,** 182 (1959). For a similar polymerization of terephthaloyl chloride and piperazine see N. C. Rose, *ibid.,* **44,** 283 (1967). Hydrolysis of Nylon 66 [W. F. Berkowitz, *ibid.,* **47,** 536 (1970)], and of latex paints [J. A. Vinson, *ibid.,* **46,** 877 (1969)] make interesting experiments. Preparation and crosslinking of an unsaturated polyester are described by M. P. Stevens, *ibid.,* **44,** 160 (1967). For experiments on polymerization kinetics see E. Senogles and L. A. Woolff, *ibid.,* **44,** 157 (1967); E. L. McCaffery, *ibid.,* **46,** 59 (1969); and *Laboratory Practice of Macromolecular Chemistry,* McGraw-Hill, New York, 1970.

529. ^{14}C Tracer Technique. Biosynthesis of an antibiotic [I. T. Glover and C. L. Limbaugh, *J. Chem. Ed.,* **46,** 866 (1969)]. An experiment with ^{14}C-labeled sodium acetate [J. C. Wright, *ibid.,* **40,** 206 (1963)].

530. Nitration of Benzoic Acid. Isomer Distribution by Isotope Dilution Technique. M. Feldman and J. W. Wheeler, *J. Chem. Ed.,* **44,** 464 (1967).

PHYSICAL ORGANIC CHEMISTRY EXPERIMENTS

531. Activation Energy Determination. A. M. Bryan and P. G. Olafsson, *J. Chem. Ed.,* **46,** 248 (1969).

532. Determination of Rate of Solvolysis of Benzhydryl Bromide. I. Horman and M. J. Strauss, *J. Chem. Ed.,* **46,** 114 (1969).

533. Methanolysis of Substituted Methyl Benzoates. I. T. Glover and C. W. Peterson, *J. Chem. Ed.,* **45,** 241 (1968).

534. Chromic Acid Oxidation of Alcohols. R. M. Lanes and D. G. Lee, *J. Chem. Ed.,* **45,** 269 (1968).

535. Alcohols to Alkyl Halides. Kinetics [J. H. Cooley, J. D. McCown, and R. M. Shill, *J. Chem. Ed.,* **44,** 280 (1967)].

536. Keto-Enol Tautomerism J. G. Dawber and M. M. Crane, *J. Chem. Ed.,* **44,** 150 (1967).

537. Kinetic Isotope Effects. J. R. Jones, *J. Chem. Ed.,* **44,** 31 (1967).

538. Activated Aromatic Nucleophilic Substitution. L. K. Dyall, *J. Chem. Ed.,* **43,** 663 (1966).

539. Acid and Base Catalyzed Isomerization of Δ^5-Cholesten-3-one, an Experiment in UV Kinetics. D. R. Dimmel and M. A. McKinney, *J. Chem. Ed.,* **49,** 373 (1972).

540. Acid-Catalyzed Hydrolysis of Sucrose. J. G. Dawber, D. R. Brown, and R. A. Reed, *J. Chem. Ed.,* **43,** 34 (1966).

541. Determination of a Reaction Rate Constant. J. O. Schreck, *J. Chem. Ed.,* **43,** 149 (1966).

542. A Second-Order Kinetics Experiment. Ethyl *p*-toluenesulfonate plus iodide ion [W. J. Teerlink, J. Asay, and J. M. Sugihara, *J. Chem. Ed.,* **41,** 161 (1964)].

543. A Pseudo First-Order–Second-Order Kinetics Experiment. An illustration of the Guggenheim method. *p*-Bromonitrosobenzene and 2,3-dimethyl-1,3-butadiene [M. Ahmad and J. Hamer, *J. Chem. Ed.,* **41,** 249 (1964)].

EXPERIMENTS INVOLVING COLUMN CHROMATOGRAPHY OR TLC

544. Nitrophenol Mixtures. I. B. Ruppel, Jr., F. L. Cuneo, and J. G. Krause, *J. Chem. Ed.,* **48,** 635 (1971).

545. Separation of Leucine and Isoleucine by TLC. K. L. Gatto and C. L. Borders, Jr., *J. Chem. Ed.,* **47,** 840 (1970).

546. Separation of Bromoaniline Isomers by Cation-Exchange Chromatography. S. J. Romano, *J. Chem. Ed.,* **47,** 478 (1970).

547. TLC Separation and Spectroscopic Analysis of *o*- and *p*-Nitroanilines. J. A. Hurlbut, *J. Chem. Ed.,* **48,** 411 (1971).

548. Relative Nucleophilic Properties of Aromatic Primary Amines. Involves tlc [A. Yeadon, *J. Chem. Ed.,* **48,** 256 (1971)].

549. Paper Chromatography. T. McCullough and A. Lechtenberg, *J. Chem. Ed.*, **47,** 141 (1970).

550. Separation by Adsorption Chromatography. S. H. Pine, *J. Chem. Ed.*, **43,** 672 (1966).

551. Paper Chromatographic Separation of 2,4-Dinitrophenylhydrazones. M. C. Burnett, *J. Chem. Ed.*, **43,** 385 (1966).

552. Competitive Aromatic Nucleophilic Displacement. D. Todd and M. Lookabaugh, *J. Chem. Ed.*, **49,** 292 (1972).

553. Optimizing Experimental Conditions, TLC. J. F. Herz, *J. Chem. Ed.*, **43,** 599 (1966). See also R. E. Bozak, *ibid.*, **43,** 73 (1966).

554. Column Chromatography Experiment Using Unknowns. S. Marmor, *J. Chem. Ed.*, **42,** 272 (1965).

555. Micro-TLC in Qualitative Organic Analysis. S. Samuels, *J. Chem. Ed.*, **43,** 145 (1966).

556. Test Tube and Glass Rod TLC. R. Ikan and E. Rapaport, *J. Chem. Ed.*, **44,** 297 (1967).

557. Talc for TLC. J. M. Walsh, *J. Chem. Ed.*, **44,** 294 (1967).

558. Quick Paper Chromatography of Amino Acids. E. P. Heimer, *J. Chem. Ed.*, **49,** 547 (1972).

559. Small Scale Thin-Layer Chromatography. U. A. Th. Brinkman and G. De Vries, *J. Chem. Ed.*, **49,** 545 (1972).

560. Partial Analysis of Vinyl-Asbestos Floor Tile. W. A. Foreman and D. R. Paulson, *J. Chem. Ed.*, **49,** 572 (1972).

561. TLC Separation of Dyes. Involves reflectance spectroscopic analysis [M. M. Frodyma and R. W. Frei, *J. Chem. Ed.*, **46,** 522 (1969)].

EXPERIMENTS INVOLVING GLC

562. The Methanolysis of Acetal. D. O. Johnston, *J. Chem. Ed.*, **44,** 33 (1967).

563. Friedel–Crafts Alkylation of Benzene. H. C. Dunathan, *J. Chem. Ed.*, **41,** 278 (1964).

564. Migratory Aptitudes. N. M. Zaczek, J. C. Ruff, A. H. Jackewitz, and D. F. Roswell, *J. Chem. Ed.*, **48**, 257 (1971). Includes pinacol syntheses and rearrangements, cleavage of pinacolones, and glc analysis of cleavage products.

565. Fractionating Column Efficiency. A. Ault, *J. Chem. Ed.*, **41**, 432 (1964).

566. The Structure and Properties of Choleic Acids. R. G. Jesaitis and A. Krantz, *J. Chem. Ed.*, **48**, 137 (1971).

567. Quantitative Gas Chromatography. H. L. Pardue, M. F. Burke, and J. R. Barnes, *J. Chem. Ed.*, **44**, 695 (1967).

568. Reactivity Ratios from Copolymerization Kinetics. A quantitative glc experiment [W. A. Mukatis and T. Ohl, *J. Chem. Ed.*, **49**, 367 (1972)].

569. The Odor of Optical Isomers. S. L. Murov and M. Pickering, *J. Chem. Ed.*, **50**, 74 (1973).

570. Oxidation of Fats: a Study Using GLC. R. J. Hamilton and M. Y. Raie, *J. Chem. Ed.*, **49**, 507 (1972).

571. The Preparation and Dehydration of 1-Benzylcycloalkanols. G. R. Newkome, J. W. Allen and G. M. Anderson, *J. Chem. Ed.*, **50**, 372 (1973).

EXPERIMENTS USING SPECTROSCOPIC METHODS

572. Structure Elucidation. B. W. Benson, E. S. Olsen, and L. A. Smeltz, *J. Chem. Ed.*, **47**, 220 (1970).

573. Experiments in NMR. G. Glaros and N. H. Cromwell, *J. Chem. Ed.*, **48**, 202 (1971).

574. An NMR Determination of Optical Purity. J. Jacobus and M. Raban, *J. Chem. Ed.*, **46**, 351 (1969).

575. Infrared Spectrometry to Study Second-Order Reaction Kinetics. B. Gastambide, J. Blanc, and Y. Allamaghy, *J. Chem. Ed.*, **41**, 613 (1964).

SYNTHETIC SEQUENCES

576. Synthetic Sequences as Laboratory Experiments. Valuable experience in synthesis is derived from the completion of a sequence of preparations in which one product becomes the starting material for the next. In a literature search for directions, it is advisable first to seek information on the final product desired. The sequence is then traced backward. Consult the instructor or assistant before attempting any of these sequences, with particular attention to quantities of starting materials.

Chapter 14 discusses the literature of organic chemistry and suggests publications of special importance for synthesis. The preparation of a report on your work is discussed in Chapter 15. In the present case you should first see if any part of your sequence has been covered in this manual. You should next check the various secondary sources given in Chapter 14, particularly *Organic Syntheses* and *Organic Reactions*. Finally a complete search for synthetic methods for your final product and for intermediate products should be made.

After approval of your plan by the instructor, write a preliminary report in your notebook, with information as to literature sources, before laboratory work is started.

A number of synthetic sequences have been worked out as experiments for the laboratory portion of courses in organic chemistry (beginning or advanced) and published in the *Journal of Chemical Education*. A number of these are included below without the accompanying references. These references should be encountered in the normal course of the literature search. In a few instances the published experiment merely suggests a number of reactions that can profitably be explored starting with some readily available chemical. These experiments are given first and the references are included in each instance.

Some of the experiments and compounds cited earlier in this section as special experiments also involve, or can be carried out as, synthetic sequences (§§506, 511, 423, 427, 434, etc.).

1. **α-Pinene, a Starting Material for a Sequence of Organic Experiments.** X. A. Dominguez and G. Leal, *J. Chem. Ed.,* **40,** 347 (1963).

2. **Conversions from Cyclohexanol.** S. B. Hanna, J. T. Wrobleski, J. T. Bohanon, and B. W. Peace, *J. Chem. Ed.,* **48,** 556 (1971).

3. **Bridgehead Reactivity.** Preparation of adamantane from cyclopentadiene dimer by aluminum chloride–catalyzed isomerization [A. Ault and R. Kopet, *J. Chem. Ed.,* **46,** 612 (1969)]. Bromination yields 1-bromoadamantane [C. W. Jefford, R. McCreadie, P. Müller, and B. Siegfried, *J. Chem. Ed.,* **48,** 708 (1971)]. 9-Bromotriptycene is obtained from 9-bromoanthracene and benzyne generated from anthranilic acid (Jefford et al.). The 9-bromoanthracene is available, but may also be prepared by bromination of anthracene [O. L. Wright and L. E. Mura, *J. Chem. Ed.,* **43,** 150 (1966)]. The reactivity of these bridgehead compounds is compared with that of *tert*-butyl bromide.

Warning: *Generation of benzyne from diazotized anthranilic acid may result in an explosion if care is not taken to make certain that the diazonium salt is decomposed completely.*

See T. F. Mich, E. J. Nienhouse, T. E. Farina, and J. J. Tufariello, *J. Chem. Ed.,* **45,** 272 (1968), and OS **48,** 12.

4. **Luminol (5-Amino-2,3-dihydro-1,4-phthalazinedione).** Phthalic anhydride, 3-nitrophthalic acid, 5-nitro-2,3-dihydrophthalazinedione. For an interesting paper on the cold light obtained from luminol see H. W. Schneider, *J. Chem. Ed.,* **47,** 519 (1970).

5. **Butyl Valerate.** *n*-Butyl alcohol, butyl bromide, butyl cyanide, valeric acid.

6. ***n*-Hexyl Alcohol.** *n*-Butyl alcohol, butyl bromide, butylmagnesium bromide: Use of ethylene oxide.

7. ***nor*-Leucine.** Chloroacetic acid, diethyl malonate, diethyl *n*-butylmalonate, caproic acid, α-bromocaproic acid.

8. **D-Arabinose.** Glucose, *d*-gluconitrile, pentaacetyl-*d*-gluconitrile.

9. **Cyclopentanecarboxaldehyde.** Cyclohexene, its bromohydrin, cyclohexene epoxide; *trans*-1,2-cyclohexanediol.

10. ***m*-Nitrophenol.** Benzene, nitrobenzene, *m*-dinitrobenzene, *m*-nitroaniline (see §459).

11. **Polystyrene.** Benzene, acetophenone, α-phenylethanol, styrene.

12. ***p*-Nitrophenylhydrazine.** Aniline, acetanilide, *p*-nitroacetanilide, *p*-nitroaniline.

13. ***m*-Bromotoluene.** *p*-Toluidine, aceto-*p*-toluidide, 4-acetamino-3-bromotoluene, 4-amino-3-bromotoluene.

14. ***p*-Aminophenylacetic acid.** Benzyl chloride, benzyl cyanide, *p*-nitrobenzyl cyanide, *p*-nitrophenylacetic acid.

15. **Congo Red.** Nitrobenzene, hydrazobenzene, benzidine. **Warning:** *This sequence is no longer suitable.* See §327.

16. **Aromatic Nitro Musk.** From *m*-xylene to 5-*tert*-butyl-1,3-dimethylbenzene. This is nitrated to 5-*tert*-butyl-1,3-dimethyl-2,4,6-trinitrobenzene (musk xylene), or acetylated and then nitrated to musk ketone.

17. **6-Aminosaccharin.** *p*-Nitrotoluene, 2-methyl-5-nitrobenzenesulfonamid, 6-nitro-1,2-benzisothiazolin-3-one 1,1-dioxide.

18. **Hexaphenylbenzene.** An extension of the benzaldehyde, benzoin, benzil sequence (Experiment 40). Dibenzyl ketone and benzil yield tetraphenylcyclopentadienone, which undergoes a Diels–Alder reaction with diphenylacetylene to give hexaphenylbenzene with loss of carbon monoxide. Diphenylacetylene may be synthesized from stilbene (Experiment 25).

19. **3-Phenylanthranil.** *o*-Benzoylbenzoic acid (§315), *o*-benzoylbenzamide, *o*-aminobenzophenone, *o*-azidobenzophenone.

20. **Quinhydrone.** Nitrobenzene, aniline, benzoquinone, hydroquinone.

21. **Triphenylguanidine.** Aniline, thiocarbanilide, phenyl mustard oil.

22. **Para Red.** A pigment. Aniline, acetanilide, *p*-nitroacetanilide, *p*-nitroaniline.

23. **Butesin (*n*-Butyl *p*-Aminobenzoate).** A local anaesthetic. *p*-Nitrotoluene, *p*-nitrobenzoic acid, *p*-aminobenzoic acid.

24. **Amidol (2,4-Diaminophenol Hydrochloride).** A photographic developing agent. Chlorobenzene, 2,4-dinitrochlorobenzene, 2,4-dinitrophenol.

25. **4-Quinazolone-7-carboxylic Acid.** *p*-Xylene, terephthalic acid, nitroterephthalic acid.

26. **Triphenylene.** Sulfanilic acid, *p*-acetaminobenzene sulfonic acid, 4-acetamino-3-bromosulfonic acid, *o*-bromoaniline.

27. **Tetraphenylethylene.** Benzophenone, dichlorodiphenylmethane.

28. **2,4-Dinitrophenylhydrazine.** Benzene, bromobenzene, 2,4-dinitrobromobenzene, 2,4-dinitrophenylhydrazine.

29. **Ninhydrin.** Dimethyl phthalate and ethyl acetate, 2-carbomethoxy-1,3-indanedione, 1,3-indanedione, 2-nitro-1,3-indanedione, 2-bromo-2-nitro-1,3-indanedione, 1,2,3-indanetrione.

30. **2,3-Diphenyl-1-indenone and Related Compounds.** Benzalphthalide, 2,3-diphenyl-1-indenone, 2,3-epoxy-2,3-diphenyl-1-indanone, dimethyl 8,9-dihydro-9-oxo-5.8-diphenyl-5,8-epoxy-5*H*-benzocycloheptene-6,7-dicarboxylate.

31. **1-Bromo-3-chloro-5-iodobenzene.** Benzene, nitrobenzene, aniline, acetanilide, *p*-bromoacetanilide, 4-bromo-2-chloroacetanilide, 4-bromo-2-chloroaniline, 4-bromo-2-chloro-6-iodoaniline, 1-bromo-3-chloro-5-iodobenzene.

32. **Rearrangement of Diazotized 4-Bromobenzophenone Hydrazone.** Ketone, hydrazone, rearrangement. The starting ketone can be synthesized. The anilides that result from rearrangement are synthesized separately by the Schotten–Bauman reaction for identification purposes. [B. L. Hawbecker, *J. Chem. Ed.*, **47,** 218 (1970).]

33. **The Ethylene Ketal Protecting Group in Organic Synthesis.** Ethyl acetoacetate, ethyl acetoacetate ethylene ketal, 4,4-diphenyl-4-hydroxy-2-butanone ethylene ketal, 4,4-diphenyl-3-buten-2-one [D. R. Paulson, A. L. Hartwig and G. F. Moran, *J. Chem. Ed.*, **50,** 216 (1973)].

Appendix

577. Vapor Pressures in Torr

Temp., °C	*Water*	*Ethyl Ether*	*n-Pentane*	*n-Heptane*	*n-Octane*	*Bromo-benzene*	*p-Di-bromo-benzene*	*α-Bromo-naphtha-lene*
0	5	185	183	11	3			
10	9	292	282	21	6			
20	17	442	420	36	10			
30	32	647	611	58	18	6		
40	55	921	873	92	31	10		
50	92	1277	1193	141	49	17		
60	149	1729	1605	209	78	28		
70	233	2296	2119	302	118	44	4	
80	355	2994	2735	427	175	66	7	
90	525	3841	3498	589	253	98	12	
91	546		3580	607	263	102	12	
92	567		3665	625	273	106	13	
93	588		3750	644	283	110	13	
94	611		3838	664	293	114	14	
95	634		3929	684	303	118	15	
96	657		4021	705	313	122	15	
97	682		4114	726	324	127	16	
98	707		4211	748	334	131	17	
99	733		4310	771	344	136	17	
100	760		4410	795	354	141	18	
110	1075			1047	482	199	28	4
120	1489			1367	646	275	42	6
130	2026				859	373	63	9
140	2711				1114	496	89	13
150	3570				1425	649	129	19
160	4636					846	178	27
170	5940					1077	234	39
180	7520					1351	316	55
190	9413					1684	407	75
200	11659						525	102
210	14305						660	134
220	17395							179
230	20978							233
240	25100							300
250	29818							382
260	35188							482
270	41261							604
280	48104							743

578. Boiling Points of Water and Naphthalene at Various Pressures

Pressure,	*Boiling Point, °C*		*Pressure,*	*Boiling Point, °C*	
torr	*Water*	*Naphthalene*	*torr*	*Water*	*Naphthalene*
570	92.1	206.0	720	98.5	215.7
580	92.6	206.7	725	98.7	216.0
590	93.1	207.4	730	98.9	216.3
600	93.5	208.1	735	99.1	216.6
610	94.0	208.7	740	99.3	216.9
620	94.4	209.4	745	99.4	217.2
630	94.8	210.1	750	99.6	217.5
640	95.3	210.8	755	99.8	217.8
650	95.7	211.4	760	100.0	218.1
660	96.1	212.1	765	100.2	218.4
670	96.5	212.7	770	100.4	218.7
680	96.9	213.3	775	100.5	219.0
690	97.3	213.9	780	100.7	219.3
700	97.7	214.5	785	100.9	
705	97.9	214.8	790	101.1	
710	98.1	215.1	795	101.3	
715	98.3	215.4	800	101.4	

579. Common Laboratory Reagents

Acid, acetic, "glacial"; density 1.05. This is industrially pure acetic acid, containing less than 1% of water.

Acid, hydrochloric, concentrated; density 1.18. An approx. 38% solution of hydrogen chloride in water; concentration about 12 *N*; colorless. The common commercial grade, usually of yellowish tint, is likely to be of slightly lower concentration and density; usually approx. 10 *N*.

Acid, nitric, concentrated; density 1.42. This is the constant-boiling mixture of 68 parts by weight of HNO_3 and 32 parts of water. Approx. 16 *N*.

Acid, nitric, (yellow) fuming; density about 1.5. This is a product that at the time of manufacture is mainly anhydrous nitric acid. By the time it reaches the purchaser it has probably decomposed slightly and may consist of about 96% HNO_3, the balance being water and oxides of nitrogen. Approx. 24 *N*.

Acid, nitric, (red) fuming; density about 1.54. This is HNO_3 plus a greater excess of oxides of nitrogen than is present in yellow fuming acid.

Acid, sulfuric, concentrated; density about 1.84. Industrially pure H_2SO_4, containing about 6% of water. Approx. 36 *N*.

Acid, sulfuric, fuming; mixtures of H_2SO_4 with varying percentages of SO_3. Should be labeled with a percentage number; for example, "20% fuming" acid is a mixture of 80 parts by weight of H_2SO_4 with 20 parts of SO_3. Often called "oleum."

Ammonium hydroxide, concentrated; density 0.90. A 28% aqueous solution of ammonia, NH_3. Approx. 15 *N*.

580. Saponification Numbers of Common Fats

Beef tallow	196–200	Linseed oil	188–195
Beeswax	88–96	Mutton tallow	195–196
Butter fat	210–230	Olive oil	185–196
Castor oil	175–183	Peanut oil	186–194
Coconut oil	253–262	Rape-seed oil	168–179
Corn oil	187–193	Sesame oil	188–193
Cottonseed oil	194–196	Sperm oil	120–137
Lard	193–203	Wool fat	82–130

See §256 for the definition of **Saponification number** and an experiment in which these numbers are used.

Index

DENSITIES AND BOILING POINTS OF LIQUID ORGANIC COMPOUNDS

(Boiling points at 600 and 10 torr are estimates based on tables and graphs published by several authors, including Stull, Dreisbach and T. E. Jordan. For laboratories at relatively high altitudes, values for the last column may be filled in on the basis of plots on Fig. 2, §35.)

Substance	Density (g/ml)	Boiling Points (°C) at Different Pressures			
		760 torr	600 torr	10 torr	____ torr
Acetic acid	1.049	118	111	18	______
Acetic anhydride	1.087	140	133	36	______
Acetone	0.792	56.5	50	−31	______
Acetyl chloride	1.105	52	46	−33	______
Aniline	1.022	184.4	175	70	______
Benzaldehyde	1.050	179	170	62	______
Benzene	0.879	80	73	(s)	______
Benzoyl chloride	1.219	197	188	73	______
Benzyl alcohol	1.050	205	196	93	______
Benzyl cyanide	1.015	234	224	106	______
Bromobenzene	1.499	156	147	40	______
Bromoform	2.890	150	142	35	______
n-Butyl acetate	0.822	126	120	23	______
n-Butyl alcohol	0.810	118	111	31	______
n-Butylbenzene	0.862	183	174	62	______
n-Butyl bromide	1.299	102	95	−1	______
n-Butyldimethylcarbinol	0.812	140	133	44	______
Carbon disulfide	1.263	47	40	−45	______
Carbon tetrachloride	1.463	77	71	−20	______
Chlorobenzene	1.107	132	125	23	______
Chloroform	1.498	61	54	−30	______
Diethyl adipate	1.009	240	232	121	______
Diethyl malonate	1.055	199	190	82	
Dimethylaniline	0.956	193	184	70	______
Ether, dibutyl	0.784	142	130	30	______
Ether, diethyl	0.714	35	28	−48	______
Ether, diisopropyl	0.726	68	61	−27	______
Ethyl acetate	0.901	77	71	−14	______
Ethyl acetoacetate	1.025	181	172	67	______
Ethyl alcohol	0.785	78	73	−2	______